Operation, Planning, and Analysis of Energy Storage Systems in Smart Energy Hubs

Behnam Mohammadi-Ivatloo • Farkhondeh Jabari
Editors

Operation, Planning, and Analysis of Energy Storage Systems in Smart Energy Hubs

Springer

Editors
Behnam Mohammadi-Ivatloo
Department of Electrical and Computer Engineering
University of Tabriz
Tabriz, Iran

Farkhondeh Jabari
Department of Electrical and Computer Engineering
University of Tabriz
Tabriz, Iran

ISBN 978-3-030-09149-1 ISBN 978-3-319-75097-2 (eBook)
https://doi.org/10.1007/978-3-319-75097-2

Softcover re-print of the Hardcover 1st edition 2018

Printed on acid-free paper

This Springer imprint is published by the registered company Springer International Publishing AG part of Springer Nature.
The registered company address is: Gewerbestrasse 11, 6330 Cham, Switzerland

Preface

Along with rapid population growth, the demand for and consumption of energy has also grown exponentially. Therefore, energy efficiency, greenhouse gas emissions, and environmental concerns are crucial concerns around the world. Meanwhile, different renewable energy sources, including solar, wind, hydro, geothermal, and tidal energy are broadly employed to provide a clean and sustainable source of electricity generation. Such energy can then be applied to air and water heating and cooling, transportation infrastructure and development, rural, off-grid energy services, and so on. The aim of the book is to robustly design and investigate a natural gas and electricity coupled energy hub as a super node that is meant to receive various energy carries, including renewable energy sources, and organize demand and supply side management with different energy storage technologies.

This book also focuses on operational issues when combining renewable energy, natural gas, and electricity into energy hubs, with an emphasis on energetic, economic, and environmental viewpoints. Moreover, the variable nature of renewable energy sources, together with traditional load and generation forecasting, introduces increased uncertainty in the operation and planning of combined renewable and natural gas electrical grids. Therefore, risk-constrained stochastic programming and robust optimization techniques are used to investigate the robustness and opportunistic aspects of optimal scheduling problems to make the risk-averse or risk-taker decisions, respectively.

The main purpose of Chap. 1 is to introduce the concept of smart energy hub. In this regard, an introduction to the concept of the smart grid, its definitions, features, and main challenges are presented. Some advantages, goals, and impacts of using energy storages in energy hub are discussed in Chap. 2.

Chapter 3 examines different technologies, structures, and the technical and operational constraints of compressed air energy storage. In addition, the impact of advanced adiabatic compressed air energy storages on day-ahead economic emission dispatch of coal and gas-fired generators is investigated using a mixed-integer, nonlinear program that utilized the GAMS software package and SBB tool.

In Chap. 4, optimal scheduling of a residential hub energy system based on the consumption and presence of solar thermal energy is presented. One model of a residential hub energy system includes equipment such as combined heat and power systems, a boiler, battery storage system, solar thermal storage, and smart appliances. In Chap. 5, short-term optimal scheduling of solar powered multi-chiller plants is presented. In Chap. 6, day-ahead economic dispatches of three different multi-chiller plants is addressed using a basic, open-source, nonlinear, mixed-integer program, using the GAMS software package. Compared with competitive heuristic algorithms, the use of the BONMIN solver in finding optimal loading points of centrifugal chillers reduces their electricity requirement significantly.

In Chap. 7 research related to demand side management programs in residential, commercial, agriculture, and industrial energy hubs is reviewed and discussed. In Chap. 8, the applicability of compressed air energy storage systems in handling the fluctuating energy generation of local renewable energy units in the hub energy system is examined. Chapter 9 investigates the capability of stochastic frameworks when dealing with energy resources scheduling problems in renewable energy hubs. The authors of Chap. 10 consider a renewable-based energy hub, which includes wind turbines, photovoltaic cells, energy storage, and boilers, to name a few. The volatile nature of renewable energy resources create new and unique problems when addressing the demand for energy. In this regard, stochastic short-term scheduling is optimal, considering the uncertainty in the supply of renewable energy generation. Chapter 11 presents grid assistance opportunities through charging and discharging of electric vehicles, and explores the technical and operational challenges in integrating this movable and changeable energy storage within the power system. The chapter discusses the development of charging load curves of electric vehicles based on mobility attributes and charging protocols. Chapter 12 optimizes the operation of a residential energy hub, which includes a combined heat and power unit, a boiler, a plug-in hybrid electric vehicle, photovoltaic panels, and a heat storage system. This is meant to provide adequate electricity and heat to a home. A two-stage stochastic model for a long-term distribution network model is proposed in Chap. 13. This is meant to solve the challenging issue of the uncertainty associated with renewable generation and the integration of electric vehicles based on the network's technical constraints. In Chap. 14, a joint energy storage and distribution system is proposed, taking into consideration voltage stability constraints. An introduction to the concept of optimal design of distributed energy systems is presented in Chap. 15. Chapter 16 endeavors to present a general modeling and optimization scheme for coupled power flow investigation in various energy networks.

In Chap. 17, the importance of different applicable pathways of sustainable power to gas is explained. Finally, Chap. 18 investigates the performance of hub energy systems from both economic and environmental viewpoints in the presence of hydrogen energy storage systems and demand response programs.

Tabriz, Iran

Behnam Mohammadi-Ivatloo
Farkhondeh Jabari

Acknowledgments

The editors would like to acknowledge the contributions of all scholars involved in this book and, more specifically, those who took part in the review process. Without their support, this book would not have become a reality. The editors would like to thank each one of the authors for the contribution of their time and expertise to this book. The editors also wish to acknowledge the valuable contributions of Dr. Mahmoud Pesaran-hajiabbas, Dr. Sayyad Nojavan, Mr. Sajad Madadi-yeganeh, and Mr. Morteza Nazari-heris. Their reviews improved the quality, coherence, and presentation of the chapters. Most of the authors also served as reviewers, which we greatly appreciate.

Contents

Chapter 1
An Introduction to Smart Energy Systems and Definition of Smart Energy Hubs

Mohammad Mohammadi, Younes Noorollahi, and Behnam Mohammadi-Ivatloo

1.1 Introduction

Energy has always been one of the most basic human needs and the main driver of the development of human societies. With the improvement of technology and the mechanization of the lifestyle, this need is increasing day by day [1]. Therefore, providing clean, affordable, safe, and sustainable energy is one of the main challenges of different countries. In the last century, the main source of energy was fossil fuel resources [2]. The fossil fuels are usually converted to the electricity at large thermal power plants, and this electricity is transmitted over long distances to reach the consumers. In these systems, primary energy is converted to electricity at a very low efficiency and sent to the consumer with high losses in transmission and distribution systems. The complexity of the transmission and distribution systems results in high system costs and making it difficult to protect and control these systems [3].

In these systems, intelligent equipment can only be found in the control, monitoring, and protection sectors that are locally installed in limited parts of the system. But today, with the development of the concept of the smart grid, power systems have become more intelligent than ever, and move towards systems that have the ability to decide and interact with adjacent systems [4]. The use of intelligent technologies, automated monitoring, and data collection and processing

---

M. Mohammadi · Y. Noorollahi (✉)
Department of Renewable Energy and Environment, Faculty of New Sciences and Technologies, University of Tehran, Tehran, Iran
e-mail: m.mohammady@ut.ac.ir; noorollahi@ut.ac.ir

B. Mohammadi-Ivatloo
Department of Electrical and Computer Engineering, University of Tabriz, Tabriz, Iran
e-mail: bmohammadi@tabrizu.ac.ir

B. Mohammadi-Ivatloo, F. Jabari (eds.), *Operation, Planning, and Analysis of Energy Storage Systems in Smart Energy Hubs*,
https://doi.org/10.1007/978-3-319-75097-2_1

systems in the smart grid leads to optimization of the power grid function and intelligent management of its various equipment. As a result, the development of the concept of smart grid leads to improved energy efficiency, reducing the need to develop energy production and transmission infrastructure and optimize the size of equipment. On the other hand, with intelligent predictions and controls, the integration of RES is facilitated and the process of development of DG is accelerated in the demand side [5].

The development of the concept of distributed generation and, more generally, the concept of distributed energy resources (DER) in recent years has led to main changes in the structure of the power system and energy markets. The development of the concept of DER (in particular demand-side resources) has led to the emergence of new energy generators in the power system and increased consumers' participation in grid management in the form of programs such as demand-side management (DSM). DSM programs incorporate multiple concepts such as demand response (DR), energy efficiency, load growth, and energy conservation. The main purpose of these programs is to coordinate the pattern of customer consumption with the needs and conditions of the network. The participating of the consumers in these programs can reduce their consumption in peak hours (high energy price periods) or shift a portion of consumption to off-peak hours (low energy price periods), and in addition to reducing their costs, can lead to a reduction in peak demand in the network and a smoother consumption curve. As a result, consumers' participation in DSM programs, in addition to better management of consumption and reduction of energy bills, can improve network stability and reliability [6].

However, the use of DER, in particular, the integration of RES and the use of energy storage systems (ESS) along with multi-generation systems and successful participation in DSM programs, requires an integrated management framework. The concept of energy hub developed in recent years for the modeling and management of multi-energy systems (MES) is a promising method for modeling future energy systems [7]. Energy hub is defined as a model in which the production, conversion, storage, and consumption of various energy carriers are carried out [8]. The hub energy is a conceptual model for controlling and managing multi-carrier and integrated energy systems [9]. So far, various studies have been carried out on this concept, and it has been shown that the integration of different energy carriers in the form of energy hub models leads to improved system performance compared to controlling and scheduling systems with an energy carrier [9]. Energy hub can be used to model various systems in different sizes such as residential houses, commercial buildings, industrial units, greenhouses, office and services buildings, which indicates that this model is a complete and comprehensive model [10]. However, the development of smart grid concept into MES and management of these systems in an intelligent framework is an issue that has not been addressed so far.

This chapter focuses on the integration of different energy systems in the form of macro energy hubs, and the advantages and challenges of these systems. The necessity of developing smart grid concept into energy hub models is discussed. Finally, with the definition of the concept of SEH, the potential of this model is discussed for the modeling of SES in the future.

1.2 Integrated Management of Energy Hubs

The energy hub is an umbrella concept that can cover all energy technologies and systems. So various energy systems with different sizes can be modeled with this concept. However, in terms of size, energy hubs can be classified into two classes of micro energy hubs and macro energy hubs. Micro hubs can be divided into four general categories of residential, commercial, industrial, and agricultural micro hubs. Clearly, in each section, the energy hub, for example, can be a residential building, a commercial building, an industrial unit, a greenhouse, etc. Integrating these micro energy hubs at a higher level will lead to the creation of a network of interconnected energy hubs called macro energy hubs. The macro hub is a collection of energy hubs that are managed and controlled in a coordinated way. Hence, large-scale energy systems such as a residential complex, an industrial area, or even a whole city can be modeled with the concept of macro energy hub.

Integrating and coordinate management of different consumption sectors in the form of a macro energy hub can bring many benefits for each of these sectors and for the entire system. For example, the pattern of energy consumption in residential and commercial buildings is usually different. The peak demand in commercial buildings occurs on a daily basis, and peak demand for residential customers usually relates to the early hours of the night. The relationship between these two sectors is usually established through electricity and gas networks. By integration and coordinated control of these two sectors, resources in one sector can be used to provide the power deficit in another sector. In the industrial sector, a significant amount of low-grade heat is produced, which can be used to heat the adjacent buildings by retrieving this waste heat. In the agricultural sector, there are usually many waste products that can be used to supply CHP-based district heating systems or to provide fuel for the transportation sector. Integrated management of these sectors together increases productivity and can reduce waste, fossil fuel consumption, environmental pollution, emissions and overall system costs.

The development of multi-generation systems has increased the opportunities to integrate various energy infrastructures such as electricity, natural gas, and district heating networks. The simplest example of this is the CHP system, where the gas purchased from the gas network can be converted to electricity and heat. These produced energies, after supplying the system demand, can be sold to electricity and district heating networks. The integration of these various energy infrastructures can be modeled in different ways, but one of the best models presented so far is the energy hub model. Modeling these systems in the form of macro energy hubs can provide optimization opportunities for the entire system. In fact, the energy hub is modeled as a large node, which connects the various energy systems and technologies. In this model, the priorities of each of the sub-systems as well as the constraints of the entire system can be included in the model. In every energy system, consumer preference is usually the reduction of bills and the energy costs while utilities, in addition to minimization of costs and increasing profits, should consider things like power quality, peak shaving, and consumption curve

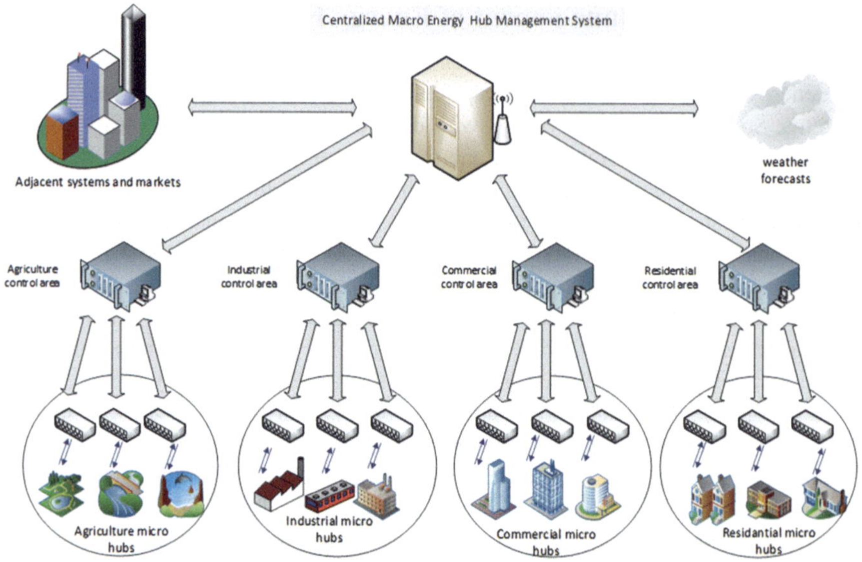

Fig. 1.1 Schematic representation of a centralized macro energy hub management system

shape. Therefore, the optimal management problem in a macro energy hub is a two-layer hierarchy problem that takes into account the priorities of all system agents simultaneously. Therefore, in macro energy hub, in addition to solving an optimization problem at the level of micro energy hubs, the macro energy hub optimization should also be considered. At the level of micro energy hubs optimization, consumers' preferences and related constraints are considered. At the macro hub level, an optimal power flow (OPF) problem or, more generally, an optimal energy flow (OEF) problem is solved. Managing and controlling a macro hub is usually done in two ways: centralized or decentralized.

In a centralized mode, an optimization problem is solved in order to optimize the performance of the macro energy hub and the optimal conditions for the operation of each of the components are determined. In this case, a central management unit is responsible for collecting and processing data and sending control signals. All decisions related to the optimal performance of the macro energy hub are taken by this unit. A demonstration of a centralized controlled macro energy hub is shown in Fig. 1.1.

The first step towards using the energy hub model in the integrated modeling of various energy infrastructures such as power grids, natural gas, and district heating networks was taken in [11]. The authors formulated and solved the OEF problem for a macro energy hub. The same authors presented the modeling of ESS in the previous model in [12] and its linearized problem in [13]. A centralized management model for a network of interconnected energy hubs using the model predictive control (MPC) method has been introduced in [14] by taking into account

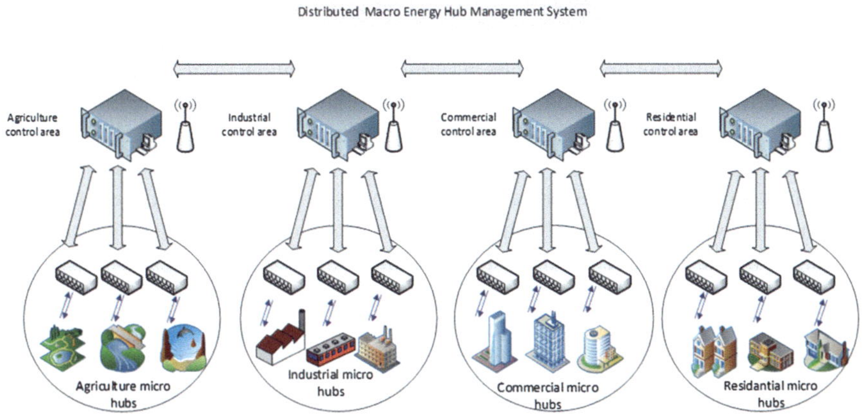

Fig. 1.2 Schematic representation of the distributed macro energy hub management system

ESS dynamics. Similar authors in [15] added the power exchange capabilities with the network to the previous model and provided a centralized controller for a network of residential energy hubs. The authors in [16] have presented a model for integrated management of interconnected energy hubs, in which the effects of considering the threshold value for subscribers' charges (defined by subscribers themselves) on the results of the optimal management problem have been evaluated. Their study results showed that entering the preferences defined by subscribers in the optimization problem led to an increase in their willingness to participate in the optimal management of the macro energy hub and to achieve a smoother consumption curve throughout the system.

Along with all the benefits of a centralized control mode, this model cannot be implemented on large-scale systems and their use is limited to small-scale systems. With the increase of components and energy carriers, the number of variables in the optimization problem as well as the data that needs to be processed increases, which leads to an increase in the volume, cost, and time of the calculations. In some cases, it even becomes difficult to reach the optimal solution at the reasonable time and the possibility of online control is lost. On the other hand, in some cases, there is no possibility to get the permission of all the agents in the system to access their information and centralized control of their equipment, and some agents do not have active participation in centralized control programs. In such cases, decentralized or distributed control can be used, whose schematic representation is shown in Fig. 1.2.

In this case, each control area is controlled by an independent controller, but the control decisions of each area and its mode of operation are exchanged between the other areas. Solving the optimization problem in distributed mode is such that the problem of optimization in a centralized state is divided into several sub-issues and solved. This mode brings a lot of benefits compared to the centralized mode. Solving the optimization problem in each sub-area and in parallel leads to less computing time and does not require a large-scale central processor. This difference in the time and load of computing on large-scale systems becomes much more apparent and the benefits of the distributed mode in these systems are palpable. On the other hand, reliability in distributed mode is more than the centralized mode, since disturbing a controller in a control area only affects the performance of the same area and has a much less impact on the performance of adjacent systems, while in the centralized schema, the existence of a problem in the process of data collection and processing can affect the performance of the entire system.

A model for distributed management of a macro energy hub has been presented in [17] using an iterative method and the OEF problem is solved for this system. Same authors in [18] have used MPC method for modeling of the optimal performance of interconnected energy hubs taking into account the ESS dynamics. The method presented in this work can be used to optimize the management of macro energy hubs in a centralized and distributed mode.

Integrating different energy infrastructure and integrated management of them in the form of energy hub models can have many technical, economic, and environmental benefits. However, in such a system, various energy technologies that have different energy carriers are interacting, and there are many connections between them. This leads to the need to process and exchange large volumes of information in order to solve the optimal management problem in such a system. In such a framework, the use of concepts similar to the smart grid is essential for collecting data and sending control signals and intelligent interactions between system controllers and agents. In this case, due to the presence of various energy carriers, a concept called smart energy systems is introduced, that is a generalized concept of the smart grid. The development of the concept of SES can have many benefits, including increasing efficiency, reducing energy consumption, reducing emissions, increasing reliability, real-time control, facilitating the integration of RES and reducing system costs. The concept of SES and the advantages of their modeling in the content of SEH models are discussed in the next section.

1.3 Smart Grid

1.3.1 Smart Grid Concept

The existence of a hierarchical structure in power systems makes it possible to distribute the power generated by centralized power plants over long distances in transmission systems and extensive distribution systems among consumers. In such a system, intelligent technologies are used only in limited parts of the system and for control purposes and special equipment protection. But today, with the development of ICT and the need for intelligent performance of the various components of the system the development of concepts such as the smart grid for the automatic and intelligent operation of the power grid is inevitable. The word "smart grid" has been used repeatedly in literature, but so far, no precise definition has been provided. The smart grid is a modern electricity grid, which utilizes ICT and energy management structures to increase reliability and improve efficiency [19]. Also, the smart grid can be considered as a system that, in addition to the functions of the power system, such as distribution, transmission, and generation of power, provides storage, decision-making, and optimal interaction possibilities [5]. In another definition, it can be said that the smart grid is seeking to use digital technology for supply and consumption of electric power [20]. Other definitions presented for the concept of the smart grid can be found in [20]. In general, the smart grid can be defined as a network in which communications, smart applications, automated monitoring, and information management are used to improve, optimize, and reconstruct various power grid infrastructures. By utilizing these technologies intelligently, a self-healing and reliable system can be found in the form of the smart grid. Improving system efficiency, reducing the cost of constructing and developing energy infrastructure, optimizing the use of different equipment, facilitating the integration of RES are some of the advantages of using the concept of the smart grid.

1.3.2 Smart Grid Components

The difference between the smart grid and a traditional power grid is in the presence of intelligent technologies and monitoring and control systems, some of which are as follows.

- ICT & smart meter (SM)
- Energy management system (EMS)
- DER
- Smart users
- DSM

1.3.2.1 ICT and SM

All activities carried out in the context of smart grid require the collection, exchange, and processing of information. This possibility is provided through the ICT [4] and SM [21] infrastructures. The existence of these systems will allow the bidirectional information exchange between subscribers and the system operator, which facilitates the collection of information, sending control signals and, consequently, real-time management over the network. These technologies can be utilized at various levels of the network from small smart home to production and transmission infrastructures. The existence of these systems allows each of the distributed components to manage their energy systems and optimize the management of DG resources and ESS to improve the overall performance of the network.

1.3.2.2 EMS

In each energy system, its successful operation requires the coordination of all components of the system and their successful operation at appropriate times and conditions. The optimal performance of an energy system is influenced by various factors, which can be used as input variables to the optimization problem. On the other hand, deciding on the work points and the optimal operational plan for all components of the system requires the analysis of system conditions and the solution of a comprehensive optimization problem. Achieving these conditions requires an EMS that can manage production, storage, and energy consumption. ESM specifies the optimal operating plan for the system by collecting required data such as forecasting demand, price and climate, and taking into account the technical constraints of the system. This system in a residential building can be introduced as a home energy management system (HEMS) in a commercial building as a building energy management system (BEMS) and in an industrial unit as an industrial energy management system (IEMS). These systems provide the optimal planning of each of these units, and even at a higher level, an EMS can manage the optimal performance of the combination of these units. ESM's successful performance leads to optimized system performance with the lowest cost and highest efficiency, active participation in network management and DSM applications, and realizing the benefits of the concept of the smart grid. On the other hand, the optimal management of an energy system in the smart grid content and using ICT and SM technologies will facilitate this process and better coordinate the energy system with other parts of the network.

1.3.2.3 DER

DER's development in recent years has created many hopes for restructuring the network and the presence of small and distributed energy producers. Small-scale renewable resources, mainly associated with the ESS, are among the most important distributed resources, that their application is increasing year by year. However, the use of these resources for local energy supply or integration of them into the main network requires appropriate forecasts of demand, market conditions, weather conditions, as well as accurate information on system conditions and the performance of other network components. This requires the gathering and processing of a large amount of information that reveals the importance of smart grid infrastructures. On the other hand, the presence of DER on the consumer side will allow them to participate in network management and achieve smart grid goals. In fact, increasing DER, especially RES, penetration, with an optimal management, in addition to providing local consumers' demand, reducing the share of fossil fuels in the energy basket, reducing emissions, can lead to the provision of ancillary services to the network and increase the sustainability and reliability of the network. On the other hand, the existence of a smart grid framework can improve the process of integration and management of RES. Further discussions can be found in [3].

In addition to all DER, plug-in electrical vehicle (PEV) is one of the most promising future technologies that, if optimally controlled, can be seen as a potential storage device and DER. Due to the high consumption of fossil fuels in the transport sector and related environmental issues, as well as the existence of international instruments to reduce greenhouse gas emissions, it is essential to move towards a clean transportation system. PEV has been widely considered as a zero-emission transportation technology (especially when using RES to charge them), and there are many research and development programs around the world on the development of this technology. At present, the technology for the manufacture and development of these vehicles is well-known among most automotive companies, and many commercial vehicles have been produced in recent years. But the main issue is the integration of them which can add new challenges to network management. Uncontrolled charging of these vehicles can lead to new peaks in the network and reliability problems. To solve this problem, methods such as the development of charging algorithms, aggregation, and vehicle-to-grid (V2G) capability have been proposed. The presence of the V2G capability provides the possibility of bidirectional power exchange with the network, and PEV can be used as a DER in network management. Generally, the optimal and programmed use of PEV along with the use of features like V2G can bring many benefits, including reducing consumers' cost, facilitating the integration of RES, reducing emission, using the benefits of participating in DSM programs, earning money by owners or aggregators, and offering ancillary services to the network and enhancing network stability. However, the exchange of information and coordination between PEVs and the network operator requires the existence of intelligent infrastructures for communication and information exchange that can be achieved within the framework of the smart grid. As a result, it can be said that the optimal use of PEV and its capabilities within the smart grid framework is achievable and the optimal PEV function as a DER can lead to improved performance of the smart grid [22].

1.3.2.4 DSM

Energy demand management involves a series of interconnected activities between the electricity industry and its customers in order to reduce network's peak demand and energy consumption, as well as level the consumption curve of the network, in order to provide more efficient and low-cost consumers' demands. In the beginning, consumption management was introduced in order to reduce peak consumption, and in fact called load management (LM) programs and gradually consumers' cost, optimal allocation of resources, and environmental pollution reduction was raised as other incentives from the DSM. By adopting these policies, the level of consumer comfort will not be reduced, but by maintaining their level of comfort and well-being, they will consume less energy or the pattern of consumption will be changed and, in addition to reducing costs, it will also be possible to earn money.

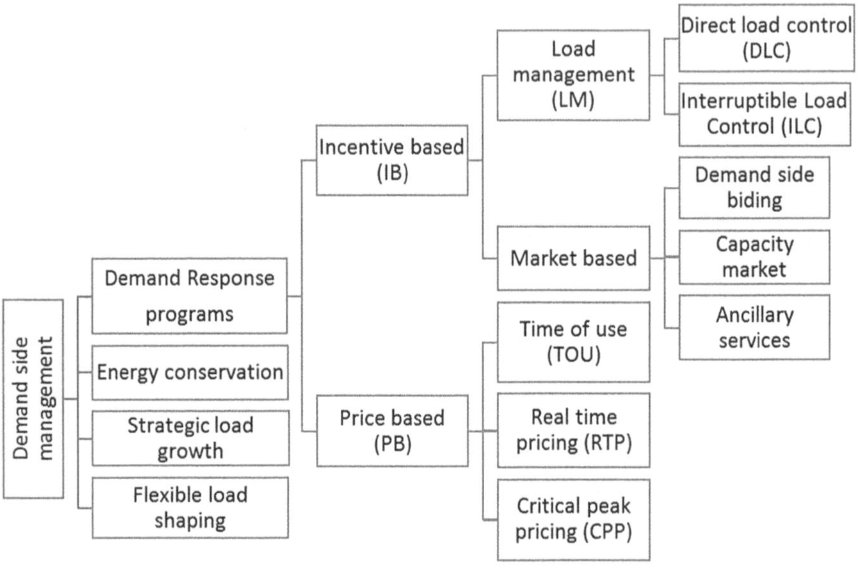

Fig. 1.3 Classification of demand-side management programs

DSM is a comprehensive concept that includes concepts such as load growth, energy saving, energy efficiency, and DR programs. Load growth refers to programs designed to increase load levels and generate more electricity in a state of emergency and strategic. The goal of energy saving is to reduce energy consumption by modifying behavior patterns and customer consumption. Reducing energy consumption in energy efficiency programs is done through specific systems on the demand side and normally without affecting the services provided. By replacing the equipment with energy efficient technologies, the same share of the service is provided to subscribers with less energy consumption. DR refers to programs to change the pattern of end-user consumption through response to a change in electricity prices over time or incentive payments to encourage a reduction in power consumption at times when the market price is high or the reliability of the system is at stake. General classification of DSM programs can be viewed in Fig. 1.3.

DR can be considered as the most important component of DSM, and many studies have been carried out on various methods of its implementation. In general, the purpose of DR is to reduce power consumption during critical hours. Critical hours are the hours when the energy price of the wholesale market is high or the system's reserve level is low due to accidental events such as outgoing transmission lines and generators, or extreme temperature conditions. Two factors that can lead to consumer responsiveness are the change in the retail price of electricity or the implementation of an incentive program to satisfy customers to reduce their consumption during critical hours.

Therefore DR programs can be divided into two types based on incentive-based (IB) and price-based (PB). IB programs include load management programs and market-based programs such as demand-side biding and capacity market. PB programs are based on dynamic pricing such as time of use (TOU) pricing, real-time pricing (RTP), and critical peak pricing (CPP). The load management (LM) program includes direct load control (DLC) and interruptible load control (ILC). The DLC usually involves residential users and refers to programs that can control a customer's load, such as home appliances, through direct operator control. ILC usually involves commercial and industrial subscribers, and it refers to programs that can reduce peak demand by interrupting the load of subscribers at peak hours with direct control of the system operator or subscribers' actions upon request by the system operator.

As a result, there are usually three ways for consumers to participate in DR programs:

- DER
- Load shifting
- Load curtailment

The existence of DER, in particular, RES and ESS, leads to consumers' demand-supply in peak hours and reduction in consumer dependency on the network. The transfer of consumer demand from peak hours to off-peak hours in the form of load shifting program will reduce consumer's costs and network's peak demand. The third ones reduce costs by reducing unnecessary loads during critical hours.

The benefits of subscriber presence in DR programs include incentive payments, reduced billing, while the improvement of system reliability and market performance. On the other hand, the need for production will be reduced, which will reduce production capacity and prevent expansion costs, investment in new infrastructure and the cost of the spinning reserve will be reduced. On the other hand, it reduces fuel consumption, especially fossil fuels, and thus reduces environmental impacts. On the other hand, by increasing the demand side's flexibility, the availability of RES also increases. Achieving these goals and benefiting from all the benefits mentioned for DSM, and in particular, DR requires an intelligent framework for managing and controlling all system components and information exchange that brings the role of the concept of the smart grid clearer. The presence of DER along with DSM in the content of an intelligent energy management system can bring more benefits and flexibility [6].

1.3.3 Smart Grid Challenges

As discussed, the concept of smart grid can provide many benefits, such as increasing productivity, reducing fossil fuel consumption, improving network sustainability, improving the quality of power, facilitating RES integration, online control, and developing concepts such as DER, DSM, and smart users.

However, overview of the different methods used in the literature to determine the economic and environmental effects of SG in [20] shows that despite the reduction of energy consumption and greenhouse gas in smart grids, investment in this area on a massive scale is not economical yet. Because of the high infrastructure investment costs the effects of cost reductions in SG are fewer than the effects of energy efficiency and emissions reduction. Other challenges of SG can be noted as follows [5]:

- Implementation obstacles, such as high investment costs and lack of knowledge about its benefits
- Physical and cyber security
- Optimal and distributed control of DER
- Consumers' tend to retain the information and their privacy
- The development of appropriate and comprehensive standards for data collection and operations within the network
- Low penetration of key technologies such as PEV

1.4 Smart Energy Hub Concept

1.4.1 The Necessity of Using Smart Energy Hub

From a macro perspective, one of the main problems of the concept of the smart grid is the focus of this concept only on the power grid. In this concept, only electricity is considered as the energy carrier in the system, and this view cannot provide a proper model of real energy systems. For example, supplying demand in a home is done through energy carriers such as electricity, natural gas, and water. Consequently, the comprehensiveness of concepts such as the smart home, which focuses only on the control of electrical equipment, can be questionable. In the other sectors of consumption, such as the commercial, industrial, and agricultural sectors, the same conditions prevail and demand for these energy systems is supplied through various energy carriers. Therefore, the smart grid cannot provide a good model of future energy systems, and this concept needs to be developed and reviewed. Since the integration of different energy systems in the presence of different energy carriers is unavoidable due to the development of multi-generation technologies, a comprehensive and realistic model of future energy systems requires the adoption of an intelligent model for multi-energy systems.

The integration of various energy technologies into the 100% renewable energy system as a model of the future smart energy system of Denmark in the future is examined in [23]. In this study, solutions such as multi-generation systems, ESS, biofuels in transport and the development of electric transport have been investigated. The results of this study showed that the integration of infrastructures such as power grids, heat, and transportation systems in a smart context could be the best way to achieve 100% renewable energy systems in the future. On a larger scale, this issue has been examined for the whole of Europe in [24]. The results of this study showed that with the optimum utilization of different energy carriers in an intelligent framework, it is possible to achieve a 100% renewable energy system in the form of smart energy systems.

By understanding the necessity of modeling different energy carriers, developing multi-generation systems and integrating various energy infrastructures, the generalization of the concept of the smart grid to SES is the only way to achieve a comprehensive model of sustainable energy systems in the future. Providing a real and comprehensive model of these systems requires an appropriate framework for integrated management of the entire system. Given the superiority of the energy hub model in the modeling of MES, it would be possible to achieve real models of SES in the future in the form of smart energy hubs.

1.4.2 Recent Research on Smart Energy Hub

In recent years, few but good research have been done to make energy hub models smarter and moving towards SES modeling. The effects of the presence of RES in an MES in both planning and operational optimization modes are evaluated in [25]. The results showed that optimal management of RES such as wind turbine and PV in the content of the smart energy system could reduce fuel consumption, energy costs, and emissions. The operational optimization of an SEH with the ability to exchange power with the network in various schemes has been considered in [26], to minimize the operating costs. The results of this study indicated that the optimum performance of smart systems in the energy hub content will lead to better performance and greater flexibility of these systems. An energy system consisting of CHP, electrical storage, boilers, responsive loads, and PEV in the form of a smart residential energy hub has been evaluated in [27]. In this study, considering the TOU program and the participation of the energy hub in the DR program, the goal is to minimize the operating costs of the system. The results showed that the optimal response to the TOU pricing plan, reduction in energy costs, and correcting the shape of the load curve are results of the presence of responsive loads in the smart residential energy hub. Same authors in [28], with the development of the previous model, have investigated the effects of the presence of PV on the optimal performance of the smart home by attending various DR programs. The results of this study showed that the presence of PV makes it possible to use more solar power in peak hours, and peak demand and system costs are reduced. By considering the V2G capability for PEV, it's possible to achieve a greater reduction in system costs. It is also concluded that the type of DR programs has a significant impact on the optimum performance of this smart residential energy hub.

The energy supply of a residential building using CCHP, along with thermal storage and PEV, has been investigated in the form of an SEH model in [29]. In this paper, the effects of the presence of responsive loads and the ability of V2G on the optimal performance of SEH have been investigated. The results showed that the existence of programmable loads would lead to the flexibility of SEH's performance in DR programs and improve its performance. Considering the capability of V2G for PEV, as well as the presence of thermal storage, will reduce peak demand, improve the consumption curve, facilitate PV integration, reduce dependency on main grid, and reduce operating costs.

The authors in [30] provided a model for the optimal operation of a residential SEH to maximize the benefit of participating in the DR program for this energy hub. In SEH, all possible connections between the various components of the energy hub are considered, in order to provide the possibility of supplying demand from different paths and increase the reliability of the system. The results of this study showed that successful participation in DR programs would improve the consumption curve and increase the sales of power to the network, thereby increasing the profit of the SEH. Also, the presence of different energy carriers and increased degrees of freedom of supply through different technologies will increase the reliability of demand-supply and improve the performance of SEH in DR programs.

A comprehensive model of a residential SEH with the ability to interact with the smart grid is formulated in [31]. SEH's operational optimization based on the model presented in this paper reduces energy costs, peak demand, and emission, taking into account consumer comfort. An empirical survey of the proposed model at a smart home in Ontario, Canada, resulted in a reduction of at least 20% in system costs and more than 50% in peak demand.

The same authors in [32] went through a similar process for a commercial SEH and provided a model for the optimal control of commercial units in the smart grid content. Implementing the proposed model on the energy system of a product storage unit has led to a reduction in energy costs and keeping system conditions on predefined values that indicate the efficiency of the SEH model for the intelligent operation of commercial energy systems.

These authors have presented a similar model in [33] for optimal performance of a greenhouse as a model of agriculture SEH. Weather and electricity prices forecasts, as well as user preferences, are included in the proposed model. The mathematical simulation of this model demonstrates the proper function of it in optimizing greenhouse's operation to reduce energy costs. Numerical results indicated a 40% reduction in energy costs in the warm months of the year and a decrease of 13% for the cold months of the year, which proves the efficiency of the proposed model for modeling the energy systems of a greenhouse in the content of the SEH model.

The use of hydrogen infrastructure in the content of a smart macro energy hub aimed at minimizing the cost of the system is examined in [34]. In this study, an urban energy system, including residential and commercial micro hubs, and a hydrogen fueling station are modeled in an intelligent framework. The simulation results of this study showed that integrated management of SEH leads to lower energy costs, increased efficiency of hydrogen infrastructure, and reduced emissions.

Modeling the implementation of the DR program in SEHS can be found in [35, 36]. These studies have shown that the concept of SEH can lead to more consumers' encouragement to participate in DR programs and improving their performance. The concept of energy hub offers freedom of action and more options for consumers. In addition to the demand shifts, these consumers can participate in DR programs by optimal resource planning and the choice between different energy carriers. These include the possibility of participating in the management of the consumption pattern of different energy carriers (and not just electricity). This new framework in this article is called integrated demand response (IDR). The mathematical modeling presented in these studies is applied to the network of six interconnected energy hubs. Numerical results showed that during peak hours, part electricity demand can shift to low energy cost hours, and the consumption curve would be smoother, and the purchase of gas would increase (to generate electricity at CHP). In peak demand hours, the purchase from the power grid will be reduced, while purchases from the gas network will increase simultaneously, but overall, subscriber charges will decrease. From the utilities point of view, electricity demand peak shaving also leads to lower costs for infrastructure development and reserve. Also, due to more gas sales, the gas company gains more profit. As a result, implementation of IDR can reduce the cost of subscribers' bills, increase utilities' profits, and increase network stability.

The subscriber's satisfaction with DR programs was studied in [37] by defining a parameter called dissatisfaction coefficient. In this study, a stochastic model has been presented for integrated management of residential SHE. The simulation results showed that optimizing the performance of residential hubs in an intelligent environment leads to lower cost of energy and peak demand, and on the other hand, increases consumers' participation in IDR programs.

A concept called dependent demand in which different energy carriers can be used to supply a demand has been introduced in [38] and its effect on the performance of an SEH for participating in DR programs has been studied. In the proposed approach, for the modeling of dependent demand, a row is added to the matrix model, which represents the contribution of each of the energy carriers to the supply of dependent demand. In this study, a scenario-based approach is used to model the uncertainty caused by consumer decisions to select different energy carriers. The results of this study showed that taking into account the dependent demand and increasing the degree of freedom on its delivery methods would reduce system costs. Also, increasing the share of loads controlled by subscribers will actually increase the impact of uncertainty of consumer behavior on the optimal system performance and increase operating costs of the energy hub.

The application of the concept of cloud computing [39] in SEH models can be found in [40]. This concept has been developed to process large volumes of information and can be very useful for improving the performance of the SEH, especially the macro energy hubs. In this study, a model for optimal management of a network of interconnected micro-SEHs is presented in the presence of DSM programs, which uses cloud computing for data processing. In this study, the game theory has been used to formulate the SEH optimization problem and the results showed that using cloud computing leads to reduced computing and processing time.

A summary of the important research on integrated management of energy hubs and SEHs is presented in Table 1.1. As you can be seen, in the micro hubs section, residential sector has the largest number of studies and less attention has been paid to commercial, industrial, and agricultural sectors. On the other hand, a lot of work has been done in the field of integrated control of macro energy hubs, which is due to the high potential of the energy hub in the modeling of large-scale MES. Finally, it's worth mentioning in Table 1.1 that the research that has been done in recent years has led to the use of intelligent frameworks for their models, which reflects the movement of energy hub models towards smart models and accessing to comprehensive models of SEH.

1.4.3 Smart Energy Hub Definition

From the discussion, it can be concluded that focusing only on an energy carrier to provide intelligent models of future energy systems cannot provide a proper perspective of future sustainable energy systems. Models of energy systems in the future need to consider different energy carriers and integration and interaction of new technologies in a smart environment. Participation in the DR program in the form of an SEH leads to improved consumer performance. Because in SEH there are various energy technologies that use different energy carriers to convert and store energy. This makes it possible to increase the freedom of operation and flexibility of the SEH in DR programs. The subscribers can participate in DR programs only by optimizing the use of different technologies and carriers, and there is no change in the level of service received and their level of comfort. The presence of responsive loads and the addition of scheduling capabilities can increase the benefits of participation in DR programs. This will increase the level of satisfaction and willingness of subscribers to participate in DR programs and network management.

Therefore, the introduction to intelligent sustainable systems in the future is to understand the need for integrated management of different energy systems in the presence of different energy carriers in the form of models such as SEH. Therefore, SEH can be defined as a multi-energy system that provides integrated management of all components, as well as optimal consumption planning, even in the presence of uncertainties, to provide a self-healing and thinker energy system. In such a system, along with a centralized control, each of the system's agents must be able to decide and optimize their tasks.

Table 1.1 Classification of research on integrated management and smart energy hub (EH)

References	Description	Publish year	Sector
[31]	Optimal operation of smart residential EHs	2012	Residential EH
[29]	Optimal energy management of a smart residential EH	2015	Residential EH
[38]	Internal and external dependency model for assessing the stochastic behavior of the demand side	2015	Residential EH
[28]	Optimal energy management of a smart residential EH	2015	Residential EH
[37]	Optimal energy management of a smart residential EH considering customer's dissatisfaction level	2016	Residential EH
[32]	Optimal operation of smart commercial EHs	2015	Commercial EH
[41]	Optimal operation of smart industrial EHs	2015	Industrial EH
[33]	Optimal operation of smart agricultural EHs	2015	Agricultural EH
[11]	Optimal power flow (OPF) of a network of EHs	2005	Macro EH
[17]	Decomposed OPF of a network of EHs	2008	Macro EH
[14]	Central controller for a network of EHs	2009	Macro EH
[15]	OPF of a network of EHs in the presence of RES and grid exchange	2010	Macro EH
[18]	Distributed controller for a network of EHs	2010	Macro EH
[42]	EH modeling for interconnected power exchange	2011	Macro EH
[43]	Decomposed OPF of a network of EHs	2014	Macro EH
[44]	Optimal power dispatch of EH under uncertainty	2014	Macro EH
[45]	Optimal optimization of a network of EHs	2015	Macro EH
[34]	Hydrogen economy evaluation in a network of EHs	2015	Macro EH
[46]	Planning a neighborhood EH	2015	Macro EH
[47]	Optimal planning of a network of EHs under uncertainty	2015	Macro EH
[48]	OPF of a network of EHs by using a generalized heuristic approach and addressing variable efficiency models	2015	Macro EH
[36]	Demand response in the context of smart EHs	2015	Macro EH
[40]	Cloud computing in a network of smart EHs	2015	Macro EH
[49]	Optimal operation of a network of EHs and its power exchange with main grid as a procumer	2015	Macro EH
[50]	Optimal planning of network of EHs	2015	Macro EH

1.5 Conclusions

This chapter has provided an introduction to integrated management and integration of different micro energy hubs in the form of macro energy hub. Also, a comprehensive assessment of the smart energy systems and moving toward smart energy hubs has been discussed.

At the level of residential, commercial, industrial, and agricultural micro energy hubs, the integrated management of various energy carriers leads to the benefits of the synergy of these carriers. From a macro perspective, the interaction of micro energy hubs and their comprehensive management in the form of macro energy hub will increase the technical, economic, and environmental benefits compared to systems that are individually modeled and managed. Achieving these benefits requires an integrated management framework for operational optimization at the level of micro energy hubs and OEF at the macro energy hub level. Also organizing concepts such as DSM, RES, and ESS also has a huge benefit to the system in the content of an integrated management.

Consequently, considering the function of the concept of energy hub, energy hub models can be considered as a promising tool for modeling, optimizing, and optimal control of multi-energy systems at various operating levels. The results of previous studies demonstrate the efficiency of the energy hub model to provide comprehensive and realistic models of residential, commercial, industrial, agricultural, and integrated energy systems. However, due to the existence of multiple energy systems and technologies that utilize different energy carriers, the multiplicity of connections between different components, energy hub models require the collection, processing, and exchange of large amounts of information, which reveals the importance of using SEH models.

As a result, future models of energy hub should be able to provide realistic models of energy systems in different sectors of consumption that can be used to solve the challenges of these sectors and improve their performance. Also, energy hub models require the use of smart technologies to achieve comprehensive and realistic models of future energy systems in the form of SEH models.

References

1. Noorollahi Y, Yousefi H, Mohammadi M (2016) Multi-criteria decision support system for wind farm site selection using GIS. Sustain Energy Technol Assess 13:38–50. https://doi.org/10.1016/j.seta.2015.11.007
2. Noorollahi Y, Itoi R, Yousefi H, Mohammadi M, Farhadi A (2017) Modeling for diversifying electricity supply by maximizing renewable energy use in Ebino city southern Japan. Sustain Cities Soc 34:371–384. https://doi.org/10.1016/j.scs.2017.06.022
3. Hossain M, Madlool N, Rahim N, Selvaraj J, Pandey A, Khan AF (2016) Role of smart grid in renewable energy: an overview. Renew Sustain Energy Rev 60:1168–1184

4. Güngör VC, Sahin D, Kocak T, Ergüt S, Buccella C, Cecati C, Hancke GP (2011) Smart grid technologies: communication technologies and standards. IEEE Trans Ind Inf 7(4):529–539
5. Tuballa ML, Abundo ML (2016) A review of the development of smart grid technologies. Renew Sustain Energy Rev 59:710–725
6. Siano P (2014) Demand response and smart grids—a survey. Renew Sustain Energy Rev 30:461–478
7. Geidl M, Andersson G (2005) Optimal power dispatch and conversion in systems with multiple energy carriers. In: Proceedings of the 15th power systems computation conference (PSCC). Citeseer
8. Geidl M, Koeppel G, Favre-Perrod P, Klockl B, Andersson G, Frohlich K (2007) Energy hubs for the future. IEEE Power Energy Mag 5(1):24
9. Mohammadi M, Noorollahi Y, Mohammadi-Ivatloo B, Yousefi H (2017) Energy hub: from a model to a concept – a review. Renew Sustain Energy Rev 80:1512–1527. https://doi.org/10.1016/j.rser.2017.07.030
10. Mohammadi M, Noorollahi Y, Mohammadi-Ivatloo B, Yousefi H, Jalilinasrabady S (2017) Optimal Scheduling of Energy Hubs in the Presence of Uncertainty-A Review. J Energy Manag Technol 1(1):1–17. https://doi.org/10.22109/jemt.2017.49432
11. Geidl M, Andersson G (2005) A modeling and optimization approach for multiple energy carrier power flow. In: 2005 IEEE Russia Power Tech, pp 1–7
12. Geidl M, Andersson G (2007) Optimal coupling of energy infrastructures. In: 2007 IEEE Lausanne Power Tech, pp 1398–1403
13. Geidl M, Andersson G (2007) Optimal power flow of multiple energy carriers. IEEE Trans Power Syst 22(1):145–155
14. Arnold M, Negenborn RR, Andersson G, De Schutter B (2009) Model-based predictive control applied to multi-carrier energy systems. In: 2009 IEEE Power & Energy Society general meeting, PES'09, pp 1–8
15. Arnold M, Andersson G (2010) Investigating renewable infeed in residential areas applying model predictive control. In: 2010 IEEE Power and Energy Society general meeting, pp 1–8
16. Rastegar M, Fotuhi-Firuzabad M, Zareipour H (2015) Centralized home energy management in multi-carrier energy frameworks. In: 2015 IEEE 15th international conference on environment and electrical engineering (EEEIC), pp 1562–1566
17. Arnold M, Andersson G (2008) Decomposed electricity and natural gas optimal power flow. In: 16th power systems computation conference (PSCC 08), Glasgow, Scotland
18. Arnold M, Negenborn R, Andersson G, De Schutter B (2010) Distributed predictive control for energy hub coordination in coupled electricity and gas networks. In: Intelligent infrastructures. Springer, Dordrecht, pp 235–273
19. Kolokotsa D (2015) The role of smart grids in the building sector. Energy Buildings 116:703
20. Moretti M, Djomo SN, Azadi H, May K, De Vos K, Van Passel S, Witters N (2016) A systematic review of environmental and economic impacts of smart grids. Renew Sustain Energy Rev 68:888
21. Sharma K, Saini LM (2015) Performance analysis of smart metering for smart grid: an overview. Renew Sustain Energy Rev 49:720–735
22. Tan KM, Ramachandaramurthy VK, Yong JY (2016) Integration of electric vehicles in smart grid: a review on vehicle to grid technologies and optimization techniques. Renew Sustain Energy Rev 53:720–732
23. Mathiesen BV, Lund H, Connolly D, Wenzel H, Østergaard PA, Möller B, Nielsen S, Ridjan I, Karnøe P, Sperling K (2015) Smart energy systems for coherent 100% renewable energy and transport solutions. Appl Energy 145:139–154
24. Connolly D, Lund H, Mathiesen B (2016) Smart energy Europe: the technical and economic impact of one potential 100% renewable energy scenario for the European Union. Renew Sustain Energy Rev 60:1634–1653

25. Rivarolo M, Greco A, Massardo A (2013) Thermo-economic optimization of the impact of renewable generators on poly-generation smart-grids including hot thermal storage. Energy Convers Manag 65:75–83
26. Teimourzadeh Baboli P, Yazdani Damavandi M, Parsa Moghaddam M, Haghifam M (2015) A mixed integer modeling of micro energy-hub system. In: 2015 IEEE Power & Energy Society general meeting, IEEE, pp 1–5
27. Rastegar M, Fotuhi-Firuzabad M, Lehtonen M (2015) Home load management in a residential energy hub. Electr Power Syst Res 119:322–328
28. Rastegar M, Fotuhi-Firuzabad M (2015) Load management in a residential energy hub with renewable distributed energy resources. Energy Buildings 107:234–242
29. Brahman F, Honarmand M, Jadid S (2015) Optimal electrical and thermal energy management of a residential energy hub, integrating demand response and energy storage system. Energy Buildings 90:65–75
30. Moghaddam IG, Saniei M, Mashhour E (2016) A comprehensive model for self-scheduling an energy hub to supply cooling, heating and electrical demands of a building. Energy 94:157–170
31. Bozchalui MC, Hashmi SA, Hassen H, Cañizares CA, Bhattacharya K (2012) Optimal operation of residential energy hubs in smart grids. IEEE Trans Smart Grid 3(4):1755–1766
32. Bozchalui MC, Cañizares CA, Bhattacharya K (2015) Optimal operation of climate control systems of produce storage facilities in smart grids. IEEE Trans Smart Grid 6(1):351–359
33. Bozchalui MC, Cañizares CA, Bhattacharya K (2015) Optimal energy management of greenhouses in smart grids. IEEE Trans Smart Grid 6(2):827–835
34. Maroufmashat A, Fowler M, Khavas SS, Elkamel A, Roshandel R, Hajimiragha A (2016) Mixed integer linear programing based approach for optimal planning and operation of a smart urban energy network to support the hydrogen economy. Int J Hydrogen Energy 41:7700
35. Bahrami S, Sheikhi A (2015) From demand response in smart grid toward integrated demand response in smart energy hub. IEEE Trans Smart Grid 7:650
36. Sheikhi A, Bahrami S, Ranjbar AM (2015) An autonomous demand response program for electricity and natural gas networks in smart energy hubs. Energy 89:490–499
37. Sheikhi A, Rayati M, Ranjbar AM (2016) Demand side management for a residential customer in multi energy systems. Sustain Cities Soc 22:63
38. Neyestani N, Yazdani-Damavandi M, Shafie-Khah M, Chicco G, Catalao JP (2015) Stochastic modeling of multienergy carriers dependencies in smart local networks with distributed energy resources. IEEE Trans Smart Grid 6(4):1748–1762
39. Fang B, Yin X, Tan Y, Li C, Gao Y, Cao Y, Li J (2016) The contributions of cloud technologies to smart grid. Renew Sustain Energy Rev 59:1326–1331
40. Sheikhi A, Rayati M, Bahrami S, Ranjbar AM, Sattari S (2015) A cloud computing framework on demand side management game in smart energy hubs. Int J Electr Power Energy Syst 64:1007–1016
41. Paudyal S, Cañizares CA, Bhattacharya K (2015) Optimal operation of industrial energy hubs in smart grids. IEEE Trans Smart Grid 6(2):684–694
42. Krause T, Kienzle F, Liu Y, Andersson G (2011) Modeling interconnected national energy systems using an energy hub approach. In: 2011 IEEE Trondheim PowerTech, pp 1–7
43. Moeini-Aghtaie M, Abbaspour A, Fotuhi-Firuzabad M, Hajipour E (2014) A decomposed solution to multiple-energy carriers optimal power flow. IEEE Trans Power Syst 29(2):707–716
44. Moeini-Aghtaie M, Dehghanian P, Fotuhi-Firuzabad M, Abbaspour A (2014) Multiagent genetic algorithm: an online probabilistic view on economic dispatch of energy hubs constrained by wind availability. IEEE Trans Sustain Energy 5(2):699–708
45. Maroufmashat A, Elkamel A, Fowler M, Sattari S, Roshandel R, Hajimiragha A, Walker S, Entchev E (2015) Modeling and optimization of a network of energy hubs to improve economic and emission considerations. Energy 93:2546–2558
46. Orehounig K, Evins R, Dorer V (2015) Integration of decentralized energy systems in neighbourhoods using the energy hub approach. Appl Energy 154:277–289

47. Salimi M, Ghasemi H, Adelpour M, Vaez-ZAdeh S (2015) Optimal planning of energy hubs in interconnected energy systems: a case study for natural gas and electricity. IET Gener Transm Distrib 9(8):695–707
48. Shabanpour-Haghighi A, Seifi AR (2015) Energy flow optimization in multicarrier systems. IEEE Trans Ind Inf 11(5):1067–1077
49. Yang H, Xiong T, Qiu J, Qiu D, Dong ZY (2016) Optimal operation of DES/CCHP based regional multi-energy prosumer with demand response. Appl Energy 167:353
50. Zhang X, Shahidehpour M, Alabdulwahab A, Abusorrah A (2015) Optimal expansion planning of energy hub with multiple energy infrastructures. IEEE Trans Smart Grid 6(5):2302–2311

Chapter 2
Impacts of Energy Storage Technologies and Renewable Energy Sources on Energy Hub Systems

Mohammad Mohammadi, Younes Noorollahi, and Behnam Mohammadi-Ivatloo

2.1 Introduction

Large-scale thermal power plants were the main source of energy in recent decades. Fossil fuels whose resources are ending are converted into other energies (mainly electricity) with very low efficiency at these plants. Transmission and distribution infrastructure over long distances are responsible for delivering this energy to consumers. However, such a structure of energy supply faces many problems. The problems caused by the fossil fuels consumption and greenhouse gases emissions have led to issues such as global warming and increasing international environmental concerns. Because of the scarcity of fossil fuels and the lack of access to the resources of this fuel in many countries in the world, it is not reasonable to use them at low-efficiency thermal power plants. On the other hand, problems such as the huge costs and losses of transmission and distribution systems, the difficulty of controlling and protecting these systems have made the current hierarchical systems not a suitable option for future energy supply. From another perspective, different energy systems were planned and managed independently [1]. But nowadays the development of technologies such as efficient multi-generation system leads to realizing the benefits of integrated energy infrastructure such as electricity, natural gas, and district heating networks, and thus a rapid movement toward multi-energy systems. In such systems, different energy carriers and systems interact together in

M. Mohammadi · Y. Noorollahi (✉)
Department of Renewable Energy and Environment, Faculty of New Sciences and Technologies, University of Tehran, Tehran, Iran
e-mail: m.mohammady@ut.ac.ir; noorollahi@ut.ac.ir

B. Mohammadi-Ivatloo
Department of Electrical and Computer Engineering, University of Tabriz, Tabriz, Iran
e-mail: bmohammadi@tabrizu.ac.ir

B. Mohammadi-Ivatloo, F. Jabari (eds.), *Operation, Planning, and Analysis of Energy Storage Systems in Smart Energy Hubs*,
https://doi.org/10.1007/978-3-319-75097-2_2

a synergistic way. However, consideration of such a concept requires a suitable tool for integrated management of the system components. Energy hub is an appropriate framework for modeling and optimal scheduling of multi-energy systems [2]. Therefore, the scarcity of fossil fuels, environmental concerns, and problems of centralized energy supply systems have led to an incentive to use energy efficient systems and alternative energy systems [3].

Nowadays, with the advent of DER in particular RES, ESS, and multi-generation systems, centralized large-scale power plants are now shifting to local and distributed energy sources. RES are one of the most commonly used distributed energy sources that their popularity is increasing day by day. From another perspective, renewable energies are much more stable than fossil fuels and have endless reserves. Therefore, these energies have also a remarkable role in sustainable development. This means that they have a much less destructive effect on the environment and can, in addition to meeting the needs of the current generation, respond to the needs of future generations and not be a threat to them. However, the main problem of these resources is their fluctuating and unpredictable nature. The production of these resources heavily depends on the location and time of their operation, which reduces the reliability of the operation of renewable systems. One of the main solutions to this problem is the use of energy storage systems. ESS, in addition to mitigating the effects of the integration of RES, can be used to provide ancillary services to energy networks and to participate effectively in demand response programs and to create a balance between energy production and demand. An energy hub can interact with different energy carriers so the energy hub can simultaneously utilize different RES and ESS. Each of these elements has some effects on the performance of the hub. The main objective of this chapter is to review and discuss the effects and the role of RES and ESS on the optimal management of energy hubs. In this regard, the role of renewable resources as inputs in energy hubs and energy storages to improve the reliability and flexibility of the energy hub is studied by reviewing previous research in this area. Finally, a model of energy hub is presented and the role of RES and electrical and thermal storage systems is discussed using numerical results.

2.2 Impact of RES on the Performance of EHs

In the energy hub models different energy carriers can interact with each other, and thus the energy hub can provide these different carriers through common sources such as electricity and natural gas networks or from renewable sources. Therefore, the range of inputs varies from fossil fuel sources to new and renewable technologies [4]. In large-scale and centralized power plants, which are major energy suppliers in many parts of the world, mostly fossil fuels are converted to low-efficiency electricity (low energy conversion efficiency at thermal power plants), and this electricity is transmitted to consumers with high losses, resulting in a large part of the primary energy is wasted in this system. For example, in a conventional coal-fired power plant, 72% of the primary energy is wasted and only 28% of it reaches the final consumer [5].

DER can be defined as systems for producing or storing energy at or near the place of consumption. The development of these systems will reduce the waste of primary energy, reduce transmission losses and thus reduce operating costs. DER have the ability to use different technologies such as fuel cells, micro gas turbines, waste heat recovery equipment, renewable technologies such as small wind turbines and PV. These types of on-site energy generation resources can be one of the main sources of energy for an energy hub. A model for optimal scheduling of a DER including renewable resources and storage systems along with power distribution networks with the goal of minimizing energy costs has been presented in [6]. Renewable resources can play an essential role in DER and their share is increasing rapidly. The inefficiency of fossil-fuel-based energy systems has led to the integration of RES with these systems and the move towards 100% renewable energy systems [7]. The use of RES, such as biomass, solar, fuel cells and the use of waste heat in the different co-generation and poly-generation technologies have been studied in [8]. The role of different technologies to achieve a 100% renewable energy system in Europe in 2050 has been discussed in [9]. The results showed that by using existing technologies there is a possibility to achieve a 100% renewable scenario, due to the possibility of optimal integration of different energy carriers. This reveals the importance of multi-energy systems and energy hub models in optimal utilization of energy resources, especially in the future renewable energy systems.

Therefore, different energy carriers used in energy hubs can be supplied only through renewable sources, and the consumption of fossil fuels in the energy hub can be zero. For example, electricity and heat of an energy hub can be generated from solar and geothermal sources. In addition to generating electricity and heat, it is also possible to produce water in a fuel cell, as one of the most promising renewable technologies. Wind power can be used to supply electricity for various applications. In this regard, the possibility of using PV along with CHP in energy hub models for centralized cooling, heating, and electricity energy supply in a residential area has been investigated in [5]. The results of this study have shown that the use of PV in addition to supplying electricity demand in district level also provides the possibility of selling excess electricity to the grid. The biomass is another renewable energy that can be used in various forms in energy hubs and can provide different energy carriers such as electricity, heat, and transport fuels. A complete model of the various components of the biomass supply chain, including electricity, heating, and gas infrastructures for modeling various biomass technologies has been offered in [10]. A comprehensive overview of the biomass energy conversion models for generating electricity, heat, and fuel, along with a discussion of the challenges in this area, can be found in [11]. An assessment has been conducted in [12] to reduce the share of fossil fuels and increase the share of RES in the form of energy hub models for a village in Switzerland. The study focuses on the development of renewable technologies such as PV, biomass-based district heating, and small hydroelectric power plants to reduce costs and emissions. By developing this model, the authors in [13] provided a model for planning a hybrid renewable energy supply system for the village in the form of different structures of energy hubs. The results show that increasing the share of RES in the current energy supply system in the framework of energy hub models will lead to increased autonomy, peak shaving, and emission reductions.

Renewable fuels such as hydrogen and ethanol can be obtained from biomass. In this regard, a model for planning the conversion of biomass to hydrogen is presented in [14] to minimize the annual cost of energy. A framework for modeling of the fuel cell, electrolyzer, and hydrogen tanks as ESS in the content of smart grid and in the presence of RES can be found in [15]. Different aspects of the use of hydrogen as a clean fuel in future transportation systems are investigated in [16]. The results show that hydrogen is a promising option for using in future energy systems and emissions reduction. A review of various hydrogen production technologies from renewable sources and related issues can be found in [17]. The use of hydrogen infrastructures in the energy hub models has been investigated in [18]. In this study, the optimal planning of hydrogen infrastructures along with infrastructure such as electricity, gas, and district heating networks has been carried out in a network of interconnected energy hubs. The results indicated a higher degree of freedom in optimizing the system and improving the overall performance of the system with the presence of hydrogen infrastructure. The effects of the presence of the hydrogen distribution system in the form of a fueling station in the structure of an energy hub have been investigated in [19]. The results showed that the optimal interaction of hydrogen fuel supply system with commercial and residential energy hubs in a smart urban energy system leads to a reduction in the cost and emission of the whole system.

Despite the many benefits of using RES, so far little attention has been paid to these energy sources as inputs in energy hub models [20]. So that most of the energy hub models presented so far have used electricity and natural gas networks as their main inputs. The most commonly used renewable sources are the wind and solar power which can be found in 20% of the energy hub models [20]. However, the use of other renewable energies, especially biomass and clean fuels such as hydrogen is very limited. The energy hub models in the future should move towards modeling sustainable energy systems. Using fossil-fuel-based energy distribution networks with many problems in their structure cannot provide a comprehensive model of future sustainable energy systems. There is a great potential for studying the effects of renewable sources in the framework of energy hub models. Therefore, energy hub models require the use of RES and the integration of these resources to meet the demand for various renewable energy systems in the future.

2.3 Impact of ESS on the Performance of EHs

As discussed in the previous section, energy systems around the world need to move towards renewable energy systems to achieve sustainable energy systems. However, one of the main problems of RES is the intermittent nature and unpredictable power generation. Consequently, in renewable systems, production control is not easy to adapt to the pattern of consumption. One of the main solutions to this problem is the use of ESS, which facilitates the integration of RES. ESS stores the energy when it is not needed and provides energy when it is needed. Using ESS in energy systems will increase system efficiency, reduce operating system costs, reduce the size of production and transmission systems, reduce fossil fuel consumption, and reduce emissions [21]. However, in addition to facilitating the integration of RES, storage systems can have various applications in the energy systems which are discussed in the following sections.

2.3.1 The Ultimate Goal of Using ESS

ESS has various applications in energy systems due to its various types. Categories of energy storage systems and various technologies are presented in Fig. 2.1.

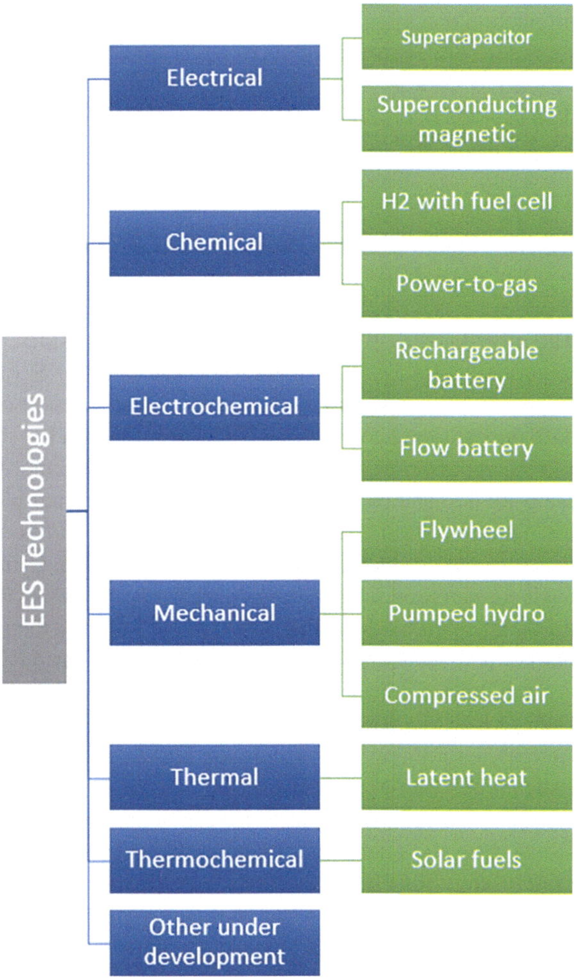

Fig. 2.1 Classification of EES technologies

Determining the proper storage system for the energy system under planning requires a complete understanding of different energy storage technologies. Various indicators can influence the choice of ESS for an energy system. These include capacity, initial cost, efficiency, lifetime, storage capacity, maturity, charging time, response time, and storage loss. Comparison and investigation of various storage technologies from the viewpoint of the above indicators can be found in [22, 23]. After recognizing the characteristics of different ESS, it is essential to determine its ultimate goal and purpose of using ESS in the energy system. Different objectives for using ESS in energy systems can be categorized into three categories [22]:

- Facilitating the integration of RES and improving system reliability [21].
- Improving system resilience and providing ancillary services.
- Increasing system flexibility and moving towards smart energy systems.

On the production side, ESS can be used to improve the pattern of RES production and align it with demand behavior. ESS can be used to provide ancillary services to the network and increase its stability. On the consumption side, ESS can be used as DER to meet the needs of subscribers and facilitate their participation in demand-side management (DSM) programs. ESS can be operated for increasing the benefit of the storage owner (merchant storage) [24].

2.3.1.1 Facilitating the Integration of RES and Improving System Reliability

The most known application of ESS can be considered as solving the problem of integration of RES. The power generated by RES varies in different times and places. On the other hand, the pattern of production of these resources may vary with the pattern of consumption. For example, in a PV system used to provide power to residential houses, peak power production occurs during midday hours, while the peak demand of the residential consumers usually occurs in the early hours of the night. Therefore, ESS can be used to store additional power and use it at peak hours for the production and consumption balance. Using ESS in an isolated system leads to increased reliability and facilitates the use of RES in these systems. In the grid-connected mode, using ESS leads to tracking the pricing pattern in the energy market and reducing system operating costs. Various ESS applications in power systems with emphasis on RES integration have been discussed in [25] and it has been proved that RES in the presence of ESS become controllable and dispensable sources. Examining the appropriate ESSs for wind power integration, as well as issues related to size determination and control systems, can be found in [23]. A survey in Europe was conducted in [26] to achieve a 100% renewable energy system focusing on the effects of storage systems. In this study, data from Germany were used to study solar and wind resources. The results showed that only solar and wind resources could supply 50% of Germany's electricity demand, and with the addition of ESS, this would increase by 80%.

Another important argument is the increasing of the system reliability in the presence of ESS. In power grid, the presence of spinning reserve can result in an appropriate response if an imbalance between production and demand is generated. Nevertheless, in off-grid systems, due to the limitations in the capacity of energy production and conversion systems, there is always no way to benefit from such a spinning reserve. In these types of systems, the ESS can be used to respond to imbalances in the system and stabilizing the system [27]. An imbalance in energy systems occurs for two reasons: the sudden drop in production or the sudden rise in demand. In systems that are separate from the network as well as in systems where the share of RES is high, this imbalance can have a huge impact on the reliability of the system. In these systems, the use of ESS is very important for responding to sudden changes in production or demand. Therefore, consideration of factors such as response times and ESS ramp rates for such purposes should be carefully checked.

In summary, the first goal of using ESS in energy systems is to improve the performance of RES and facilitate integration in order to increase the reliability of the system and create a balance between production and demand in renewable energy systems and so renewable energy hubs.

2.3.1.2 Improving System Resilience and Providing Ancillary Services

In power grid and more generally in energy systems, ESS can be used to increase system stability. In this case, ESS is used to reduce uncertainties, improve power quality, provide ancillary services to the network, and improve its conditions. Some of the applications of ESS for this purpose are frequency regulation, spinning reserve, voltage regulation, network inertia, volatility reduction, black starter energy supply, network synchronization, direct voltage supply in fault conditions, and equipment capacity optimization. In addition, the use of ESS for demand shifting in different periods of energy prices will lead to peak shaving and smoother demand curve. This will reduce the cost of production and transmission of electricity for the power grid, as well as improve its stability in peak hours. Further discussions in this area can be found in [28].

2.3.1.3 Increasing System Flexibility and Moving Towards Smart Energy Systems

The third objective for using ESS is to focus on their applications on demand side. Installing ESS on the consumer side, in addition to providing consumer energy, enables their active participation in DSM programs and benefits from the smart grid advantages. The effects of the final consumer storage systems in the content of the smart grid in the presence of RES and demand response (DR) programs have been investigated in [29]. The results show that the presence of ESS leads to better tracking of energy prices, increased productivity, and improved performance of other system equipment, especially distributed generation (DG) sources such as CHP. Thus, it can be said that ESS in the smart grid content can lead to a balance between production and energy consumption, smoothing the consumption curve, benefiting from the advantages of DG especially RES, and to improve the overall system performance and increasing its productivity.

One of the most promising smart technologies that have recently been considered as potential storage system is plug-in electric vehicle (PEV). These vehicles are mainly on the consumer side and have the ability for bi-directional power transfer with the network. The advanced technology used in the batteries of these vehicles has increased their charging and discharge rates. This has caused their potential applications in providing network-side services such as the spinning reserve, frequency regulation, and network stability. However, one vehicle alone cannot provide such a service, because participation in services such as frequency regulation requires high power capacity and fast dynamic response. So, as a rule, a large number of these vehicles are controlled centrally by the aggregators, so that, with the optimum control of the charge and discharge of these vehicles, in addition to lowering the costs of subscribers, they can provide ancillary services to the network [30]. In a study to optimize PEV charging program in the residential micro grid, three different technologies used in the battery of PEVs were investigated in presence of RES [31]. The results showed that optimal control of PEV leads to its successful operation for peak shaving purposes.

In general, we can say that the presence of ESS on the demand side can provide the possibility to benefit from the advantages of the smart grid. In some cases, such as PEV, with the coordinated control of the storages of these vehicles, in addition to meeting requests and reducing the cost of owners of these vehicles, they can be used as an energy storage system for the entire system and for realizing the concept of the smart grid.

2.3.2 Optimal Scheduling of ESS in EHs

When the main purpose of the ESS application was identified, then the optimal planning is important in the next step. At this stage, the goal is to determine the appropriate size for the storage system, which includes items such as power capacity and should be selected based on various parameters. Various parameters such as resource capacity, the pattern of consumption, climatic conditions, etc. affect the proper selection of ESS. Various parameters such as resource capacity, the pattern of consumption, climatic conditions, etc. affect the proper size selection of ESS. In the hub energy models, the type of connection and ESS installation location must also be carefully checked. In the energy hub, ESS can be installed at the place of production or purchase from the network and/or consumption side, so that the energy carrier can be stored on the input side, or after being transformed into a qualified energy carrier be stored at the demand. The choice of this installation location should be based on the desired indicators in the objective function and make the maximum controllability for ESS.

Storage systems can also be installed or controlled in a distributed or aggregated manner [32]. In distributed mode, each storage system is individually connected to the system and controlled, but in the aggregated scheme, a large group of storages is managed by a central control system. Therefore, the choice of an appropriate control strategy for ESS should also be considered. To design a charging and discharging controller for ESS, there are a lot of things to consider, including technical constraints, resource capacity or forecasting of production capacity in RES resources, energy pricing and market conditions, patterns of consumption, climatic conditions, and so on [33]. A model is presented in [34] to optimize the performance of a hybrid renewable energy system with a combination of a wind turbine, diesel and biomass generators in the presence of ESS. The presence of diesel generator with low inertia and variations in wind power production leads to voltage and frequency disturbances in this system. The results indicate that the use of ESS along with a suitable controller in the short term is essential for maintaining the system's stability and power quality, and in the long term, it will improve system performance and smooth the demand curve. Given the little work that has been done on ESS control systems in the energy hub, and this fact the models presented so far have used a simple charging strategies, there is a good potential for designing and studying the effects of different control strategies on the optimal performance of the energy hubs.

2.3.3 ESS Performance in EHs

This section reviews the recent research done on the application of ESS in multi-energy systems and energy hubs. The authors in [35] examined the effects of the thermal storage size on the performance of a multi-energy system for generating electricity, heat, and cooling in the presence of RES. The results of this study showed that the use of thermal storage leads to optimization of equipment capacity (reducing the need for production of heat and reducing boiler capacity), reducing primary energy consumption and increasing system efficiency. Feasibility study of pit thermal storage, to capture the waste heat produced in a biomass poly-generation system, with district heating and cooling networks in residential buildings is done in [36]. The results indicate that pit storage is a suitable method for combination with a biomass power plant which increases the annual efficiency of the system. The effects of adding a pump storage system to the energy system of a touristic resort, a system for supplying electricity, heat, and water, have been investigated in [37]. The results indicate that adding new storage system will reduce the discharge power of the available battery, increase its lifetime, and reduce system's costs. In another study, the effects of ESS and RES in a combined cooling, heating and power (CCHP) production system have been investigated in [38]. In this study, it has been shown that increasing the contribution of RES to electricity production reduces the need for heat generation by CCHP and directly affects the capacity of the thermal storage and reduces its capacity. The results indicate that RES and ESS have interactions on each other, even if they do not have a direct connection. Same authors have optimized the performance of a system for supplying electricity, heat, and cooling, taking into account different storage combinations in [21]. The results of the study of the effects of different storage systems have shown that the use of thermal storage along with CHP leads to a decrease in the dependence of the system on the main grid and an increase in power generated by CHP. Optimal design and operation of advance compressed air energy storage and air source heat pumps in CCHP systems is studied in [39]. Impact of battery energy storage system on operation of renewable energy based CCHP system is studied in [40]. The effect of thermal storage on a poly-generation system in the presence of a ground source heat pump has been investigated in [41]. In the designed system, electricity demand is purchased directly from the grid, the electricity produced by the CHP is consumed by the heat pump, and the excess heat generated by the heat pump is stored in the thermal storage. This combination allows the thermal storage to react to changes in the price of electricity and acts as an electrical storage for the system, while the initial cost of a thermal storage is much lower than an electrical storage. As a result, with this combination, the use of thermal storage leads to a demand shift to off-peak hours, reduces the capacity of the equipment, and reduces operating costs of the system.

Due to the various features and applications of ESS, an optimal combination of storage systems can be used to achieve the desired result or to meet different goals in a system. The idea of combining different storage systems with their application can be found in [22]. In this work, a diagram of possible combinations of different ESS is presented to minimize costs considering technical constraints. Considering the complementary features of various storage systems and considering the design and optimal management of hybrid storage systems have been done in [42].

In the energy hub models, as previously mentioned, ESS can be embedded in different places and have different effects. The impact of thermal energy market on operation of energy hub with heat and electrical storage is studied in [43]. The effects of various ESS such as electrical, gas, and heat storages from the perspective of operating costs on the performance of the energy hub are investigated in [44]. Also, the effect of different parameters such as the horizon of prediction and ESS size on the optimal performance of the energy hub has been studied in [45]. The results of this study showed that, in addition to the size of the ESS which affects system costs, an increase in simulation horizons could also reduce system costs, and even its impact can be more than increasing storage size. Therefore, in modeling, there should be a balance between increasing the computational time due to increased forecast horizons as well as the size of the storage system in order to achieve optimal performance and cost of the system.

As discussed, the energy storage system is one of the main systems in the energy hub, but unfortunately, so far, little research has been done on the effects of this system and its optimal control in the framework of energy hub models. Optimal planning and placement, as well as designing an appropriate control strategy, are potential fields for studying ESS in energy hubs that require more research and studies. Energy hub models provide the ability to use different ESSs and even different combinations of them, which could be the subject of future research in this area.

2.4 Case Studies

In this section, the performance of RES and ESS in the optimal energy hub management problem is studied by modeling an energy hub and evaluating numerical results. For this purpose, a complete model of the energy hub is used which has different inputs, converters, storages, and outputs to meet different demands. A schematic representation of the studied energy hub can be seen in Fig. 2.2. In this model of the energy hub, the wind turbine is used as a source of renewable energy production. This energy hub is powered by electricity and natural gas networks. Transformer, converters, CHP, and boiler have been used to convert various energy carriers. Electrical and thermal storages are also used as energy storage systems. On demand side, given the usual demands for energy systems, electricity, heat, and natural gas demands are considered for this energy hub model.

2.4.1 Energy Hub Modeling

In the energy hub, various objective functions can be considered. The proposed objective function is formulated based on the cost of purchasing energy (electricity and natural gas), electricity sales to the grid, the cost of charging and discharging electrical and thermal storages, emission costs and reliability indicators. The objective function is formulated in a deterministic environment of wind speed, demand, and hourly price of the electricity market. This objective function is optimized to minimize operational costs in a one-day time horizon subject to different constraints. The objective function of the optimal management problem of the proposed energy hub can be considered as follows:

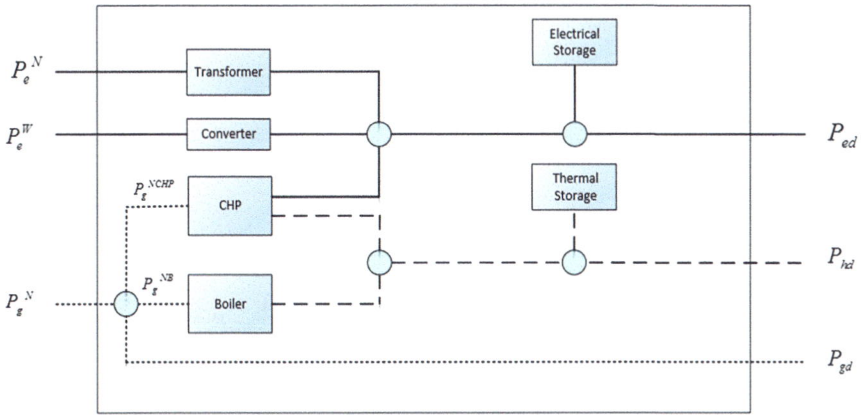

Fig. 2.2 Schematic representation of the proposed energy hub

minimize

$$
\begin{aligned}
\mathrm{TC} = & \sum_{t=1}^{24} \left[\pi_e{}^N(t) P_e{}^N(t)\right] + \left[\pi_e{}^W P_e{}^W(t)\right] \\
& + \left[\pi_g{}^N P_g{}^N(t)\right] + \left[\pi_e{}^S \left(P_e{}^{\mathrm{ch}}(t) + P_e{}^{\mathrm{dis}}(t)\right)\right] \\
& + \left[\pi_h{}^S \left(P_h{}^{\mathrm{ch}}(t) + P_h{}^{\mathrm{dis}}(t)\right)\right] + \left[\pi_e{}^{\mathrm{ENS}} P_e{}^{\mathrm{ENS}}(t)\right] + \left[\pi_h{}^{\mathrm{ENS}} P_h{}^{\mathrm{ENS}}(t)\right] \\
& + \left[\sum_{t=1}^{24} \sum_{\mathrm{em}=1}^{3} \pi_{\mathrm{em}} \left(\mathrm{EF}_{\mathrm{em}}{}^N P_e{}^N(t) + \mathrm{EF}_{\mathrm{em}}{}^{\mathrm{NCHP}} P_g{}^{\mathrm{NCHP}}(t) + \mathrm{EF}_{\mathrm{em}}{}^{\mathrm{NB}} P_g{}^{\mathrm{NB}}(t)\right)\right]
\end{aligned}
\tag{2.1}
$$

Subject to

$$
\begin{aligned}
P_{\mathrm{ed}}(t) = & \left[A^N \eta_T P_e{}^N(t)\right] + \left[A^{\mathrm{CHP}} \eta_{\mathrm{echp}} P_g{}^{\mathrm{NCHP}}(t)\right] \\
& + \left[A^W \eta_C P_e{}^W(t)\right] + \left[P_e{}^{\mathrm{dis}}(t) - P_e{}^{\mathrm{ch}}(t)\right] \\
& + \left[P_e{}^{\mathrm{shdo}}(t) - P_e{}^{\mathrm{shup}}(t)\right] + \left[P_e{}^{\mathrm{ENS}}(t)\right]
\end{aligned}
\tag{2.2}
$$

$$
\begin{aligned}
P_{\mathrm{hd}}(t) = & \left[A^{\mathrm{CHP}} \eta_{\mathrm{hchp}} P_g{}^{\mathrm{NCHP}}(t)\right] + \left[\eta_B P_g{}^{\mathrm{NB}}(t)\right] \\
& + \left[P_h{}^{\mathrm{dis}}(t) - P_h{}^{\mathrm{ch}}(t)\right] + \left[P_h{}^{\mathrm{ENS}}(t)\right]
\end{aligned}
\tag{2.3}
$$

$$
P_{\mathrm{gd}}(t) = \left[P_g{}^N(t)\right] - \left[P_g{}^{\mathrm{NCHP}}(t)\right] - \left[P_g{}^{\mathrm{NB}}(t)\right] \tag{2.4}
$$

$$
-P_e{}^{N\max} \le P_e{}^N(t) \le P_e{}^{N\max} \tag{2.5}
$$

$$
0 \le P_g{}^N(t) \le P_g{}^{N\max} \tag{2.6}
$$

$$
\eta_T P_e{}^N(t) \le P^T \tag{2.7}
$$

$$
\eta_{\mathrm{echp}} P_g{}^{\mathrm{NCHP}}(t) \le P^{\mathrm{CHP}} \tag{2.8}
$$

$$
\eta_B P_g{}^{\mathrm{NB}}(t) \le P^B \tag{2.9}
$$

$$
P_e{}^S(t) = P_e{}^S(t-1) + \eta_e{}^{\mathrm{ch}} P_e{}^{\mathrm{ch}}(t) - \frac{P_e{}^{\mathrm{dis}}(t)}{\eta_e{}^{\mathrm{dis}}} - P_e{}^{\mathrm{loss}}(t) \tag{2.10}
$$

$$
P_e{}^{\mathrm{loss}}(t) = \alpha_e{}^{\mathrm{loss}} P_e{}^S(t) \tag{2.11}
$$

$$\alpha_e^{\min} P_e^{SC} \leq P_e^S(t) \leq \alpha_e^{\max} P_e^{SC} \tag{2.12}$$

$$\alpha_e^{\min} P_e^{SC} I_e^{ch}(t) \leq P_e^{ch}(t) \leq \alpha_e^{\max} P_e^{SC} I_e^{ch}(t) \tag{2.13}$$

$$\alpha_e^{\min} P_e^{SC} I_e^{dis}(t) \leq P_e^{dis}(t) \leq \alpha_e^{\max} P_e^{SC} I_e^{dis}(t) \tag{2.14}$$

$$0 \leq I_e^{ch}(t) + I_e^{dis}(t) \leq 1 \tag{2.15}$$

$$P_h^S(t) = P_h^S(t-1) + \eta_h^{ch} P_h^{ch}(t) - \frac{P_h^{dis}(t)}{\eta_h^{dis}} - P_h^{loss}(t) \tag{2.16}$$

$$P_h^{loss}(t) = \alpha_h^{loss} P_h^S(t) \tag{2.17}$$

$$\alpha_h^{\min} P_h^{SC} \leq P_h^S(t) \leq \alpha_h^{\max} P_h^{SC} \tag{2.18}$$

$$\alpha_h^{\min} P_h^{SC} \frac{1}{\eta_h^{ch}} I_h^{ch}(t) \leq P_h^{ch}(t) \leq \alpha_h^{\max} P_h^{SC} \frac{1}{\eta_h^{ch}} I_h^{ch}(t) \tag{2.19}$$

$$\alpha_h^{\min} P_h^{SC} \eta_h^{dis} I_h^{dis}(t) \leq P_h^{dis}(t) \leq \alpha_h^{\max} P_h^{SC} \eta_h^{dis} I_h^{dis} \tag{2.20}$$

$$0 \leq I_h^{ch}(t) + I_h^{dis}(t) \leq 1 \tag{2.21}$$

In this objective function, the first term is related to the value of power exchange with the electricity network. The second term is the cost of providing electric power from the wind turbine. The third term also refers to the cost of purchasing natural gas from the network. Fourth and fifth terms are included in order to take into account the operating costs of charging and discharging storages. Fifth and sixth terms are related to the electrical and thermal energies not supplied cost. And finally, the last term is related to environmental costs of different greenhouse gas emissions. In the above relations, Eqs. (2.2)–(2.4) are related to power equilibrium, so that at each step of the simulation demand is equal to the total energy generation.

Equations (2.5) and (2.6) are defined to take into account the technical and contractual limitations of gas and electricity networks and assume a maximum amount of the power exchanged with these networks. Hub components are installed with limited production capacity and for considering this maximum capacity for the transformer, CHP and boiler, Eqs. (2.7)–(2.9) are defined respectively. Equations (2.10)–(2.15) are related to the operational constraints of electrical storage. Equation

(2.10) refers to the state of charge of storage. The content of the storage at any time is a function of the storage content in the previous step, as well as the amount of charge, the amount of discharge, and the amount of storage loss in that time step. According to (2.11), the amount of electrical storage losses is defined as a certain percentage of its charge content. The allowed amount of storage content in each step is specified by (2.12). Binary variables $I_e^{ch}(t)$ and $I_e^{dis}(t)$ are defined in such a way that storage charging and discharging do not occur simultaneously. So at any time step, only one of the binary variables can have a value of 1. This constraint is applied through (2.15). Thermal storage constraints are defined in the same way in (2.16)–(2.21).

2.4.2 Simulation Results

Five cases are considered to evaluate the effects and roles of RES and ESS in the energy hub operation. The case study results are compared and analyzed based on operational cost, emission, and reliability. These five cases are categorized in Table 2.1. In case 1, the energy hub also has CHP and boiler in addition to the possibility of purchasing energy from electricity and gas networks. This case is considered as the base case of the model. In case 2, the electrical storage (ES) is added to the base case. Case 3 uses a heat storage (HS) instead of an electrical storage to balance production and demand. However, the case 4 energy hub uses an on-site wind turbine (WT) for clean electric power generation. Finally, a combination of electrical and thermal storages alongside wind turbine and CHP is evaluated in case 5. The demand of energy hub for electricity, heat, and natural gas can be seen in Fig. 2.3. Also, hourly electricity prices and hourly wind speed are shown in Figs. 2.4 and 2.5, respectively.

The values of the input parameters and other assumptions for the optimal energy hub management problem can be found in Table 2.2.

The proposed optimal management problem for energy hub is an MILP model that has been solved in the GAMS software using the CPLEX solving algorithm. Due to the linearity of the objective function and the convexity of the solution space, the solutions obtained from the problem are the optimal global solutions. These results are discussed for different cases in the following sections.

Table 2.1 Defined cases for the energy hub optimal operation

Cases	Energy hub structure
Case 1 (base case)	Base
Case 2	Base + ES
Case 3	Base + HS
Case 4	Base + WT
Case 5	Base + WT + HS + ES

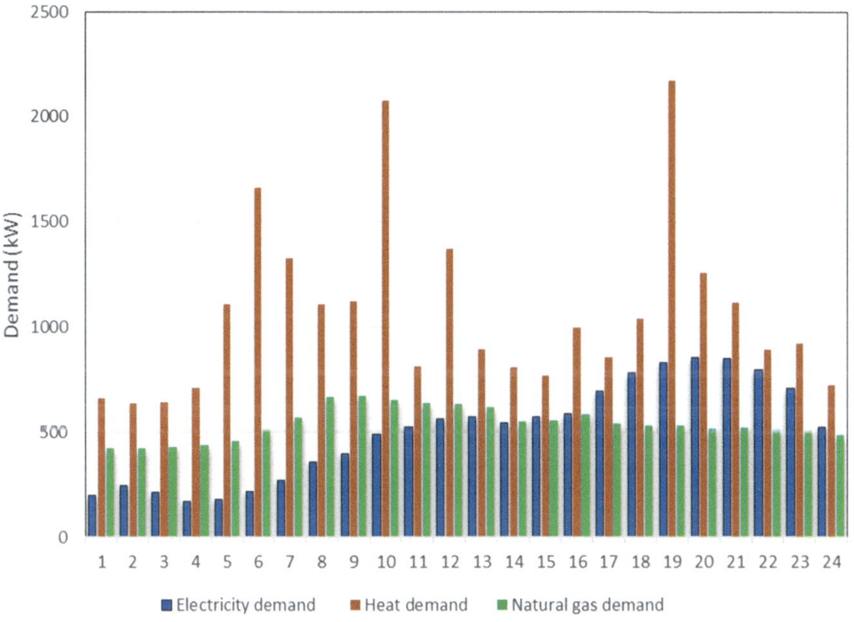

Fig. 2.3 The hourly electricity, heat, and natural gas demand

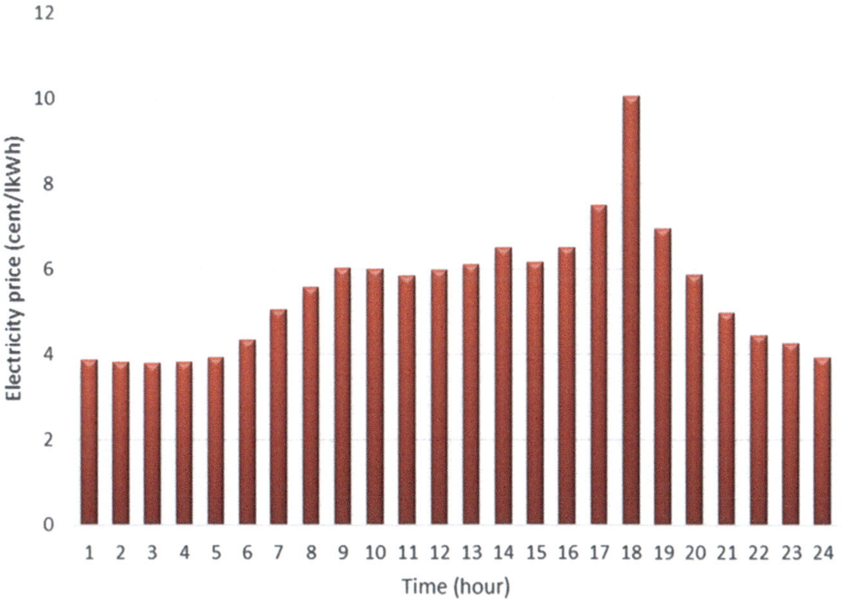

Fig. 2.4 The hourly electricity price

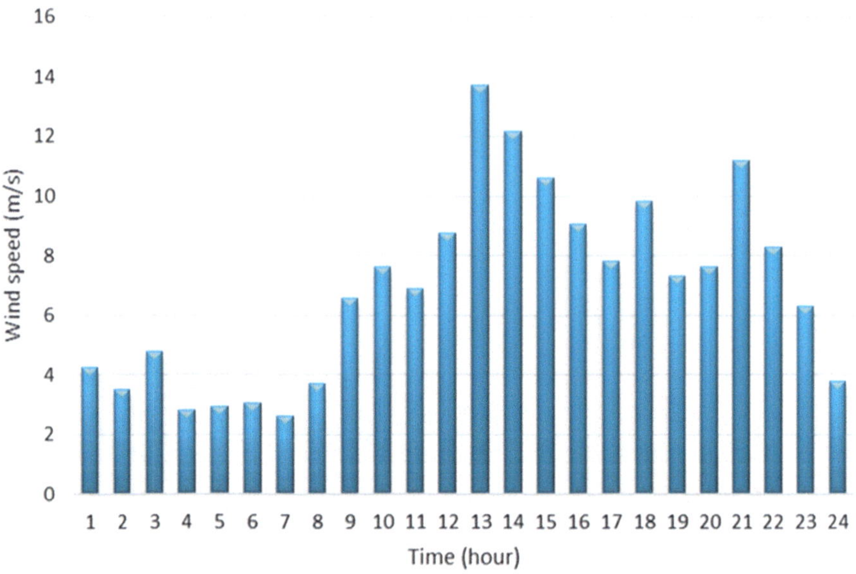

Fig. 2.5 The hourly wind speed

2.4.2.1 Case 1

In this case, CHP is the main supplier of electrical and thermal demand, and electricity and boiler networks are considered as a backup system for CHP. If demand is not fully met by CHP, the remainder of the demand will be supplied from the grid. In the case of heat demand, the boiler is also responsible for the thermal power deficit. The numerical results of this case can be seen in Table 2.3.

As shown in the table above, CHP is responsible for supplying electrical and thermal energy with maximum capacity during the day. This is due to lower gas price than electricity and the possibility of supplying heat demand simultaneously. In cases where the demand increases and the CHP is unable to meet this demand, the fraction of this power is purchased from the main network, which leads to an increase in operating costs of the system. The existence of CHP in the system allows for the sale of excess electricity to the network. In the table above, the negative values for power exchanged with the network are sales of this power to the network. So, in the early hours of the day, the energy hub can earn money by selling excess electricity to the network. However, it is observed that part of the electrical demand does not come at peak times, which reduces the reliability of the energy hub. The schematic representation of the above concepts can be seen in Fig. 2.6. This figure shows how to supply the energy demand by the energy hub.

In the case of heat demand, 252 kW of heat demand is produced by CHP and the rest of the demand is provided by burning gas in the boiler. In this case, the current structure of energy hub in the heat demand peak hours (hours 10 and 19) is not able

Table 2.2 Energy hub input data and parameters

Parameter	Unit	Value	Parameter	Unit	Value
α_e^{loss}	–	0.02	$EF_{em}^{CHPCO_2}$	kg/kWh	0.412
α_h^{loss}	–	0.02	$EF_{em}^{CHPSO_2}$	kg/kWh	0.008
α_e^{min}	–	0.1	$EF_{em}^{CHPNO_2}$	kg/kWh	0.000112
α_e^{max}	–	0.9	$EF_{em}^{BCO_2}$	kg/kWh	0.617
α_h^{min}	–	0.1	$EF_{em}^{BSO_2}$	kg/kWh	0.011
α_h^{max}	–	0.9	$EF_{em}^{BNO_2}$	kg/kWh	0.000284
η_e^{ch}	–	0.9	$eELF_{max}$	kg/kWh	0.05
η_e^{dis}	–	0.9	$EF_{em}^{NCO_2}$	kg/kWh	0.424
η_h^{ch}	–	0.9	$EF_{em}^{NSO_2}$	kg/kWh	0.00226
η_h^{dis}	–	0.9	$EF_{em}^{NNO_2}$	kg/kWh	0.000925
η_C	–	0.9	π_e^{ENS}	₵/kWh	20
η_T	–	0.9	π_h^{ENS}	₵/kWh	20
η_{eCHP}	–	0.4	π_e^S	₵/kWh	2
η_{hCHP}	–	0.35	π_e^W	₵/kWh	0
η_B	–	0.85	$\pi_{em}^{CO_2}$	₵/kg	0.014
A^{CHP}	–	0.96	$\pi_{em}^{SO_2}$	₵/kg	0.99
A_e^N	–	0.99	$\pi_{em}^{NO_2}$	₵/kg	4.2
A^W	–	0.96	π_g^N	₵/kWh	1.838
$P_e^{N\,max}$	kW	600	π_h^S	₵/kWh	2
$P_g^{N\,max}$	kW	4000	P^{CHP}	kW	300
P^T	kW	600	P_e^{SC}	kW	300
P^B	kW	1800	P_h^{SC}	kW	300

to provide all the demand and part of this heat demand is not provided. In this case, the energy hub faces an energy not supplied penalty and increases operating costs. The total operational cost of the energy hub, in this case, is 190,734.8 Euro cents.

2.4.2.2 Case 2

In this case, an electrical storage is added to the system so that during the excess electricity production period, some of this additional power is stored and used at times required to meet the demand. The numerical results of this case are summarized in Table 2.4. In this case, the electrical storage is charged at times when the energy price is low and it provides part of the electrical demand at peak hours. How to exchange electrical energy in the energy hub in the presence of an electrical storage can be seen in Fig. 2.7.

The existence of the electrical storage leads to a reduction in the electrical energy sold to the network from 519.3 kWh in the base case to 417.7 kWh in this case. The reason for this can be attributed to spending some of that energy on charging electrical storage at low-cost energy hours. So that even the charging of the storage at 4:00 am leads to the purchase of energy from the network. However, the purchase

Table 2.3 Optimal operational plan for energy hub in the first case

t	P_e^N	P_e^{CHP}	P_e^{ENS}	P_g^N	P_g^{NB}	P_g^{NCHP}	P_{gd}	P_h^B	P_h^{CHP}	P_h^{ENS}
1	−100.9	288	0	1656.3	483.7	750	422.7	411.1	252	0
2	−47.9	288	0	1632	455.6	750	426.4	387.3	252	0
3	−87	288	0	1636.1	458.4	750	427.7	389.6	252	0
4	−135	288	0	1728.8	542.4	750	436.4	461	252	0
5	−120.9	288	0	2214.5	1007.1	750	457.4	856	252	0
6	−81.4	288	0	2922.6	1660.5	750	512	1411.5	252	0
7	−18.2	288	0	2586.3	1267.8	750	568.5	1077.6	252	0
8	81.6	288	0	2426.9	1009.7	750	667.2	858.2	252	0
9	124.7	288	0	2444.8	1024.2	750	670.6	870.6	252	0
10	227.1	288	0	3519	2117.6	750	651.3	1800	252	25.5
11	265.8	288	0	2046.1	658.7	750	637.4	559.9	252	0
12	309.3	288	0	2702.4	1320.2	750	632.1	1122.2	252	0
13	319.2	288	0	2126.9	759.9	750	617	645.9	252	0
14	284	288	0	1952.6	654.3	750	548.3	556.2	252	0
15	316.9	288	0	1914	610.5	750	553.6	518.9	252	0
16	335.4	288	0	2213.5	877.8	750	585.8	746.1	252	0
17	453	288	0	2000.4	710.3	750	540.1	603.8	252	0
18	552.4	288	0	2210.9	928.4	750	532.5	789.1	252	0
19	600	288	9.9	3397.4	2117.6	750	529.7	1800	252	120.4
20	600	288	33.1	2456.2	1189.3	750	516.9	1010.9	252	0
21	600	288	28.3	2291.2	1021.2	750	520	868	252	0
22	572.9	288	0	2016.1	755.5	750	510.7	642.2	252	0
23	471.1	288	0	2047.4	790	750	507.4	671.5	252	0
24	264.6	288	0	1794.6	555.7	750	489	472.3	252	0
$\sum_t$	5786.7	6912	71.3	53,937	22,976.4	18,000	12,960.7	19,529.9	6048	145.9

of energy occurs in the hours when energy price is low. The addition of the electrical storage device results in a significant reduction in the amount of electrical energy not supplied, and only a small amount of electrical energy is not provided at 7:00 pm. The most discharge amount occurs at an hour when the price of electrical energy is at its highest (6:00 pm). This causes the electrical storage to have the greatest impact in reducing the operating costs of the energy hub. In the thermal behavior of the energy hub, there is no change and its operational plan is similar to the base case for thermal demand. The set of these factors will reduce the total operating cost of the energy hub by 189,930.7 cents. Therefore, it can be said that addition of an electrical storage in addition to reducing operating costs leads to increase of reliability of the system in the field of supply of electrical demand.

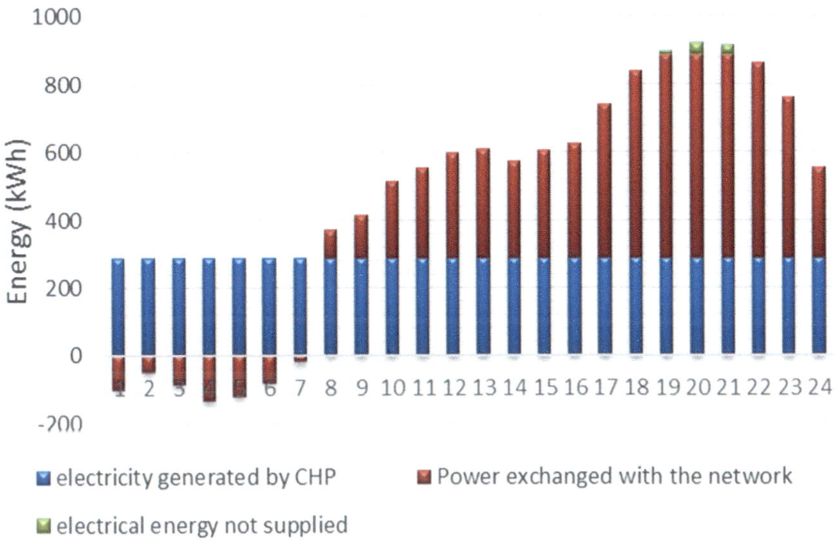

Fig. 2.6 Supply of electricity demand by the energy hub in case 1

2.4.2.3 Case 3

In this case, in order to compare the effectiveness of the heat storage in the optimal performance of the energy hub, a thermal storage is added to the base state. Figure 2.8 shows how to supply the heat demand by the energy hub in this case.

By adding a thermal storage, some thermal energy is stored in the non-peak hours of thermal demand in this storage and it is used in peak hours. Charging the heat storage in the first hour leads to an increase in gas purchases from the grid and an increase in boiler production compared to the base case. The same thing can be seen at 6:00 pm. An increase in the state of charge of the heat storage at 6:00 pm will result in the heat demand deficit being compensated at 7:00 pm (peak hour), and this demand will be fully met at this hour. This will increase the reliability of the system. In terms of cost, the total operating cost of the energy hub, in this case, is 189,059.6 cents, which is lower than both previous cases. If we consider that the value of each kilowatt-hour of electrical and thermal energy for the consumer be same, it can be said that the thermal storage system creates a greater reduction in the amount of unmet energy and creates better conditions for reliability than electrical storage. The electrical operation plan of the energy hub, in this case, is similar to the two previous cases and has not changed. The performance of various components of the energy hub in each time step can be seen in Table 2.5 separately.

Table 2.4 Optimal operational plan for energy hub in case 2

t	P_e^N	P_e^{CHP}	P_e^{ch}	P_e^{dis}	P_e^S	P_e^{ENS}	P_h^{CHP}	P_h^B	P_h^{ENS}
1	−62.3	288	34.3	0	30.6	0	252	411.1	0
2	−47.9	288	0	0	30.3	0	252	387.3	0
3	−87	288	0	0	30	0	252	389.6	0
4	167.6	288	269.7	0	270	0	252	461	0
5	−120.9	288	0	0	267.3	0	252	856	0
6	−81.4	288	0	0	264.7	0	252	1411.5	0
7	−18.2	288	0	0	262.1	0	252	1077.6	0
8	81.6	288	0	0	259.5	0	252	858.2	0
9	124.7	288	0	0	256.9	0	252	870.6	0
10	227.1	288	0	0	254.4	0	252	1800	25.5
11	265.8	288	0	0	251.8	0	252	559.9	0
12	309.3	288	0	0	249.3	0	252	1122.2	0
13	319.2	288	0	0	246.9	0	252	645.9	0
14	284	288	0	0	244.4	0	252	556.2	0
15	316.9	288	0	0	242	0	252	518.9	0
16	335.4	288	0	0	239.6	0	252	746.1	0
17	453	288	0	0	237.2	0	252	603.8	0
18	415.9	288	0	121.6	101.1	0	252	789.1	0
19	600	288	0	0	100.1	9.9	252	1800	120.4
20	600	288	0	33.1	62.7	0	252	1010.9	0
21	600	288	0	28.3	30.9	0	252	868	0
22	572.9	288	0	0	30.6	0	252	642.2	0
23	471.1	288	0	0	30.3	0	252	671.5	0
24	264.6	288	0	0	30	0	252	472.3	0
$\sum t$	5991.4	6912	304	183	4022.7	9.9	6048	19,529.9	145.9

2.4.2.4 Case 4

In this case, the energy hub uses wind turbine as a renewable energy source, in addition to CHP, this technology will also be used to generate electrical power. The effects of adding a wind turbine to the optimal operational program of energy hub can be seen in Table 2.6.

By adding a wind turbine to the energy hub, local power supplies can be provided from RES and energy hub will be able to sell more energy to the grid. How to supply electricity demand and exchange energy with the network in the presence of a wind turbine is shown in Fig. 2.9. As can be seen, unlike previous cases that energy sales were only made in the early hours of the day, this scenario would allow the sale of energy during the day and even in the afternoon, when the price of electricity in the market is higher than the early hours of the day. As a result, energy sales are higher and also at higher prices, which results in higher energy hub revenues. The amount of energy sales in this scenario reaches 864.1 kWh, which is significantly higher than previous ones. On the other hand, the amount of electricity purchased

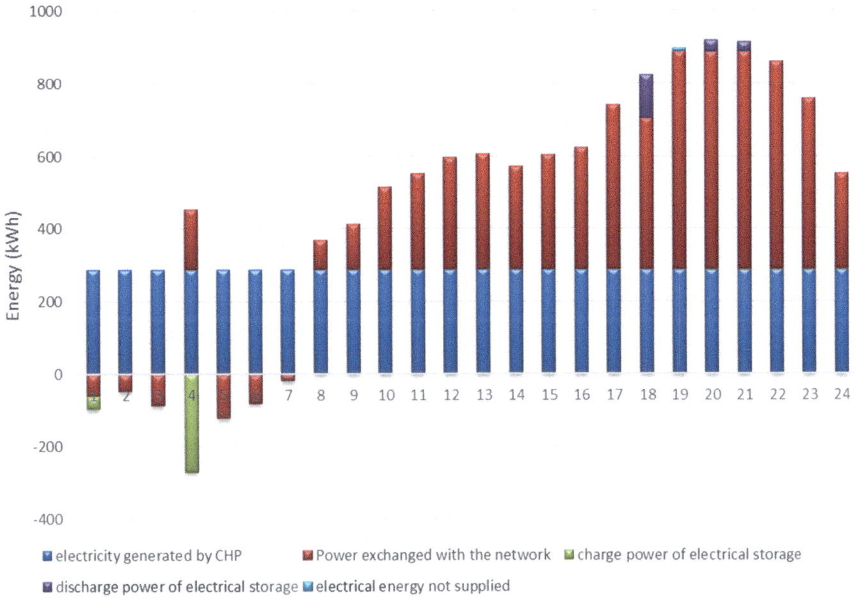

Fig. 2.7 Supply of electricity demand by the energy hub in case 2

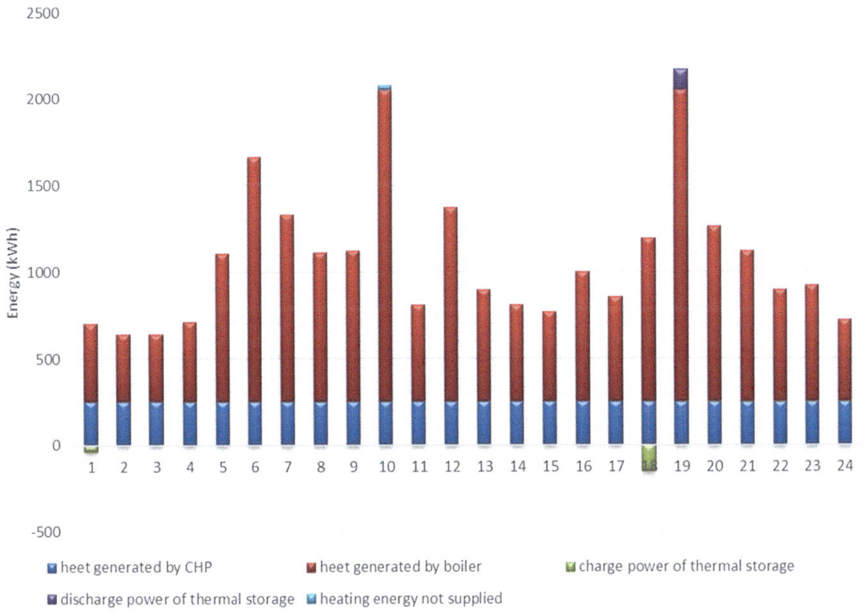

Fig. 2.8 Supply of electricity demand by the energy hub in case 3

Table 2.5 Optimal operational plan for energy hub in case 3

t	P_e^N	P_e^{CHP}	P_e^{ENS}	P_g^N	P_h^{CHP}	P_h^B	P_h^{ch}	P_h^{dis}	P_h^S	P_h^{ENS}
1	−100.9	288	0	1702.8	252	450.6	39.5	0	35.2	0
2	−47.9	288	0	1632	252	387.3	0	0	34.8	0
3	−87	288	0	1636.1	252	389.6	0	0	34.5	0
4	−135	288	0	1728.8	252	461	0	0	34.1	0
5	−120.9	288	0	2214.5	252	856	0	0	33.8	0
6	−81.4	288	0	2922.6	252	1411.5	0	0	33.5	0
7	−18.2	288	0	2586.3	252	1077.6	0	0	33.1	0
8	81.6	288	0	2426.9	252	858.2	0	0	32.8	0
9	124.7	288	0	2444.8	252	870.6	0	0	32.5	0
10	227.1	288	0	3519	252	1800	0	0	32.2	25.5
11	265.8	288	0	2046.1	252	559.9	0	0	31.8	0
12	309.3	288	0	2702.4	252	1122.2	0	0	31.5	0
13	319.2	288	0	2126.9	252	645.9	0	0	31.2	0
14	284	288	0	1952.6	252	556.2	0	0	30.9	0
15	316.9	288	0	1914	252	518.9	0	0	30.6	0
16	335.4	288	0	2213.5	252	746.1	0	0	30.3	0
17	453	288	0	2000.4	252	603.8	0	0	30	0
18	552.4	288	0	2390.4	252	941.7	152.6	0	165.7	0
19	600	288	9.9	3397.4	252	1800	0	120.4	31.5	0
20	600	288	33.1	2456.2	252	1010.9	0	0	31.2	0
21	600	288	28.3	2291.2	252	868	0	0	30.9	0
22	572.9	288	0	2016.1	252	642.2	0	0	30.6	0
23	471.1	288	0	2047.4	252	671.5	0	0	30.3	0
24	264.6	288	0	1794.6	252	472.3	0	0	30	0
$\sum t$	5786.7	6912	71.3	54,163	6048	19,722	192.1	120.4	903	25.5

in this scenario is 2979.1 kWh, which is less than half that for the base case. The combination of these factors leads to a reduction in the operating costs of the energy hub to a value of 165,831.83 cents. On the other hand, with the addition of a wind turbine, it is possible to provide all the electricity demand, and the reliability of the energy hub is remarkably improved. There is no change in the optimal energy hub plan for the supply of thermal demand. In total, it can be said that the addition of the wind turbine results in better performance of energy hub than the previous cases. Therefore, in the next case, the effect of adding storage systems in the presence of wind turbine is investigated.

2.4.2.5 Case 5

A combination of the wind turbine, thermal and electrical storages is added to the base case to study the effect of this structure on the optimal performance of the energy hub. The numerical results of this case can be found in Table 2.7.

Table 2.6 Optimal operational plan for energy hub in case 4

t	P_e^N	P_e^{CHP}	P_e^W	P_e^{ENS}	P_g^N	P_h^{CHP}	P_h^B	P_h^{ENS}
1	−107.5	288	6.8	0	1656.3	252	411.1	0
2	−47.9	288	0	0	1632	252	387.3	0
3	−109.8	288	23.4	0	1636.1	252	389.6	0
4	−135	288	0	0	1728.8	252	461	0
5	−120.9	288	0	0	2214.5	252	856	0
6	−81.4	288	0	0	2922.6	252	1411.5	0
7	−18.2	288	0	0	2586.3	252	1077.6	0
8	81.6	288	0	0	2426.9	252	858.2	0
9	18.5	288	109.5	0	2444.8	252	870.6	0
10	53.2	288	179.3	0	3519	252	1800	25.5
11	139.4	288	130.3	0	2046.1	252	559.9	0
12	44.8	288	272.8	0	2702.4	252	1122.2	0
13	−68.6	288	400	0	2126.9	252	645.9	0
14	−103.8	288	400	0	1952.6	252	556.2	0
15	−71	288	400	0	1914	252	518.9	0
16	42.1	288	302.5	0	2213.5	252	746.1	0
17	263.6	288	195.3	0	2000.4	252	603.8	0
18	183	288	380.9	0	2210.9	252	789.1	0
19	457.1	288	158.9	0	3397.4	252	1800	120.4
20	461.1	288	181.5	0	2456.2	252	1010.9	0
21	243.9	288	400	0	2291.2	252	868	0
22	346.6	288	233.4	0	2016.1	252	642.2	0
23	379.6	288	94.3	0	2047.4	252	671.5	0
24	264.6	288	0	0	1794.6	252	472.3	0
$\sum t$	2115	6912	3868.9	0	53,937	6048	19,529.9	145.9

Energy hub primarily uses CHP and wind turbine to provide electrical demand. In times of capacity shortage, this amount is purchased from the power grid. The electrical storage is responsible for the coordination of production with the pattern of consumption, and especially the price pattern of the electricity market. At hour that the lowest electricity prices and the lowest electricity demand occur (4:00 am), this storage is charged and at hour, which has the highest rates for electricity price (6:00 pm), it is discharged and in addition to compensating the electricity generation deficit, provides the possibility of electricity sales to the network in this hour. This leads to more revenue and lower operating costs. With this operational plan, all electrical demand will be provided at the lowest operating cost. Such an operation is also used to provide heat demand and much of this demand is provided with a minimum operating cost. Therefore, operating costs of the energy hub are expected to decrease in this scenario. The amount, in this case, is 163,870.2 cents, which is the least amount among all examined cases. So, case 5 has the best performance in terms of operational costs and reliability.

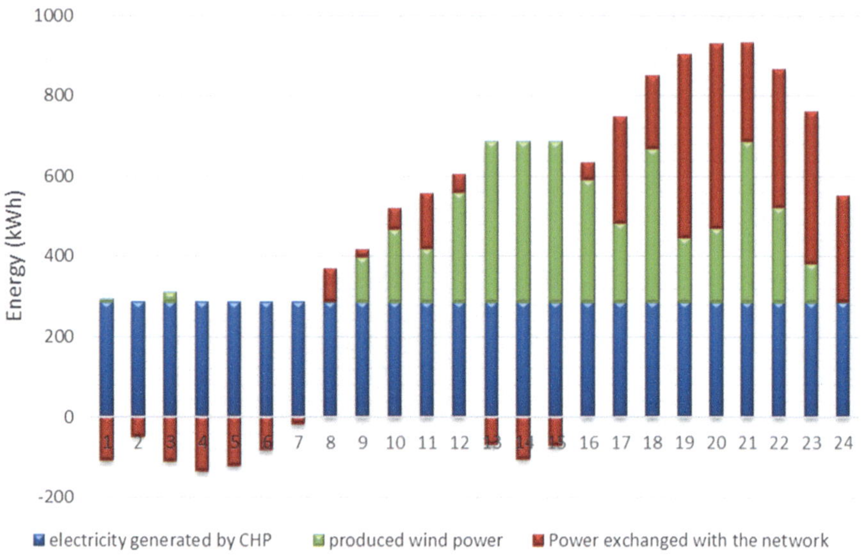

Fig. 2.9 Supply of electricity demand by the energy hub in case 4

2.5 Conclusion

In this chapter, the effects and the role of RES and ESS have been evaluated on the performance of the energy hub. Increasing demand for energy, along with limited fossil fuel storage and growing concern about the environmental problems caused by fossil fuel consumption, increases the need to use RES such as the wind and solar and increases their penetration in the energy systems. The main inputs of the energy hubs models are electrical and gas networks that are mainly based on the use of fossil fuel energies. So, future energy hub models should move towards using RES to generate energy and supply different demands of the energy hub. Increasing the share of these resources, especially in the form of DES, and the unpredictable nature of the production of these resources can lead to imbalances in supply and demand of energy and reduce the stability of the system. Utilization of ESS for energy hubs can reduce the effects of the integration of renewable sources and increase the reliability of the system. ESS can be used to provide ancillary services to the network and improve the quality of power and reduce system stability problems, as well as intelligent demand-side performance and the goals of demand-side management programs. In this chapter, in order to numerically investigate the effects of RES and ESS, energy hub has been modeled in the presence of wind turbine and electrical and thermal storages and numerical results have been discussed. The results indicate that adding RES will reduce the dependence of the energy hub on the fossil fuel networks and increase the sales of renewable energy to these networks, thereby reducing the operating costs of the energy hub. The use of ESS will increase the reliability of the system by reducing the amount of energy not supplied, as well as increasing the flexibility of the energy hub in dealing with different pricing plans for energy.

Table 2.7 Optimal operational plan for energy hub in case 5

t	P_e^N	P_e^{CHP}	P_e^W	P_e^{ch}	P_e^{dis}	P_e^{ENS}	P_g^N	P_h^{ch}	P_h^{dis}	P_h^{ENS}
1	−68.9	288	6.8	34.3	0	0	1702.8	39.5	0	0
2	−47.9	288	0	0	0	0	1632	0	0	0
3	−109.8	288	23.4	0	0	0	1636.1	0	0	0
4	167.6	288	0	269.7	0	0	1728.8	0	0	0
5	−120.9	288	0	0	0	0	2214.5	0	0	0
6	−81.4	288	0	0	0	0	2922.6	0	0	0
7	−18.2	288	0	0	0	0	2586.3	0	0	0
8	81.6	288	0	0	0	0	2426.9	0	0	0
9	18.5	288	109.5	0	0	0	2444.8	0	0	0
10	53.2	288	179.3	0	0	0	3519	0	0	25.5
11	139.4	288	130.3	0	0	0	2046.1	0	0	0
12	44.8	288	272.8	0	0	0	2702.4	0	0	0
13	−68.6	288	400	0	0	0	2126.9	0	0	0
14	−103.8	288	400	0	0	0	1952.6	0	0	0
15	−71	288	400	0	0	0	1914	0	0	0
16	42.1	288	302.5	0	0	0	2213.5	0	0	0
17	263.6	288	195.3	0	0	0	2000.4	0	0	0
18	−24.2	288	380.9	0	184.6	0	2390.4	152.6	0	0
19	457.1	288	158.9	0	0	0	3397.4	0	120.4	0
20	461.1	288	181.5	0	0	0	2456.2	0	0	0
21	243.9	288	400	0	0	0	2291.2	0	0	0
22	346.6	288	233.4	0	0	0	2016.1	0	0	0
23	379.6	288	94.3	0	0	0	2047.4	0	0	0
24	264.6	288	0	0	0	0	1794.6	0	0	0
$\sum_t$	2249	6912	3868.9	304	184.6	0	54,163	192.1	120.4	25.5

Nomenclature

Indices

B	Boiler
C	Converter
ch	Charge
dis	Discharge
e	Electricity
ed	Electricity demand
em	Emission CO_2,SO_2,NO_2
es	Electrical storage
g	Gas
gd	Gas demand
h	Heat

hd	Heat demand
hs	Heat storage
N	Network
t	Time
T	Transformer

Parameters

α_e^{loss}	Loss efficiency of electrical storage
α_h^{loss}	Loss efficiency of thermal storage
$\alpha_e^{\min}$	Minimum factor of electrical storage
$\alpha_e^{\max}$	Maximum factor of electrical storage
$\alpha_h^{\min}$	Minimum factor of thermal storage
$\alpha_h^{\max}$	Maximum factor of thermal storage
η_e^{ch}	Electrical storage charge efficiency
η_e^{dis}	Electrical storage discharge efficiency
η_h^{ch}	Thermal storage charge efficiency
η_h^{dis}	Thermal storage discharge efficiency
η_C	Electricity efficiency of converter
η_T	Electricity efficiency of transformer
η_{eCHP}	Gas to electricity efficiency of CHP
η_{hCHP}	Gas to heat efficiency of CHP
η_B	Gas to heat efficiency of boiler
π_e^{DR}	DR operation cost
π_e^{ENS}	Electricity energy not supplied cost
π_h^{ENS}	Heating energy not supplied cost
$\pi_e^N(t)$	Hourly electricity price
π_e^S	Electrical storage operation cost
π_e^W	Produced wind power cost
π_{em}	Cost of CO_2 SO_2 and NO_2 emissions
π_g^N	Gas price
π_h^S	Thermal storage operation cost
A^{CHP}	CHP availability
A_e^N	Electricity network availability
A^W	Wind turbine availability
A, B, C	Parameters of wind turbine characteristic curve
EF_{em}	Emission factor for electricity network, CHP and boiler
$\text{eELF}_{\max}$	Electricity maximum equivalent loss factor
LPF^{shup}	Load participation factor for shifting up the electricity demand
LPF^{shdo}	Load participation factor for shifting down the electricity demand
P^B	Boiler capacity
P^{CHP}	CHP capacity
$P_{\text{ed}}(\text{t})$	Hourly electricity demand

$P_{gd}(t)$	Hourly gas demand
$P_{hd}(t)$	Hourly heat demand
$P_e^{N\max}$	Maximum capacity of electricity network
$P_g^{N\max}$	Maximum capacity of gas network
P_r	Rated power of wind turbine
P_e^{SC}	Electrical storage capacity
P_h^{SC}	Thermal storage capacity
P^T	Transformer capacity
$P_e^W(t)$	Produced wind power
TC	Objective function's variable
$V_w(t)$	Hourly wind speed
V_r	Rated wind speed of wind turbine
V_{ci}	Cut-in wind speed of wind turbine
V_{co}	Cut out wind speed of wind turbine

Variables

EC	Emission cost
eELF	Equivalent loss factor
eLOEE	Loss of energy expected
eLPSP	Loss of power supply probability
hELF	Equivalent loss factor
hLOEE	Loss of energy expected
hLPSP	Loss of power supply probability
$I_e^{ch}(t)$	Binary variable for charge state of electrical storage
$I_e^{dis}(t)$	Binary variable for discharge state of electrical storage
$I_h^{ch}(t)$	Binary for charge state of thermal storage
$I_h^{dis}(t)$	Binary variable for discharge state of thermal storage
$I_e^{shup}(t)$	Binary variable for shift up of the electricity demand
$I_e^{shdo}(t)$	Binary variable for shift down of the electricity demand
$P_e^{ch}(t)$	Charged power of electrical storage
$P_e^{dis}(t)$	Discharged power of electrical storage
$P_e^{ENS}(t)$	Electrical energy not supplied
$P_e^N(t)$	Power exchanged with the network
$P_e^{loss}(t)$	Loss power of electrical storage
$P_e^S(t)$	Electrical storage capacity
$P_e^{shup}(t)$	Shifted up power via DR
$P_e^{shdo}(t)$	Shifted down power via DR
$P_g^N(t)$	Imported gas power from network
$P_g^{NCHP}(t)$	Purchased gas for CHP from network
$P_g^{NB}(t)$	Purchased gas for boiler from network
$P_h^{ch}(t)$	Charged power of thermal storage
$P_h^{dis}(t)$	Discharged power of thermal storage

$P_h^{\mathrm{loss}}(t)$	Loss power of thermal storage
$P_h^S(t)$	Thermal storage capacity
$P_e^{\mathrm{CHP}}(\mathrm{t})$	Electricity generated by CHP
$P_h^{\mathrm{CHP}}(\mathrm{t})$	Heat generated by CHP
$P_h^B(\mathrm{t})$	Heat generated by boiler
$P_h^{\mathrm{ENS}}(t)$	Heating energy not supplied

References

1. Dolatabadi A, Mohammadi-Ivatloo B (2017) Stochastic risk-constrained scheduling of smart energy hub in the presence of wind power and demand response. Appl Therm Eng 123:40–49
2. Mohammadi M, Noorollahi Y, Mohammadi-Ivatloo B, Yousefi H, Jalilinasrabady S (2017) Optimal Scheduling of Energy Hubs in the Presence of Uncertainty-A Review. J Energy Manag Technol 1(1):1–17. https://doi.org/10.22109/jemt.2017.49432
3. Hossain M, Madlool N, Rahim N, Selvaraj J, Pandey A, Khan AF (2016) Role of smart grid in renewable energy: an overview. Renew Sust Energ Rev 60:1168–1184
4. Maniyali Y, Almansoori A, Fowler M, Elkamel A (2013) Energy hub based on nuclear energy and hydrogen energy storage. Ind Eng Chem Res 52(22):7470–7481
5. Ondeck AD, Edgar TF, Baldea M (2015) Optimal operation of a residential district-level combined photovoltaic/natural gas power and cooling system. Appl Energy 156:593–606
6. Yang Y, Zhang S, Xiao Y (2015) Optimal design of distributed energy resource systems coupled with energy distribution networks. Energy 85:433–448
7. Noorollahi Y, Itoi R, Yousefi H, Mohammadi M, Farhadi A (2017) Modeling for diversifying electricity supply by maximizing renewable energy use in Ebino City southern Japan. Sustain Cities Soc 34:371
8. Raj NT, Iniyan S, Goic R (2011) A review of renewable energy based cogeneration technologies. Renew Sust Energ Rev 15(8):3640–3648
9. Connolly D, Lund H, Mathiesen B (2016) Smart energy Europe: the technical and economic impact of one potential 100% renewable energy scenario for the European Union. Renew Sust Energ Rev 60:1634–1653
10. Van Dyken S, Bakken BH, Skjelbred HI (2010) Linear mixed-integer models for biomass supply chains with transport, storage and processing. Energy 35(3):1338–1350
11. Mafakheri F, Nasiri F (2014) Modeling of biomass-to-energy supply chain operations: applications, challenges and research directions. Energy Policy 67:116–126
12. Orehounig K, Evins R, Dorer V, Carmeliet J (2014) Assessment of renewable energy integration for a village using the energy hub concept. Energy Procedia 57:940–949
13. Orehounig K, Evins R, Dorer V (2015) Integration of decentralized energy systems in neighbourhoods using the energy hub approach. Appl Energy 154:277–289
14. Woo Y-B, Cho S, Kim J, Kim BS (2016) Optimization-based approach for strategic design and operation of a biomass-to-hydrogen supply chain. Int J Hydrog Energy 41(12):5405–5418
15. Nojavan S, Zare K, Mohammadi-Ivatloo B (2017) Application of fuel cell and electrolyzer as hydrogen energy storage system in energy management of electricity energy retailer in the presence of the renewable energy sources and plug-in electric vehicles. Energy Convers Manag 136:404–417. https://doi.org/10.1016/j.enconman.2017.01.017
16. Sharma S, Ghoshal SK (2015) Hydrogen the future transportation fuel: from production to applications. Renew Sust Energ Rev 43:1151–1158
17. Hosseini SE, Wahid MA (2016) Hydrogen production from renewable and sustainable energy resources: promising green energy carrier for clean development. Renew Sust Energ Rev 57:850–866

18. Hajimiragha A, Canizares C, Fowler M, Geidl M, Andersson G (2007) Optimal energy flow of integrated energy systems with hydrogen economy considerations. In: 2007 IREP symposium, bulk power system dynamics and control-VII. Revitalizing operational reliability, IEEE, pp 1–11
19. Maroufmashat A, Fowler M, Khavas SS, Elkamel A, Roshandel R, Hajimiragha A (2015) Mixed integer linear programing based approach for optimal planning and operation of a smart urban energy network to support the hydrogen economy. Int J Hydrog Energy 41:7700
20. Mohammadi M, Noorollahi Y, Mohammadi-Ivatloo B, Yousefi H (2017) Energy hub: from a model to a concept–a review. Renew Sust Energ Rev 80:1512–1527
21. Barberis S, Rivarolo M, Traverso A, Massardo A (2016) Thermo-economic analysis of the energy storage role in a real polygenerative district. J Energy Storage 5:187–202
22. Palizban O, Kauhaniemi K (2016) Energy storage systems in modern grids—matrix of technologies and applications. J Energy Storage 6:248
23. Zhao H, Wu Q, Hu S, Xu H, Rasmussen CN (2015) Review of energy storage system for wind power integration support. Appl Energy 137:545–553
24. Shafiee S, Zareipour H, Knight AM, Amjady N, Mohammadi-Ivatloo B (2017) Risk-constrained bidding and offering strategy for a merchant compressed air energy storage plant. IEEE Trans Power Syst 32(2):946–957
25. Koohi-Kamali S, Tyagi V, Rahim N, Panwar N, Mokhlis H (2013) Emergence of energy storage technologies as the solution for reliable operation of smart power systems: a review. Renew Sust Energ Rev 25:135–165
26. Weitemeyer S, Kleinhans D, Vogt T, Agert C (2015) Integration of renewable energy sources in future power systems: the role of storage. Renew Energy 75:14–20
27. Moradi MH, Eskandari M, Hosseinian SM (2016) Cooperative control strategy of energy storage systems and micro sources for stabilizing microgrids in different operation modes. Int J Electr Power Energy Syst 78:390–400
28. Lucas A, Chondrogiannis S (2016) Smart grid energy storage controller for frequency regulation and peak shaving, using a vanadium redox flow battery. Int J Electr Power Energy Syst 80:26–36
29. Del Granado PC, Pang Z, Wallace SW (2016) Synergy of smart grids and hybrid distributed generation on the value of energy storage. Appl Energy 170:476–488
30. Alipour M, Mohammadi-Ivatloo B, Moradi-Dalvand M, Zare K (2017) Stochastic scheduling of aggregators of plug-in electric vehicles for participation in energy and ancillary service markets. Energy 118:1168–1179
31. Coelho VN, Coelho IM, Coelho BN, Cohen MW, Reis AJ, Silva SM, Souza MJ, Fleming PJ, Guimarães FG (2016) Multi-objective energy storage power dispatching using plug-in vehicles in a smart-microgrid. Renew Energy 89:730–742
32. Shotorbani AM, Ghassem-Zadeh S, Mohammadi-Ivatloo B, Hosseini SH (2017) A distributed secondary scheme with terminal sliding mode controller for energy storages in an islanded microgrid. Int J Electr Power Energy Syst 93:352–364
33. Abbaspour M, Satkin M, Mohammadi-Ivatloo B, Hoseinzadeh Lotfi F, Noorollahi Y (2013) Optimal operation scheduling of wind power integrated with compressed air energy storage (CAES). Renew Energy 51:53–59
34. Wang C, Liu Y, Li X, Guo L, Qiao L, Lu H (2016) Energy management system for stand-alone diesel-wind-biomass microgrid with energy storage system. Energy 97:90–104
35. Chesi A, Ferrara G, Ferrari L, Magnani S, Tarani F (2013) Influence of the heat storage size on the plant performance in a Smart User case study. Appl Energy 112:1454–1465
36. Dominković DF, Ćosić B, Medić ZB, Duić N (2015) A hybrid optimization model of biomass trigeneration system combined with pit thermal energy storage. Energy Convers Manag 104:90–99
37. Stoppato A, Benato A, Destro N, Mirandola A (2016) A model for the optimal design and management of a cogeneration system with energy storage. Energ Buildings 124:241
38. Rivarolo M, Greco A, Massardo A (2013) Thermo-economic optimization of the impact of renewable generators on poly-generation smart-grids including hot thermal storage. Energy Convers Manag 65:75–83

39. Jabari F, Nojavan S, Ivatloo BM (2016) Designing and optimizing a novel advanced adiabatic compressed air energy storage and air source heat pump based μ-combined cooling, heating and power system. Energy 116:64–77
40. Jabari F, Nojavan S, Mohammadi-Ivatloo B, Sharifian MB (2016) Optimal short-term scheduling of a novel tri-generation system in the presence of demand response programs and battery storage system. Energy Convers Manag 122:95–108
41. Liu W, Chen G, Yan B, Zhou Z, Du H, Zuo J (2015) Hourly operation strategy of a CCHP system with GSHP and thermal energy storage (TES) under variable loads: a case study. Energ Buildings 93:143–153
42. Bocklisch T (2016) Hybrid energy storage approach for renewable energy applications. J Energy Storage 8:311
43. Vahid-Pakdel MJ, Nojavan S, Mohammadi-Ivatloo B, Zare K (2017) Stochastic optimization of energy hub operation with consideration of thermal energy market and demand response. Energy Convers Manag 145:117–128
44. Pazouki S, Haghifam M-R (2014) Impact of energy storage technologies on multi carrier energy networks. In: Smart Grid Conference (SGC), 2014, IEEE, pp 1–6
45. Adamek F, Arnold M, Andersson G (2014) On decisive storage parameters for minimizing energy supply costs in multicarrier energy systems. IEEE Trans Sustainable Energy 5(1): 102–109

Chapter 3
Robust Economic Emission Dispatch of Thermal Units and Compressed Air Energy Storages

Farkhondeh Jabari and Behnam Mohammadi-Ivatloo

3.1 Introduction

Nowadays, different renewable energy sources such as hydro, wind, solar, geothermal, ocean waves, and tidal are attracting world's attention due to the major concerns about the excessive CO_2 emissions and global energy crisis [1–4]. Therefore, a secure, efficient, and economic scheduling of modern power systems is an arduous challenge because of the volatility nature of renewables. Meanwhile, energy storage systems are used to reshape daily electricity demand profile and add more flexibility on power system operation. Figure 3.1 summarizes the power rating versus discharge time at rated power for different energy storages.

As obvious from this figure, large-scale storage facilities such as pumped hydro storages and compressed air plants are able to participate in bulk power management with higher power rating for long discharge time period [5].

3.2 Types of Compressed Air Energy Storages

In the conventional diabatic compressed air energy storage (CAES) shown in Fig. 3.2, surplus electrical energy purchased from the local power grid over the low-cost or off-peak demand periods drives an air compressor with an intercooling

F. Jabari (✉) · B. Mohammadi-Ivatloo
Department of Electrical and Computer Engineering, University of Tabriz, Tabriz, Iran
e-mail: f.jabari@tabrizu.ac.ir; bmohammadi@tabrizu.ac.ir

B. Mohammadi-Ivatloo, F. Jabari (eds.), *Operation, Planning, and Analysis of Energy Storage Systems in Smart Energy Hubs*,
https://doi.org/10.1007/978-3-319-75097-2_3

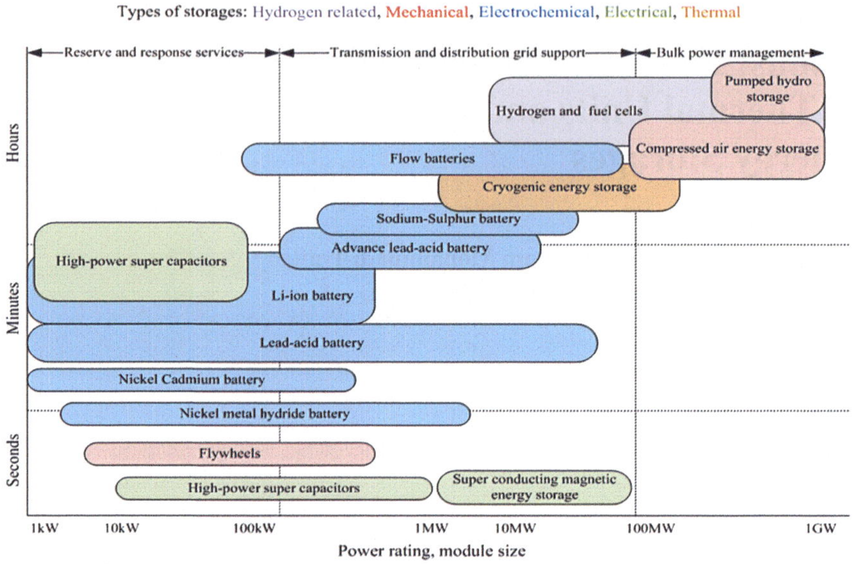

Fig. 3.1 A comparison between different energy storage technologies from power rating and discharge time perspectives

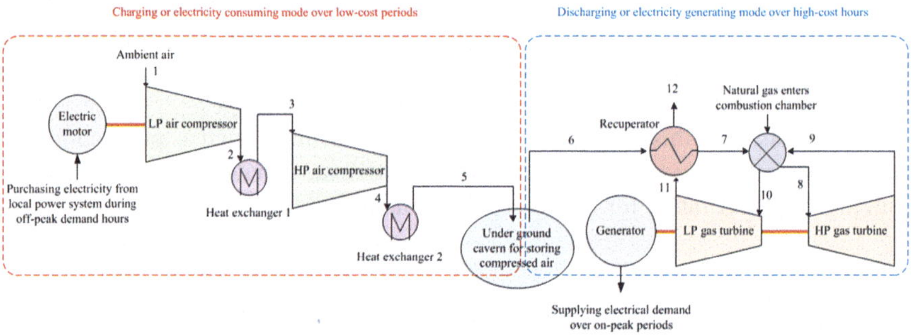

Fig. 3.2 Schematic diagram of conventional diabatic compressed air energy storage

process. After the compression stage, the compressed air is stored in an underground cavern. During the mid-peak and on-peak time intervals with higher electricity rates, a recuperator is employed to preheat the stored air before heating it up by a combustion chamber operating with natural gas. Then, the high-temperature supplied air enters to an expansion valve and generates electricity. Use of a recuperator for air preheating increases the overall system efficiency by 15% [6].

The advantages and the main drawbacks of the conventional CAES plants can be summarized as follows [7, 8]:

Advantages:

- Fast reaction time
  - Ramp rate from 0% to 100% of rated capacity in approximately 10 min
  - From 10% to 100% of rated capacity in less than 4 min
  - From 50% to 100% of rated capacity in almost 15 s
- Large-scale energy supplier
- Frequent start-ups and shut-downs
- Used for ancillary services such as frequency regulation, load following, and voltage control due to their capability for frequent start-up and shut-down.

Disadvantages:

- Heating energy losses: The heating flow of compression stage should be dissipated to avoid from the deterioration of underground cavern and the pressure drop of stored air. Hence, natural gas is used to heat up the stored air for discharging or power generating process to prevent from the extremely low temperature of gas turbine.
- Fossil fuels requirements such as natural gas
- Reliance on natural geological characteristics of the installation region as a major disadvantage
- Overall efficiency less than 45%

The interest for the development of other CAES technologies such as advanced adiabatic CAES [9, 10], adsorption enhanced CAES [11], near-isothermal CAES [12], and underwater CAES [13] plants was reignited by increasing fuel prices and greenhouse gas emissions. For example, Fig. 3.3 illustrates the operating principle of advanced adiabatic compressed air energy storages (AA-CAES). Obviously, the additional heat storage is installed for reserving the heat released in the low- and high-pressure compressors. It causes that overall efficiency of adiabatic type reaches up to 70% with near-zero carbon footprints.

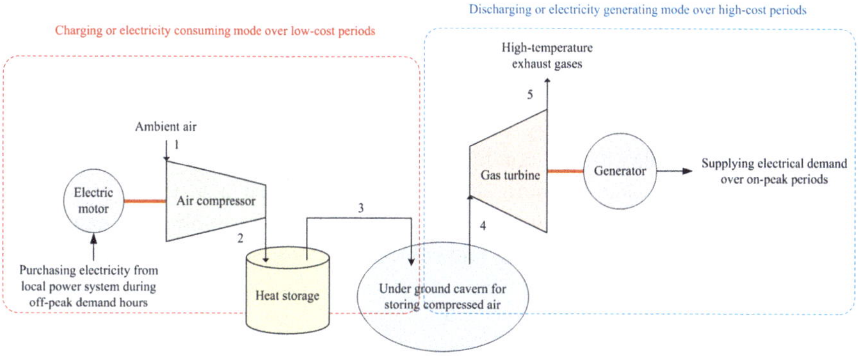

Fig. 3.3 Schematic diagram of advanced adiabatic compressed air energy storage

3.3 Literature Review on CAES

Recently, several remarkable efforts have been carried out on designing, modeling, and scheduling of different integrated energy hubs using compressed air energy storages from energetic-exergetic-economic-environmental viewpoints. In this context, Wang et al. [14] designed and analyzed a multi-level underwater compressed air energy storage to improve the overall efficiency between 62% and 81% in comparison with conventional diabatic CAES systems. This system is composed of a CAES, rock bed thermal storage unit, and battery energy storage. The advantages of this system are listed as follows:

- Flexible operation for storing variable renewable energy sources and meet fluctuating electricity demand of end-consumers
- No preexisting bathymetry requirements
- Battery pack acts as auxiliary electricity storage: When the produced electrical power is more than the electricity requirement of the compressor train, it will be stored in the battery pack.

Perazzelli and Anagnostou [15] investigated the technical feasibility of shallow lined rock cavern tunnels or shafts under a wide range of geotechnical conditions. A compressed CO_2 energy storage system with higher density and round-trip efficiency than that of CAES is introduced in [16]. To decrease the energy consumption of compression train in the charging process, and enhance the round trip efficiency, a CAES combined with an air conditioner is developed by Chen et al. [17]. In this system, the air conditioner is utilized to precool the inlet air of compressor, decrease the energy consumption of compression train in the charging process, and improve the round trip efficiency by more than 3%. A marine current turbine farm is used in [18] as the main energy supply for a stand-alone island. To compensate the impacts of tidal phenomenon on marine current turbine farm power variation, an ocean compressed air energy storage is established for facilitating real-time power management under uncertain operation conditions such as different load levels, different tidal speed levels, one-day cases, and one-week case. In [19], a combined wind-solar-compressed air energy storage system is developed to transform the fluctuating wind and solar power into a stable electricity and hot water, and increase the storage efficiency, round trip efficiency, and exergy efficiency up to 87.7%, 61.2%, and 65.4%, respectively. Zhao et al. [20] combined an advanced adiabatic compressed air energy storage system with a flywheel energy storage to operate under variable ambient conditions, inlet temperature of compressor, storage cavern temperature, and pressure. Lv et al. [21] utilized a CAES for cooling, heating, and electricity generation. The cooling demand is supplied by expanding the compressed air, and the heating one is recovered in the process of compression and storage. It is found that annual monetary cost saving is about 53.9%. Energy-exergy based performance analysis of an integrated micro gas turbine, CAES, and solar dish concentrator is presented in [22]. It is revealed that the round trip efficiency increases as the difference between minimum and maximum pressure of air cavern increases.

In [23], a novel CAES based combined cooling, heating, and power system based system is designed. This trigeneration microgrid composes of a gas engine, supplemental heat exchangers, and an ammonia-water absorption chiller. An evolutionary multi-objective algorithm is then applied to increase overall exergy efficiency up to 53.04% and reduce total product cost to 20.54 ¢/kWh. Jabari et al. [9] designed and scheduled a novel air to air heat pump based trigeneration plant for residential summer cooling and winter heating applications. An advanced adiabatic CAES unit is modeled with injected and produced power limits, storage power and energy capacities, air balance, and operational constraints to reduce total energy procurement cost by 21.79% and 22.36% in cooling and heating cycles, respectively. Li et al. [24] developed a novel energy storage technology that stores the excessive energy as compressed air and thermal flux for cooling and heating of domestic households. The cooling demand is supplied by direct expanding of compressed air instead of installing an absorption chiller. Besides, sensible heat of solar thermal and geothermal energy are stored in a storage medium and released for building heating in the winter. In [25], a CAES is integrated with a pneumatic motor for cogeneration facilities. As the amount of generated power is greater than demand, the surplus electricity is utilized to compress ambient air. During on-peak hours, the stored air is extracted to drive the pneumatic motor and generate electricity. Moreover, the exhausted air from pneumatic motor is utilized as a cold storage medium to satisfy the cooling demand. Bagdanavicius and Jenkins [26] combined a compressed air storage with a thermal storage unit to store the waste heat recovered from the compression stage for district heating applications. It is demonstrated that use of thermal energy storage and waste heat recovery reduces total exergy cost from 13.89 ¢/kWh to 11.20 ¢/kWh. Shafiee et al. [27] proposed an information gap decision theory (IGDT) based risk constrained bidding/offering strategy for a merchant CAES plant for participation in the day-ahead electricity market taking into account the price uncertainties. The IGDT technique assesses the robustness and opportunistic aspects of the optimal self-scheduling scenarios in the presence of price uncertainty to make the risk-averse and risk-taker decisions, respectively. In the literature, many scholars have focused on design and characteristics analysis [28–30], optimization, and analysis of system components, such as turbine, electromotor, air storage cavern, and thermal energy storage [31–35] of CAES systems.

3.4 Mathematical Modeling of Advanced Adiabatic Compressed Air Energy Storage

In this section, mathematical modeling of optimal charge and discharge decisions of adiabatic CAES is presented. As mentioned, the adiabatic CAES utilizes the electrical power for compressing ambient air over the off-peak demand hours and generates electricity by expanding it via a gas turbine as shown in Fig. 3.3. In this chapter, constraint (3.1)–(3.7) is used to model the operational limits of the adiabatic CAES [9]:

$$V_t^{\text{inj}} = \alpha^{\text{inj}} P_{\text{Comp},t} \tag{3.1}$$

$$P_{\text{Gen},t} = \alpha^{\text{pump}} V_t^{\text{pump}} \tag{3.2}$$

$$V_{\min}^{\text{inj}} u_t^{\text{inj}} \le V_t^{\text{inj}} \le V_{\max}^{\text{inj}} u_t^{\text{inj}} \tag{3.3}$$

$$V_{\min}^{\text{pump}} u_t^{\text{pump}} \le V_t^{\text{pump}} \le V_{\max}^{\text{pump}} u_t^{\text{pump}} \tag{3.4}$$

$$u_t^{\text{pump}} + u_t^{\text{inj}} \le 1 \tag{3.5}$$

$$\text{SOC}_{t-1} = \text{SOC}_t + V_t^{\text{inj}} - V_t^{\text{pump}} \tag{3.6}$$

$$\text{SOC}_{\min} \le \text{SOC}_t \le \text{SOC}_{\max} \tag{3.7}$$

where V_t^{inj} is the energy content of injected air to the underground cavern at time horizon t; α^{inj} is the efficiency of injecting process; $P_{\text{Comp},\,t}$ is the electricity requirement of compressor train at time horizon t; $P_{\text{Gen},\,t}$ is the electricity generated by CAES at time t; α^{pump} is the efficiency of generating mode; V_t^{pump} is the energy content of pumped air to gas turbine at time t; $V_{\min}^{\text{inj}}, V_{\max}^{\text{inj}}$ is the minimum and maximum amounts of injected air to the underground cavern; $u_t^{\text{inj}} = \begin{cases} 1 & \text{at injecting mode} \\ 0 & \text{Otherwise} \end{cases}$; $V_{\min}^{\text{pump}}(t), V_{\max}^{\text{pump}}(t)$ is the minimum and maximum values of pumped air to the gas turbine; $u_t^{\text{pump}} = \begin{cases} 1 & \text{at pumping mode} \\ 0 & \text{Otherwise} \end{cases}$; SOC is the state of charge of adiabatic CAES at time t; $\text{SOC}_{\min}, \text{SOC}_{\max}$ is the minimum and maximum state of charge of adiabatic CAES.

3.5 Economic Environmental Dispatch of Thermal Units

Dynamic economic emission dispatch (DEED) problem determines the generation levels for the committed units focusing on two main objectives. The first objective is maximization of total benefits or total money received from selling energy minus total fuel costs, and the second one is minimization of pollutant emissions. Traditionally, total fuel cost curve of coal and gas fired plants composed of a quadratic function and a sinusoidal (valve-point effect) term which can be given by Eq. (3.8) [36–38]. Conventional thermal generating units consist of multiple valves, which are installed for controlling their power output. As the steam admission valves are opened for the first time, a rapid increase in losses occurred resulting in ripples in equivalent cost function which can be modeled as a sinusoidal cost function [39, 40]. Note that λ_t and $P_{i,\,t}$ are the forecasted electricity market price and the power generation of thermal unit i at time horizon t, respectively. Moreover, a_i, b_i, and c_i represent the no-load, linear, and the quadratic cost coefficients of thermal generator i, respectively, while e_i and f_i model the valve point effect of generating unit i.

$$\text{Max}\left[\sum_{t=1}^{T}\sum_{i=1}^{N_{\text{Gen}}}\lambda_t P_{i,t}-\sum_{t=1}^{T}\sum_{i=1}^{N_{\text{Gen}}}\left(a_i+b_iP_{i,t}+c_iP_{i,t}^2+\left|e_i\sin\left(f_i\left(P_i^{\min}-P_i\right)\right)\right|\right)\right] \tag{3.8}$$

As obvious from Eq. (3.9), the second objective of DEED problem is to mitigate total greenhouse gas emissions which generally consist of three major pollutants including sulfur oxides (SO_x), nitrogen oxides (NO_x), and carbon dioxide (CO_2) [41, 42], where constants γ_i, η_i, δ_i, ξ_i, and φ_i indicate the emission characteristics.

$$\text{Min}\left[\sum_{t=1}^{T}\sum_{i=1}^{N_{\text{Gen}}}\left(\gamma_i+\eta_iP_{i,t}+\delta_iP_{i,t}^2+\xi_i\exp\left(\varphi_iP_{i,t}\right)\right)\right] \tag{3.9}$$

Aggregating two objectives, DEED problem can be mathematically formulated as a nonlinear optimization process, which can be maximized as a single-objective optimization problem using the weighting factor, $\omega \in [0, 1]$, as follows:

Objective function

$$\begin{aligned}&= \underset{P_{i,t}}{\text{Max}}\left\{\omega\times\sum\nolimits_{t=1}^{T}\sum\nolimits_{i=1}^{N_{\text{Gen}}}\left(\lambda_tP_{i,t}-a_i-b_iP_{i,t}-c_iP_{i,t}^2-\left|e_i\sin\left(f_i\left(P_i^{\min}-P_{i,t}\right)\right)\right|\right)\right.\\&\quad\left.-\frac{(1-\omega)\times B_{\max}}{E_{\max}}\times\sum\nolimits_{t=1}^{T}\sum\nolimits_{i=1}^{N_{\text{Gen}}}\left(\gamma_i+\eta_iP_{i,t}+\delta_iP_{i,t}^2+\xi_i\exp\left(\varphi_iP_{i,t}\right)\right)\right\}\end{aligned} \tag{3.10}$$

$$E_{\max}=\sum\nolimits_{i=1}^{N_{\text{Gen}}}\left(\gamma_i+\eta_iP_i^{\max}+\delta_iP_i^{\max^2}+\xi_i\exp\left(\varphi_iP_i^{\max}\right)\right) \tag{3.11}$$

$$B_{\max} = \sum_{i=1}^{N_{\text{Gen}}} \left[\lambda_t P_i^{\max} - a_i - b_i P_i^{\max} - c_i P_i^{\max^2} - \left| e_i \sin\left(f_i \left(P_i^{\min} - P_i^{\max} \right) \right) \right| \right] \tag{3.12}$$

Subject to

- Power balance criterion: Total active power output of thermal generating units should meet the electricity demand, P_d^t, and transmission losses, P_{Loss}^t, which can be fulfilled by Eq. (3.13). As formulated by Eq. (3.14), total transmission losses can be calculated using *B*-coefficients.

$$\sum_{i=1}^{N_{\text{Gen}}} P_{i,t} + P_{\text{Gen},t} = P_d^t + P_{\text{Loss}}^t + P_{\text{Comp},t} \quad \forall t = 1, 2, \ldots, T \tag{3.13}$$

$$P_{\text{Loss}}^t = \sum_{i=1}^{N_{\text{Gen}}} \sum_{i=1}^{N_{\text{Gen}}} P_{i,t} B_{ij} P_{j,t} \quad \forall t = 1, 2, \ldots, T \tag{3.14}$$

- Electrical power generation limits: The active power output of gas and coal fired thermal units should be limited by lower and upper boundaries as inequality constraint (3.15), where $P_i^{\min}$ and $P_i^{\max}$ refer to the minimum and maximum power generation of *i*th thermal unit, respectively.

$$P_i^{\min} \leq P_{i,t} \leq P_i^{\max} \quad \forall t = 1, 2, \ldots, T; \forall i = 1, 2, \ldots, N_{\text{Gen}} \tag{3.15}$$

- Ramp up and down rates of thermal units: The most important constraint of thermal generators is ramp up and down limitations. A thermal generation unit can increase or decrease its production by maximum and minimum constant ramp rates, which can be stated as follows:

$$\begin{gathered} \text{Max}\left(P_i^{\min}, P_{i,t-1} - \text{DR}_i \right) \leq P_{i,t} \leq \text{Min}\left(P_i^{\max}, P_{i,t-1} + \text{UR}_i \right) \\ \forall t = 1, 2, \ldots, T; \forall i = 1, 2, \ldots, N_{\text{Gen}} \end{gathered} \tag{3.16}$$

3.6 Robust Optimization Method for Modeling Wind Generation Uncertainty

Equations (3.17)–(3.19) formulate the standard nonlinear programming (NLP) problem:

$$\underset{x_j}{\text{Minimize}} \, f(x_1, x_2, \ldots, x_n) \tag{3.17}$$

Subject to

$$g_m(x_1, x_2, \ldots, x_n) \leq b_m, m = 1, \ldots, M \tag{3.18}$$

$$x_j \geq 0, \ \ j = 1, \ldots, n \tag{3.19}$$

For building a robust nonlinear programming (RNLP), it is assumed that d_j refers to the deviance from nominal value of coefficient of the decision variable x_j. The integer parameter Γ_0 controls the robustness level of the objective function, which belongs to $[0, |J_0|]$, where $J_0 = \{j|\ d_j > 0\}$. The robustness index Γ_0 should be set to zero in order to ignore the effect of cost deviations on the objective function. If $\Gamma_0 = |J_0|$, all deviations of cost function have been considered. Therefore, an RNLP can be formulated as (3.20)–(3.25).

$$\underset{x_j, q_{oj}, y_j, z_0}{\text{Minimize}}\ f(x_1, x_2, \ldots, x_n) + z_0\Gamma_0 + \sum_{j=1}^{n} q_{oj} \tag{3.20}$$

Subject to constraints (3.1)–(3.7) and (3.13)–(3.16)

$$z_0 + q_{oj} \geq d_j y_j, \ \ j \in J_0 \tag{3.21}$$

$$q_{oj} \geq 0, \ \ j = 1, \ldots, n \tag{3.22}$$

$$y_j \geq 0, \ \ j = 1, \ldots, n \tag{3.23}$$

$$z_0 \geq 0 \tag{3.24}$$

$$x_j \leq y_j, \ \ j = 1, \ldots, n \tag{3.25}$$

In which, variables z_0 and q_{oj} are dual variables of the optimization (3.17)–(3.19) used to take into account the known bounds of coefficient of the decision variable x_j. The auxiliary variable y_j is used to achieve the corresponding linear declarations. Eventually, RNLP can be implemented on economic emission dispatch of thermal-wind-CAES units as (3.26)–(3.31) considering the upper and lower bounds of the wind productions for uncertainty modeling.

$$\underset{\substack{P_{i,t}, P_{\text{Gen},t} \\ P_{\text{Comp},t}, q_{oi}, y_i \\ \forall i; z_0}}{\text{Max}} \left\{ \omega \times \sum_{t=1}^{T} \sum_{i=1}^{N_{\text{Gen}}} \left(\lambda_t P_{i,t} - a_i - b_i P_{i,t} - c_i P_{i,t}^2 - \left| e_i \sin \left(f_i \left(P_i^{\min} - P_{i,t} \right) \right) \right| \right) \right.$$
$$\left. - z_0 \Gamma_0 - \sum_{i=1}^{T} q_{oi} - \frac{(1-\omega) \times B_{\max}}{E_{\max}} \times \sum_{t=1}^{T} \sum_{i=1}^{N_{\text{Gen}}} \left(\gamma_i + \eta_i P_{i,t} + \delta_i P_{i,t}^2 + \xi_i \exp \left(\varphi_i P_{i,t} \right) \right) \right\} \tag{3.26}$$

Subject to constraints (3.1)–(3.7) and (3.13)–(3.16)

$$z_0 + q_{oi} \geq d_i y_i, \quad i = 1, \ldots, T \tag{3.27}$$

$$q_{oi} \geq 0, \quad i = 1, \ldots, T \tag{3.28}$$

$$y_i \geq 0, \quad i = 1, \ldots, T \tag{3.29}$$

$$z_0 \geq 0 \tag{3.30}$$

$$\lambda_i \leq y_i, \quad i = 1, \ldots, T \tag{3.31}$$

The proposed interval robust optimization algorithm for modeling the wind production uncertainties comprises the following steps:

1. Set wind production $P_{\text{WT}_i} = P_{\text{WT}_i}^{\min}$ $(i = 1, \ldots, T)$, and $\Gamma_0 = T$ to consider all possible deviations of wind speeds.
2. Set $d_i^k = G^k \left(P_{\text{WT}_i}^{\max} - P_{\text{WT}_i}^{\min} \right), \quad (i = 1, \ldots, T)$, where G^k is a coefficient that uses increasing values within [0, 1] and k is the iteration counter. Using the parameter G^k, a consecutive nested subinterval can be derived as $\left[P_{\text{WT}_i}^{\min}, P_{\text{WT}_i}^{\min} + d_i^k \right]$.
3. RNLP optimization (3.1)–(3.7), (3.13)–(3.16), and (3.26)–(3.31) is solved to obtain total fuel cost and emissions at the iteration k.
4. For covering all ranges of coefficient G^k, steps 2 and 3 should be repeated iteratively (categorized by k).

The wind production, $P_{\text{WT}_i}^k = P_{\text{WT}_i}^{\min} + d_i^k, \quad (i = 1, \ldots, T)$ are calculated in each iteration k by using the energy procurement level per time period.

3.7 Simulation Results and Discussions

3.7.1 Input Data

In this section, DEED problem is applied on a 10 generators standard power system. Simulations are implemented as a mixed integer nonlinear programming (MINLP) problem over a 24-h time interval in general algebraic modeling system (GAMS) (https://www.gams.com/) software package and solved by SBB (https://www.gams.com/latest/docs/solvers/sbb/index.html) tool running on a Lenovo with 2.10 GHz CPU, 4 GB RAM. Detailed specifications and technical parameters of 10-unit test system can be found in [43]. Figure 3.4 illustrates the 24-h load curve with 2150 MW on-peak and 1036 MW off-peak electricity demand which respectively occurred at $t = 12^{\text{noon}}$ and $t = 1^{\text{a. m.}}$. Moreover, the forecasted electricity market prices are depicted in Fig. 3.5. Other technical characteristics of generators and B-coefficients are reported in Tables 3.1 and 3.2, respectively. In this chapter, an AA-CAES is considered with operational characteristics as Table 3.3.

3.7.2 Results of DEED Without and with Participation of CAES

Figure 3.6 shows hourly generation schedules obtained from solving DEED under GAMS environment with and without considering CAES. Total fuel cost and emissions in Pareto optimal solutions obtained from solving DEED problem with

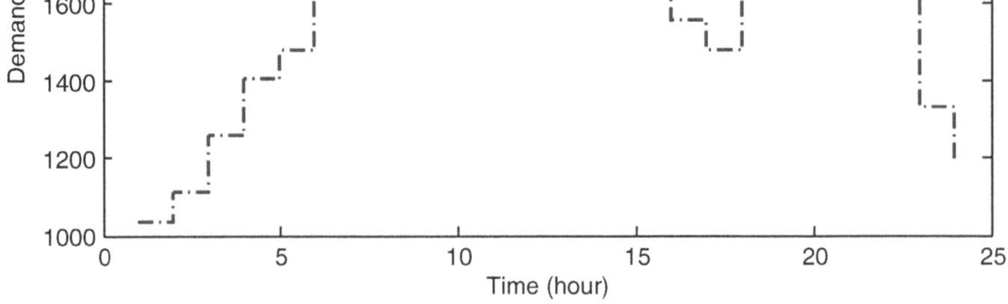

Fig. 3.4 Total electricity demand over a 24-h study horizon

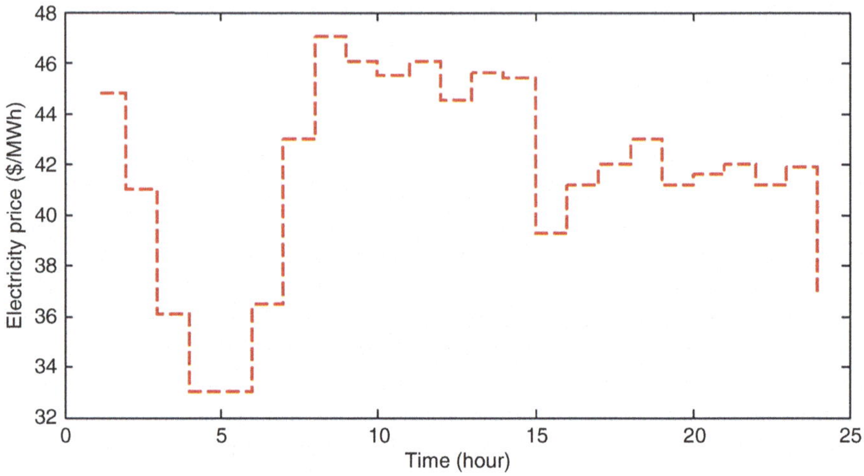

Fig. 3.5 Hourly forecasted electricity prices

Table 3.1 Technical specification of thermal units

Unit	P^{min}	P^{max}	UR	DR	a	b	c
1	150	470	80	80	786.8	38.54	0.1524
2	135	470	80	80	451.33	46.159	0.1058
3	73	340	80	80	1050	40.397	0.028
4	60	300	50	50	1243.5	38.306	0.0354
5	73	243	50	50	1658.6	36.328	0.0211
6	57	160	50	50	1356.7	38.27	0.0179
7	20	130	30	30	1450.7	36.51	0.0121
8	47	120	30	30	1450.7	36.51	0.0121
9	20	80	30	30	1455.6	39.58	0.109
10	10	55	30	30	1469.4	40.541	0.1295

Unit	γ	η	δ	ξ	φ
1	103.39	−2.4444	0.0312	0.5035	0.0207
2	103.39	−2.4444	0.0312	0.5035	0.0207
3	300.39	−4.0695	0.0509	0.4968	0.0202
4	300.39	−4.0695	0.0509	0.4968	0.0202
5	320	−3.8132	0.0344	0.4972	0.02
6	320	−3.8132	0.0344	0.4972	0.02
7	330.01	−3.9023	0.0465	0.5163	0.0214
8	330.01	−3.9023	0.0465	0.5163	0.0214
9	350.01	−3.9524	0.0465	0.5475	0.0234
10	360	−3.9864	0.047	0.5475	0.0234

Table 3.2 Transmission loss *B*-coefficients

$B =$	0.000049	0.000014	0.000015	0.000015	0.000016	0.000017	0.000017	0.000018	0.000019	0.000020
	0.000014	0.000045	0.000016	0.000016	0.000017	0.000015	0.000015	0.000016	0.000018	0.000018
	0.000015	0.000016	0.000039	0.000010	0.000012	0.000012	0.000014	0.000014	0.000016	0.000016
	0.000016	0.000016	0.000010	0.000040	0.000014	0.000010	0.000011	0.000012	0.000014	0.000015
	0.000016	0.000017	0.000012	0.000014	0.000035	0.000011	0.000013	0.000013	0.000015	0.000016
	0.000017	0.000015	0.000012	0.000010	0.000011	0.000036	0.000012	0.000012	0.000014	0.000015
	0.000017	0.000015	0.000014	0.000011	0.000013	0.000012	0.000038	0.000016	0.000016	0.000018
	0.000018	0.000016	0.000014	0.000012	0.000013	0.000012	0.000016	0.000040	0.000015	0.000016
	0.000019	0.000018	0.000016	0.000014	0.000015	0.000014	0.000016	0.000015	0.000042	0.000019
	0.000020	0.000018	0.000016	0.000015	0.000016	0.000015	0.000018	0.000016	0.000019	0.000044

Table 3.3 Technical characteristics of CAES

α^{inj}, α^{pump}	V_{min}^{inj}, V_{min}^{pump}	V_{max}^{inj}, V_{max}^{pump}	SOC_{min}	SOC_{max}
0.7	0	75	0	210

different weighing factors are shown in Fig. 3.7. As expected, it is obvious that total fuel cost will dramatically be decreased as the value of weighing factor ω increases and vice versa. Besides, when the amount of weighing factor increases, emission productions gradually increase. In other words, total fuel costs and emissions are conflicting in nature when optimizing a multi-objective problem and one objective is usually improved, while other is getting worse. Figure 3.8 depicts total fuel costs minus total money received from selling energy. According to Figs. 3.7 and 3.8, if a decision maker or power system operator desires lower fuel cost or higher profit, more emissions will be produced and vice versa. Moreover, total fuel cost and emission functions in case "with CAES" are less than case "without CAES." For example, optimal charge and discharge decisions of CAES for $\omega = 0.5$ are illustrated in Fig. 3.9.

3.7.3 Robust DEED with Participation of CAES

In this section, a case study is used to show the application of the robust optimization approach for DEED of thermal-wind-CAES units. Considering these input parameters, RNLP problem is solved for five iterations (G^k is changed by the steps as: $G = [0, 0.27, 0.64, 0.79, 1]$) to produce the required data for constructing DEED strategies. The lower and upper bounds of the wind power data for the study horizon have been shown in Fig. 3.10. Starting from lower to upper level, RNLP problem is solved for each interval within the wind power bound to achieve the robust cost. The power output of 1st and 2nd thermal generating units, charge and discharge power of CAES, fuel costs, and emission productions for five iterations have been depicted in Figs. 3.11, 3.12, and 3.13, respectively. As expected, if the value of wind power decreases (smaller k), the amount of robust cost will be increased resulting in a higher emission production and higher fuel cost and vice versa.

3.8 Conclusion

This chapter expressed different technologies, structures, technical and operational constraints of compressed air energy storages, comprehensively. Mathematical framework was presented to model CAES and formulate a dynamic economic environmental dispatch problem for conventional thermal plants. In addition, impact of advanced adiabatic compressed air energy storages on day-ahead economic emission dispatch of coal and gas-fired generators is investigated by solving a mixed

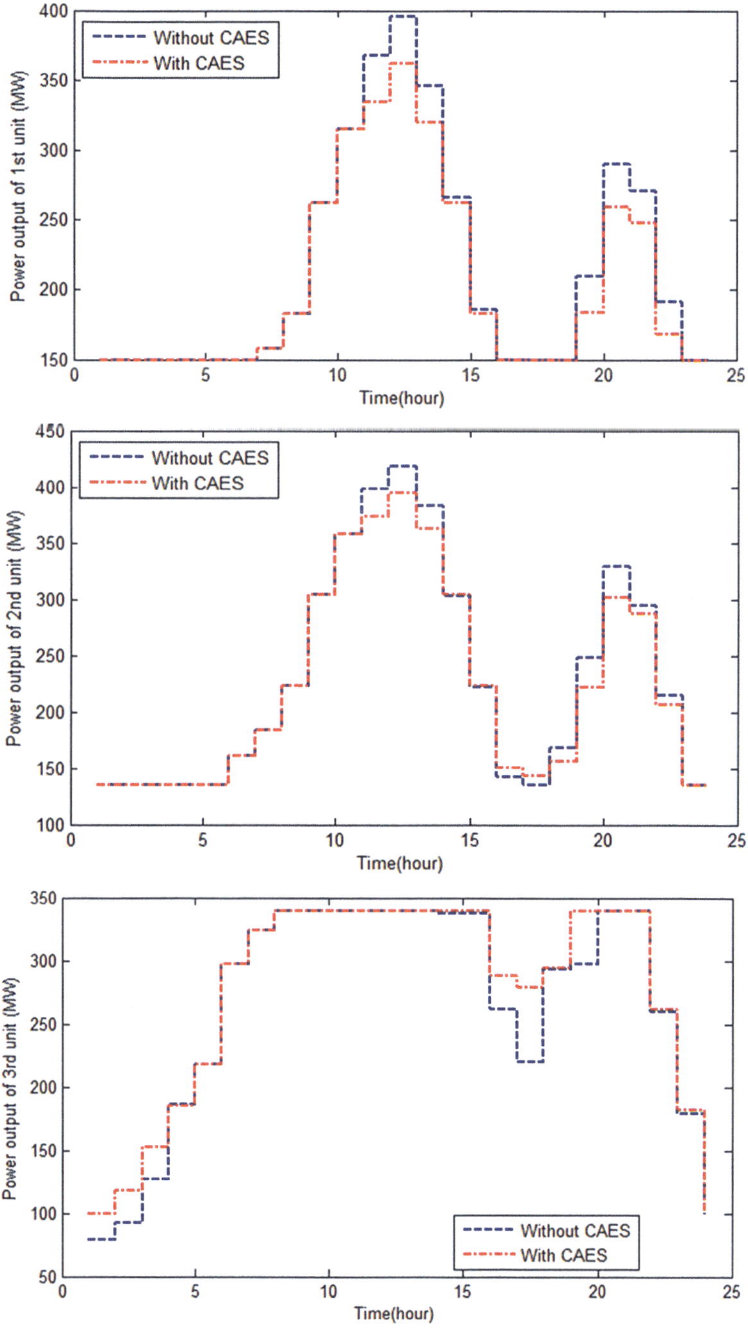

Fig. 3.6 Hourly generation schedules for $\omega = 0.5$

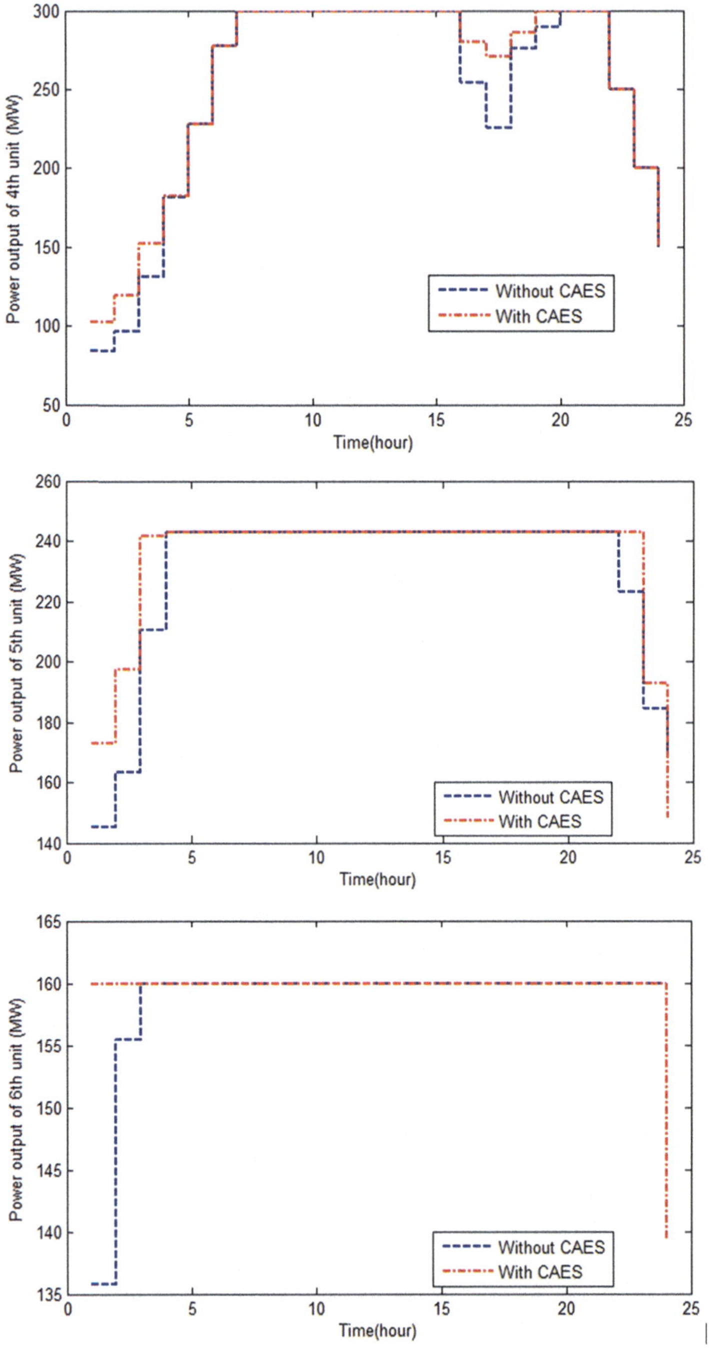

Fig. 3.6 (continued)

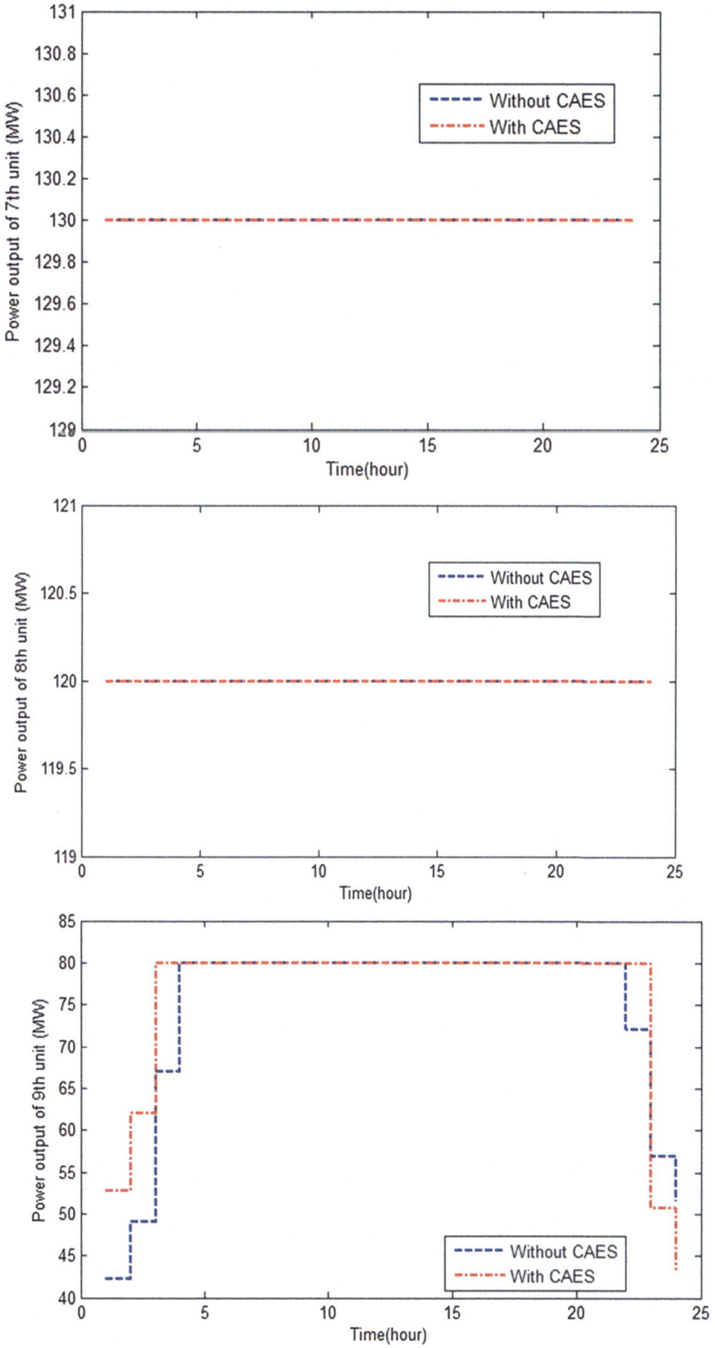

Fig. 3.6 (continued)

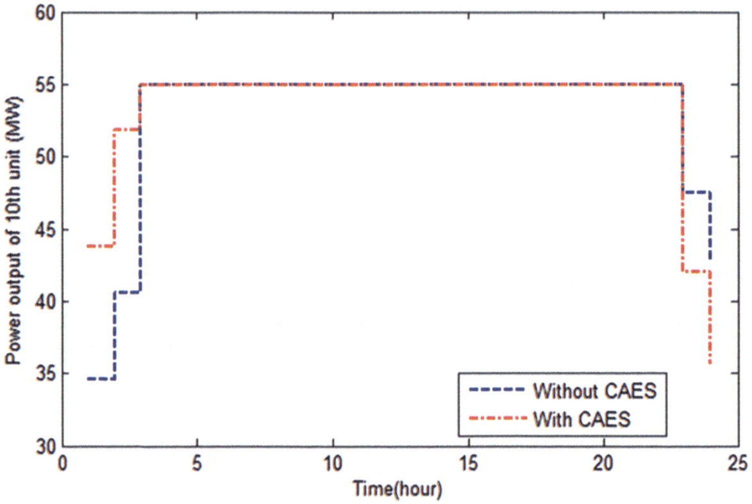

Fig. 3.6 (continued)

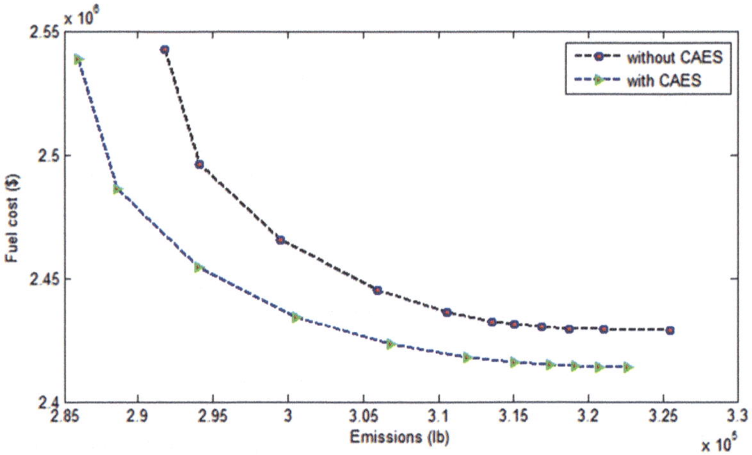

Fig. 3.7 Pareto optimal solutions obtained by varying weights ω

integer nonlinear programming problem in GAMS software package under SBB tool. Numerical results revealed that use of advanced adiabatic CAESs and robust DEED of thermal-wind-CAES units reduces total fuel cost and pollutant emission footprints, significantly.

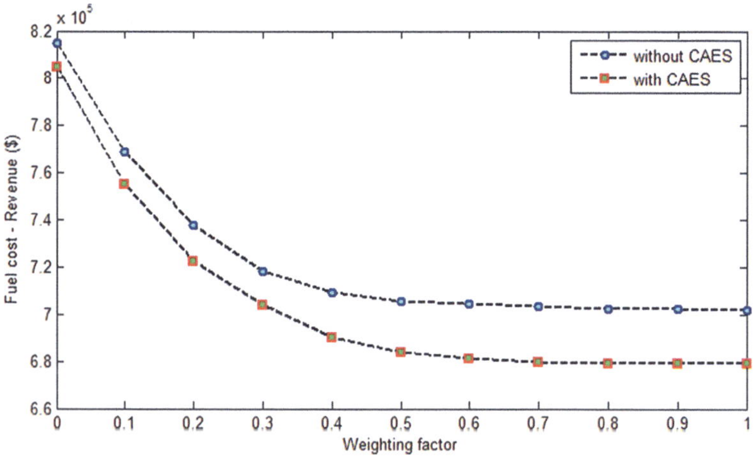

Fig. 3.8 Total fuel costs minus total money received from selling energy in different weights ω

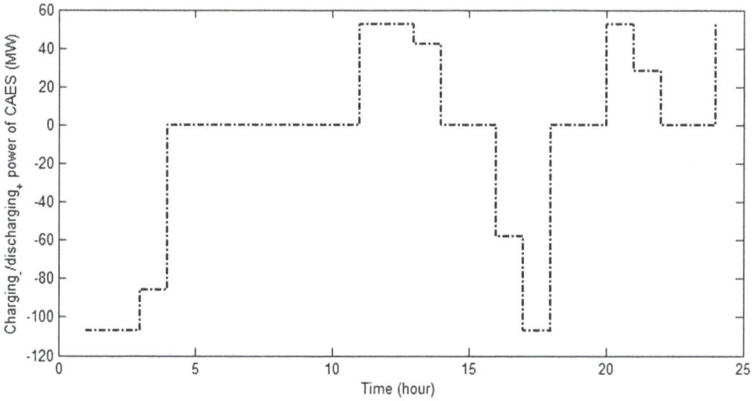

Fig. 3.9 Optimal charge and discharge decisions of CAES for $\omega = 0.5$

Nomenclature

α^{inj}	Efficiency of injected power
α^{pump}	Efficiency of produced power
$\gamma_i, \eta_i, \delta_i, \xi_i, \varphi_i$	Emission characteristics of thermal generation station i
λ_t	Electricity price at time horizon t [$/MWh]
ω	Weighting factor $\in[0, 1]$
a_i	No-load cost coefficient of thermal generator i [$]
b_i	Linear cost coefficient of thermal generator i [$/MWh]
B_{ij}	ijth element of loss coefficient square matrix of size N_{Gen}[1/MW]
c_i	Quadratic cost coefficient of thermal generator i [$/MW^{2}h]

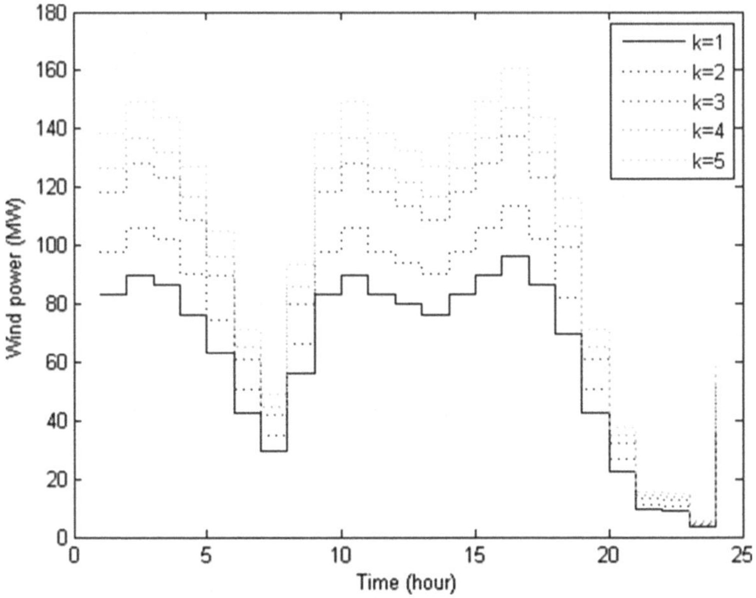

Fig. 3.10 Nested wind generation intervals

e_i, f_i	Cost coefficient of thermal generator i reflecting the valve point effect respectively in [$] and [1/MW]
$P_{\text{Comp}, t}$	Power consumed by CAES at time t for compressing and injecting air [MWh]
$P_{\text{Gen}, t}$	Power output of CAES at time horizon t [MWh]
$P_{i, t}$	Active power output of thermal generation station i at time t [MW]
$P_i^{\min}, P_i^{\max}$	Minimum and maximum power generation of ith thermal unit, respectively
P_d^t	Electrical demand at time t [MW]
P_{Loss}^t	Active power losses of transmission lines at time t [MW]
SOC_t	Amount of stored energy at time t [MWh]
$\text{SOC}_{\min}$	Minimum level of storage [MWh]
$\text{SOC}_{\max}$	Maximum level of storage [MWh]
u_t^{inj}	Binary variable, which is equal to 1 if air injected by CAES at time t, and 0 otherwise
u_t^{pump}	Binary variable, which is equal to 1 if air pumped by CAES at time t, and 0 otherwise
V_t^{inj}	Energy equivalent of injected air to storage at time t [MW/h]
$V_{\min}^{\text{inj}}$	Minimum level of injected air into storage [MW/h]
$V_{\max}^{\text{inj}}$	Minimum level of injected air into storage [MW/h]

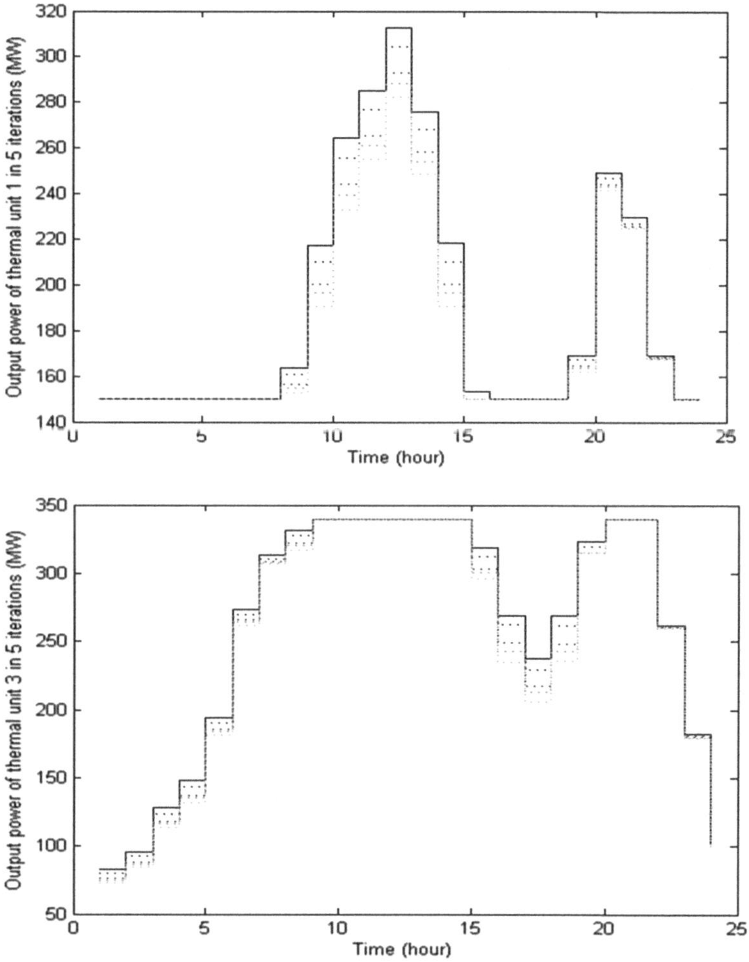

Fig. 3.11 Output power of each thermal generating unit for $\omega = 0.5$

V_t^{pump}	Energy equivalent of pumped air to combustion chamber at time t [MWh]
$V_{\text{min}}^{\text{pump}}$	Minimum level of pumped air from storage to combustion chamber [MW/h]
$V_{\text{max}}^{\text{pump}}$	Maximum level of pumped air from storage to combustion chamber [MW/h]

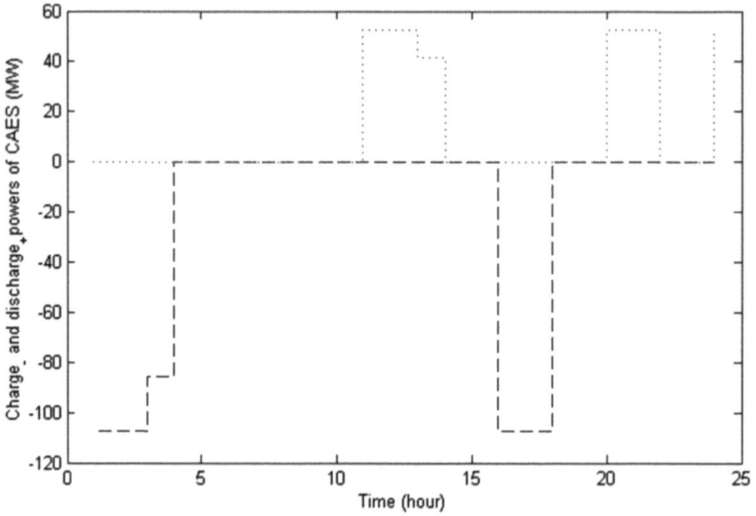

Fig. 3.12 Optimal charge and discharge decisions of CAES for $\omega = 0.5$

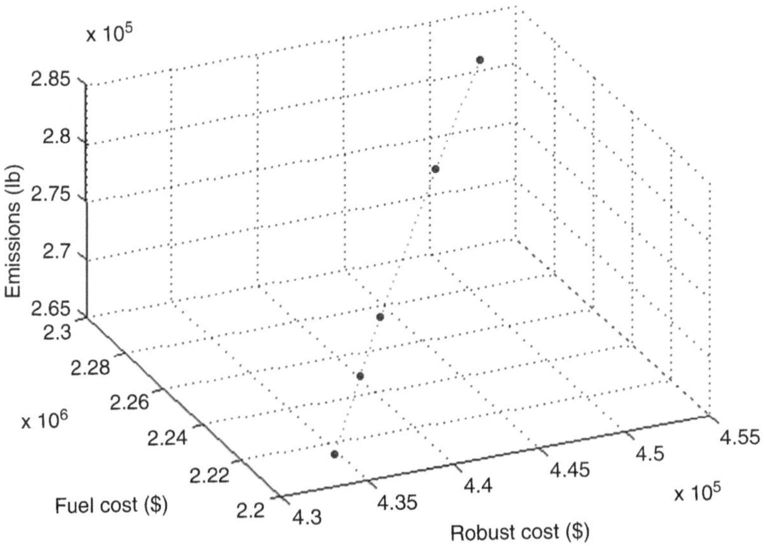

Fig. 3.13 Fuel costs and emission productions for five iterations and $\omega = 0.5$

References

1. Jabari F, Nojavan S, Mohammadi Ivatloo B, Sharifian MBB (2016) Optimal short-term scheduling of a novel tri-generation system in the presence of demand response programs and battery storage system. Energy Convers Manag 122:95–108
2. Jabari F, Nojavan S, Mohammadi-Ivatloo B, Ghaebi H, Mehrjerdi H (2018) Risk-constrained scheduling of solar Stirling engine based industrial continuous heat treatment furnace. Appl Therm Eng 128:940–955
3. Jabari F, Seyedi H, Najafi Ravadanegh S (2015) Large-scale power system controlled islanding based on backward elimination method and primary maximum expansion areas considering static voltage stability. Int J Electr Power Energy Syst 67:368–380
4. Jabari F, Seyedi H, Najafi Ravadanegh S, Mohammadi-Ivatloo B (2017) Multi-objective optimal preventive islanding based on stochastic backward elimination strategy considering uncertainties of loads and wind farms. Int Trans Electr Energy Syst 27:e2451
5. Chen L, Zheng T, Mei S, Xue X, Liu B, Lu Q (2016) Review and prospect of compressed air energy storage system. J Mod Power Syst Clean Energy 4:529–541
6. Karellas S, Tzouganatos N (2014) Comparison of the performance of compressed-air and hydrogen energy storage systems: Karpathos island case study. Renew Sust Energ Rev 29:865–882
7. Gonzalez A, Ó'Gallachóir B, McKeogh E, Lynch K (2004) Study of electricity storage technologies and their potential to address wind energy intermittency in Ireland. Final report. Sustainable Energy Research Group and Rockmount Capital Partners Cork. RE/HC/03/001. Published May 2004
8. Saidur R, Rahim NA, Hasanuzzaman M (2010) A review on compressed-air energy use and energy savings. Renew Sust Energ Rev 14:1135–1153
9. Jabari F, Nojavan S, Mohammadi Ivatloo B (2016) Designing and optimizing a novel advanced adiabatic compressed air energy storage and air source heat pump based μ-Combined Cooling, heating and power system. Energy 116(Part 1):64–77
10. Zhang Y, Yang K, Li X, Xu J (2014) Thermodynamic analysis of energy conversion and transfer in hybrid system consisting of wind turbine and advanced adiabatic compressed air energy storage. Energy 77:460–477
11. Havel TF (2012) Adsorption-enhanced compressed air energy storage. Google Patents, 2012
12. Odukomaiya A, Abu-Heiba A, Gluesenkamp KR, Abdelaziz O, Jackson RK, Daniel C et al (2016) Thermal analysis of near-isothermal compressed gas energy storage system. Appl Energy 179:948–960
13. Wang Z, Xiong W, Ting DSK, Carriveau R, Wang Z (2016) Conventional and advanced exergy analyses of an underwater compressed air energy storage system. Appl Energy 180:810–822
14. Wang Z, Ting DSK, Carriveau R, Xiong W, Wang Z (2016) Design and thermodynamic analysis of a multi-level underwater compressed air energy storage system. J Energy Storage 5:203–211
15. Perazzelli P, Anagnostou G (2016) Design issues for compressed air energy storage in sealed underground cavities. J Rock Mech Geotech Eng 8:314–328
16. Liu H, He Q, Borgia A, Pan L, Oldenburg CM (2016) Thermodynamic analysis of a compressed carbon dioxide energy storage system using two saline aquifers at different depths as storage reservoirs. Energy Convers Manag 127:149–159
17. Chen L-X, Hu P, Sheng C-C, Xie M-N (2017) A novel compressed air energy storage (CAES) system combined with pre-cooler and using low grade waste heat as heat source. Energy 131:259–266
18. Sheng L, Zhou Z, Charpentier JF, Benbouzid MEH (2017) Stand-alone island daily power management using a tidal turbine farm and an ocean compressed air energy storage system. Renew Energy 103:286–294
19. Ji W, Zhou Y, Sun Y, Zhang W, An B, Wang J (2017) Thermodynamic analysis of a novel hybrid wind-solar-compressed air energy storage system. Energy Convers Manag 142:176–187

20. Zhao P, Dai Y, Wang J (2014) Design and thermodynamic analysis of a hybrid energy storage system based on A-CAES (adiabatic compressed air energy storage) and FESS (flywheel energy storage system) for wind power application. Energy 70:674–684
21. Lv S, He W, Zhang A, Li G, Luo B, Liu X (2017) Modelling and analysis of a novel compressed air energy storage system for trigeneration based on electrical energy peak load shifting. Energy Convers Manag 135:394–401
22. Mohammadi A, Mehrpooya M (2016) Exergy analysis and optimization of an integrated micro gas turbine, compressed air energy storage and solar dish collector process. J Clean Prod 139:372–383
23. Yao E, Wang H, Wang L, Xi G, Maréchal F (2017) Multi-objective optimization and exergoeconomic analysis of a combined cooling, heating and power based compressed air energy storage system. Energy Convers Manag 138:199–209
24. Li Y, Wang X, Li D, Ding Y (2012) A trigeneration system based on compressed air and thermal energy storage. Appl Energy 99:316–323
25. Liu J-L, Wang J-H (2015) Thermodynamic analysis of a novel tri-generation system based on compressed air energy storage and pneumatic motor. Energy 91:420–429
26. Bagdanavicius A, Jenkins N (2014) Exergy and exergoeconomic analysis of a compressed air energy storage combined with a district energy system. Energy Convers Manag 77:432–440
27. Shafiee S, Zareipour H, Knight AM, Amjady N, Mohammadi-Ivatloo B (2017) Risk-constrained bidding and offering strategy for a merchant compressed air energy storage plant. IEEE Trans Power Syst 32:946–957
28. Jubeh NM, Najjar YSH (2012) Power augmentation with CAES (compressed air energy storage) by air injection or supercharging makes environment greener. Energy 38:228–235
29. Zunft S, Jakiel C, Koller M, Bullough C (2006) Adiabatic compressed air energy storage for the grid integration of wind power. In: Sixth international workshop on large-scale integration of wind power and transmission networks for offshore windfarms, 26–28 October 2006, Delft, The Netherlands, pp 346–351
30. Grazzini G, Milazzo A (2008) Thermodynamic analysis of CAES/TES systems for renewable energy plants. Renew Energy 33(9):1998–2006
31. Hartmann N, Vöhringer O, Kruck C, Eltrop L (2012) Simulation and analysis of different adiabatic compressed air energy storage plant configurations. Appl Energy 93:541–548
32. Proczka JJ, Muralidharan K, Villela D, Simmons JH, Frantziskonis G (2013) Guidelines for the pressure and efficient sizing of pressure vessels for compressed air energy storage. Energy Convers Manag 65:597–605
33. Raju M, Kumar Khaitan S (2012) Modeling and simulation of compressed air storage in caverns: a case study of the Huntorf plant. Appl Energy 89(1):474–481
34. Wolf D, Berthold S, Dötsch C, Lopez JU, Erkan T, Span R (2009) Dynamic simulation of possible heat management solutions for adiabatic compressed air energy storage. In: 11th international conference on thermal energy storage. Stockholm, Sweden, 2009
35. Grazzini G, Milazzo A (2012) A thermodynamic analysis of multistage adiabatic CAES. Proc IEEE 100:461–472
36. Ma H, Yang Z, You P, Fei M (2017) Multi-objective biogeography-based optimization for dynamic economic emission load dispatch considering plug-in electric vehicles charging. Energy 135:101
37. Zou D, Li S, Li Z, Kong X (2017) A new global particle swarm optimization for the economic emission dispatch with or without transmission losses. Energy Convers Manag 139:45–70
38. Jabari F, Shamizadeh M, Mohammadi-ivatloo B (2017) Dynamic economic generation dispatch of thermal units incorporating aggregated plug-in electric vehicles. In: 3rd international conference of technology and energy management. Shahid Beheshti University, Tehran
39. Jadoun VK, Gupta N, Niazi K, Swarnkar A (2015) Modulated particle swarm optimization for economic emission dispatch. Int J Electr Power Energy Syst 73:80–88

40. Elaiw A, Xia X, Shehata A (2013) Hybrid DE-SQP and hybrid PSO-SQP methods for solving dynamic economic emission dispatch problem with valve-point effects. Electr Power Syst Res 103:192–200
41. Shilaja C, Ravi K (2017) Optimization of emission/economic dispatch using euclidean affine flower pollination algorithm (eFPA) and binary FPA (BFPA) in solar photo voltaic generation. Renew Energy 107:550–566
42. Chen F, Zhou J, Wang C, Li C, Lu P (2017) A modified gravitational search algorithm based on a non-dominated sorting genetic approach for hydro-thermal-wind economic emission dispatching. Energy 121:276–291
43. Basu M (2008) Dynamic economic emission dispatch using nondominated sorting genetic algorithm-II. Int J Electr Power Energy Syst 30:140–149

Chapter 4
Solar Thermal Energy Storage for Residential Sector

Afshin Najafi-Ghalelou, Sayyad Nojavan, Majid Majidi, Farkhondeh Jabari, and Kazem Zare

4.1 Introduction

An energy hub system is a multi-generation system where multiple energy carriers are converted, stored [1], and distributed to meet heat and electrical demand [2]. Converter devices can be solar thermal storage, CHP, and boiler. Solar thermal storage is used to convert solar irradiation to heat. Other technologies such as CHP and boiler are also suggested to convert natural gas to electricity and heat.

4.1.1 Literature Review

In this chapter, energy management system inside a residential energy hub system has been investigated. In order to minimize discomfort and operation costs, a new efficient algorithm for energy management system inside a residential energy hub system is presented in [3]. Stochastic programming is implemented in [4] for modeling optimal scheduling of energy hub systems. To minimize energy cost based on availability of each expected demand, resources, and prices, a new optimal management algorithm for optimal management of distributed energy resources in facilities with energy hub systems is provided in [5]. A day-ahead dynamic optimal

A. Najafi-Ghalelou · S. Nojavan (✉) · M. Majidi · K. Zare
Faculty of Electrical and Computer Engineering, University of Tabriz, Tabriz, Iran
e-mail: afshin.najafi95@ms.tabrizu.ac.ir; sayyad.nojavan@tabrizu.ac.ir; majidmajidi95@ms.tabrizu.ac.ir; kazem.zare@tabrizu.ac.ir

F. Jabari
Department of Electrical and Computer Engineering, University of Tabriz, Tabriz, Iran
e-mail: f.jabari@tabrizu.ac.ir

B. Mohammadi-Ivatloo, F. Jabari (eds.), *Operation, Planning, and Analysis of Energy Storage Systems in Smart Energy Hubs*,
https://doi.org/10.1007/978-3-319-75097-2_4

operation and dispatch strategies of energy hub systems are presented in [6] to minimize daily operation cost. A stochastic scheduling for wind integrated smart energy hub system problem is presented in [7].

Energy management systems inside smart home have been investigated as follows: In [8], prioritizing operation of controllable appliances from the customer's viewpoint has been investigated to minimize customer energy costs. An energy management solution is presented in [9] to combine and describe advantages and features of both energy hub system framework and demand side management methods. In order to tackle the household load scheduling problem with uncertain ambient temperature and hot water demand, an interval number optimization method is provided in [10]. An optimization problem is proposed in [11] which simultaneously selects, sizes, and determines optimal operation of residential heating systems. In order to reduce electricity price of smart buildings and manage battery storage and temperature of thermal appliances, a new control algorithm is presented in [12]. In [13], various optimization techniques applied to demand side management system have been reviewed.

Literature review about solar thermal storage can be expressed as follows: Performance of a liquid thermocline and a packed bed are compared with each other in [14] for an off-shore wind-TP system. To reduce heat pumps operational temperature differences, application of hybrid pumped thermal electricity storage is studied and provided in [15]. To determine operational state of power generation unit based on thermal and electric demand, a new thermal storage strategy is provided in [16]. Economic impact of designing thermal energy storage system is analyzed and provided in [17]. Technologies about high temperature solar receivers associated with power tower systems and power dish are compared and provided in [18]. A new distributed energy resources customer adoption model of thermal energy storage is provided in [19] to improve tracking of losses based on temperature of ambient and storage. Summary of different thermal energy storage systems and solar thermal storage materials is provided and compared in [20].

4.1.2 Novelty and Contributions of This Research

According to our knowledge, there is no research available about optimal energy consumption scheduling of a residential energy hub system in the presence of solar thermal storage system. So, in this chapter, a residential hub energy system model containing CHP generator, boiler, electrical storage, solar thermal storage, and smart appliance is proposed. Two cases studied are used to assess the impacts of solar thermal energy storage on operation cost of residential energy hub system. According to the above information, the novelty and contributions of this paper are presented below:

- Energy management of a residential hub energy systems is proposed in the presence of solar thermal energy storage.

- Scheduling optimal performance of all equipment within the residential hub energy systems.
- Scheduling and prioritizing performance of smart appliances in the presence of distributed energy sources with the aim of minimizing total operation cost of residential hub energy systems.
- Employing mixed-integer programming (MIP) to guarantee global optimal.

4.1.3 Chapter Organization

The rest of the proposed chapter is categorized as follows: The mathematical model has been presented in Sect. 4.2. Input data, case study, and the results are provided in Sect. 4.3. Discussion and conclusions are presented in Sect. 4.4.

4.2 Problem Formulation

As shown in Fig. 4.1, the proposed residential hub energy system model contains CHP generator, boiler, battery storage system, solar thermal storage, and smart appliances. Optimal energy consumption scheduling of a residential energy hub system has been formulated in this section. The objective function includes operation cost of equipment in a residential energy hub system which can be presented as:

$$\mathrm{OBJ} = \left\{ \begin{array}{l} \sum_{j=1}^{J}\sum_{t=1}^{T} \frac{\left(\lambda^{\mathrm{Gas}} \times P_{j,t}^{\mathrm{CHP}}\right)}{\eta^{\mathrm{CHP}}} + \sum_{j=1}^{J}\sum_{t=1}^{T} \frac{\left(\lambda^{\mathrm{Gas}} \times P_{j,t}^{\mathrm{Boiler}}\right)}{\eta^{\mathrm{Boiler}}} \\ + \sum_{j=1}^{J}\sum_{t=1}^{T} \mathrm{MC}^{\mathrm{elec}} \times \mathrm{DR}^{\mathrm{elec}} + \sum_{j=1}^{J}\sum_{t=1}^{T} \lambda_t^{\mathrm{Grid}} \times P_{j,t}^{\mathrm{Import}} \\ - \sum_{j=1}^{J}\sum_{t=1}^{T} P_{j,t}^{\mathrm{export}} \times \lambda^{\mathrm{export}} \end{array} \right\} \times \Delta t \tag{4.1}$$

The objective function includes operation cost of CHP, boiler, battery storage system, cost of purchased power from grid, and profit of selling power to grid.

4.2.1 Combined Heat and Power (CHP) Generator

The output power of CHP generator should not exceed its designed capacity which is presented as [21]:

$$P_{j,t}^{\mathrm{CHP}} \leq \mathrm{CAP}^{\mathrm{CHP}} \tag{4.2}$$

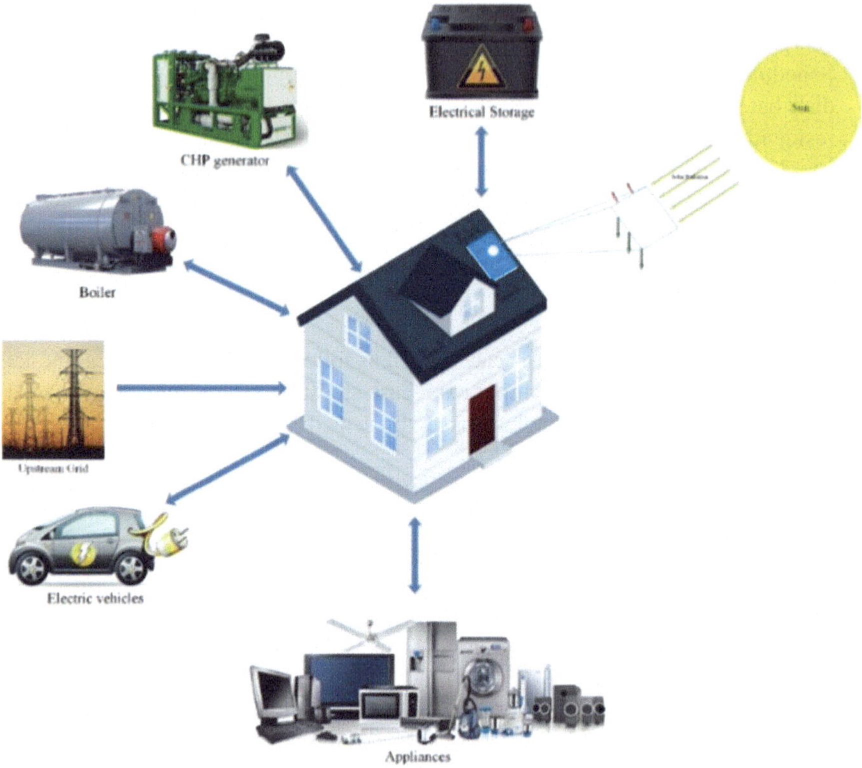

Fig. 4.1 Schematic diagram of proposed residential hub energy system model

4.2.2 Boiler

The output power of boiler as well as CHP generator should not exceed its designed capacity. In this regard, Eq. (4.3) is presented [21].

$$P_{j,t}^{\text{Boiler}} \leq \text{CAP}^{\text{boiler}} \tag{4.3}$$

4.2.3 Battery Storage System

The model of central battery storage system which is available for all residential hub energy system sector is obtained from [21]. The technical constraints related to battery storage system are described as follows:

The output power of battery storage system as well as other equipment should not exceed its designed capacity and therefore Eq. (4.4) is presented.

$$\sum_{j=1}^{J} \mathrm{SOC}_{j,t}^{\mathrm{elec}} \leq \mathrm{CAP}^{\mathrm{elec}} \tag{4.4}$$

The state of charge of battery storage system at time t is equal to the state of charge of battery storage system at time $t-1$ plus the charged amount at time t minus the discharge power at time t. Also, discharge rate of battery storage at time t should not exceed state of charge of battery storage at time $t-1$. Mathematical formulation of mentioned statements is provided as follows:

$$\mathrm{SOC}_{j,t}^{\mathrm{elec}} = \mathrm{SOC}_{j,t+1}^{\mathrm{elec}} + \left(\mathrm{CR}_{j,t}^{\mathrm{elec}} \times \eta^{\mathrm{elec}} \times \Delta t\right) - \left(\frac{\mathrm{DR}_{j,t}^{\mathrm{elec}} \times \Delta t}{\eta^{\mathrm{elec}}}\right) \tag{4.5}$$

$$\frac{\mathrm{DR}_{j,t}^{\mathrm{elec}} \times \Delta t}{\eta^{\mathrm{elec}}} \leq \mathrm{SOC}_{j,t-1}^{\mathrm{elec}} \tag{4.6}$$

The charge and discharge rate of battery storage system should not exceed charge and discharge limits of battery storage:

$$\mathrm{CR}_{j,t}^{\mathrm{elec}} \leq M^{\mathrm{elec}} \times B_{j,t}^{\mathrm{elec}} \tag{4.7}$$

$$\mathrm{DR}_{j,t}^{\mathrm{elec}} \leq M^{\mathrm{elec}} \times \left(1 - B_{j,t}^{\mathrm{elec}}\right) \tag{4.8}$$

The total state of charge of the battery storage system at each time period is equal to the sum of state of charge of sub-batteries storage system at each residential hub energy system sector:

$$\mathrm{SOC}_{t}^{\mathrm{Totalelec}} = \sum_{j=1}^{J} \mathrm{SOC}_{j,t}^{t} \tag{4.9}$$

In order to avoid net accumulation, state of charge of the battery storage system at the end of the each sample day should be equal to the initial value of battery storage. In the proposed model, the initial state of charge of battery is set as variable to determine the best initial state of charge for one day utilization [21]. Otherwise, it can be set as parameter which is obtained from the end of previous day.

$$\mathrm{SOC}_{1}^{\mathrm{Totalelec}} = \mathrm{SOC}_{48}^{\mathrm{Totalelec}} = S^{\mathrm{elec}} \tag{4.10}$$

The charge and discharge rate of battery storage system should not exceed the charge and discharge limits of battery storage which are defined by the battery manufacturer. For this reason, Eqs. (4.11) and (4.12) are presented.

$$\sum_{j=1}^{J} \mathrm{CR}_{j,t}^{\mathrm{elec}} \leq \mathrm{CRL}^{\mathrm{elec}} \tag{4.11}$$

$$\sum_{j=1}^{J} \mathrm{DR}_{j,t}^{\mathrm{elec}} \leq \mathrm{DRL}^{\mathrm{elec}} \tag{4.12}$$

4.2.4 *Exchanged Power Between the Residential Energy Hub System and Upstream Grid*

The imported/exported power from/to grid at each period of time is calculated as follows:

$$P_{j,t}^{\mathrm{Import}} \leq M^{\mathrm{Grid}} \times B_{j,t}^{\mathrm{Grid}} \tag{4.13}$$

$$P_{j,t}^{\mathrm{export}} \leq M^{\mathrm{Grid}} \times \left(1 - B_{j,t}^{\mathrm{Grid}}\right) \tag{4.14}$$

4.2.5 *Appliances*

Household appliances can be noted as fridge, washing machine, dishwasher, etc. The appliances should be ON between the specific time periods which is determined by the owner of residential energy hub system. Also, each appliance must be active continuously (θ) based on the predefined length of time ($P_{j,\,i}$) within the determined time period which is determined by the owner of residential energy hub system and for this reason, Eq. (4.15) which is obtained from [21] is provided as follows:

$$\sum_{t=T_{j,i}^{\mathrm{Start}}}^{T_{j,i}^{\mathrm{Finish}}-P_{j,i}} \omega_{j,i,t-\theta} \tag{4.15}$$

4.2.6 *Solar Thermal Storage*

The model of solar thermal storage is obtained from [19]. Solar thermal storage converts solar irradiation to thermal which is used directly or stored in the thermal storage system to be used in other periods. The technical constraints related to thermal storage are described as follows:

The state of charge of solar thermal storage at time t is equal to the state of charge of solar thermal storage at time $t-1$ plus the charged heat minus the discharged heat and loss of heat at time t. Also, state of charge of solar thermal storage should not exceed its designed capacity. In this regard, Eqs. (4.16) and (4.17) are presented.

$$H_t^{\text{stored}} = H_{t-1}^{\text{stored}} + \eta^{\text{ch,Ther}} \times H_t^{\text{ch,Ther}} - \sum_j^J \frac{H_{j,t}^{\text{dch,Ther}}}{\eta^{\text{dch,Ther}}} - H_t^{\text{loss,Ther}} \tag{4.16}$$

$$H_t^{\text{stored}} \le \text{CAP}^{\text{Ther}} \tag{4.17}$$

The imported/exported power from/to solar thermal storage at each period of time is limited by discharge/charge rate. For this reason Eqs. (4.18) and (4.19) are presented.

$$H_t^{\text{ch,Ther}} \le B_t^{\text{ch,ther}} \times \text{CAP}^{\text{Ther}} \times H^{\text{ch,Ther,max}} \tag{4.18}$$

$$H_{j,t}^{\text{dch,Ther}} \le B_{j,t}^{\text{dch,ther}} \times \text{CAP}^{\text{Ther}} \times H^{\text{dch,Ther,max}} \tag{4.19}$$

Heat losses of solar thermal storage depend on the capacity of solar thermal storage, ambient temperature, and amount of stored energy in the solar thermal storage. So, the heat losses can be formulated as follows:

$$H_t^{\text{loss,Ther}} = H_{t-1}^{\text{stored}} \times \theta^{\text{storage}} + \theta^{\text{static}} \times E_t^{\text{unuse}} \tag{4.20}$$

The unused energy of solar thermal storage system can be calculated based on the minimum/maximum temperature of solar thermal storage, ambient temperature, and the capacity of solar thermal storage as follows:

$$E_t^{\text{unuse}} = \text{CAP}^{\text{Ther}} \times \frac{T^{\min} - T_t^{\text{amb}}}{T^{\max} - T^{\min}} \tag{4.21}$$

The amount of converted solar irradiation to the heat at each period of time depends on the solar irradiation, efficiency, and the surface area of thermal energy storage panel. Also, the charge rate of solar thermal storage at each period is limited by the converted amount of solar irradiation to heat. For this reason, Eqs. (4.22) and (4.23) are presented.

$$Q_t = \varphi_t^{\text{solar}} \times A^{\text{app}} \times \gamma^{\text{Ther}} \tag{4.22}$$

$$Q_t \ge H_t^{\text{ch,Ther}} \tag{4.23}$$

4.2.7 Energy Balances

Energy balance constraint between the production and consumption power can be written as:

$$\sum_{j=1}^{J}\sum_{i=1}^{I}\sum_{\theta=1}^{P_{j,i}-1} P_i^{\text{Consump}} \times \omega_{j,i,t-\theta} = P_{j,t}^{\text{CHP}} + \text{DR}_{j,t}^{\text{elec}} - \text{CR}_{j,t}^{\text{elec}} + P_{j,t}^{\text{Import}} - P_{j,t}^{\text{export}} \tag{4.24}$$

4.2.8 Thermal Balances

The heat balance constraint between producers and consumers can be written as:

$$H_{j,t}^{\text{Demand}} = \alpha^{\text{CHP}} \times P_{j,t}^{\text{CHP}} + P_{j,t}^{\text{Boiler}} \tag{4.25a}$$

With considering the solar thermal storage, Eq. (4.25a) will be updated as:

$$H_{j,t}^{\text{Demand}} = \alpha^{\text{CHP}} \times P_{j,t}^{\text{CHP}} + P_{j,t}^{\text{Boiler}} + H_{j,t}^{\text{dch,Ther}} \tag{4.25b}$$

4.3 Numerical Simulation

The proposed residential hub energy system model contains CHP generator, boiler, battery storage, solar thermal storage, and smart appliances. The entire time horizon of case study is 24 h with time interval of 30 min. The starting time of case study is from 8 AM and the ending time is 8 AM of the next morning. The proposed optimization problem has been studied in two case studies with and without considering the effect of solar thermal storage on total operation cost of residential hub energy system.

Case study 1 is related to the optimal scheduling of residential hub energy system consumption without considering the effect of solar thermal storage. In this case, the objective is to minimize total energy cost of residential hub energy system (4.1) subject to constraints (4.2)–(4.15) and (4.25a). In the second study, the effect of solar thermal storage is considered in which the objective is to minimize the total energy cost of residential hub energy system (4.1) subject to constraints (4.2)–(4.24) and (4.25b).

4.3.1 Input Data

Technical information of solar thermal storage is presented in Table 4.1 [19]. Consumption power and operation time length of each appliance are presented in Table 4.2 [21]. All appliances except the washing machine, dish washer, and tumble dryer have constant power consumption rate during the operation time while the electrical profiles for washing machine, dish washer, and tumble dryer are presented in Fig. 4.2 [22]. Technical information of CHP, boiler, and battery storage system are provided in Table 4.3 [21]. The earliest starting time of appliances and latest finishing time of appliances are presented in Tables 4.4 and 4.5, respectively. Market price and ambient temperature are presented in Figs. 4.3 and 4.4 [23, 24], respectively. Solar irradiation is presented in Fig. 4.5 [25]. Heat demands for each residential hub energy system sector are presented in Figs. 4.6 and 4.7, respectively [21]. Natural gas price is considered to be 2.7 p/kWh and the cost of selling power to the upstream grid is set to be 1 p/kWh$_e$ [21]. It should be mentioned that the developed MIP model is implemented using CPLEX [26] in GAMS software [27].

Table 4.1 Technical information of solar thermal storage [19]

Parameters	Value	Parameters	Value
A^{app}	50 m^2	$\theta^{storage}$	5.7%
γ^{Ther}	95%	θ^{static}	5.6%
CAP^{Ther}	100 kW	$H^{dch, Ther, max}$	25%
$\eta^{ch, Ther}$	95%	$H^{ch, Ther, max}$	25%
$\eta^{dch, Ther}$	95%	T^{max}	65 °C
T^{min}	36 °C		

Table 4.2 Power consumption and length of operation time of each appliance

Appliances	Power consumption (kW) [21]	Length of operation time (h) [21]
Washing machine	Fig. 4.2	2
Dish washer	Fig. 4.2	2
Tumble dryer	Fig. 4.2	1.5
Cooker hob	3	0.5
Cooker oven	5	0.5
Microwave	1.7	0.5
Interior lighting	0.84	6
Laptop	0.1	2
Desktop	0.3	3
Vacuum cleaner	1.2	0.5
Fridge	0.3	24
Electrical car	3.5	3

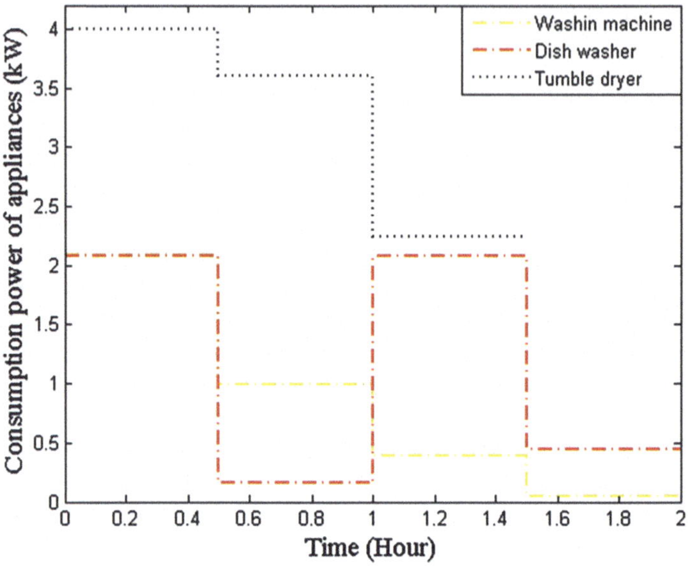

Fig. 4.2 Electricity utilization profiles of washing machine, dish washer, and tumble dryer

Table 4.3 Technical information of CHP, boiler, battery storage system, and thermal storage system

Parameter	Value
CHP	
η^{CHP}	35%
CAP^{CHP}	4 kW_e
α^{CHP}	1.3
Boiler	
η^{Boiler}	85%
CAP^{boiler}	24 kW_{th}
Battery storage system	
η^{elec}	95%
CAP^{elec}	4 kW_eh
MC^{elec}	0.005 p/kWh_e
M^{elec}	2 kW_e
CRL^{elec}	4 kW_e
DRL^{elec}	4 kW_e

4.3.2 Simulation Results

In this section the effect of solar thermal energy storage system has been investigated in two cases. In case 1, total operation cost of residential hub energy system without considering effect of solar thermal storage has been solved. In order to show the effect of solar thermal energy storage, the same problem is been solved in case 2 with considering the effect of solar thermal storage system. With comparing the results of cases 1 and 2, it can be seen that operation cost in case 2 is decreased

Table 4.4 Earliest starting time of appliances (h) [21]

Smart homes	1	2	3	4	5	6	7	8	9	10
Washing machine	12	11	–	13	–	18	14	16	11	–
Dish washer	16	14	–	11	–	22	22	20	16	–
Tumble dryer	19	17	–	14	–	1	1	23	19	–
Cooker hob	15	10	–	13	10	14	18	11	10	–
Cooker oven	11	15	–	20	13	13	–	–	19	20
Microwave	21	13	–	20	12	17	–	18	20	10
Interior lighting	18	–	20	20	22	19	–	17	20	21
Laptop	19	–	17	17	19	21	–	18	19	19
Desktop	17	–	16	–	14	19	20	22	20	–
Vacuum cleaner	18	–	19	–	20	16	22	21	21	21
Fridge	0	–	0	–	0	0	0	–	0	0
Electrical car	21	–	20	–	19	18	17	–	21	19

Table 4.5 Latest finishing time of appliances (h) [21]

Smart homes	1	2	3	4	5	6	7	8	9	10
Washing machine	20	18	–	19	–	23	18	20	15	–
Dish washer	19	16	–	14	–	1	24	22	18	–
Tumble dryer	24	21	–	17	–	6	3	1	20	–
Cooker hob	16	11	–	15	13	17	23	15	15	–
Cooker oven	12	16	–	22	16	16	–	–	24	1
Microwave	22	14	–	22	15	20	–	20	21	11
Interior lighting	24	–	2	2	4	1	–	23	2	3
Laptop	1	–	22	20	24	3	–	22	24	24
Desktop	23	–	20	–	19	1	1	1	24	–
Vacuum cleaner	2	–	23	–	1	22	4	4	4	5
Fridge	24	–	24	–	24	24	24	–	24	24
Electrical car	7	–	3	–	23	2	1	–	6	5

about 16.88%. In case 2, solar energy storage system is used to meet heat demand instead of boiler. So, the cost of gas consumption is reduced and this causes the reduction of operation cost of residential hub energy system. Comparison results of two cases related to the operation cost of residential hub energy system are studied and presented in Table 4.6.

Output power of CHP and boiler are presented in Figs. 4.8 and 4.9, respectively. In the second case study, the output power of CHP is decreased 33.60 kW. The produced heat by boiler after 12 PM has become zero and instead of boiler, heat produced by solar thermal storage is used to meet heat demand.

Charge and discharge rates and state of charge of battery storage system are provided in Figs. 4.10 and 4.11, respectively. Battery storage in the second case study is charged 6.77 kW more in comparison with case one. Also, battery storage system is discharged 6.33 kW more compared to case one. So, in the second case

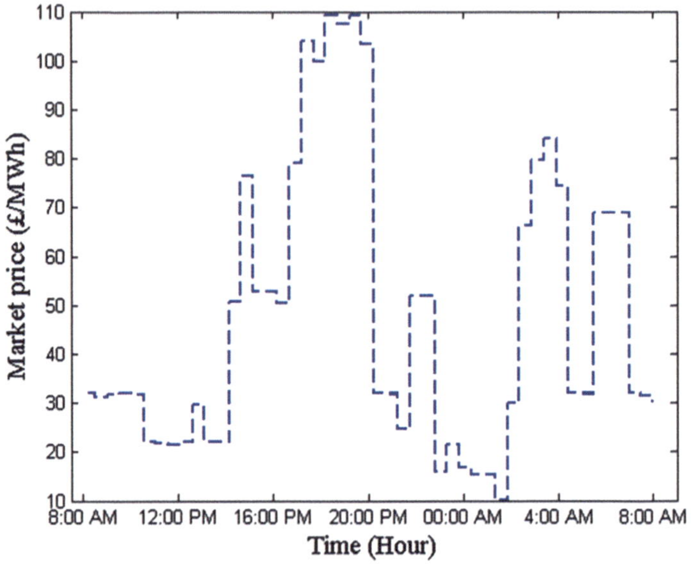

Fig. 4.3 Market price (£/MWh) [23]

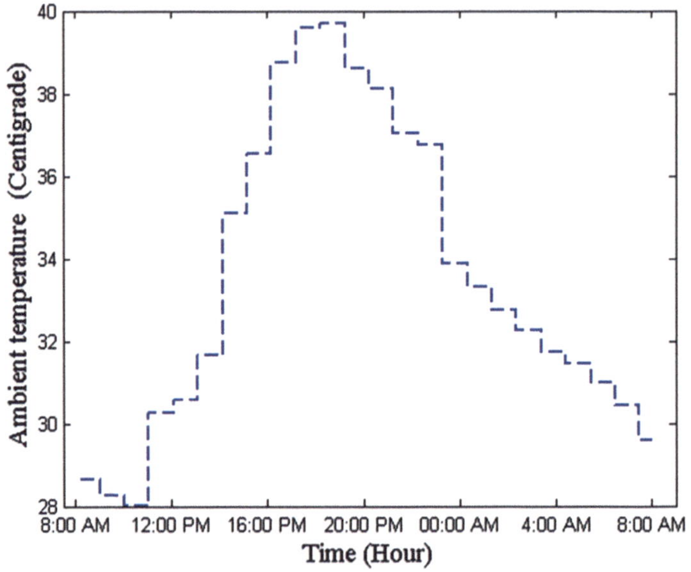

Fig. 4.4 Ambient temperature (°C) [24]

study, the state of charge of battery is 2.61 kW less in comparison with case one. Also, it can be observed that with considering the effect of solar thermal storage, battery storage system is charged and discharged more.

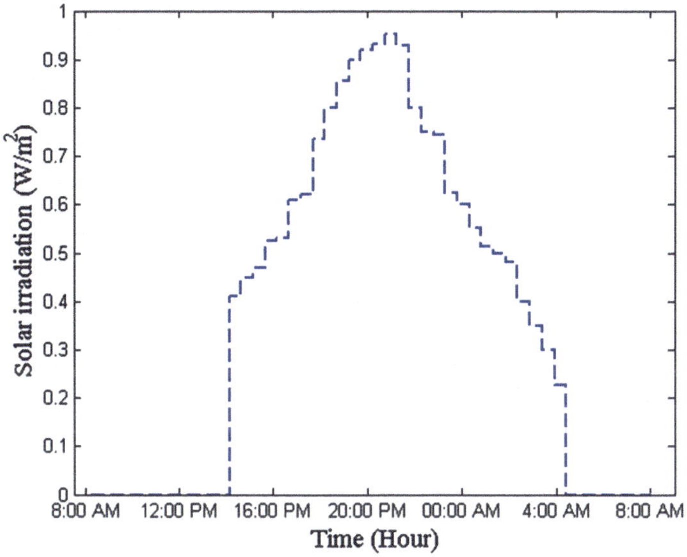

Fig. 4.5 Solar irradiation (W/m^2)

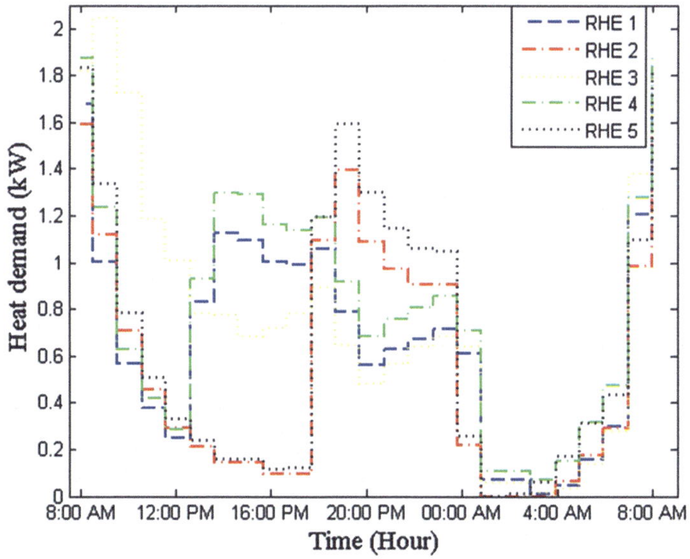

Fig. 4.6 Heat demand of sectors 1–5 in residential hub energy system [21]

The charge, discharge, state of charge, and thermal losses of solar thermal storage are presented in Fig. 4.12. It can be observed that the solar thermal storage produced 337.52 kW heat with converting the solar irradiation and discharged 164.61 kW to

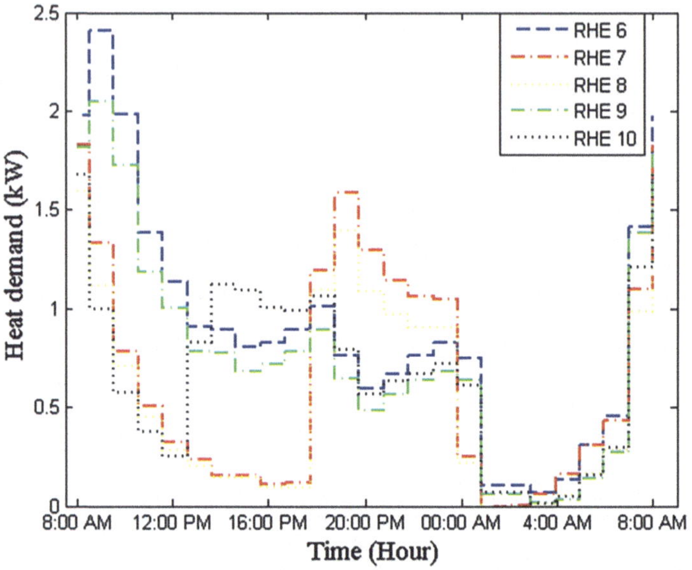

Fig. 4.7 Heat demand of sectors 6–10 in residential hub energy system [21]

Table 4.6 Total operation cost of residential hub energy system

Case 1:Without considering effect of solar thermal storage	13.9483 £
Case 2: With considering effect of solar thermal storage	11.5943 £
Cost reduction in comparison with case 1	16.88%

meet heat demand of residential hub energy system. Total losses of solar thermal storage are 147.38 kW during the 24 h study case.

Imported/exported power from/to the grid is presented in Fig. 4.13. In case 1, imported power from grid is 472.84 kW and exported power to grid is 1.9 kW. In case 2, imported power from grid is 506.89 kW. So, imported power from grid is increased 7.2% in comparison with case 1 and exported power remained constant in comparison with case 1.

The activation time of each appliance in each residential hub energy system sector for cases 1 and 2 is presented in Tables 4.7 and 4.8, respectively. It should be mentioned that each appliance is active continuously (θ) within the determined time period ($P_{j,\,i}$) by the owner of smart home. With comparing the obtained results from two case studies, it can be understood that the activation time of some appliances is only shifted in small time intervals.

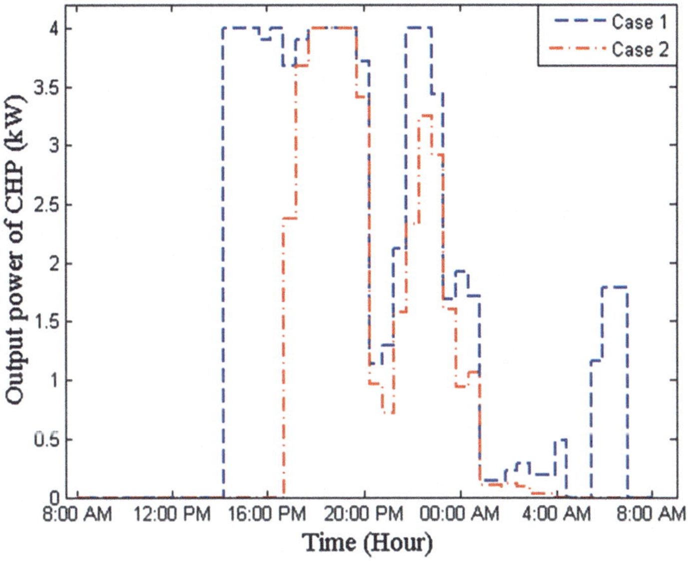

Fig. 4.8 Output power of CHP (kW)

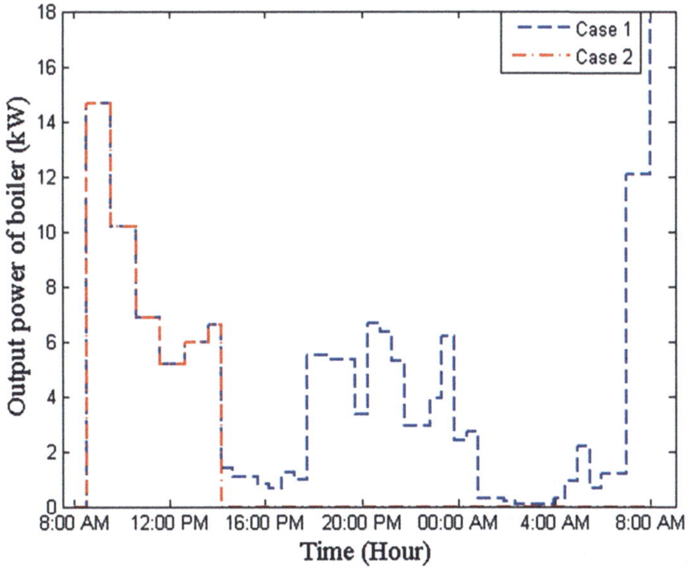

Fig. 4.9 Output power of boiler (kW)

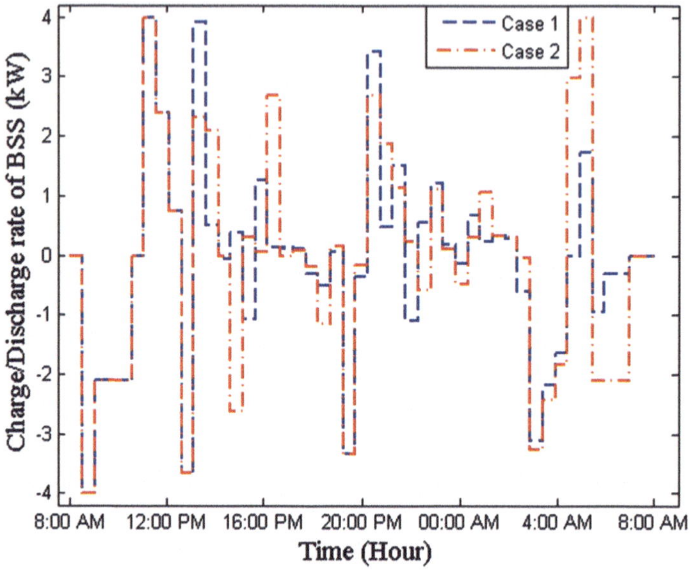

Fig. 4.10 Charge/discharge rate of battery storage system (kW)

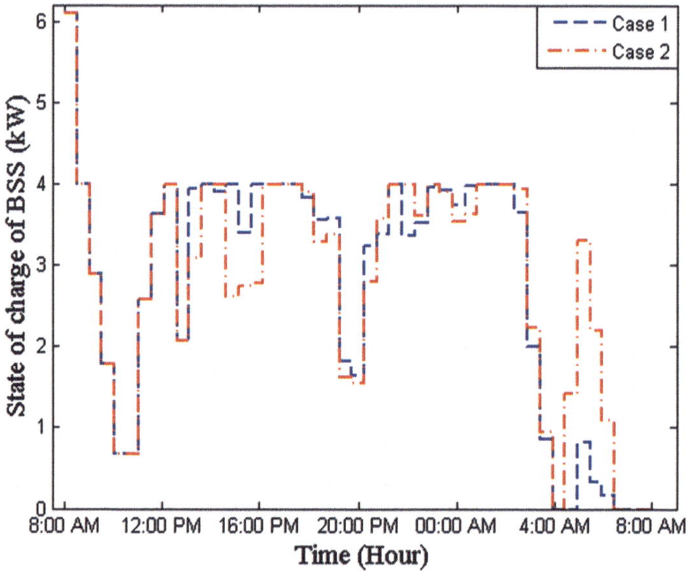

Fig. 4.11 State of charge of battery storage system (kW)

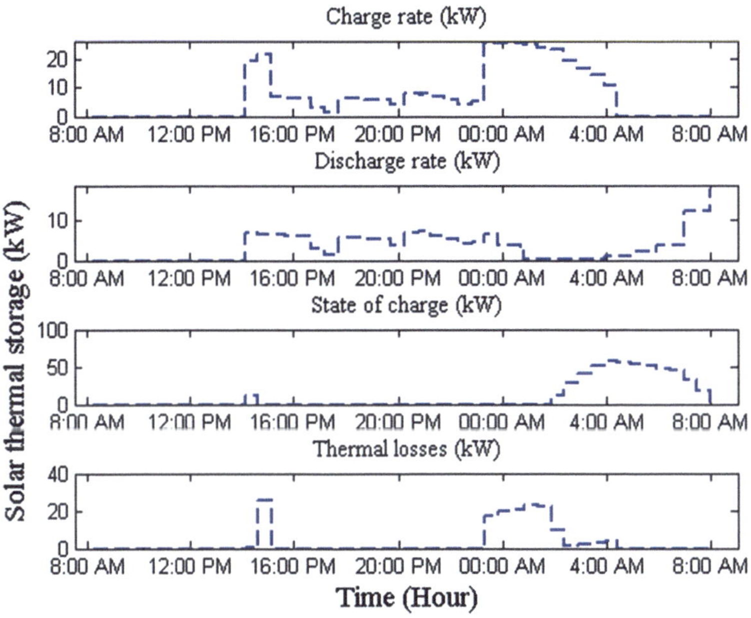

Fig. 4.12 Solar thermal storage (kW)

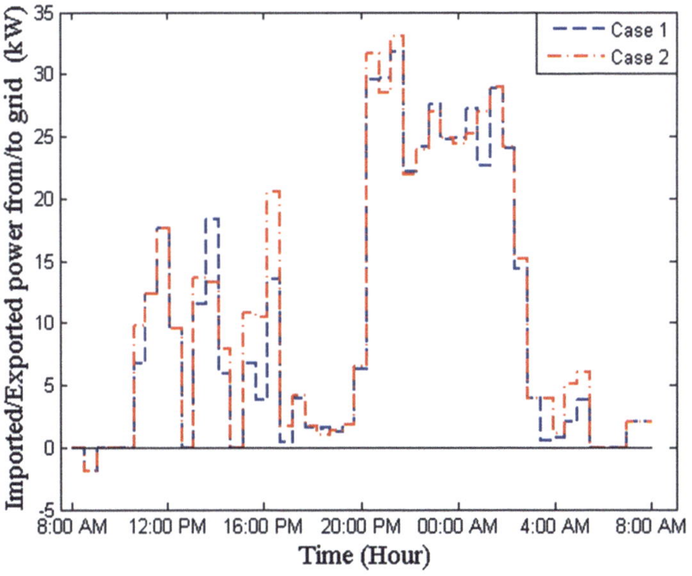

Fig. 4.13 Solar thermal storage (kW)

Table 4.7 Activation time of appliances without considering the effect of solar thermal storage (h)

Sectors	1	2	3	4	5	6	7	8	9	10
Washing machine	12–14	11–13	–	13–15	–	18–20	14–16	16–18	11–13	–
Dish washer	16–18	14–16	–	11–13	–	23–1	22:30–00:30	20–22	16–18	–
Tumble dryer	21:30–23	20–21:30	–	15–16:30	–	1–2:30	1–2:30	23:30–1	19–20:30	–
Cooker hob	15–15:30	10:30–11	–	13:30–14	11:30–12	14–14:30	23–23:30	11:30–12	13–13:30	–
Cooker oven	11:30–12	16–16:30	–	21–21:30	13:30–14	13:30–14	–	–	23–23:30	20:30–21
Microwave	21–21:30	13:30–14	–	20:30–21	12–12:30	17:30–18	–	19:30–20	21–21:30	10:30–11
Interior lighting	18–24	–	20–2	20–2	22–4	19–1	–	17–23	20–2	21–3
Laptop	20–22	–	19:30–21:30	18:30–20:30	19:30–21:30	23–1	–	20–22	20–22	19:30–21:30
Desktop	20–23	–	16–19	–	14–17	22–1	22–1	22:30–1:30	20–23	–
Vacuum cleaner	20:30–21	–	22:30–23	–	1–1:30	16–16:30	00:30–1	1–1:30	23:30–24	1–1:30
Fridge	1–24	–	1–24	–	1–24	1–24	1–24	–	1–24	1–24
Electrical car	23–2	–	23:30–2:30	–	20–23	20–23	20–23	–	24–3	21:30–00:30

Table 4.8 Activation time of appliances with considering the effect of solar thermal storage (h)

Sectors	1	2	3	4	5	6	7	8	9	10
Washing machine	12–14	11–13	–	13–15	–	18–20	15–17	16–18	11–13	–
Dish washer	16–18	14–16	–	11–13	–	23–1	22:30–00:30	20–22	16–18	–
Tumble dryer	21:30–23	20–21:30	–	15–16:30	–	1–2:30	1–2:30	23:30–1	19–20:30	–
Cooker hob	15:30–16	10:30–11	–	13:30–14	11:30–12	14–14:30	23–23:30	11:30–12	10:30–11	–
Cooker oven	11:30–12	16–16:30	–	21–21:30	13–13:30	13:30–14	–	–	23–23:30	20:30–21
Microwave	21–21:30	13–13:30	–	20–20:30	12–12:30	17:30–18	–	19:30–20	21–21:30	10:30–11
Interior lighting	18–23:30	–	20–2	20–2	22–4	19–1	–	17–23	20–2	21–2:30
Laptop	20–22	–	20–22	18:30–20:30	22–24	23–1	–	20–22	21–23	20–22
Desktop	20–23	–	16–19	–	14–17	22–1	22–1	22:30–1:30	20–22:30	–
Vacuum cleaner	20–20:30	–	21–21:30	–	1–1:30	16–16:30	00:30–1	1–1:30	24–00:30	1–1:30
Fridge	1–24	–	1–24	–	1–24	1–24	1–24	–	1–24	1–24
Electrical car	23–2	–	23:30–2:30	–	20–23	20–23	20–23	–	00:30–3:30	22–1

4.4 Conclusion

In this chapter, optimal energy consumption scheduling of a residential hub energy system containing CHP unit, boiler, battery storage, solar thermal storage, and smart appliances is proposed. In this chapter, effect of solar thermal storage has been analyzed in two cases. By comparing the obtained results, it can be found that operation cost of residential hub energy system with considering effect of solar thermal storage is decreased 16.88%, imported power from grid is increased 7.2%, and battery storage system is charged and discharged 28.8 and 21.17% more, respectively. Output power of boiler is decreased 43.28% and output power of CHP is decreased 55.13%. It should be mentioned that the developed MIP model is implemented using CPLEX in GAMS software. Finally, risk-based optimal energy consumption scheduling of a residential energy hub system in the presence of solar thermal storage system can be modeled using information gap decision theory framework and robust optimization approach as a future work.

Nomenclature

Index

j	Residential hub energy system sector
t	Time period index
i	Appliances index
θ	Operation period of appliances index

Parameter

$\mathrm{CAP}^{\mathrm{CHP}}$	Capacity of CHP generator ($\mathrm{kW_e}$)
α^{CHP}	Heat to power ratio of CHP
$\mathrm{CAP}^{\mathrm{boiler}}$	Capacity of boiler ($\mathrm{kW_{th}}$)
$\mathrm{CAP}^{\mathrm{elec}}$	Capacity of battery storage system ($\mathrm{kWh_e}$)
η^{elec}	Battery storage charge/discharge efficiency (%)
M^{elec}	Maximum capacity of battery storage system ($\mathrm{kW_e}$)
$\mathrm{DRL}^{\mathrm{elec}}$	Discharge limit of battery storage system ($\mathrm{kW_e}$)
$\mathrm{CRL}^{\mathrm{elec}}$	Charge limit of battery storage system ($\mathrm{kW_e}$)
$P_{i,\theta}^{\mathrm{Consump}}$	Consumption power of ith appliance at the operation period θ ($\mathrm{kW_e}$)
$P_{j,i}$	Processing time of ith smart appliance at jth residential hub energy system sector (h)

$T_{j,i}^{\mathrm{Start}}$	Latest finishing time of *i*th smart appliance at *j*th residential hub energy system sector (h)
$T_{j,i}^{\mathrm{Finish}}$	Earliest starting time of *i*th smart appliance at *j*th residential hub energy system sector (h)
M^{Grid}	Maximum capacity of bought power from grid (kW_e)
$\eta^{\mathrm{ch,\ Ther}}$	Charge efficiency of solar thermal storage system (%)
$\eta^{\mathrm{dch,\ Ther}}$	Discharge efficiency of solar thermal storage system (%)
$\mathrm{CAP}^{\mathrm{Ther}}$	Capacity of solar thermal storage system (kWh_{th})
$T^{\max}$	Maximum operation temperature (°C)
$T^{\min}$	Minimum operation temperature (°C)
T_t^{amb}	Ambient temperature (°C)
$H^{\mathrm{ch,\ Ther,\ max}}$	Maximum charge rate of solar thermal storage (kW_{th})
$H^{\mathrm{dch,\ Ther,\ max}}$	Maximum discharge rate of solar thermal storage (kW_{th})
$\theta^{\mathrm{storage}}$	Coefficient of solar thermal storage loss (scalar number)
θ^{static}	Coefficient of static solar thermal storage loss (Scalar number)
E_t^{unuse}	Unusable energy due to temperature limitation (kWh)
$\varphi_t^{\mathrm{solar}}$	Solar irradiation (W/m^2)
A^{app}	Surface of solar thermal panel (m^2)
γ^{Ther}	Efficiency of solar thermal panel (%)
$H_{j,t}^{\mathrm{Demand}}$	Heat demand (kW_{th})
Δt	Time interval duration (h)
$\lambda_t^{\mathrm{Grid}}$	Price of imported power from upstream grid (£/kWh_e)
$\lambda^{\mathrm{export}}$	Cost of selling power to the upstream grid (£/kWh_e)
λ^{Gas}	Natural gas price (£/kWh)

Variables

$P_{j,t}^{\mathrm{CHP}}$	Output power of CHP (kWe)
$P_{j,t}^{\mathrm{Boiler}}$	Output power of boiler (kW_{th})
$\mathrm{SOC}_{j,t}^{\mathrm{elec}}$	State of charge of sub-batteries storage system (kWh_e)
$\mathrm{SOC}_t^{\mathrm{Totalelec}}$	Total state of charge of battery storage system (kWh_e)
$\mathrm{CR}_{j,t}^{\mathrm{elec}}$	Charge rate of battery storage system (kW_e)
$\mathrm{DR}^{\mathrm{elec}}$	Discharge rate of battery storage system (kW_e)
$\mathrm{MC}^{\mathrm{elec}}$	Maintenance cost of battery storage system (£/kWh_e)
$P_{j,t}^{\mathrm{Import}}$	Imported power from grid (kW_e)
$P_{j,t}^{\mathrm{export}}$	Exported power to grid (kW_e)
H_t^{stored}	State of charge of solar thermal storage (kWh_{th})
$H_t^{\mathrm{ch,Ther}}$	Charge rate of solar thermal storage (kW_{th})
$H_{j,t}^{\mathrm{dch,Ther}}$	Discharge rate of solar thermal storage (kW_{th})
$H_t^{\mathrm{loss,Ther}}$	Heat loss rate of solar thermal storage (kW_{th})
Q_t	Amount of converted solar irradiation to heat (kW_{th})

Binary Variable

$B_{j,t}^{\text{elec}}$ Binary variable: equal to 1 if battery storage is charged at time t; otherwise 0

$\omega_{j,i,t}$ Binary variable: equal to 1 if ith appliances at jth residential hub energy system sector is ON at time t; otherwise 0

$B_{j,t}^{\text{Grid}}$ Binary variable: equal to 1 if power is bought from grid at time t; otherwise 0

$B_{t}^{\text{ch,ther}}$ Binary variable: equal to 1 if solar thermal storage is charged at time t; otherwise 0

$B_{j,t}^{\text{dch,ther}}$ Binary variable: equal to 1 if solar thermal storage is discharged at time t; otherwise 0

References

1. Ghalelou AN, Fakhri AP, Nojavan S, Majidi M, Hatami H (2016) A stochastic self-scheduling program for compressed air energy storage (CAES) of renewable energy sources (RESs) based on a demand response mechanism. Energy Convers Manag 120:388–396
2. Majidi M, Nojavan S, Zare K (2017) A cost-emission framework for hub energy system under demand response program. Energy 134:157–166
3. Kamyab F, Bahrami S (2016) Efficient operation of energy hubs in time-of-use and dynamic pricing electricity markets. Energy 106:343–355
4. Vahid-Pakdel M, Nojavan S, Mohammadi-ivatloo B, Zare K (2017) Stochastic optimization of energy hub operation with consideration of thermal energy market and demand response. Energy Convers Manag 145:117–128
5. Roldán-Blay C, Escrivá-Escrivá G, Roldán-Porta C, Álvarez-Bel C (2017) An optimisation algorithm for distributed energy resources management in micro-scale energy hubs. Energy 132:126–135
6. Ma T, Wu J, Hao L (2017) Energy flow modeling and optimal operation analysis of the micro energy grid based on energy hub. Energy Convers Manag 133:292–306
7. Dolatabadi A, Mohammadi-Ivatloo B (2017) Stochastic risk-constrained scheduling of smart energy hub in the presence of wind power and demand response. Appl Therm Eng 123:40–49
8. Rastegar M, Fotuhi-Firuzabad M, Zareipour H (2016) Home energy management incorporating operational priority of appliances. Int J Electr Power Energy Syst 74:286–292
9. Batić M, Tomašević N, Beccuti G, Demiray T, Vraneš S (2016) Combined energy hub optimisation and demand side management for buildings. Energy Buildings 127:229–241
10. Wang J, Li Y, Zhou Y (2016) Interval number optimization for household load scheduling with uncertainty. Energy Buildings 130:613–624
11. Patteeuw D, Helsen L (2016) Combined design and control optimization of residential heating systems in a smart-grid context. Energy Buildings 133:640–657
12. Shakeri M, Shayestegan M, Abunima H, Reza SS, Akhtaruzzaman M, Alamoud A, Sopian K, Amin N (2017) An intelligent system architecture in home energy management systems (HEMS) for efficient demand response in smart grid. Energy Buildings 138:154–164
13. Esther BP, Kumar KS (2016) A survey on residential demand side management architecture, approaches, optimization models and methods. Renew Sustain Energy Rev 59:342–351
14. Davenne T, Garvey S, Cardenas B, Simpson M (2017) The cold store for a pumped thermal energy storage system. J Energy Storage 14:295–310

15. Frate GF, Antonelli M, Desideri U (2017) A novel pumped thermal electricity storage (PTES) system with thermal integration. Appl Therm Eng 121:1051–1058
16. Zheng C, Wu J, Zhai X, Wang R (2017) A novel thermal storage strategy for CCHP system based on energy demands and state of storage tank. Int J Electr Power Energy Syst 85:117–129
17. Seitz M, Johnson M, Hübner S (2017) Economic impact of latent heat thermal energy storage systems within direct steam generating solar thermal power plants with parabolic troughs. Energy Convers Manag 143:286–294
18. Dutta P (2017) High temperature solar receiver and thermal storage systems. Appl Therm Eng 124:624–632
19. Steen D, Stadler M, Cardoso G, Groissböck M, DeForest N, Marnay C (2015) Modeling of thermal storage systems in MILP distributed energy resource models. Appl Energy 137:782–792
20. Alva G, Liu L, Huang X, Fang G (2017) Thermal energy storage materials and systems for solar energy applications. Renew Sustain Energy Rev 68:693–706
21. Zhang D, Liu S, Papageorgiou LG (2014) Fair cost distribution among smart homes with microgrid. Energy Convers Manag 80:498–508
22. Nistor S, Wu J, Sooriyabandara M, Ekanayake J (2011) Cost optimization of smart appliances. In: 2011 2nd IEEE PES International Conference and Exhibition on Innovative Smart Grid Technologies (ISGT Europe). IEEE, pp 1–5
23. Balancing mechanism reporting system. The new electricity trading arrangements. Available online: https://www.bmreports.com. 20 May 2017
24. Climate Information for every country in the world. https://en.tutiempo.net/
25. National Solar Radiation Data Base. http://rredc.nrel.gov/solar/old_data/nsrdb/
26. The GAMS Software Website (2012) https://www.gams.com/
27. Brooke A, Kendrick D, Meeraus A (1990) GAMS user's guide. The Scientific Press, Redwood City

Chapter 5
Optimal Short-Term Scheduling of Photovoltaic Powered Multi-chiller Plants in the Presence of Demand Response Programs

Farkhondeh Jabari and Behnam Mohammadi-Ivatloo

5.1 Motivation

During the extremely hot weather or sudden transient heat waves, air-conditioning systems are the most common energy consumers in the different residential, commercial, industrial, and administrative buildings especially in the tropical regions. As obvious from Fig. 5.1, currently 30% of total electrical demand is assigned to cooling air-conditioning applications.

In the meantime, use of solar radiations as primary energy resource in a multi-chiller plant not only increases the economic savings in using non-renewable petroleum products and mitigates pollutant emission productions of electric chillers, but also supplies the heating demand of solar assisted absorption chillers and reduces total electricity requirements of central air-conditioners significantly.

5.2 Literature Review

In the literature, some scholars have focused on optimal performance investigation of multi-chiller plants using different evolutionary algorithms. In this context, an improved ripple bee swarm optimization algorithm is proposed in [1, 2] to obtain the economic chiller loading points. Using the features of biological communities, some movement models are developed to minimize total energy requirements of cooling towers and pumps within the feasible solution space. References [3–6] solved the economic chiller dispatching problem using the particle swarm optimization

F. Jabari (✉) · B. Mohammadi-Ivatloo
Department of Electrical and Computer Engineering, University of Tabriz, Tabriz, Iran
e-mail: f.jabari@tabrizu.ac.ir; bmohammadi@tabrizu.ac.ir

B. Mohammadi-Ivatloo, F. Jabari (eds.), *Operation, Planning, and Analysis of Energy Storage Systems in Smart Energy Hubs*,
https://doi.org/10.1007/978-3-319-75097-2_5

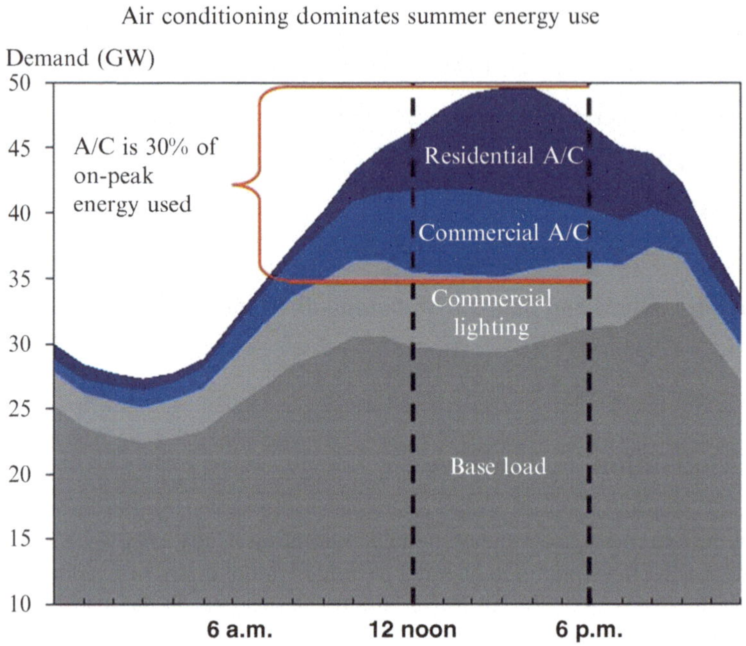

Fig. 5.1 Typical electrical loads during extremely hot summer days

technique. A day-ahead optimal chiller dispatching problem is solved by Powell et al. [7] and implemented on a benchmark district cooling system with and without considering a thermal energy storage. In [8], differential cuckoo search algorithm (DCSA) [9] based on obligate brood-parasitic behavior of some cuckoo species is introduced to optimize the chiller loading design problem. Reference [10] simplifies the complicated evolution process of the genetic algorithm (GA) for solving optimal chiller loading using the evolution strategy (ES). Other search approaches such as GA [11–14], simulated annealing (SA) [15, 16], differential evolution (DE) [17], gradient method (GM) [18], Lagrangian method [19], empirical model [20], artificial neural network (ANN) [21–25], firefly algorithm [26] have also been proposed in the literature.

5.3 Problem Formulation

5.3.1 Multiple-Chiller Plant

As illustrated in Fig. 5.2, a multi-chiller plant consists of two or more chillers connected in parallel or series piping to a distribution system [7, 17].

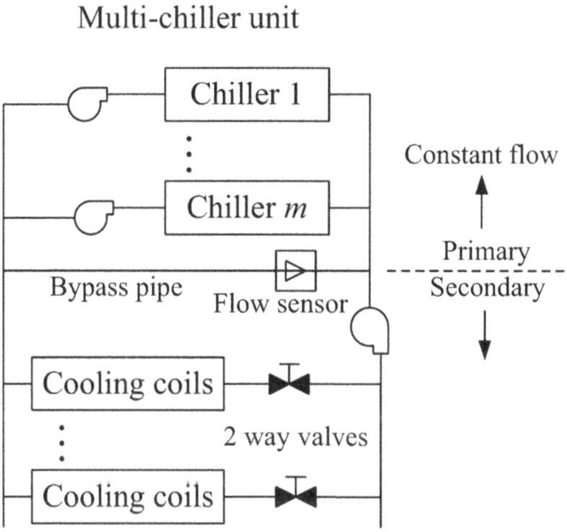

Fig. 5.2 Schematic presentation of a typical CCHP system

In the short-term economic dispatch of the multi-chiller plant, the total electrical power consumed by the centrifugal chillers can be calculated as Eq. (5.1):

$$P_t^{\mathrm{ch}} = \sum_{i=1}^{N} U_i^t \times \left(\alpha_i + \beta_i \times \mathrm{PLR}_i^t + \gamma_i \times \mathrm{PLR}_i^{t^2} + \zeta_i \times \mathrm{PLR}_i^{t^3}\right) \tag{5.1}$$

where N is the number of chillers; U_i^t is a binary decision variable that will be equal to 1, if ith chiller is on; otherwise it will be 0; α_i, β_i, γ_i, ζ_i are the coefficients related to the operating characteristic of chiller i; PLR_i^t is the partial load ratio (PLR) of chiller i at time horizon t that is defined as relation (5.2).

$$\mathrm{PLR}_i^t = \frac{\text{Cooling load of chiller } i \text{ at time } t}{\text{Power consumption of chiller } i \text{ at time } t} \tag{5.2}$$

Subject to:

Power balance criterion which can be stated by Eq. (5.3):

$$\sum_{i=1}^{N} \left(U_i^t \times \mathrm{PLR}_i^t \times \mathrm{RT}_i\right) = \mathrm{CL}_t; \quad \forall t = 1, 2, \ldots, T \tag{5.3}$$

where RT_i is the Capacity of chiller i; CL_t is the Total cooling demand at time t.

5.3.2 Solar Photovoltaic Cells

In the last decade, use of solar collectors such as flat plat collectors and evacuated tube collectors for thermally driven solar cooling systems and photovoltaic cells to generate electricity for vapor compressing in air-conditioners is rapidly gaining popularity due to nearly-zero carbon footprints. This chapter aims to present optimal short-term dispatching of solar photovoltaic based multi-chiller plants in the presence of time-of-use cooling-demand response programs. Use of solar irradiance as primary energy source during extremely-hot summer days not only mitigate total carbon footprints, but also reduces total energy consumptions of electrical chillers from fossil fuels based non-renewable energy sources, specially by applying peak clipping and valley filling demand response strategies on cooling demand. The power output of a photovoltaic module can be calculated from Eq. (5.4) [27].

$$P_t^{\mathrm{pv}} = \eta S \Phi_t \left[1 - 0.005 \times \left(T_t^a - 25\right)\right] \tag{5.4}$$

where P_t^{pv} is the Power output of a photovoltaic panel; η is the Conversion coefficient of a photovoltaic panel; S is the Array area of a photovoltaic module; Φ_t is the Solar irradiance; T_t^a is the Ambient temperature at time t.

5.3.3 Demand Response Programs

Demand response programs (DRPs) are defined as effective and practical solution to change electrical energy utilization of consumers with respect to their usual power consumption pattern [28]. The US Department of Energy (DOE) defined DRPs as the capability of changing power consumption pattern of industrial, residential, and commercial consumers considering changes in electrical energy price or incentive payments [29]. Application of DRPs to electrical energy systems takes advantages of modifying of market clearing price (MCP) [30], avoiding raising of power market price over production cost as well as improving the performance of the markets [31]. Moreover, employing DRPs is effective in helping the system reliability by decreasing the rate of forced outages of the system [31]. In addition, the industrial loads can rival in power market by incorporating demands in the market. DRPs are mainly classified into time-based programs and incentive-based programs, where the first category involves the programs based on power market pricing and the second one aims to participate in the programs through financial incentives. Applying time-of-use DRPs, end-users shift their electricity consumptions from on-peak high-price hours to off-peak low-price periods. Time-of-use DRPs are illustrated in Fig. 5.3. The dashed section of demand profile doesn't participate in DRPs, while the other one shifts a part of electricity load from mid-peak or on-peak time intervals to off-peak hours.

$$\mathrm{CL}_t = \mathrm{CL}_t^0 \left(1 - \mathrm{DR}_t\right) + \mathrm{ldr}_t \tag{5.5}$$

$$\mathrm{CL}_t^0 - \mathrm{CL}_t = \mathrm{ldr}_t = \mathrm{DR}_t \times \mathrm{CL}_t^0 \tag{5.6}$$

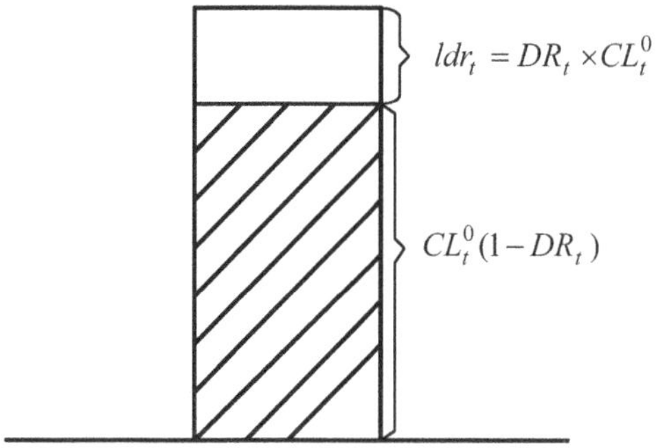

Fig. 5.3 Participation of cooling demand in time-of-use DRP

$$\sum_{t=1}^{T} \mathrm{ldr}_t = \sum_{t=1}^{T} \mathrm{DR}_t \times \mathrm{CL}_t^0 \tag{5.7}$$

$$\mathrm{CL}_t^{\mathrm{inc}} \leq \mathrm{inc}_t \times \mathrm{CL}_t^0 \tag{5.8}$$

$$\mathrm{DR}_t \leq \mathrm{DR}_{\mathrm{max}} \tag{5.9}$$

$$\mathrm{inc}_t \leq \mathrm{inc}_{\mathrm{max}} \tag{5.10}$$

where CL_t^0 is the Initial demand which participates in time-of-use DRPs; CL_t is the Cooling demand after implementation of DRPs at time horizon t; DR_t is the Percentage of participation in DRPs at time t; ldr_t is the Shifted demand at time t; $\mathrm{CL}_t^{\mathrm{inc}}$ is the Increased demand at time t; inc_t is the Amount of load increase at time t; $\mathrm{DR}_{\mathrm{max}}$ is the Maximum value of load participation in DRP; $\mathrm{inc}_{\mathrm{max}}$ is the Maximum value of load increase.

5.3.4 Objective Function and Constraints

In this chapter, total electricity procurement cost of a multi-chiller plant over the study horizon should be minimized as follows:

$$\mathrm{Min} \sum_{t=1}^{T} \lambda_t P_t^{\mathrm{grid}} \tag{5.11}$$

Subject to:

- Electrical power balance constraint

$$\begin{aligned} & P_t^{\text{grid}} + N_{\text{pv}} P_t^{\text{pv}} = P_t^{\text{ch}}; \quad \forall t = 1, 2, \ldots, T \\ & \text{Constraints (5.1)–(5.10)} \end{aligned} \tag{5.12}$$

where P_t^{grid} is the Purchased electrical power from upstream grid; N_{pv} is the Number of photovoltaic panels.

5.4 Illustrative Examples

5.4.1 Plant 1 with Six Chillers

In this section, four cases are studied for optimal dispatching of multi-chiller plants 1 and 2 in the presence of solar photovoltaic panels and demand response programs as follows:

- Case 1: Without PVs and DRPs
- Case 2: With DRPs
- Case 3: With PVs
- Case 4: With PVs and DRPs

The problem is modeled as a mixed integer nonlinear program (MINLP) and solved using SBB solver under general algebraic mathematical system (GAMS) environment [32]. Figures 5.4 and 5.5 depict total cooling demand of a semiconduc-

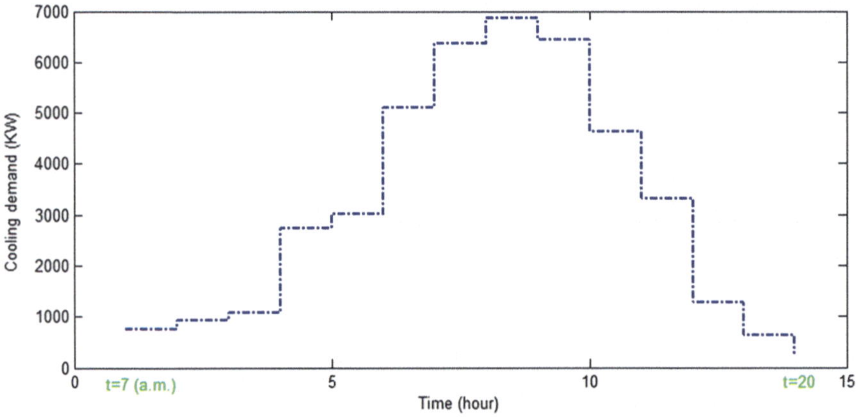

Fig. 5.4 Cooling demand of a semiconductor factory located at Hsinchu Scientific Garden (Taiwan) [11]

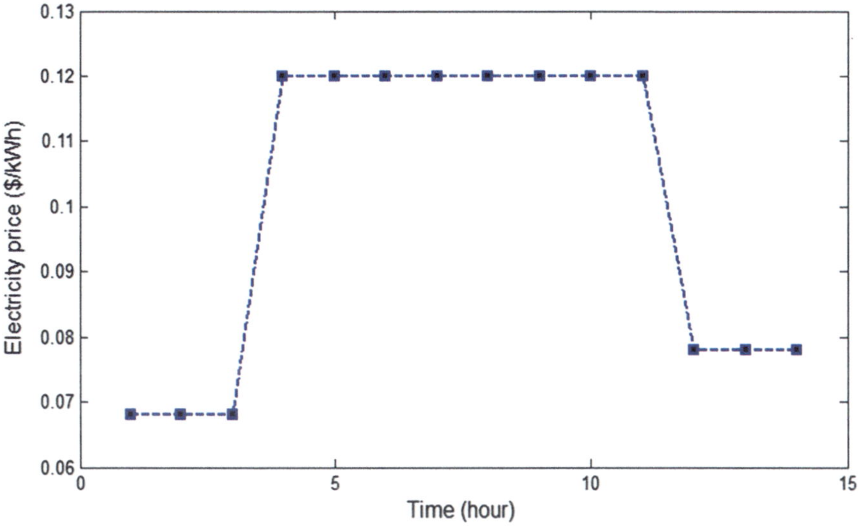

Fig. 5.5 Hourly electricity rates over the study horizon from $t = 7^{\text{a.m.}}$ to $t = 20$

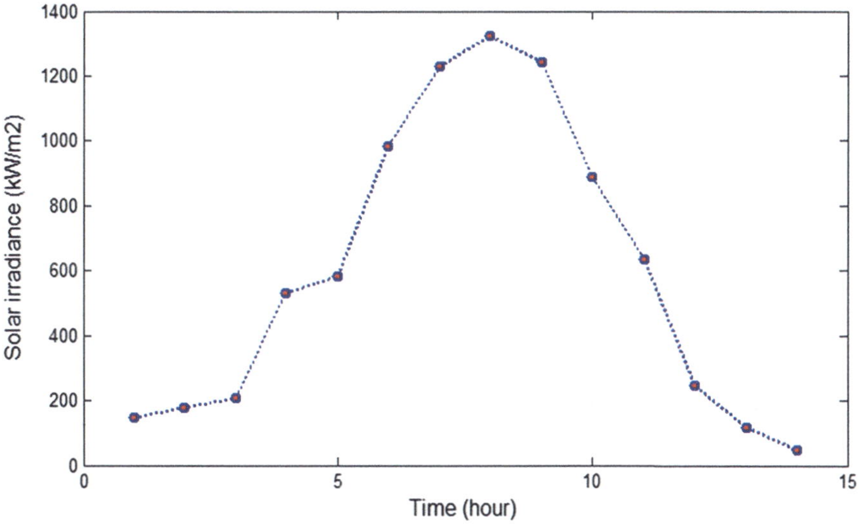

Fig. 5.6 Solar irradiance during a sample extremely-hot summer day

tor factory located at Hsinchu Scientific Garden (Taiwan) [11] and hourly electricity rates [33, 34], respectively. In addition, Figs. 5.6, 5.7, and 5.8 illustrate the variations of solar irradiance, ambient temperature, and power output of PV panels during a sample extremely-hot summer day from $t = 7^{\text{a.m.}}$ to $t = 20$ [35]. Tables 5.1 and 5.2, respectively, present all coefficients related to PV panels and operating characteristic of six chillers plant 1 which participate in supplying the cooling demand.

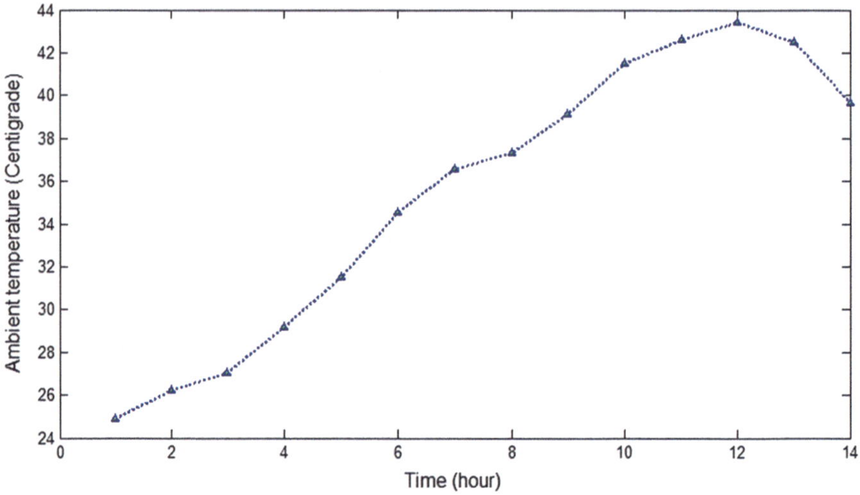

Fig. 5.7 Variations of ambient temperature from $t = 7^{\text{a.m.}}$ to $t = 20$

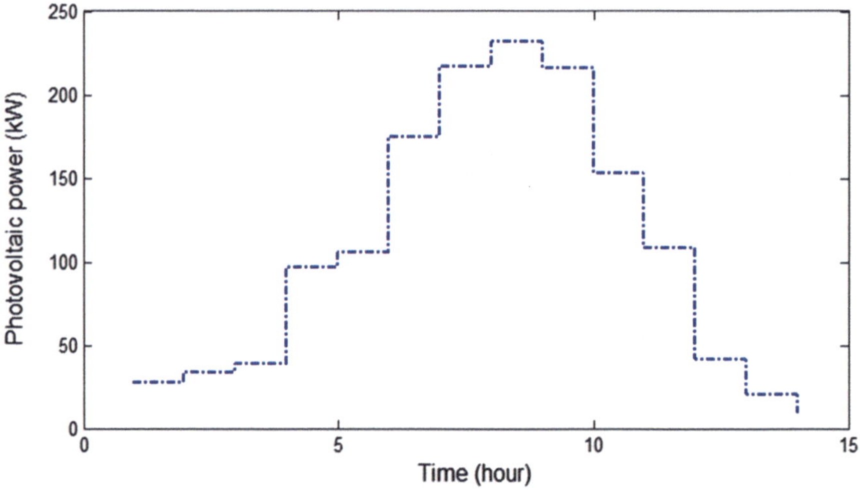

Fig. 5.8 Power output of photovoltaic panels from $t = 7^{\text{a.m.}}$ to $t = 20$

Table 5.1 Parameters of PV panels [27]

N_{pv}	η	S
400	0.187	2.5

Table 5.3 summarizes total electricity requirements of six chillers in their optimum operating points. Considering $DR_{max} = inc_{max} = 0.15$, the optimum operating points of these chillers in cases 2 and 4 vary as reported in Table 5.4. Moreover, the

Table 5.2 Chiller data for six units [11]

Chiller	α_i	β_i	γ_i	ζ_i	Chiller capacity (RT)
1	399.345	−122.12	770.46	0	1280
2	287.116	80.04	700.48	0	1280
3	−120.505	1525.99	−502.14	0	1280
4	−19.121	898.76	−98.15	0	1280
5	−95.029	1202.39	−352.16	0	1280
6	191.750	224.86	524.4	0	1280

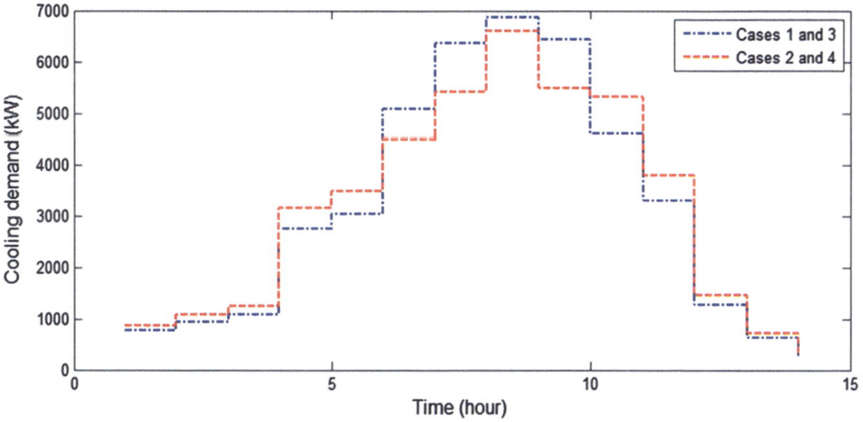

Fig. 5.9 Variations of cooling demand before and after participation in time-of-use DRPs

variations of cooling load in four cases before and after implementation of time-of-use DRPs are shown in Fig. 5.9.

As obvious from Fig. 5.9, time-of-use DRPs shift the cooling demand from on-peak hours to other mid-peak and off-peak periods. Moreover, total energy procurement cost of this multi-chiller plant in four cases with and without participation of PVs and DRPs can be reported as Table 5.5. As expected, using the photovoltaic panels and implementing the time-of-use DRPs on cooling demand reduces total energy cost of multiple-chiller plants.

5.4.2 Plant 2 with Four Chillers

In this subsection, same cases are studied on another multi-chiller plant with four units. The operating characteristics of four centrifugal chillers and total cooling demand of a benchmark hotel building located in Ahvaz, Iran have, respectively, been shown in Table 5.6 and Fig. 5.10. Solar radiations, ambient air temperature, and electrical power generated by 400 photovoltaic cells during a severe-hot summer day in Ahvaz, Iran are shown in Figs. 5.11, 5.12, and 5.13, respectively.

Table 5.3 Economic loading points of six chillers in cases 1 and 3

CL^t (kW)	Chiller	PLR_i^t	CL^t (kW)	Chiller	PLR_i^t
$t = 7^{a.m.}$	1	0.937	$t = 14$	1	0.794
762 (kW)	2	0.463	6858 (kW)	2	0.729
	3	0.081		3	1
	4	0.021		4	1
	5	0.081		5	1
	6	0.412		6	0.836
$t = 8^{a.m.}$	1	0.939	$t = 15$	1	0.703
933 (kW)	2	0.451	6445.8 (kW)	2	0.629
	3	0.081		3	1
	4	0.021		4	1
	5	0.081		5	1
	6	0.545	6		0.703
$t = 9^{a.m.}$	1	0.941	$t = 16$	1	0.720
1080 (kW)	2	0.375	4618.6 (kW)	2	0.713
	3	0.081		3	0.081
	4	0.048		4	1
	5	0.081		5	1
	6	0.634		6	0.814
$t = 10^{a.m.}$	1	0.720	$t = 17$	1	0.720
2752.5 (kW)	2	0.640	3304.5 (kW)	2	0.667
	3	0.081		3	0.081
	4	1		4	1
	5	0.416		5	0.081
	6	0.653		6	0.753
$t = 11^{a.m.}$	1	0.720	$t = 18$	1	0.937
3024.6 (kW)	2	0.640	1275 (kW)	2	0.597
	3	0.363		3	0.081
	4	1		4	0.834
	5	1		5	0.081
	6	0.605		6	0.542
$t = 12^{noon}$	1	0.665	$t = 19$	1	0.935
5092.9 (kW)	2	0.587	622 (kW)	2	0.406
	3	0.081		3	0.081
	4	1		4	0.021
	5	1		5	0.081
	6	0.646		6	0.303
$t = 13$	1	0.688	$t = 20$	1	0.931
6375.9 (kW)	2	0.613	264.5 (kW)	2	0.485
	3	1		3	0.081
	4	1		4	0.045
	5	1		5	0.081
	6	0.681		6	0.605

Table 5.4 Economic loading points of six chillers in cases 2 and 4

CL^t (kW)	Chiller	PLR_i^t	CL^t (kW)	Chiller	PLR_i^t
$t = 7^{a.m.}$	1	0.951	$t = 14$	1	0.737
876.3 (kW)	2	0.504	6597.784 (kW)	2	0.666
	3	0.081		3	1
	4	0.021		4	1
	5	0.081		5	1
	6	0.501		6	0.752
$t = 8^{a.m.}$	1	0.952	$t = 15$	1	0.749
1072.95 (kW)	2	0.508	5478.93 (kW)	2	0.68
	3	0.081		3	0.081
	4	0.041		4	1
	5	0.081		5	1
	6	0.635		6	0.77
$t = 9^{a.m.}$	1	0.946	$t = 16$	1	0.712
1242 (kW)	2	0.559	5311.39 (kW)	2	0.639
	3	0.081		3	0.081
	4	0.808		4	1
	5	0.081		5	1
	6	0.561		6	0.716
$t = 10^{a.m.}$	1	0.695	$t = 17$	1	0.639
3165.375 (kW)	2	0.424	3800.175 (kW)	2	0.559
	3	0.081		3	0.081
	4	0.921		4	1
	5	1		5	0.081
	6	0.47		6	0.609
$t = 11^{a.m.}$	1	0.738	$t = 18$	1	0.693
3478.29 (kW)	2	0.38	1466.25 (kW)	2	0.52
	3	0.081		3	0.081
	4	1		4	0.983
	5	1		5	0.081
	6	0.636		6	0.535
$t = 12^{noon}$	1	0.72	$t = 19$	1	0.952
4478.744 (kW)	2	0.666	715.3 (kW)	2	0.547
	3	0.081		3	0.081
	4	1		4	0.021
	5	1		5	0.081
	6	0.752		6	0.375
$t = 13$	1	0.737	$t = 20$	1	0.951
5421.637 (kW)	2	0.666	304.175 (kW)	2	0.215
	3	0.081		3	0.081
	4	1		4	0.076
	5	1		5	0.081
	6	0.752		6	0.605

Table 5.5 Total energy cost of six chillers in four cases

Case study	Energy cost ($)
1	3212.22
2	3138.83
3	3043.40
4	2970.02

Table 5.6 Chiller data for four units of plant 2

Chiller	α_i	β_i	γ_i	ζ_i	Chiller capacity (RT)
1	104.09	166.57	−430.13	512.53	850
2	−67.15	1177.79	−2174.53	1456.53	1200
3	384.71	−779.13	1151.42	−63.2	1630
4	541.63	413.48	−3626.5	4021.41	1850

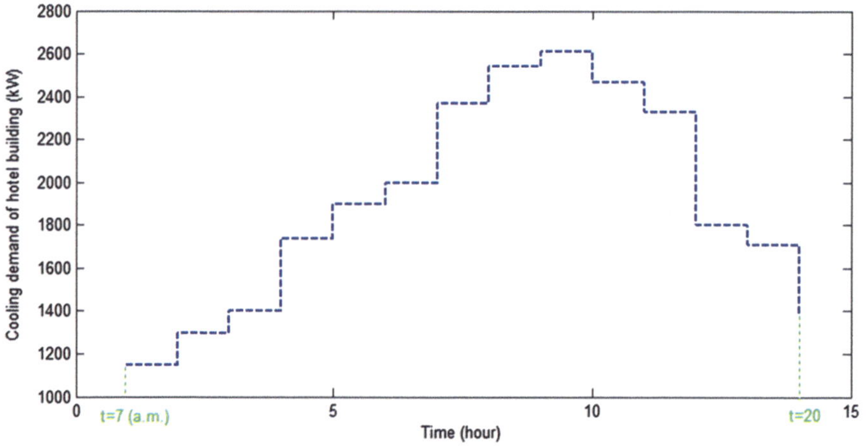

Fig. 5.10 [!t] Cooling demand of a hotel building located in Ahvaz, Iran before participation in time-of-use DRPs

Table 5.7 summarizes total electricity requirements of four chillers in their optimum operating points. Considering $DR_{max} = inc_{max} = 0.2$, the optimum operating points of these chillers in cases 2 and 4 vary as reported in Table 5.8. Moreover, the variations of hotel cooling load in four cases before and after implementation of time-of-use DRPs are shown in Fig. 5.14.

As obvious from Fig. 5.14, time-of-use DRPs shift the cooling demand from on-peak hours to other mid-peak and off-peak periods. Moreover, total energy procurement cost of this multi-chiller plant in four cases with and without participation of PVs and DRPs can be reported as Table 5.9. As expected, using the photovoltaic cells and implementing the time-of-use DRPs on hotel cooling demand reduces total energy cost of multiple-chiller plants.

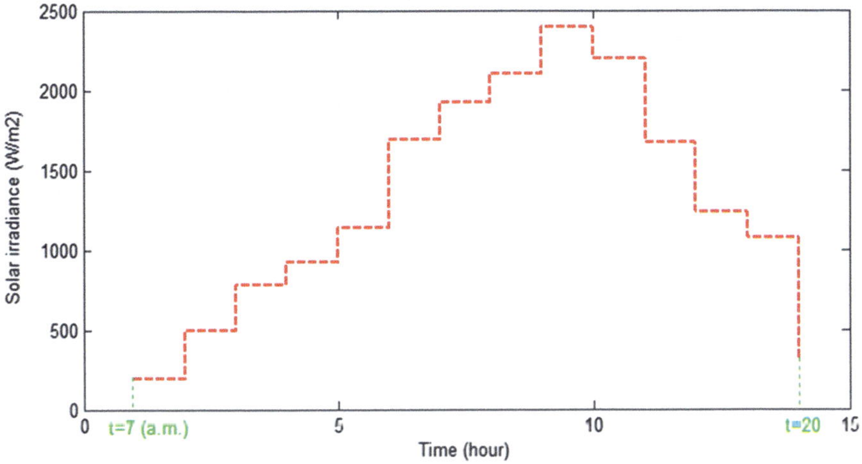

Fig. 5.11 Solar radiations during a severe-hot summer day in Ahvaz, Iran

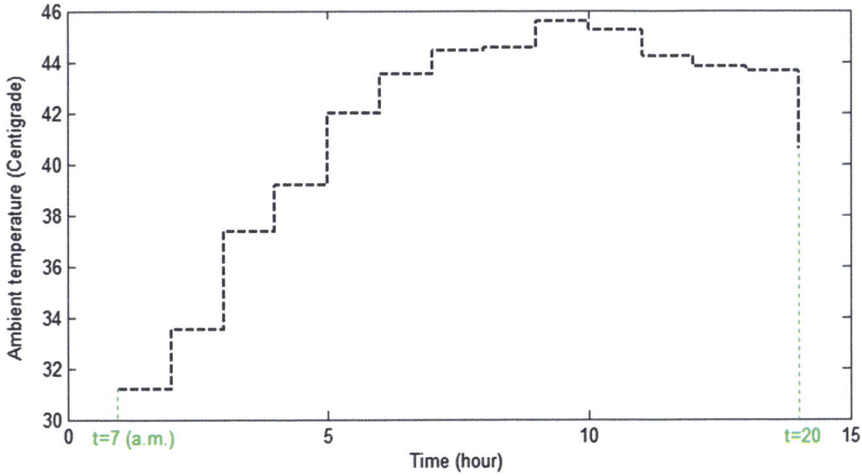

Fig. 5.12 Climatic conditions over a sample summer day in Ahvaz, Iran

5.5 Concluding Remarks

In this chapter, short-term optimal scheduling of solar powered multi-chiller plants was presented. As we know, total cooling demand directly depends on solar irradiations in a way that when solar irradiance increases, the value of building cooling demand in different residential, commercial, and industrial sectors will be increased. Hence, use of solar energy for supplying total electricity requirement of chillers will be a cost-effective way in comparison with other energy resources. This is an interesting result indicating that if solar photovoltaic panels are employed to

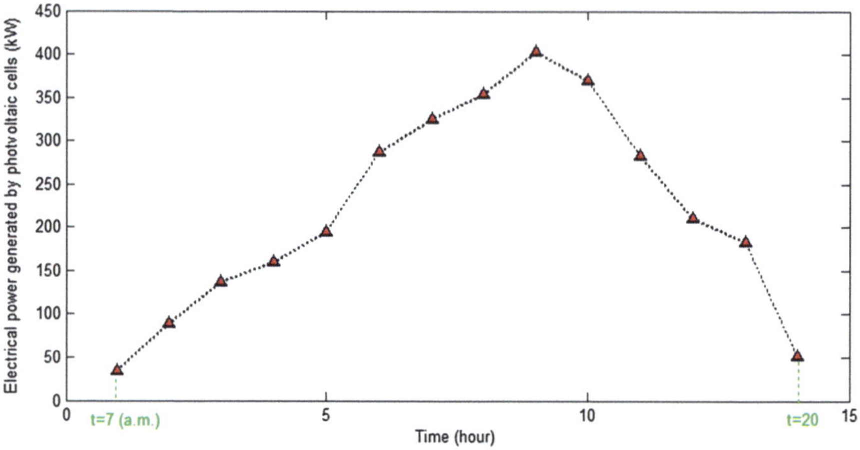

Fig. 5.13 Power output of photovoltaic panels from $t = 7^{\text{a.m.}}$ to $t = 20$

Table 5.7 Optimum loading points of four chillers in cases 1 and 3

CL^t (kW)	Chiller	PLR_i^t	CL^t (kW)	Chiller	PLR_i^t
$t = 7^{\text{a.m.}}$	2	0.064	$t = 14$	2	0.906
1150 (kW)	3	0.658	2540 (kW)	3	0.891
$t = 8^{\text{a.m.}}$	2	0.064	$t = 15$	2	0.920
1300 (kW)	3	0.750	2610 (kW)	3	0.924
$t = 9^{\text{a.m.}}$	2	0.064	$t = 16$	2	0.893
1400 (kW)	3	0.812	2470 (kW)	3	0.858
$t = 10^{\text{a.m.}}$	2	0.715	$t = 17$	2	0.864
1740 (kW)	3	0.541	2330 (kW)	3	0.793
$t = 11^{\text{a.m.}}$	2	0.762	$t = 18$	2	0.733
1900 (kW)	3	0.605	1800 (kW)	3	0.564
$t = 12^{\text{noon}}$	2	0.788	$t = 19$	2	0.706
2000 (kW)	3	0.647	1710 (kW)	3	0.529
$t = 13$	2	0.873	$t = 20$	2	0.064
2370 (kW)	3	0.812	1380 (kW)	3	0.799

produce electricity for driving chiller equipment, higher coefficient of performance for chillers will be attained and lower electricity cost will be paid while increasing the amount of cooling demand. Moreover, it is demonstrated that use of photovoltaic panels as renewable based power generation facilities and time-of-use demand response programs for peak clipping reduces total electricity cost significantly.

Table 5.8 Optimum loading points of four chillers in cases 2 and 4

CL^t (kW)	Chiller	PLR_i^t	CL^t (kW)	Chiller	PLR_i^t
$t = 7^{\text{a.m.}}$	2	0.064	$t = 14$	2	0.802
1437.5 (kW)	3	0.799	2036.4 (kW)	3	0.671
$t = 8^{\text{a.m.}}$	2	0.064	$t = 15$	2	0.810
1569.2 (kW)	3	0.910	2036.4 (kW)	3	0.685
$t = 9^{\text{a.m.}}$	2	0.064	$t = 16$	2	0.802
1569.2 (kW)	3	0.932	2036.4 (kW)	3	0.671
$t = 10^{\text{a.m.}}$	2	0.802	$t = 17$	2	0.802
2036.4 (kW)	3	0.671	2036.4 (kW)	3	0.671
$t = 11^{\text{a.m.}}$	2	0.802	$t = 18$	2	0.827
2036.4 (kW)	3	0.671	2250 (kW)	3	0.716
$t = 12^{\text{noon}}$	2	0.802	$t = 19$	2	0.801
2036.4 (kW)	3	0.671	2137.5 (kW)	3	0.669
$t = 13$	2	0.802	$t = 20$	2	0.064
2036.4 (kW)	3	0.671	1445 (kW)	3	0.853

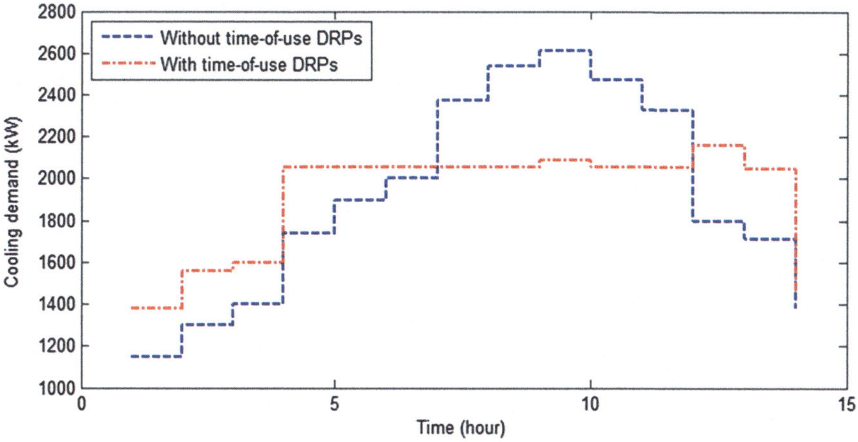

Fig. 5.14 Variations of cooling demand before and after participation in time-of-use DRPs

Table 5.9 Total energy cost of four chillers in four cases

Case study	Energy cost ($)
1	874.82
2	818.15
3	537.36
4	480.70

Nomenclature

α_i, β_i, γ_i, ζ_i	Coefficients related to the operating characteristic of chiller i
η	Conversion coefficient of a photovoltaic panel

Φ_t	Solar irradiance
CL_t	Cooling demand after implementation of DRPs at time horizon t
CL_t^0	Initial demand which participates in time-of-use DRPs
N	Number of chillers
N_{pv}	Number of photovoltaic panels
P_t^{ch}	Total electrical power consumed by all centrifugal chillers at time t
P_t^{pv}	Power output of a photovoltaic panel
P_t^{grid}	Purchased electrical power from upstream grid
PLR_i^t	Partial load ratio (PLR) of chiller i at time horizon t
RT_i	Capacity of chiller i
S	Array area of a photovoltaic module
U_i^t	A binary decision variable that will be equal to 1, if ith chiller is on; otherwise it will be 0
T_t^a	Ambient temperature at time t

References

1. Lo C-C, Tsai S-H, Lin B-S (2016) Economic dispatch of chiller plant by improved ripple bee swarm optimization algorithm for saving energy. Appl Therm Eng 100:1140–1148
2. Niknam T, Golestaneh F (2013) Enhanced bee swarm optimization algorithm for dynamic economic dispatch. IEEE Syst J 7(4):754–762
3. Lee W-S, Lin L-C (2009) Optimal chiller loading by particle swarm algorithm for reducing energy consumption. Appl Therm Eng 29(8):1730–1734
4. Beghi A, Cecchinato L, Cosi G, Rampazzo M (2012) A PSO-based algorithm for optimal multiple chiller systems operation. Appl Therm Eng 32:31–40
5. Askarzadeh A, dos Santos Coelho L (2015) Using two improved particle swarm optimization variants for optimization of daily electrical power consumption in multi-chiller systems. Appl Therm Eng 89:640–646
6. Ardakani AJ, Ardakani FF, Hosseinian S (2008) A novel approach for optimal chiller loading using particle swarm optimization. Energy Build 40(12):2177–2187
7. Powell KM, Cole WJ, Ekarika UF, Edgar TF (2013) Optimal chiller loading in a district cooling system with thermal energy storage. Energy 50:445–453
8. dos Santos Coelho L, Klein CE, Sabat SL, Mariani VC (2014) Optimal chiller loading for energy conservation using a new differential cuckoo search approach. Energy 75:237–243
9. Valian E, Tavakoli S, Mohanna S, Haghi A (2013) Improved cuckoo search for reliability optimization problems. Comput Ind Eng 64(1):459–468
10. Chang Y-C, Lee C-Y, Chen C-R, Chou C-J, Chen W-H, Chen W-H (2009) Evolution strategy based optimal chiller loading for saving energy. Energy Convers Manag 50(1):132–139
11. Chang Y-C (2005) Genetic algorithm based optimal chiller loading for energy conservation. Appl Therm Eng 25(17):2800–2815
12. Chow T, Zhang G, Lin Z, Song C (2002) Global optimization of absorption chiller system by genetic algorithm and neural network. Energy Build 34(1):103–109
13. Chang Y-C, Lin J-K, Chuang M-H (2005) Optimal chiller loading by genetic algorithm for reducing energy consumption. Energy Build 37(2):147–155
14. Beghi A, Cecchinato L, Rampazzo M (2011) A multi-phase genetic algorithm for the efficient management of multi-chiller systems. Energy Convers Manag 52(3):1650–1661
15. Chang Y-C (2006) An innovative approach for demand side management – optimal chiller loading by simulated annealing. Energy 31(12):1883–1896

16. Chang Y-C, Chen W-H, Lee C-Y, Huang C-N (2006) Simulated annealing based optimal chiller loading for saving energy. Energy Convers Manag 47(15):2044–2058
17. Lee W-S, Chen Y-T, Kao Y (2011) Optimal chiller loading by differential evolution algorithm for reducing energy consumption. Energy Build 43(2):599–604
18. Chang Y-C, Chan T-S, Lee W-S (2010) Economic dispatch of chiller plant by gradient method for saving energy. Appl Energy 87(4):1096–1101
19. Chang Y-C (2006) An outstanding method for saving energy-optimal chiller operation. IEEE Trans Energy Convers 21(2):527–532
20. Wang H (2017) Empirical model for evaluating power consumption of centrifugal chillers. Energy Build 140:359–370
21. Frey P, Fischer S, Drück H (2014) Artificial Neural Network modelling of sorption chillers. Sol Energy 108:525–537
22. Wei X, Xu G, Kusiak A (2014) Modeling and optimization of a chiller plant. Energy 73:898–907
23. Labus J, Hernández J, Bruno J, Coronas A (2012) Inverse neural network based control strategy for absorption chillers. Renew Energy 39(1):471–482
24. Congradac V, Kulić F (2012) Recognition of the importance of using artificial neural networks and genetic algorithms to optimize chiller operation. Energy Build 47:651–658
25. Chang Y-C, Chen W-H (2009) Optimal chilled water temperature calculation of multiple chiller systems using Hopfield neural network for saving energy. Energy 34(4):448–456
26. dos Santos Coelho L, Mariani VC (2013) Improved firefly algorithm approach applied to chiller loading for energy conservation. Energy Build 59:273–278
27. Nguyen DT, Le LB (2014) Optimal bidding strategy for microgrids considering renewable energy and building thermal dynamics. IEEE Trans Smart Grid 5(4):1608–1620
28. Kamyab F, Amini M, Sheykhha S, Hasanpour M, Jalali MM (2016) Demand response program in smart grid using supply function bidding mechanism. IEEE Trans Smart Grid 7(3):1277–1284
29. Qdr Q (2006) Benefits of demand response in electricity markets and recommendations for achieving them. US Department of Energy
30. Vlachos AG, Biskas PN (2013) Demand response in a real-time balancing market clearing with pay-as-bid pricing. IEEE Trans Smart Grid 4(4):1966–1975
31. Strbac G (2008) Demand side management: benefits and challenges. Energy Policy 36(12):4419–4426
32. Brook A, Kendrick D, Meeraus A (1988) GAMS, a user's guide. ACM Signum Newsl 23(3–4):10–11
33. Jabari F, Nojavan S, Ivatloo BM (2016) Designing and optimizing a novel advanced adiabatic compressed air energy storage and air source heat pump based μ-combined cooling, heating and power system. Energy 116:64–77
34. Jabari F, Nojavan S, Ivatloo BM, Sharifian MB (2016) Optimal short-term scheduling of a novel tri-generation system in the presence of demand response programs and battery storage system. Energy Convers Manag 122:95–108
35. Jabari F, Masoumi A, Mohammadi-ivatloo B (2017) Long-term solar irradiance forecasting using feed-forward back-propagation neural network. In: 3rd international conference of IEA technology and energy management. Shahid Beheshti University, Tehran

Chapter 6
Basic Open-Source Nonlinear Mixed Integer Programming Based Dynamic Economic Dispatch of Multi-chiller Plants

Farkhondeh Jabari and Behnam Mohammadi-Ivatloo

6.1 Literature Review

In the literature, some scholars have focused on optimal performance investigation of multi-chiller plants using different evolutionary algorithms. In this context, an improved ripple bee swarm optimization algorithm is proposed in [1, 2] to obtain the economic chiller loading points. Using the features of biological communities, some movement models are developed to minimize total energy requirements of cooling towers and pumps within the feasible solution space. References [3–6] solved the economic chiller dispatching problem using the particle swarm optimization technique. A day-ahead optimal chiller dispatching problem is solved by Powell et al. [7] and implemented on a benchmark district cooling system with and without considering a thermal energy storage. In [8], differential cuckoo search algorithm (DCSA) [9] based on obligate brood-parasitic behavior of some cuckoo species is introduced to optimize the chiller loading design problem. Reference [10] simplifies the complicated evolution process of the genetic algorithm (GA) for solving optimal chiller loading using the evolution strategy (ES). Other search approaches such as GA [11–14], simulated annealing (SA) [15, 16], differential evolution (DE) [17], gradient method (GM) [18], Lagrangian method [19], empirical model [20], artificial neural network (ANN) [21–25], firefly algorithm [26] have also been proposed.

This chapter develops a computationally efficient mixed integer nonlinear programming framework for solving economic multi-chiller loading problem using basic open-source nonlinear mixed integer programming (BONMIN) [27] solver under general algebraic modeling system (GAMS) [28] environment.

F. Jabari (✉) · B. Mohammadi-Ivatloo
Department of Electrical and Computer Engineering, University of Tabriz, Tabriz, Iran
e-mail: f.jabari@tabrizu.ac.ir; bmohammadi@tabrizu.ac.ir

B. Mohammadi-Ivatloo, F. Jabari (eds.), *Operation, Planning, and Analysis of Energy Storage Systems in Smart Energy Hubs*,
https://doi.org/10.1007/978-3-319-75097-2_6

6.2 Problem Formulation

In the economic dispatching problem of a multi-chiller plant, total electrical power consumed by all centrifugal chillers should be minimized as Eq. (6.1):

$$\text{Objective} = \text{Min}\left\{\sum_{t=1}^{T}\sum_{i=1}^{N} U_i^t \times \left(\alpha_i + \beta_i \times \text{PLR}_i^t + \gamma_i \times \text{PLR}_i^{t2} + \zeta_i \times \text{PLR}_i^{t3}\right)\right\} \tag{6.1}$$

where N is the number of chillers; U_i^t is a binary decision variable that will be equal to 1, if ith chiller is on; otherwise it will be 0; α_i, β_i, γ_i, ζ_i are the coefficients related to the operating characteristic of chiller i; PLR_i^t is the partial load ratio (PLR) of chiller i at time horizon t that is defined as relation (6.2).

$$\text{PLR}_i^t = \frac{\text{Cooling load of chiller } i \text{ at time } t}{\text{Power consumption of chiller } i \text{ at time } t} \tag{6.2}$$

Subject to power balance criterion which can be stated by Eq. (6.3):

$$\sum_{i=1}^{N}\left(U_i^t \times \text{PLR}_i^t \times \text{RT}_i\right) = \text{CL}^t; \quad \forall t = 1, 2, \ldots, T \tag{6.3}$$

where RT_i is the capacity of chiller i; CL^t is the total cooling demand at time t.

6.3 Case Studies

In this section, three case studies are illustrated to demonstrate the speed and the effectiveness of BONMIN solver in finding the optimum operating point of a multiple-chiller system. Tables 6.1, 6.2, and 6.3 present all coefficients related to the operating characteristic of chillers which are used in order to satisfy the cooling demand of a semiconductor factory located at Hsinchu Scientific Garden (Taiwan) [11, 13]. Multi-chiller plants of cases 1–3 composed of six 1280RT units, four units

Table 6.1 Chiller data for six units of case study 1

Chiller	α_i	β_i	γ_i	ζ_i	Chiller capacity (RT)
1	399.345	−122.12	770.46	0	1280
2	287.116	80.04	700.48	0	1280
3	−120.505	1525.99	−502.14	0	1280
4	−19.121	898.76	−98.15	0	1280
5	−95.029	1202.39	−352.16	0	1280
6	191.750	224.86	524.4	0	1280

Table 6.2 Chiller data for four units of case study 2

Chiller	α_i	β_i	γ_i	ζ_i	Chiller capacity (RT)
1	104.09	166.57	−430.13	512.53	450
2	−67.15	1177.79	−2174.53	1456.53	450
3	384.71	−779.13	1151.42	−63.2	1000
4	541.63	413.48	−3626.5	4021.41	1000

Table 6.3 Chiller data for three units of case study 3

Chiller	α_i	β_i	γ_i	ζ_i	Chiller capacity (RT)
1	100.95	818.61	−973.43	788.55	800
2	66.598	606.34	−380.58	275.95	800
3	130.09	304.5	14.377	99.8	800

with two 450RT and two 1000RT chillers, and three 800RT units. A comprehensive comparison between simulation results obtained from BONMIN solver and different optimization techniques is presented in Tables 6.4, 6.5, 6.6, 6.7, 6.8, and 6.9. As obvious from Tables 6.5, 6.7, and 6.9, total electricity requirements of multi-chiller systems in three cases reduce significantly as these optimization problems are solved using BONMIN tool under GAMS software.

6.4 Conclusion

In this chapter, day-ahead economic dispatch of three different multi-chiller plants is solved using basic open-source nonlinear mixed integer programming solver under GAMS software package. Compared with competitive heuristic algorithms, use of BONMIN solver in finding optimal loading points of centrifugal chillers reduces their electricity requirement significantly.

Nomenclature

α_i, β_i, γ_i, ζ_i	Coefficients related to the operating characteristic of chiller i
CL^t	Cooling demand after implementation of DRPs at time horizon t
N	Number of chillers
P_t^{ch}	Total electrical power consumed by all centrifugal chillers at time t
PLR_i^t	Partial load ratio (PLR) of chiller i at time horizon t
RT_i	Capacity of chiller i
U_i^t	A binary decision variable that will be equal to 1, if ith chiller is on; otherwise it will be 0

Table 6.4 Simulation results of case study 1 obtained from BONMIN solver and different optimization approaches

CL^t (kW)	Chiller	PLR_i^t					
		ALM [11]	SA [15]	PSO [6]	ES [10]	DCSA [8]	BONMIN
6858	1	0.9	0.7789	0.8026	0.82	0.812726	0.794
	2	0.9	0.7587	0.7799	0.75	0.749619	0.710
	3	0.9	0.9791	0.9996	1	1	0.627
	4	0.9	0.9781	0.9998	1	1	0.753
	5	0.9	0.9820	0.9999	1	1	0.753
	6	0.9	0.9265	0.8183	0.83	0.838559	0.729
6477	1	0.85	0.8051	0.7606	0.74	0.727731	0.637
	2	0.85	0.6056	0.6555	0.64	0.656132	0.545
	3	0.85	0.9689	1	1	1	0.687
	4	0.85	0.9941	1	1	1	0.559
	5	0.85	0.9866	1	1	1	1
	6	0.85	0.7432	0.6835	0.72	0.716524	1
6096	1	0.8	0.5635	0.6591	0.64	0.642735	1
	2	0.8	0.5743	0.5798	0.55	0.562645	1
	3	0.8	0.9675	0.9991	1	1	1
	4	0.8	0.9798	0.9979	0.9980	1	1
	5	0.8	0.9845	0.9921	1	1	1
	6	0.8	0.7338	0.5710	0.61	0.594490	1
5717	1	0.75	0.6140	0.7713	0.57	0.843697	1
	2	0.75	0.4429	0.7177	0.46	0.783794	1
	3	0.75	0.9891	0.3	1	0.000001	1
	4	0.75	0.8867	0.9991	1	1	1
	5	0.75	0.9841	1	1	1	1
	6	0.75	0.5878	0.7187	0.47	0.883049	1
5334	1	0.7	0.6265	0.6418	0.63	0.749969	1
	2	0.7	0.7403	0.6621	0.6	0.682477	0.836
	3	0.7	0.3093	0.3301	0.3	0.000012	0.713
	4	0.7	0.9546	0.9906	1	1	0.591
	5	0.7	0.9511	0.9990	1	1	0.780
	6	0.7	0.6250	0.5806	0.67	0.776363	0.609

Table 6.5 Comparison between objective function of case 1 obtained from BONMIN solver and other search algorithms

CL^t (kW)	Objective function (kW)					
	ALM [11]	SA [15]	PSO [6]	ES [10]	DCSA [8]	BONMIN
6858 ($t = 12^{\text{noon}}$)	4916.93	4777.03	4739.53	4738.76	4738.575	4690.546
6477 ($t = 1^{\text{p.m.}}$)	4635.22	4453.67	4423.04	4422.06	4421.649	4382.031
6096 ($t = 2^{\text{p.m.}}$)	4358.71	4178.73	4147.69	4144.12	4143.706	4111.756
5717 ($t = 3^{\text{p.m.}}$)	4087.42	3925.51	3921.07	3906.19	3840.055	3798.091
5334 ($t = 4^{\text{p.m.}}$)	3821.34	3675.34	3642.55	3627.46	3507.270	3504.009
Total	21,820	21,010	20,874	20,839	20,651	20,486

Table 6.6 Simulation results of case study 2 obtained from BONMIN solver and different optimization approaches

CL^t (kW)	Chiller	PLR_i^t				
		GA [11]	DE [17]	PSO [3]	DCSA [8]	BONMIN
2610 ($t = 12^{noon}$)	1	0.99	0.99	0.99	0.990988	0.946
	2	0.95	0.91	0.91	0.905473	0.813
	3	1	1	1	1	0.710
	4	0.74	0.76	0.76	0.756593	0.730
2320 ($t = 1^{p.m.}$)	1	0.86	0.83	0.83	0.828756	0.781
	2	0.81	0.81	0.81	0.805457	0.781
	3	0.88	0.90	0.90	0.896722	0.878
	4	0.69	0.69	0.69	0.687883	0.796
2030 ($t = 2^{p.m.}$)	1	0.66	0.73	0.73	0.773478	0.730
	2	0.76	0.74	0.74	0.739801	0.782
	3	0.76	0.72	0.72	0.721146	0.064
	4	0.64	0.65	0.65	0.627878	0.064
1740 ($t = 3^{p.m.}$)	1	0.6	0.6	0.6	0.767678	1
	2	0.7	0.66	0.66	0.004531	0.868
	3	0.57	0.56	0.56	0.746317	0.698
	4	0.59	0.61	0.61	0.646189	0.727
1450 ($t = 4^{p.m.}$)	1	0.6	0.61	0.61	0.515832	0.739
	2	0.36	0	0	0.000001	0.515
	3	0.44	0.57	0.57	0.610547	0.737
	4	0.58	0.61	0.61	0.607328	0.682
1160 ($t = 5^{p.m.}$)	1	0.33	0	0	0	0.643
	2	0.32	0	0	0.000014	0.650
	3	0.32	0.56	0.56	0.570369	0.653
	4	0.54	0.6	0.6	0.589625	0.593

Table 6.7 Comparison between objective function of case 2 obtained from BONMIN solver and other search algorithms

CL^t (kW)	Objective function (kW)				
	GA [11]	DE [17]	PSO [3]	DCSA [8]	BONMIN
2610 ($t = 12^{noon}$)	1862.18	1857.3	1857.3	1857.299	1767.305
2320 ($t = 1^{p.m.}$)	1457.23	1455.66	1455.66	1455.665	1404.411
2030 ($t = 2^{p.m.}$)	1183.8	1178.14	1178.14	1178.137	1147.274
1740 ($t = 3^{p.m.}$)	1001.62	998.53	998.53	942.183	980.747
1450 ($t = 4^{p.m.}$)	907.72	820.07	820.07	753.004	807.006
1160 ($t = 5^{p.m.}$)	856.3	651.07	651.07	583.923	630.303
Total	7268.9	6960.8	6960.8	6770.2	6737

Table 6.8 Simulation results of case study 3 obtained from BONMIN solver and different optimization approaches

CL^t (kW)	Chiller	PLR_i^t					
		GM [18]	GA [13]	DE [17]	PSO [3]	DCSA [8]	BONMIN
2610 ($t = 12^{noon}$)	1	0.73	0.81	0.73	0.73	0.725258	0.709
	2	0.97	0.93	0.97	0.97	0.974742	0.649
	3	1	0.96	1	1	1	0.587
1920 ($t = 1^{p.m.}$)	1	0.66	0.7	0.66	0.66	0.659065	0.643
	2	0.86	0.8	0.86	0.86	0.858458	0.643
	3	0.88	0.9	0.88	0.88	0.882477	0.527
1680 ($t = 2^{p.m.}$)	1	0.6	0.69	0.6	0.6	0.6	0.946
	2	0.75	0.68	0.74	0.74	0.74	0.841
	3	0.76	0.73	0.76	0.76	0.76	0.728
1440 ($t = 3^{p.m.}$)	1	0.53	0.52	0	0	0	0.869
	2	0.62	0.74	0.89	0.89	0.896314	0.728
	3	0.65	0.54	0.91	0.91	0.903686	0.641
1200 ($t = 4^{p.m.}$)	1	–	0.49	0	0	0	0.991
	2	–	0.44	0.74	0.74	0.743026	0.862
	3	–	0.57	0.76	0.76	0.756974	0.743
960 ($t = 5^{p.m.}$)	1	–	0.31	0	0	0	0.895
	2	–	0.32	0.57	0.57	0.536846	0.742
	3	–	0.58	0.63	0.63	0.663154	0.649

Table 6.9 Comparison between objective function of case 3 obtained from BONMIN solver and other search algorithms

CL^t (kW)	Objective function (kW)					
	GM [18]	GA [13]	DE [17]	PSO [3]	DCSA [8]	BONMIN
2610 ($t = 12^{noon}$)	1583.81	1590.96	1583.81	1583.81	1583.807	1549.365
1920 ($t = 1^{p.m.}$)	1403.2	1406.02	1403.2	1403.2	1403.196	1376.431
1680 ($t = 2^{p.m.}$)	1244.32	1250.06	1244.32	1244.32	1244.32	1223.531
1440 ($t = 3^{p.m.}$)	1102.26	1107.75	993.6	993.6	993.602	972.879
1200 ($t = 4^{p.m.}$)	–	971.21	832.33	832.33	832.325	817.496
960 ($t = 5^{p.m.}$)	–	842.18	692.25	692.25	692.251	738.462
Total	3749.8	5577.2	5165.7	5165.7	5165.7	5128.8

References

1. Lo C-C, Tsai S-H, Lin B-S (2016) Economic dispatch of chiller plant by improved ripple bee swarm optimization algorithm for saving energy. Appl Therm Eng 100:1140–1148
2. Niknam T, Golestaneh F (2013) Enhanced bee swarm optimization algorithm for dynamic economic dispatch. IEEE Syst J 7(4):754–762
3. Lee W-S, Lin L-C (2009) Optimal chiller loading by particle swarm algorithm for reducing energy consumption. Appl Therm Eng 29(8):1730–1734
4. Beghi A, Cecchinato L, Cosi G, Rampazzo M (2012) A PSO-based algorithm for optimal multiple chiller systems operation. Appl Therm Eng 32:31–40

5. Askarzadeh A, dos Santos Coelho L (2015) Using two improved particle swarm optimization variants for optimization of daily electrical power consumption in multi-chiller systems. Appl Therm Eng 89:640–646
6. Ardakani AJ, Ardakani FF, Hosseinian S (2008) A novel approach for optimal chiller loading using particle swarm optimization. Energy Build 40(12):2177–2187
7. Powell KM, Cole WJ, Ekarika UF, Edgar TF (2013) Optimal chiller loading in a district cooling system with thermal energy storage. Energy 50:445–453
8. dos Santos Coelho L, Klein CE, Sabat SL, Mariani VC (2014) Optimal chiller loading for energy conservation using a new differential cuckoo search approach. Energy 75:237–243
9. Valian E, Tavakoli S, Mohanna S, Haghi A (2013) Improved cuckoo search for reliability optimization problems. Comput Ind Eng 64(1):459–468
10. Chang Y-C, Lee C-Y, Chen C-R, Chou C-J, Chen W-H, Chen W-H (2009) Evolution strategy based optimal chiller loading for saving energy. Energy Convers Manag 50(1):132–139
11. Chang Y-C (2005) Genetic algorithm based optimal chiller loading for energy conservation. Appl Therm Eng 25(17):2800–2815
12. Chow T, Zhang G, Lin Z, Song C (2002) Global optimization of absorption chiller system by genetic algorithm and neural network. Energy Build 34(1):103–109
13. Chang Y-C, Lin J-K, Chuang M-H (2005) Optimal chiller loading by genetic algorithm for reducing energy consumption. Energy Build 37(2):147–155
14. Beghi A, Cecchinato L, Rampazzo M (2011) A multi-phase genetic algorithm for the efficient management of multi-chiller systems. Energy Convers Manag 52(3):1650–1661
15. Chang Y-C (2006) An innovative approach for demand side management – optimal chiller loading by simulated annealing. Energy 31(12):1883–1896
16. Chang Y-C, Chen W-H, Lee C-Y, Huang C-N (2006) Simulated annealing based optimal chiller loading for saving energy. Energy Convers Manag 47(15):2044–2058
17. Lee W-S, Chen Y-T, Kao Y (2011) Optimal chiller loading by differential evolution algorithm for reducing energy consumption. Energy Build 43(2):599–604
18. Chang Y-C, Chan T-S, Lee W-S (2010) Economic dispatch of chiller plant by gradient method for saving energy. Appl Energy 87(4):1096–1101
19. Chang Y-C (2006) An outstanding method for saving energy-optimal chiller operation. IEEE Trans Energy Convers 21(2):527–532
20. Wang H (2017) Empirical model for evaluating power consumption of centrifugal chillers. Energy Build 140:359–370
21. Frey P, Fischer S, Drück H (2014) Artificial neural network modelling of sorption chillers. Sol Energy 108:525–537
22. Wei X, Xu G, Kusiak A (2014) Modeling and optimization of a chiller plant. Energy 73: 898–907
23. Labus J, Hernández J, Bruno J, Coronas A (2012) Inverse neural network based control strategy for absorption chillers. Renew Energy 39(1):471–482
24. Čongradac V, Kulić F (2012) Recognition of the importance of using artificial neural networks and genetic algorithms to optimize chiller operation. Energy Build 47:651–658
25. Chang Y-C, Chen W-H (2009) Optimal chilled water temperature calculation of multiple chiller systems using Hopfield neural network for saving energy. Energy 34(4):448–456
26. dos Santos Coelho L, Mariani VC (2013) Improved firefly algorithm approach applied to chiller loading for energy conservation. Energy Build 59:273–278
27. https://www.gams.com/latest/docs/S_BONMIN
28. Brooke A, Kendrick D, Meeraus A, Raman R, America U (1998) The general algebraic modeling system. GAMS Development Corporation, vol 1050

Chapter 7
Demand Response Participation in Renewable Energy Hubs

Mohammad Mohammadi, Younes Noorollahi, and Behnam Mohammadi-Ivatloo

7.1 Introduction

The optimal performance of a power system is based on the equivalence of power production with power consumption and losses. In the power systems, operating point is continuously changing. Therefore, the production level of power generator units should be changed to balance production and consumption. These systems have been the main power supply systems of the last century. In such structures, responses to demand changes are made through a change in production plan and supply side management. However, the use of large-scale and costly production units in order to meet peak demands during the year is not a good choice. Because, in addition to the high cost of these systems, this additional capacity is used only in limited periods throughout the year and actually leads to capacity waste [1].

Nowadays, other options such as distributed energy resources (DER), especially renewable energy sources (RES), energy storage systems (ESS), and DSM have been introduced to replace current energy supply systems and have led to changes in the structure of energy systems. The development of RES, especially as distributed generation (DG) resources, has some serious problems, such as high investment costs, geographical location dependency, fluctuating nature and precarious planning of production, and the need for expensive equipment such as control systems [2]. ESS, despite their many advantages, such as facilitating the integration of RES, are

M. Mohammadi · Y. Noorollahi (✉)
Department of Renewable Energy and Environment, Faculty of New Sciences and Technologies, University of Tehran, Tehran, Iran
e-mail: m.mohammady@ut.ac.ir; noorollahi@ut.ac.ir

B. Mohammadi-Ivatloo
Department of Electrical and Computer Engineering, University of Tabriz, Tabriz, Iran
e-mail: bmohammadi@tabrizu.ac.ir

B. Mohammadi-Ivatloo, F. Jabari (eds.), *Operation, Planning, and Analysis of Energy Storage Systems in Smart Energy Hubs*,
https://doi.org/10.1007/978-3-319-75097-2_7

still considered as expensive and under-study technology. On the other hand, all of the above options are also based on changing the production plan based on the demand and so supply side management. But among the above options, DSM is a practical approach and an immediate solution that focuses on matching demand with available resources.

In this regard, this chapter provides a brief introduction of DR programs. Subsequently, an investigation is carried out on the implementation of DR programs in the energy hub models in the literature. Finally, a mathematical modeling and simulation of an energy hub are presented in the presence of DR programs and the effects of these programs on the operational program of the energy hub have been investigated.

7.2 DSM Concept

Demand-side energy management involves a series of interconnected activities between the electricity utility and its customers in order to reduce network peaks and energy consumption, as well as smoothing the consumption curve of the network in order to provide more efficient and low cost energy for consumers. In the beginning, consumption management was introduced as load management (LM) program to reduce peak consumption. Gradually reducing consumer costs, resource allocation, and reducing environmental pollution were introduced as other incentives for DSM. By adapting these policies, the level of consumer comfort will not be reduced. In fact by maintaining consumers' level of comfort and well-being, they will consume less energy, and/or the pattern of consumption will be changed. This leads to a reduction in operation costs and it will also be possible to earn money. DSM is a widespread concept that includes things like load growth, energy savings, energy efficiency, and DR programs. Load growth refers to programs that are used to increase the level of load through electricity supply in a strategic state. Energy savings refers to measures that reduce energy consumption. Energy efficiency refers to programs aimed at reducing energy consumption through specific systems on the consumption side and typically without affecting the services provided. These programs reduce total electricity consumption by replacing equipment with energy efficient technologies in existing systems to produce the same share of services for subscribers.

According to the definition of the American Energy Academy, DR refers to programs to change the pattern of electricity consumption by changing the pattern of final consumers' consumption by responding to a change in electricity prices over time or incentive payments to encourage a reduction in electricity consumption at a time when the wholesale market price is high or reliability of the system is at risk [3]. DR programs lead to lower energy consumption at times when the electricity price is high and it changes the pattern of consumer consumption and, consequently, the load curve. Advantages of the presence of costumers in DR programs include incentive payments, reduced billing, while enhancing system reliability, better market performance, and lower infrastructure costs.

7.2.1 DR Programs

In traditional systems, consumption management programs were used to overcome some power system problems. DR programs were also part of these programs. But after the restructuring of power systems, these programs were gradually set aside due to inconsistencies with the nature of the market. But after some time due to problems such as instability of price, re-implementation of consumption management programs was again considered. This time these programs were modified in such a way to fit the restructured power system. After restructuring of the power system, DR programs have been one of the main parts of the consumption management programs. Today, these programs are considered as an appropriate solution to some of the problems of the power system.

DR includes approaches from DSM that are used to change customer consumption due to price changes in the market. It should be noted that some such programs were used in the traditional electricity system in the form of multi-tariff meters. Different pricing for electricity causes two long-term and short-term changes in consumption patterns. In the long term, the high price of electricity will lead to savings in power consumption. If the tariffs difference between peak and off-peak hours is high, consumers will be encouraged to install energy storage and efficient devices in order to prevent the use of electrical energy during high-price hours. Therefore, in the long term, the creation of various tariffs for power consumption will increase the energy efficiency of the consumer. Also, in the short term, some customers have the opportunity to reduce their consumption or shift this consumption to low-cost hours. For example, an industrial consumer, if the production at peak hours is not profitable due to the high price of electricity, decision makers would be discouraged from producing goods during these hours.

In general, the main objective of DR is to reduce power consumption during critical hours. Critical hours are the hours when the price of the wholesale market is high or the system's resiliency level is low due to accidental events such as outgoing transmission lines and generators, or severe weather conditions. Two factors that can lead to consumer responsiveness are the change in retail price of electricity or the implementation of an incentive program to satisfy customers to reduce their consumption during critical hours. This incentive is different from the normal price paid for electricity. This incentive can be a payment to the consumer to reduce the consumption, set a penalty for not reducing the load, or both. DR is in fact a change in the behavior of loads in response to a stimulus. DR can appear in the form of reprogramming of an industrial consumer production plan, reorient a commercial customer's heating system, or direct control of the electricity company on the domestic water heater system. DR programs are widely used for lowering cost of electric power utilities [4], increasing profit of retailers [5], increasing power system voltage stability margin [6], increasing profit of energy hub owner [7], and many other applications. Further discussions in the literature review section will be presented separately.

7.2.2 DR Programs Classification

DR programs can be divided into two main types including incentive-based (IB) and price-based (PB). IB programs include LM programs and market-based programs such as demand side bidding, capacity market, and ancillary services. PB programs are based on dynamic pricing such as time of use (TOU) pricing, real time pricing (RTP), and critical peak pricing (CPP). LM includes direct load control (DLC) and interruptible load control (ILC). The DLC usually involves residential users and refers to programs that can control a costumer load, such as home appliances, through direct operator control. The ILC usually involves commercial and industrial subscribers, and it refers to programs that can reduce peak load by interrupting customers' load at peak hours with direct control of the operator or subscribers' actions upon request from the operator. The above classification can be seen in Fig. 7.1.

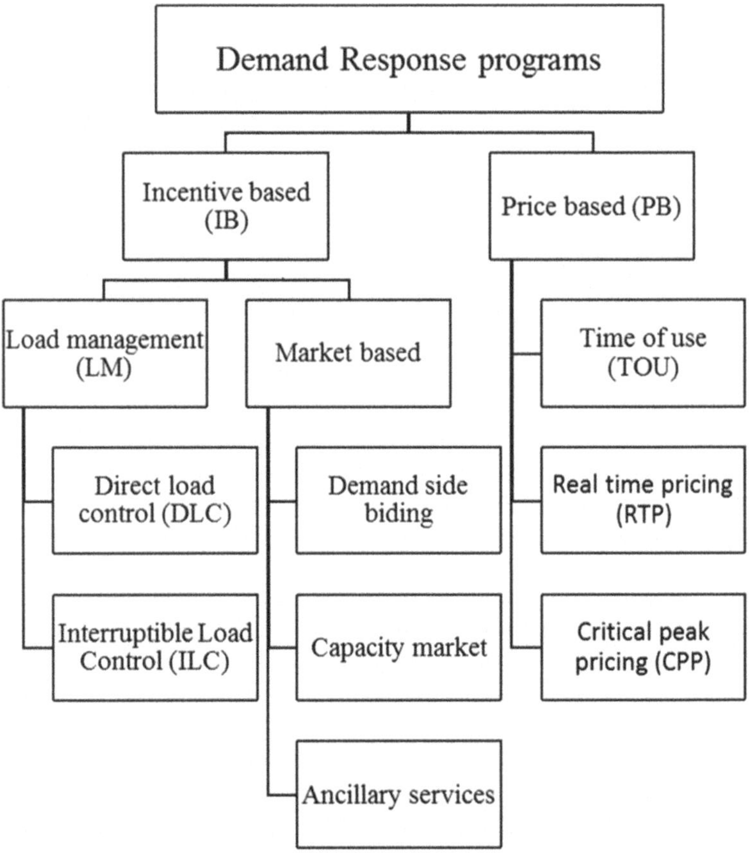

Fig. 7.1 Classification of DR programs

7.3 Literature Review

7.3.1 DR in Residential Energy Hubs

In many countries, the residential sector has a significant share of energy consumption. However, due to significant losses in energy distribution systems, as well as lack of responsive loads and energy efficiency equipment, energy management in the residential sector is not appropriate and a significant share of energy in this sector is wasted. For example, the share of the residential sector in United States energy consumption is 22%, which 47% of this energy consumption are electricity related losses in this sector [8]. On the other hand, energy prices are rising due to issues such as energy resource constraints, increased demand, and the restructuring of energy markets. Therefore, this section requires the use of alternative systems and programs to increase efficiency.

In the residential sector, DSM can be one of the main options for the energy management in buildings. However, the effective participation of residential loads in DSM programs requires the management and optimal control of various equipment in a residential building. Accordingly, we can classify residential loads into two main groups of responsive loads and non-responsive loads, which can be seen in Fig. 7.2.

In a residential building there are loads that have a large share in energy consumption and, by an optimal controlling their consumption can be optimized without affecting the consumer's comfort. In the above category, equipment such as a washing machine and a dishwasher that has a specific operating interval and their operation can be shifted to another period of time, are considered as

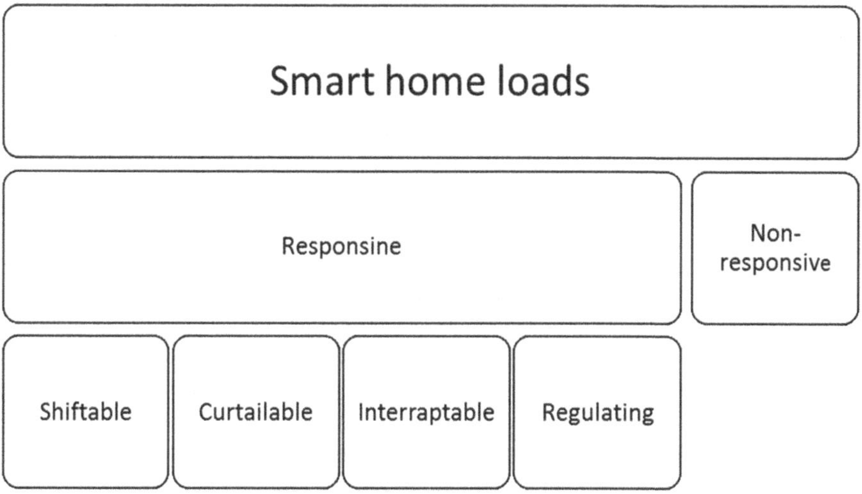

Fig. 7.2 Classification of the smart home loads

shiftable loads. Loads such as lights can be dimmed in critical situations, and can be considered as curtailable loads. Another part of the loads can be disconnected in critical conditions, and the consumption of them can be transferred to another period, which is said to be interruptible loads. The plugin electric vehicle (PEV) charging is considered as an interruptible load. The regulating loads are loads that can be programmed to operate within a range of operations with a slight deviation from the defined level. Heating and cooling loads are considered as examples of this category. However, depending on the operating range and the simulation steps, the classification can vary for each of these loads [9].

Optimizing a residential building with CHP and PEV in the form of an energy hub to participate in DR programs with TOU pricing has been investigated in [10]. The effects of the presence of responsive and non-responsive loads are studied in this energy hub. The results showed that when the energy hub is not responsive, optimal energy production planning is the only way to participate in the DR program. In this case, during high energy price hours, CHP production will be increased to reduce electricity purchases. In the presence of responsive loads, the possibility of load shifting can also be provided and the operating costs of subscribers will be reduced. A similar optimal operational plan for an energy hub within the framework of a combined cooling, heat and power (CCHP) system for participating in DR programs can also be found in [11].

One of the technologies that can lead to more effective participation of the energy hubs in DR programs is energy storage system, and in particular potential storage devices such as PEVs. Options such as aggregation, two-way communication with the network and vehicle to grid (V2G), in addition to solving the problems of charging electric cars, will enable PEVs to be used to improve the network's sustainability. A model for optimizing the performance of a residential energy hub with PV, PEV, and ESS has been presented in [12]. Optimization results in this study showed that optimal control of PV and responsive loads in the presence of ESS and taking into account the ability of V2G for PEV lead to peak shaving and lower operating costs for the energy hub. Optimization of the performance of a CCHP in the form of an energy hub model for participation in the DR program is investigated in [13]. The results of this study showed that the use of thermal storage and PEV for participation in the DR program lead to peak shaving, increasing the CCHP's contribution in demand supply and thus reducing the operating costs of the energy hub.

A comprehensive modeling framework for optimal management of a smart energy hub with the goal of minimizing energy costs, peak demand and emission has been presented in [14]. In this study, consumer preferences and also weather and electricity price forecasts are considered in modeling. The results of this study showed that optimizing the performance of a residential house in Ontario, Canada has led to a 15% drop in peak demand, and a reduction of 45% in total energy costs.

In summary, home energy management systems can be modeled in the framework of energy hub models, which improve the performance of these systems and benefiting from the benefits of concepts such as DSM, DER, and smart energy systems. The integrated management of home energy system in the framework of the energy hub facilitates the participation of houses in DR programs and also, integration with multi-energy systems (MES) and smart grid becomes easier, which improves the performance and efficiency of home energy system [15].

7.3.2 DR in Commercial Energy Hubs

A large part of the world's energy is consumed in buildings. Buildings worldwide account for 40% of total primary energy consumption and are responsible for one third of greenhouse gas emissions [16]. Therefore, energy efficiency in this sector can have a major impact on reducing energy consumption and emissions. Commercial buildings that can include service, office, and entertainment buildings are among the largest buildings in the world that have significant energy consumption. The energy supply of this building in the past was mainly done through purchasing from energy networks. However, today factors such as rising energy prices, restructuring in energy markets, and finding efficient DG resources have led to an increase in the share of DER in supplying energy demand for commercial buildings. RES and multi-generation systems have been most used in commercial buildings.

One of the main features of commercial buildings is that they have large energy consumers and most of their energy consumption is related to lighting, heating and ventilation and air conditioning (HVAC) systems. Therefore, controlling these loads can greatly improve the energy efficiency of the building [17]. On the other hand, the pattern of consumption of these major loads is usually predictable. The energy consumption in commercial buildings follows a daily, monthly, and even seasonal pattern. For example, energy consumption in an office building has two patterns of weekday and weekend patterns, which energy consumes for the whole year follow from these two patterns. This feature represents the high potential of energy systems in commercial buildings to manage demand and participate in DR programs. Also, as mentioned, the existence of DER such as CHP, PV, wind turbine, fuel cell will improve the performance of commercial buildings in DR programs.

On the other hand, ESS performance in commercial buildings can be better than residential buildings. Peak demand in commercial buildings occurs throughout the day, and DER such as PV can produce their most energy during these hours. In this case, the use of ESS is minimized, and the active participation in DSM programs and the energy stored in them can be used to sell to the network and generate revenue.

An energy management system for the optimal operation of a commercial building in the framework of energy hub with the aim of minimizing the cost, increasing efficiency, and reducing emissions has been presented in [18]. The results showed that by combining different technologies and integrated management of the building energy systems, building's peak demand and energy cost is reduced while the lower emissions are obtained.

In the commercial buildings, an integrated system for energy management is required to make optimal scheduling, successful participation in the DSM programs and benefit from the smart grids advantages. This integrated management leads to lower cost, peak shaving, reducing environmental impact and successful participation in DR programs which reveals a great potential for application of energy hubs models in this sector.

7.3.3 DR in Industrial Energy Hubs

The industrial sector is the driving force behind the development of each country, and it is usually responsible for the most energy consumption and greenhouse gas emissions. According to the EIA report, in 2010, 52% of the world's energy consumption has been happened in the industrial sector and this energy consumption rises by an annual rate of 1.4% from 2010 to 2040 [19]. Demand in the industrial sector, with the growth of industrialization, the rise of new industrial countries and consumption growth in the developing countries is quickly growing.

Therefore, energy management in this sector is essential and can have a huge impact on reducing consumption and optimizing the use of existing resources. Implementing DSM programs, especially DR, in the industrial sector can be a good option for energy efficiency improvement in this sector, but the implementation of these programs in this section is always faced with serious challenges. The existence of various flows, such as the raw materials along with the energy flow, as well as the need to maintain the balance of production with the demand and quality of manufactured goods, and the problems of maintaining the competitiveness of the production unit in the industrial sector are the main problems of implementing DR programs.

In this regard, an integrated management framework is essential for the optimal management of various systems and technologies in an industrial unit. The use of RES in industrial units and their contribution to DSM programs have been studied in [20]. In this study, wind turbines provide part of the energy demand of two industrial units, and these units have the ability to manage their demand in the content of various DSM programs. The results of this study showed that the participation of industrial units in DR programs has led to an increase in the share of wind power in supplying energy and the shift in demand to low energy price hours. Short-term scheduling of industrial cogeneration system in the presence of DR programs is studied in [21] without consideration of uncertainties. Optimization of the performance of a cogeneration industrial unit is formulated in [22] for participation in DR programs and the exchange of electricity and heat with adjacent systems in a stochastic environment. The risk constrained scheduling of industrial heat and power systems is proposed in [23]. Stochastic scheduling of industrial CHP systems with DR and RES is provided in [24]. An operational optimization model for an oxygen production unit for participation in DR programs and the day-ahead electricity market was presented in [25]. The results showed that optimal participation in DR programs result in a load shifting to low energy price hours that leads to a reduction in unit's costs and an improvement in network reliability. In this study, only the implementation of Passive DR programs is considered, and the implementation of such models on a large number of business subscribers can lead to create new peak demand in the network.

In a passive DR program usually power supply companies provide a general model for energy pricing regardless of the type of consumers and their consumption behavior information. This can lead to a transmission of a large part of consumer demand to off-peak and low prices periods and creation of new peaks in the system which reduces the benefits of DR. To avoid this issue, consumption patterns of each consumption sectors must be considered and DR programs should be presented based on a detailed categories of consumers which results in an active DR program. On the other hand, the number of industry consumers is usually less than other sectors and the assessment of their behavior and so the implementation of active DR in the industrial sector can be easier. A model of the active DR program based on dynamic pricing for industrial consumers has been provided in [26]. An agent-based approach for modeling the behavior of consumers to participate in DR programs and choose the best behavioral pattern considering various industrial processes for the three cement plant has been presented. The results showed that the implementation of active DR results in lower peak demand in the industrial units and in the whole system load curve is more smoother than passive DR.

Therefore, it can be said that to optimize the performance of industrial units several purposes can be considered such as lower costs, increase profits, reduce primary energy consumption and raw materials, reduce waste, reduce environmental impacts, and so on. These objectives can be achieved by different methods. One solution is the use of distributed energy resources such as DG, especially cogeneration and RES, as well as ESS. Moreover, demand scheduling can be used to manage energy, raw material consumption and waste. In addition, DR can be used to reduce system costs and participation in energy markets. Planning all this together requires an integrated management system in the form of an energy hub model.

7.3.4 DR in Agricultural Energy Hubs

Energy in agriculture plays an important role, and energy efficiency in the agricultural sector has a direct impact on sustainability and food security in each country. Therefore, optimum use of energy in agriculture plays a major role in agricultural sector productivity [27]. On the other hand, increased demand for food and the growth of the use of mechanized equipment in agriculture have led to an increase in demand for energy in agriculture. A significant portion of this energy demand is supplied through fossil fuels. However, agricultural farms are usually located in remote areas and there is a high potential for using DER and especially RES in these areas. The use of DER in the agricultural sector can prevent the expansion of energy distribution systems to these areas that leads to lower costs and energy losses. Also, the use of RES, in addition to supplying energy demand of farms, can also help to solve environmental problems.

In addition to the need to expand the use of DER and, in particular, the RES in the agricultural sector, intelligent technologies can be used to optimize farm management. Using intelligent technologies in the content of a smart energy system can be achieved by optimal measurement, control and management, which can increase the efficiency, reliability, and optimal utilization of energy sources.

On the other hand, energy consumption in agriculture is increasing due to factors such as lack of productivity, growth of technology, and automated mechanism and the need for more food in the world.

Despite the high potential in the farms for energy supply from RES, unfortunately, still the largest share of energy production in the agricultural sector is related to the fossil fuels. The waste of agricultural products is a good source for the production of renewable fuels such as biogas, bioethanol, and biodiesel which can be used in CHP to supply energy of farm and even selling excess power. Compared to other sectors, less attention has been paid to the use of RES in the agricultural sector. While agricultural farms are usually in remote areas and there is a great potential for the use of RES to avoid the high costs and losses of energy transfer to these areas.

In addition to the RES integration, one important strategy that has attracted a lot of attention recently is the use of smart technology especially smart meters in the agricultural sector. Smart grid, with improved monitoring, control and metering, improves the efficiency, sustainability, reliability, and optimal distribution of the energy resources. Using smart technology is not limited to the smart grid and the development of these concepts in the agricultural sector led to the creation of concepts like smart farms, smart agriculture, and precision agriculture. So in the agricultural sector, the information and communication technology can be used for the monitoring and collecting information about the physical condition of products and farm in order to classify the process, decision-making, statistical reports, forecasting, etc. A variety of the wireless technologies and their applications in various agricultural sectors can be found in [28]. A framework for the application of smart systems in the form of wireless systems for smart farms in order to help the farmers for real-time control of the production, reporting, statistical information in the context of the smart grid is presented in [29]. A model for the integration of the smart grids and smart farm with an emphasis on information management systems has been studied in [30]. It is a model of a smart farm that includes a waste-based CHP that can exchange power with the smart electricity grid. The use of the precision agriculture features and communication between the farm and smart grid results in improved farm planning, optimal exchanges with the power grid and even participate in energy markets and the DSM programs. According to our knowledge, there is no comprehensive model in literature for optimal performance of smart farms to manage their energy systems and energy exchange with neighboring systems particularly in the content of DSM programs. Due to the ability of the energy hub in the communicating between energy systems in the presence of various energy carriers, there is a good potential for smart farms modeling in the form of smart agricultural energy hubs.

7.4 Energy Hub Modeling

In order to investigate the effects of DR programs in energy hub models, this study attempts to use a complete model of energy hub that has inputs, converters, and storages to meet different demands. A schematic representation of the studied energy hub can be seen in Fig. 7.3. This energy hub is powered by electricity and natural gas networks and wind power. Transformer, converters, CHP, and boiler are used to convert different energy carriers. Electrical and thermal storages are selected as ESS in this energy hub. In demand side, demands for electricity, heat, and natural gas is considered for this hub. Also DR capabilities for electricity demand are included in this model.

7.4.1 *Objective Function*

In the model presented for energy hubs, cost, reliability, and environmental indicators are included in the objective function. The proposed objective function is formulated based on the cost of purchasing energy from various networks, revenue from electricity sales to the network, cost for charging and discharging electrical and thermal storages, DR costs, emission costs, and reliability indicators. In this chapter, the objective function in a deterministic environment is optimized to minimize operational costs in a one-day time horizon under the influence of different constraints. The objective function of the optimal management problem of the proposed energy hub can be considered as follows:

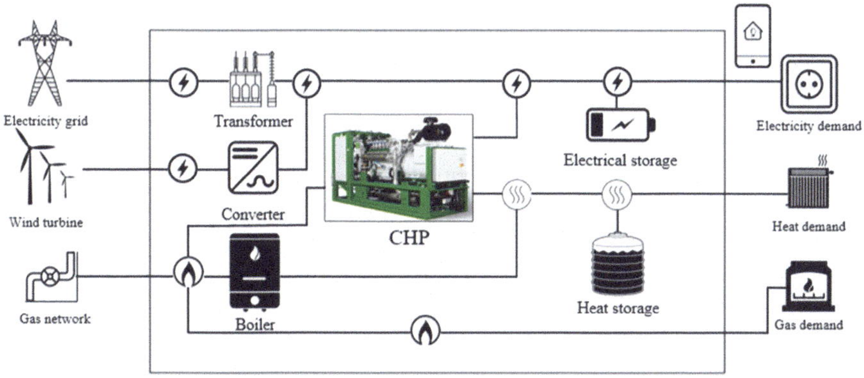

Fig. 7.3 A schematic representation of the studied energy hub

$$
\begin{aligned}
&\text{minimization of TC} \\
&= \sum_{t=1}^{24} \left[\pi_e{}^N(t) P_e{}^N(t)\right] + \left[\pi_e{}^W P_e{}^W(t)\right] + \left[\pi_g{}^N P_g{}^N(t)\right] \\
&+ \left[\pi_e{}^S \left(P_e{}^{\text{ch}}(t) + P_e{}^{\text{dis}}(t)\right)\right] + \left[\pi_h{}^S \left(P_h{}^{\text{ch}}(t) + P_h{}^{\text{dis}}(t)\right)\right] \\
&+ \left[\pi_e{}^{\text{DR}} \left(P_e{}^{\text{shup}}(t) + P_e{}^{\text{shdo}}(t)\right)\right] + \left[\pi_e{}^{\text{ENS}} P_e{}^{\text{ENS}}(t)\right] + \left[\pi_h{}^{\text{ENS}} P_h{}^{\text{ENS}}(t)\right] \\
&+ \left[\sum_{t=1}^{24} \sum_{\text{em}=1}^{3} \pi_{\text{em}} \left(\text{EF}_{\text{em}}{}^N P_e{}^N(t) + \text{EF}_{\text{em}}{}^{\text{NCHP}} P_g{}^{\text{NCHP}}(t) + \text{EF}_{\text{em}}{}^{\text{NB}} P_g{}^{\text{NB}}(t)\right)\right]
\end{aligned}
\tag{7.1}
$$

In the above objective function, energy purchases from different networks (electrical energy and natural gas) are included in the respective costs $\pi_e{}^N(t)$ and π_g^N respectively, for electricity and natural gas networks. Also, in this objective function, $P_e{}^N(t)$ can have positive and negative values and the energy hub has the potential to sell power and earn money. The cost of purchasing wind power $P_e{}^W(t)$ is at a price of $\pi_e{}^W$, which in this study the wind turbine is belonging to the energy hub and the purchase price of zero for wind power is considered. In the case of energy storage devices, the operating costs for charging $P_e{}^{\text{ch}}(t)$, $P_h{}^{\text{ch}}(t)$ and discharging $P_e{}^{\text{dis}}(t)$, $P_h{}^{\text{dis}}(t)$ are considered with the corresponding unit costs $\pi_e{}^S$ and $\pi_h{}^S$ for electric and thermal storages, respectively. In addition, operating costs $P_e{}^{\text{shup}}(t)$ and $P_e{}^{\text{shdo}}(t)$ for the load shift are modeled at unit cost $\pi_e{}^{\text{DR}}$ in the form of another operating cost for the energy hub. The penalty for not-supplied demand $P_e{}^{\text{ENS}}(t)$ and $P_h{}^{\text{ENS}}(t)$ is priced with $\pi_e{}^{\text{ENS}}$ and $\pi_h{}^{\text{ENS}}$ for electrical and thermal demands, respectively, to include reliability indicators in the management of energy hub. Finally, the cost of CO_2, SO_2, and NO_2 emissions for the grid, boiler, and CHP is taken into account in the final statement of the objective function. The above objective function is affected by various constraints, which are described below.

7.4.2 *Operational Constraints*

The optimal performance of an energy hub requires consideration of various constraints that their satisfaction will provide the optimal operation for the system. Each of these constraints is stated separately below.

7.4.2.1 Demand Constraints

The power equilibrium constraints for supplying electrical, thermal, and natural gas demands can be formulated as follows:

$$
\begin{aligned}
P_{\text{ed}}(t) &= \left[A^N \eta_T P_e{}^N(t)\right] + \left[A^{\text{CHP}} \eta_{\text{echp}} P_g{}^{\text{NCHP}}(t)\right] + \left[A^W \eta_C P_e{}^W(t)\right] \\
&+ \left[P_e{}^{\text{dis}}(t) - P_e{}^{\text{ch}}(t)\right] + \left[P_e{}^{\text{shdo}}(t) - P_e{}^{\text{shup}}(t)\right] + \left[P_e{}^{\text{ENS}}(t)\right]
\end{aligned}
\tag{7.2}
$$

$$
\begin{aligned}
P_{\mathrm{hd}}(t) &= \left[A^{\mathrm{CHP}} \eta_{\mathrm{hchp}} P_g{}^{\mathrm{NCHP}}(t)\right] + \left[\eta_B P_g{}^{\mathrm{NB}}(t)\right] \\
&+ \left[P_h{}^{\mathrm{dis}}(t) - P_h{}^{\mathrm{ch}}(t)\right] + \left[P_h{}^{\mathrm{ENS}}(t)\right]
\end{aligned}
\tag{7.3}
$$

$$
P_{\mathrm{gd}}(t) = \left[P_g{}^N(t)\right] - \left[P_g{}^{\mathrm{NCHP}}(t)\right] - \left[P_g{}^{\mathrm{NB}}(t)\right] \tag{7.4}
$$

Electricity demand $P_{\mathrm{ed}}(t)$ in Eq. (7.2) is achieved by using wind power $P_e{}^W(t)$ through a wind turbine with accessibility A^W, [31] and its converter efficiency η_C. CHP provides electricity by burning gas with an efficiency of η_{echp}. Part of the electric demand can be reduced ($P_e{}^{\mathrm{shdo}}(t)$) or increased ($P_e{}^{\mathrm{shdo}}(t)$) with demand shifts to other timescales, which allows the participation of energy hub in DR programs. An electrical storage device is used to store additional power $P_e{}^{\mathrm{ch}}(t)$ and discharge it $P_e{}^{\mathrm{dis}}(t)$ to supply the demand. The remaining electrical demand is provided by purchasing electricity from the network $P_e{}^N(t)$ with availability A^{CHP}, [31] and transformer efficiency η_T.

Thermal demand in the Eq. (7.3) will be provided through CHP with availability A^{CHP} [32] and the gas-to-heat conversion efficiency η_{hchp}. Boiler with efficiency η_B is used as a backup system to provide heat demand. Thermal storage is considered for charging $P_h{}^{\mathrm{ch}}(t)$ in case of excess production and discharge $P_h{}^{\mathrm{dis}}(t)$ to meet demand. The gas demand $P_{\mathrm{gd}}(t)$ for energy hub in Eq. (7.4) is achieved by direct purchasing from the gas network $P_g{}^N(t)$ and after deducting the amount of gas consumed in CHP, $P_g{}^{\mathrm{NCHP}}(t)$, and boiler $P_g{}^{\mathrm{NB}}(t)$.

7.4.2.2 Network Constraints

In the purchase from electricity and gas networks, the restrictions are considered which can be formulated with Eqs. (7.5) and (7.6).

$$
-P_e{}^{N\max} \leq P_e{}^N(t) \leq P_e{}^{N\max} \tag{7.5}
$$

$$
-P_e{}^{N\max} \leq P_e{}^N(t) \leq P_e{}^{N\max} \tag{7.6}
$$

7.4.2.3 Converters Constraints

The components of the energy hub in this study have certain installed capacities. Converters such as CHP, boiler, and transformer have their own output limitations, which are imposed through Eqs. (7.7)–(7.9) to the objective function.

$$
\eta_T P_e{}^N(t) \leq P^T \tag{7.7}
$$

$$
\eta_T P_e{}^N(t) \leq P^T \tag{7.8}
$$

$$\eta_B P_g{}^{\mathrm{NB}}(t) \le P^B \tag{7.9}$$

7.4.2.4 ESS Constraints

In the proposed energy hub model, the storage systems are modeled accurately, taking into account cases such as power losses and operational constraints [33]. Such modeling can be seen in Eqs. (7.10)–(7.15).

$$P_e{}^S(t) = P_e{}^S(t-1) + \eta_e{}^{\mathrm{ch}} P_e{}^{\mathrm{ch}}(t) - \frac{P_e{}^{\mathrm{dis}}(t)}{\eta_e{}^{\mathrm{dis}}} - P_e{}^{\mathrm{loss}}(t) \tag{7.10}$$

$$P_e{}^{\mathrm{loss}}(t) = \alpha_e{}^{\mathrm{loss}} P_e{}^S(t) \tag{7.11}$$

$$\alpha_e{}^{\min} P_e{}^{\mathrm{SC}} \le P_e{}^S(t) \le \alpha_e{}^{\max} P_e{}^{\mathrm{SC}} \tag{7.12}$$

$$\alpha_e{}^{\min} P_e{}^{\mathrm{SC}} I_e{}^{\mathrm{ch}}(t) \le P_e{}^{\mathrm{ch}}(t) \le \alpha_e{}^{\max} P_e{}^{\mathrm{SC}} I_e{}^{\mathrm{ch}}(t) \tag{7.13}$$

$$\alpha_e{}^{\min} P_e{}^{\mathrm{SC}} I_e{}^{\mathrm{dis}}(t) \le P_e{}^{\mathrm{dis}}(t) \le \alpha_e{}^{\max} P_e{}^{\mathrm{SC}} I_e{}^{\mathrm{dis}}(t) \tag{7.14}$$

$$0 \le I_e{}^{\mathrm{ch}}(t) + I_e{}^{\mathrm{dis}}(t) \le 1 \tag{7.15}$$

Equation (7.10) indicates the state of charging of the energy storage in each time step, which is a function of its charging level in the previous step, the charge rate, discharge rate, and the amount of losses in that time step. The amount of losses of the electrical storage in each step can be obtained from Eq. (7.11). The minimum and maximum limits for the charge level of the storage are applied through Eq. (7.12). The limits for charging and discharging power can be found in Eqs. (7.13) and (7.14). The binary variables $I_e{}^{\mathrm{ch}}(t)$ and $I_e{}^{\mathrm{dis}}(t)$ are defined for the non-simultaneous charging and discharging at each time step, so at any time maximum one of them can accept a value of 1. This constraint is applied through Eq. (7.15).

By applying the same constraints to the thermal storage, Eqs. (7.16)–(7.21) are formulated as follows:

$$P_h{}^S(t) = P_h{}^S(t-1) + \eta_h{}^{\mathrm{ch}} P_h{}^{\mathrm{ch}}(t) - \frac{P_h{}^{\mathrm{dis}}(t)}{\eta_h{}^{\mathrm{dis}}} - P_h{}^{\mathrm{loss}}(t) \tag{7.16}$$

$$P_h{}^{\mathrm{loss}}(t) = \alpha_h{}^{\mathrm{loss}} P_h{}^S(t) \tag{7.17}$$

$$\alpha_h^{\min} P_h^{SC} \le P_h^{S}(t) \le \alpha_h^{\max} P_h^{SC} \tag{7.18}$$

$$\alpha_h^{\min} P_h^{SC} \frac{1}{\eta_h^{ch}} I_h^{ch}(t) \le P_h^{ch}(t) \le \alpha_h^{\max} P_h^{SC} \frac{1}{\eta_h^{ch}} I_h^{ch}(t) \tag{7.19}$$

$$\alpha_h^{\min} P_h^{SC} \eta_h^{dis} I_h^{dis}(t) \le P_h^{dis}(t) \le \alpha_h^{\max} P_h^{SC} \eta_h^{dis} I_h^{dis}(t) \tag{7.20}$$

$$0 \le I_h^{ch}(t) + I_h^{dis}(t) \le 1 \tag{7.21}$$

7.4.2.5 DR Constraints

Responsive load can be shifted from peak demand and higher-priced periods to non-peak and low cost periods. In this regard, DR program for shifting part of the energy hub electrical load in the presence of a real-time pricing schedule can be formulated as follows [34]:

$$\sum_{t=1}^{24} P_e^{shup}(t) = \sum_{t=1}^{24} P_e^{shdo}(t) \tag{7.22}$$

$$0 \le P_e^{shup}(t) \le \mathrm{LPF}^{shup} P_{ed}(t) I_e^{shup}(t) \tag{7.23}$$

$$0 \le P_e^{shdo}(t) \le \mathrm{LPF}^{shdo} P_{ed}(t) I_e^{shdo}(t) \tag{7.24}$$

$$0 \le I_e^{shup}(t) + I_e^{shdo}(t) \le 1 \tag{7.25}$$

The demand shift program in Eq. (7.22) for shifting up $P_e^{shup}(t)$ or shifting down $P_e^{shdo}(t)$ a part of (LPF^{shup}, LPF^{shdo}) the electrical demand $P_{ed}(t)$ should be applied in such a way that the sum of the shift up and shift down loads is equal in the simulation horizon. Binary variables $I_e^{shup}(t)$ and $I_e^{shup}(t)$ are defined in order to avoid the simultaneous reduction or increase in load at any time step.

7.4.3 Wind Turbine Model

In wind turbines, the power generated from them is a function of wind speed. The power output of a wind turbine can be determined by its power curve. This curve shows the amount of power produced at different wind speeds. An example of this power curve can be seen in Fig. 7.4. A wind turbine starts producing power at cut-in

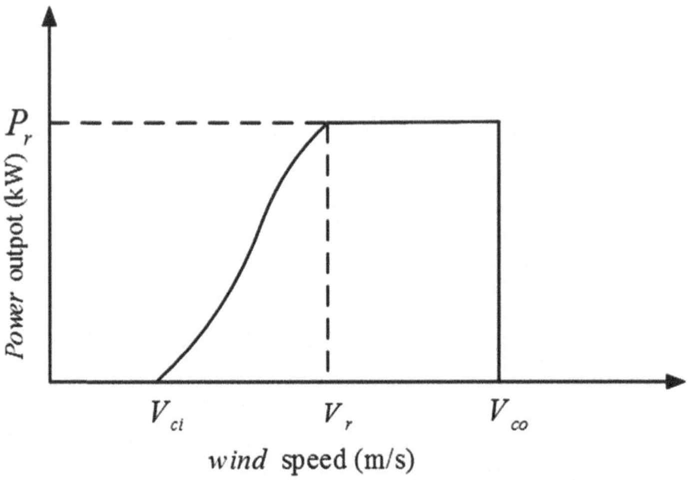

Fig. 7.4 Power curve of wind turbine

wind speed V_{ci} and its power is stopped at cut-out wind speeds V_{co} because of safety issues. The nominal power P_r is produced when the wind speed is within the rated speed V_r range to the upper limit of the speed. A nonlinear relationship between power generation and wind speed can be found in the range of cut-in wind speed up to the rated speed.

Therefore, the following relation can be derived for the wind turbine output by wind speed [35].

$$W_{\text{av}}(t) = P_r{}^* \begin{cases} A + B^* V_w(t) + C^* V_w(t)^2 & V_{\text{ci}} \le V_w(t) \le V_r \\ 1 & V_r \le V_w(t) \le V_{\text{co}} \\ 0 & \text{otherwise} \end{cases} \tag{7.26}$$

In the above relation, the coefficients A, B, and C can be obtained from the following relationships.

$$A = \frac{1}{(V_{\text{ci}} - V_r)^2} \left\{ V_{\text{ci}} (V_{\text{ci}} + V_r) - 4 V_{\text{ci}} V_r \left[\frac{V_{\text{ci}} + V_r}{2 V_r} \right]^3 \right\} \tag{7.27}$$

$$B = \frac{1}{(V_{\text{ci}} - V_r)^2} \left\{ 4 (V_{\text{ci}} + V_r) \left[\frac{V_{\text{ci}} + V_r}{2 V_r} \right]^3 - (3 V_{\text{ci}} + V_r) \right\} \tag{7.28}$$

$$C = \frac{1}{(V_{\text{ci}} - V_r)^2} \left\{ 2 - 4 \left[\frac{V_{\text{ci}} + V_r}{2 V_r} \right]^3 \right\} \tag{7.29}$$

7.5 Simulation Results

The energy hub presented in Fig. 7.3 is optimized based on the minimization of operating costs. In this study, the effects of DR programs are evaluated in five different cases. In the first and second cases the impact of the presence of the energy hub in the DR program is assessed subject to a relatively high price for natural gas. In the third and fourth cases, low prices are assumed for natural gas, and optimal operating schedule for the energy hub is provided with and without participating in DR programs. In the fifth scenario, CHP is eliminated from the energy hub, and the effect of the interaction of different carriers is examined in the presence of DR program. Various cases are listed in Table 7.1.

Different demands for electricity, heat and natural gas for the energy hub are shown in Fig. 7.5. Also, hourly electricity price and hourly wind speed are shown in Figs. 7.6 and 7.7, respectively.

The values of the input parameters and other assumptions for the optimal energy hub management problem can be found in Table 7.2.

Table 7.1 Different cases for studding the participation of the energy hub in DR program

Cases	Energy hub structure
Case 1	High gas price—without DR
Case 2	High gas price—with DR
Case 3	Low gas price—without DR
Case 4	Low gas price—with DR
Case 5	Energy hub without CHP

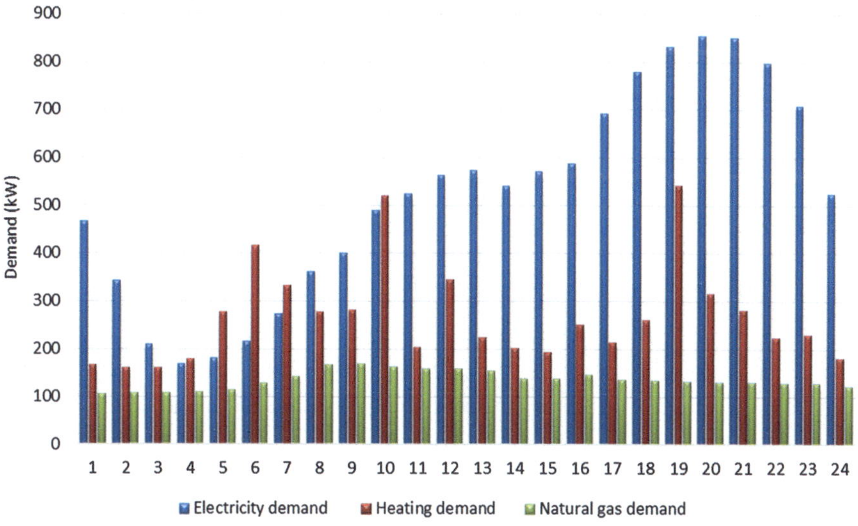

Fig. 7.5 The energy hub electricity, heat and natural gas demands

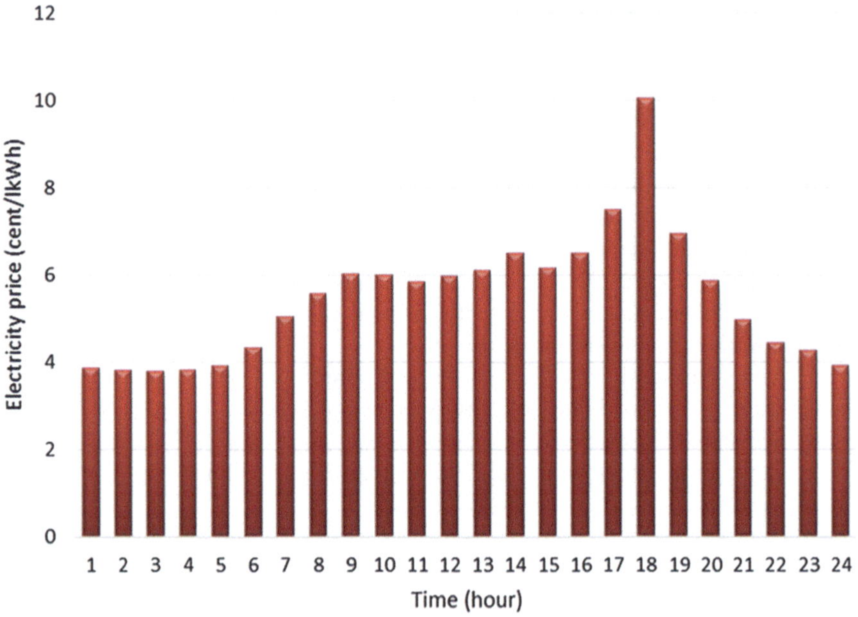

Fig. 7.6 The hourly electricity price

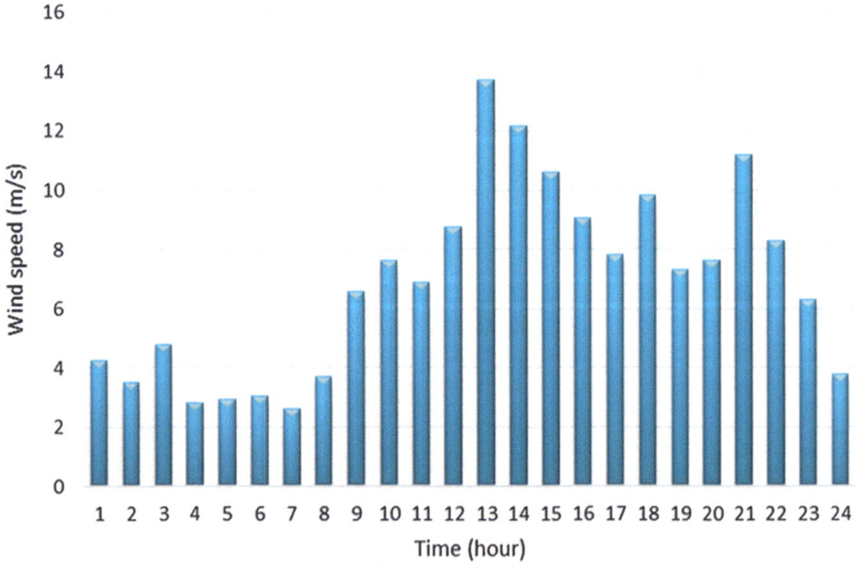

Fig. 7.7 The hourly wind speed

The problem of optimizing the energy hub in the presence of DR program is a MILP model that has been solved in the GAMS software using the CPLEX solving algorithm. Numerical results for different cases separately are discussed in the following sections.

Table 7.2 The values of the input parameters and other assumptions for the optimal energy hub management problem

Parameter	Unit	Value	Parameter	Unit	Value
α_e^{loss}	–	0.02	EF_{em}^{CHPCO2}	kg/kWh	0.412
α_h^{loss}	–	0.02	EF_{em}^{CHPSO2}	kg/kWh	0.008
α_e^{min}	–	0.1	EF_{em}^{CHPNO2}	kg/kWh	0.000112
α_e^{max}	–	0.9	EF_{em}^{BCO2}	kg/kWh	0.617
α_h^{min}	–	0.1	EF_{em}^{BSO2}	kg/kWh	0.011
α_h^{max}	–	0.9	EF_{em}^{BNO2}	kg/kWh	0.000284
η_e^{ch}	–	0.9	$eELF_{max}$	kg/kWh	0.05
η_e^{dis}	–	0.9	EF_{em}^{NCO2}	kg/kWh	0.424
η_h^{ch}	–	0.9	EF_{em}^{NSO2}	kg/kWh	0.00226
η_h^{dis}	–	0.9	EF_{em}^{NNO2}	kg/kWh	0.000925
η_C	–	0.9	π_e^{ENS}	/kWh¢	4000
η_T	–	0.9	π_h^{ENS}	/kWh¢	4000
η_{eCHP}	–	0.4	π_e^S	/kWh¢	2
η_{hCHP}	–	0.35	π_e^W	/kWh¢	0
η_B	–	0.85	π_{em}^{CO2}	/kg¢	0.014
A^{CHP}	–	0.96	π_{em}^{SO2}	/kg¢	0.99
A_e^N	–	0.99	π_{em}^{NO2}	/kg¢	4.2
A^W	–	0.96	π_g^N	/kWh¢	2, 6
LPF^{shup}	–	0.3	π_h^S	/kWh¢	2
LPF^{shdo}	–	0.3	π_e^{DR}	/kWh¢	2
A	–	0.0311	P_e^{Nmax}	kW	400
B	–	−0.0776	P_g^{Nmax}	kW	2000
C	–	0.0174	P^T	kW	600
P_r	kW	300	P^B	kW	1800
V_r	m/s	10	P^{CHP}	kW	300
V_{ci}	m/s	4	P_e^{SC}	kW	300
V_{co}	m/s	22	P_h^{SC}	kW	300

7.5.1 Numerical Results for Case 1

In this case, the price of natural gas is considered to be relatively expensive, and the energy hub should pay 6 cents for each kilowatt of purchased gas from the network. In this case the energy hub has CHP, which is the main supplier of electrical and thermal demands and the electricity and boiler network are considered as a backup system for CHP. The numerical results of this case can be seen in Table 7.3. This table actually represents the optimal operating schedule for the performance of the energy hub during a day.

In this case, the energy hub has two independent power generation technologies (CHP and wind turbines). The existence of CHP in the system allows for the sale of excess electricity to the network. In the table above, the negative values for the power exchanged with the network is the sale of this power to the network. However, at some hours, because of the high price of natural gas, the energy hub does not use CHP and purchasing electricity from the grid provides power deficit.

Table 7.3 Optimal operation schedule of the energy hub in case 1

t	P_e^N	P_e^{CHP}	P_e^W	P_e^{ch}	P_e^{dis}	P_g^N	P_h^{ch}	P_h^{dis}	P_h^B
1	400	139.1	6.8	34	0	561.8	35.7	0	79.8
2	383.2	0	0	0	0	294.6	0	0	159.8
3	265	0	23.4	45.9	0	295.6	0	0	160.4
4	400	0	0	188.7	0	318.8	0	0	178.3
5	239.7	0	0	33.3	0	440.2	0	0	277
6	−189.2	384	0	0	0	1222	0	0	79.9
7	−121.3	379.9	0	0	0	1131.4	0	0	0
8	−26.1	384	0	0	0	1166.8	58.4	0	0
9	−89.3	384	109.5	0	0	1167.6	55.4	0	0
10	−54.5	384	179.3	0	0	1274.3	0	88.6	94.8
11	202.3	232	130.3	0	0	763.4	0	0	0
12	−63	384	272.8	0	0	1166.9	0	0	7.6
13	−33.3	256.5	400	0	0	822.3	0	0	0
14	−39.8	230.9	400	0	0	738.4	0	0	0
15	42.4	220.3	400	33.3	0	712	0	0	0
16	−51.3	371.2	302.5	0	0	1113	75.2	0	0
17	155.9	384	195.3	0	0	1135	122.1	0	0
18	−159.6	384	380.9	0	209.3	1133.1	75.7	0	0
19	349.3	384	158.9	0	0	1132.4	0	207.1	0
20	379.4	360.8	181.5	0	0	1068.9	0	0	0
21	208	320	400	0	0	963.4	0	0	0
22	383.1	255.5	233.4	0	0	793	0	0	0
23	364	302	94.3	0	0	913.2	33.3	0	0
24	400	167.4	0	0	0	559.3	0	33.6	1
$\sum t$	3344.9	6307.6	3868.9	335.2	209.3	20,887.4	455.8	329.3	1038.6

The existence of a wind turbine in the energy hub will allow the generation of local power from RES and the possibility to sell more energy to the network will provided in the energy hub. Given the limited capacity of CHP, in the absence of a turbine, the sale of electrical energy to the grid is only possible in the early hours of the day when demand is low. However, with the addition of a wind turbine, it is possible to sell energy during the day and even in the afternoon, when the price of electricity in the market is higher than it in the early hours of the day. As a result, energy sales go higher and with higher price, resulting in higher energy hub revenues.

An electrical storage has been used alongside the wind turbine to reduce the intermittent effects of this source on the performance of the hub energy and to better match the power production with the pattern of consumption. The electrical storage charges at off-peak hours and discharge at a time when there is the highest electricity price in the electricity market (18:00) and provides a significant part of the electricity demand at this hour. So CHP and wind turbine produce maximum power at peak hours, and the highest electrical storage discharges occur in these hours. The set of these factors makes the energy hub cover all its demand at the network peak hour and even sell excess energy to the network at the highest price.

Table 7.4 Optimal operation schedule of the energy hub in case 2

t	P_e^N	P_e^{CHP}	P_e^{ch}	P_e^{dis}	P_e^{shup}	P_e^{shdo}	P_g^N	P_h^{ch}	P_h^{dis}	P_h^B
1	400	230.3	34	0	91.2	0	705.4	35.7	0	0
2	400	87.4	0	0	102.4	0	432.3	0	0	83.3
3	400	0	103.1	0	63.1	0	295.6	0	0	160.4
4	400	29.5	167.9	0	50.3	0	365.3	0	0	152.4
5	263	0	0	0	54.1	0	440.2	0	0	277
6	−116.7	384	0	0	64.6	0	1222	0	0	79.9
7	−69.9	379.9	0	0	45.8	0	1131.4	0	0	0
8	−26.1	384	0	0	0	0	1166.8	58.4	0	0
9	−89.3	384	0	0	0	0	1167.6	55.4	0	0
10	−54.5	384	0	0	0	0	1274.3	0	88.6	94.8
11	202.3	232	0	0	0	0	763.4	0	0	0
12	−63	384	0	0	0	0	1166.9	0	0	7.6
13	−33.3	256.5	0	0	0	0	822.3	0	0	0
14	−39.8	230.9	0	0	0	0	738.4	0	0	0
15	5	220.3	0	0	0	0	712	0	0	0
16	−39.5	360.6	0	0	0	0	1085.6	66	0	0
17	−77	384	0	0	0	207.5	1135	122.1	0	0
18	−394.6	384	0	184.6	0	234	1133.1	75.7	0	0
19	69	384	0	0	0	249.8	1132.4	0	207.1	0
20	379.4	360.8	0	0	0	0	1068.9	0	0	0
21	400	320	0	0	171.1	0	963.4	0	0	0
22	400	255.5	0	0	15.1	0	793	0	0	0
23	400	263.9	0	0	0	6	814	0	0	0
24	400	206.9	0	0	39.6	0	661.2	0	0	0
$\sum t$	3115	6506.5	305	184.6	697.3	697.3	21,190.5	413.3	295.7	855.4

In order to meet the demand for heat, the boiler provides deficits in the hours when the CHP is unable to meet all demand. Also, in times when the demand for heat is not high, the excess energy is stored in the storage, and at peak periods the storage supplies the part of the heat demand by discharging, resulting in lower operating costs for the energy hub. The total operating cost of the energy hub in this case is 15,5767.3 Euro cents. In the next case, the effects of adding DR capabilities to the energy hub are assessed.

7.5.2 Numerical Results for Case 2

The difference between this case and the previous case is in the energy hub capability to participate in DR program. The numerical results of this case are presented in Table 7.4.

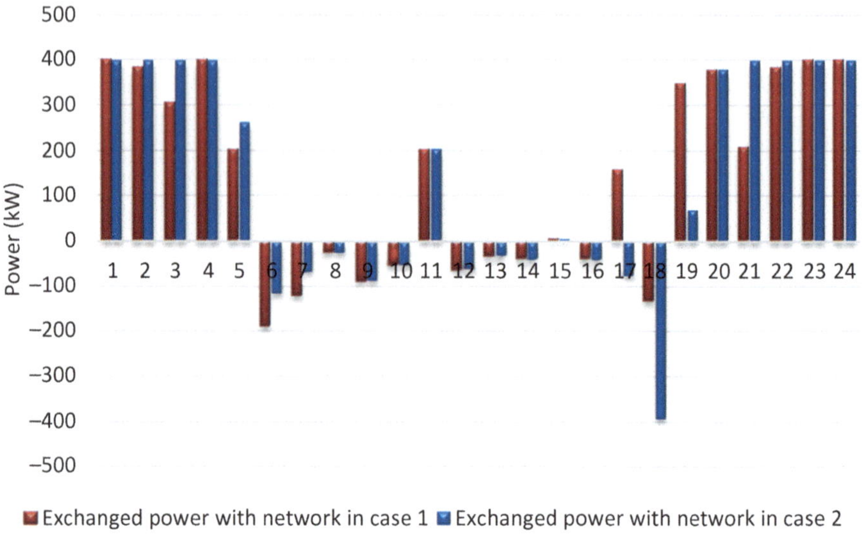

Fig. 7.8 Power exchanged with the network in cases 1 and 2

In this case, the possibility of shifting the load makes the energy hub shift part of its demand from peak hours to off-peak hours. As a result, the demand of energy hub decreases during peak hours and increases in off-peak hours. This makes the energy hub buy less electricity from the network at peak hours and even with the available energy generation resources, energy hub can sell its extra energy to the network at some peak hours. On the other hand, this shift of load will increase the use of CHP in off-peak hours and part of the added demand in these hours can be provided through CHP. This will increase gas purchases from the network. As shown in Tables 7.3 and 7.4, the amount of gas purchased from the network has increased from 30,521.1 kW in case 1 to 30,911.2 kW in case 2. The schematic representation of the above concepts can be seen in Figs. 7.8 and 7.9. These figures represent the power exchanged with the grid and the amount of power produced by the CHP.

As can be seen, due to the demand shift to off-peak hours, the amount of CHP production increased during hours 1, 2, 4, and 24. The effects of this load shift on the power exchanged with the network curve can be seen as increasing the purchase from the network at hours 2, 3, 21, and 22, as well as reducing sales to the network at hours 6 and 7. The demand shift and demand drop in peak hours led to the sale of power to the network at hour 17, increased sales to the network at hour 18, and reduced purchases from the network at hour 19. The set of these factors cause the operating costs of the hub energy to be 15,270.2 Euro cents that is less than the first case.

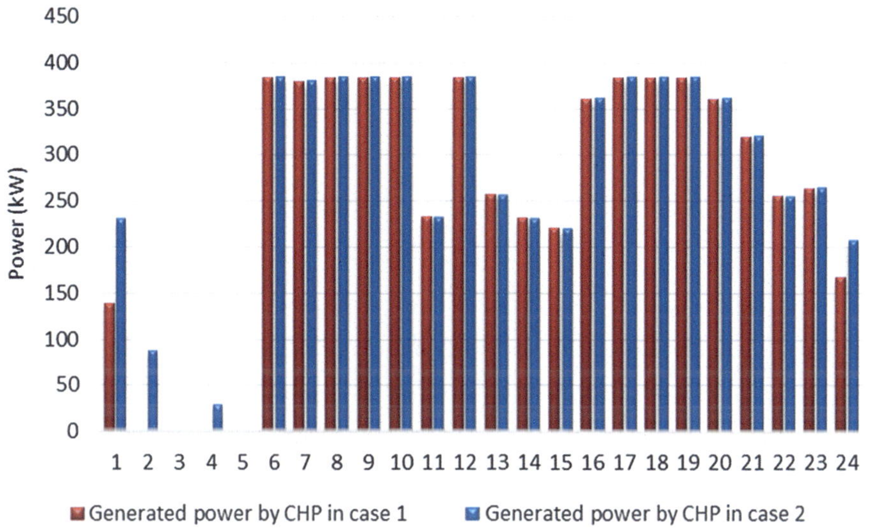

Fig. 7.9 Power produced by the CHP in cases 1 and 2

7.5.3 Numerical Results for Case 3

In this case, the price of natural gas is lower than electricity in all hours of the day. In addition, the energy hub does not have the ability to participate in DR programs and all demand must be met according to the requested pattern. The optimal operating schedule for the energy hub is shown in Table 7.5.

A remarkable point about this case compared to case 1 is the significant increase in the use of CHP due to the low price of natural gas compared to electricity. In this case, the energy hub with the ability of using different energy carriers uses cheaper energy carriers to supply a specific demand. As can be seen in the table, the amount of electricity produced by CHP in this case is 7823.7 kW, which has increased 1516.1 kW compared to case 1. This increase in the use of CHP leads to an increase in gas purchases from the network about 2799 kW compared to the first case. This increase can be seen in Fig. 7.10. As seen in the figure, the amount of gas purchased from the network has increased over most hours of the day.

Another point is the reduction of boiler production. With the increase of CHP production, the need for auxiliary source is reduced and the use of boilers is minimized. Also, with the increase in thermal energy produced by CHP, the use of thermal storage also increases sharply. This is due to matching the production of CHP with the thermal demand pattern. In this case, the heat storage is recharged at different times of the day when the CHP production is more than heat demand and, when required, by discharging provides a significant part of the thermal demand and minimizes the use of boilers. As can be seen in Tables 7.5 and 7.3, the charge and discharge power of the heat storage is more than twice as large as the cases 1. The

Table 7.5 Optimal operation schedule of the energy hub in case 3

t	P_e^N	P_e^{CHP}	P_e^W	P_e^{ch}	P_e^{dis}	P_g^N	P_h^{ch}	P_h^{dis}	P_h^B
1	125.5	384	6.8	34.3	0	1105.7	170.2	0	0
2	178.2	182.7	0	0	0	582.3	0	0	0
3	7.7	183.3	23.4	0	0	584.3	0	0	0
4	262.2	203.7	0	269.7	0	639.6	0	0	0
5	−228.6	384	0	0	0	1114.4	59	0	0
6	−189.2	384	0	0	0	1128	0	79.9	0
7	−121.3	379.9	0	0	0	1131.4	0	0	0
8	−26.1	384	0	0	0	1166.8	58.4	0	0
9	−89.3	384	109.5	0	0	1167.6	55.4	0	0
10	−54.5	384	179.3	0	0	1226	0	129.7	53.7
11	202.3	232	130.3	0	0	763.4	0	0	0
12	−63	384	272.8	0	0	1166.9	0	0	7.6
13	1.3	225.7	400	0	0	742	0	27	0
14	−211.6	384	400	0	0	1137.1	134	0	0
15	−171.2	377.2	400	0	0	1120.8	137.4	0	0
16	244.1	108	302.5	0	0	427.6	0	155.1	0
17	155.9	384	195.3	0	0	1135	122.1	0	0
18	−131.9	384	380.9	0	184.6	1133.1	75.7	0	0
19	349.3	384	158.9	0	0	1132.4	0	207.1	0
20	379.4	360.8	181.5	0	0	1068.9	0	0	0
21	136.2	384	400	0	0	1130	56	0	0
22	238.8	384	233.4	0	0	1127.7	112.5	0	0
23	285.9	371.5	94.3	0	0	1094.3	94.2	0	0
24	355.6	206.9	0	0	0	661.2	0	0	0
$\sum t$	1635.7	7823.7	3868.9	304	184.6	23,686.5	1074.9	598.8	61.3

shift of energy carrier to natural gas and its further use will reduce the dependence on the electricity grid and make less purchase from the electricity network. For a better comparison, see Fig. 7.11. This figure shows how to exchange electrical power with the network in cases 1 and 3.

It is clearly seen in this figure that purchases from the electricity grid have decreased, especially in the early and last hours of the day. This is due to the fact that at case 1 at these hours the cost of producing power from natural gas was higher than electricity, and the energy hub purchased electricity from the electricity grid. But in case 3, with the decline in natural gas prices, demand for these hours is provided through CHP and a fraction of this demand is purchased from the network. More power generation by CHP makes it possible to sell electricity to the grid even during the mid-day hours. Reducing dependence on the power grid, increasing the use of more affordable energy carrier, and better performance of ESS in this case has led to a dramatic reduction in the operating costs of the energy hub and reaching it to 65,094.2 Euro cents. In the next case, the effect of DR participation is investigated in the presence of a low-cost energy carrier (natural gas).

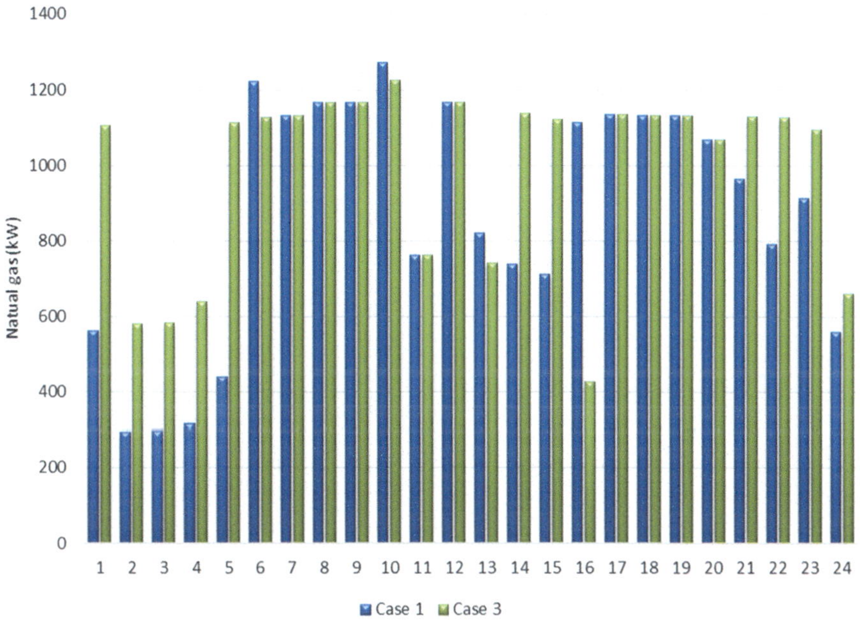

Fig. 7.10 The amount of natural gas purchased from the network in cases 1 and 3

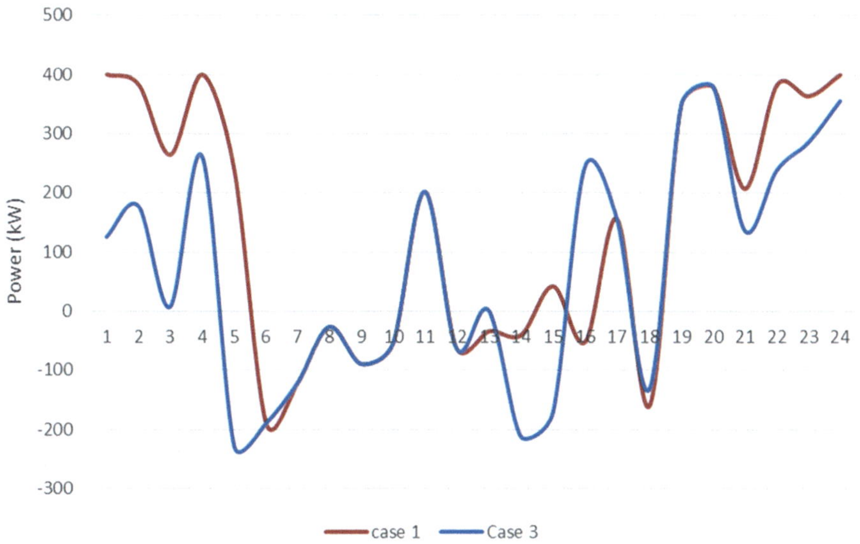

Fig. 7.11 Power exchanged with the network in cases 1 and 3

Table 7.6 Optimal operation schedule of the energy hub in case 4

t	P_e^N	P_e^{CHP}	P_e^{ch}	P_e^{dis}	P_e^{shup}	P_e^{shdo}	P_g^N	P_h^{ch}	P_h^{dis}	P_h^B
1	282.9	384	34.3	0	140.2	0	1105.7	170.2	0	0
2	293.1	182.7	0	0	102.4	0	582.3	0	0	0
3	78.5	183.3	0	0	63.1	0	584.3	0	0	0
4	318.7	203.7	269.7	0	50.3	0	639.6	0	0	0
5	−167.9	384	0	0	54.1	0	1114.4	59	0	0
6	−116.7	384	0	0	64.6	0	1128	0	79.9	0
7	−121.3	379.9	0	0	0	0	1131.4	0	0	0
8	−26.1	384	0	0	0	0	1166.8	58.4	0	0
9	−89.3	384	0	0	0	0	1167.6	55.4	0	0
10	−54.5	384	0	0	0	0	1226	0	129.7	53.7
11	202.3	232	0	0	0	0	763.4	0	0	0
12	−63	384	0	0	0	0	1166.9	0	0	7.6
13	1.3	225.7	0	0	0	0	742	0	27	0
14	−211.6	384	0	0	0	0	1137.1	134	0	0
15	−171.2	377.2	0	0	0	0	1120.8	137.4	0	0
16	167.5	108	0	0	0	68.3	427.6	0	155.1	0
17	−77	384	0	0	0	207.5	1135	122.1	0	0
18	−394.6	384	0	184.6	0	234	1133.1	75.7	0	0
19	69	384	0	0	0	249.8	1132.4	0	207.1	0
20	379.4	360.8	0	0	0	0	1068.9	0	0	0
21	136.2	384	0	0	0	0	1130	56	0	0
22	400	384	0	0	143.6	0	1127.7	112.5	0	0
23	400	371.5	0	0	101.6	0	1094.3	94.2	0	0
24	400	206.9	0	0	39.6	0	661.2	0	0	0
$\sum t$	1635.7	7823.7	304	184.6	759.5	759.6	23,686.5	1074.9	598.8	61.3

7.5.4 Numerical Results for Case 4

In this case an affordable price is considered for natural gas, but the DR capability is only applied to electrical demand. The optimal operating schedule for the energy hub in this case can be found in Table 7.6.

In this case, as in case 2, part of the electricity demand is reduced in network peak hours and is transmitted to off-peak hours. The effects of demand shifts on the demand curve for electrical energy in cases 2 and 4 can be seen in Fig. 7.12.

The demand shifts reduce the purchase from the network at peak hours and even sell more electrical energy to the network at some of these hours. The electric power exchange curve with the network can be seen in Fig. 7.13. As shown in this figure, the demand reduction in peak hours as well as the existence of distributed generation sources such as CHP and wind turbine will make it possible for the energy hub not only to reduce purchase from the network during the peak hours, but also to sell excess electric power to the network and earn revenue. This will reduce the

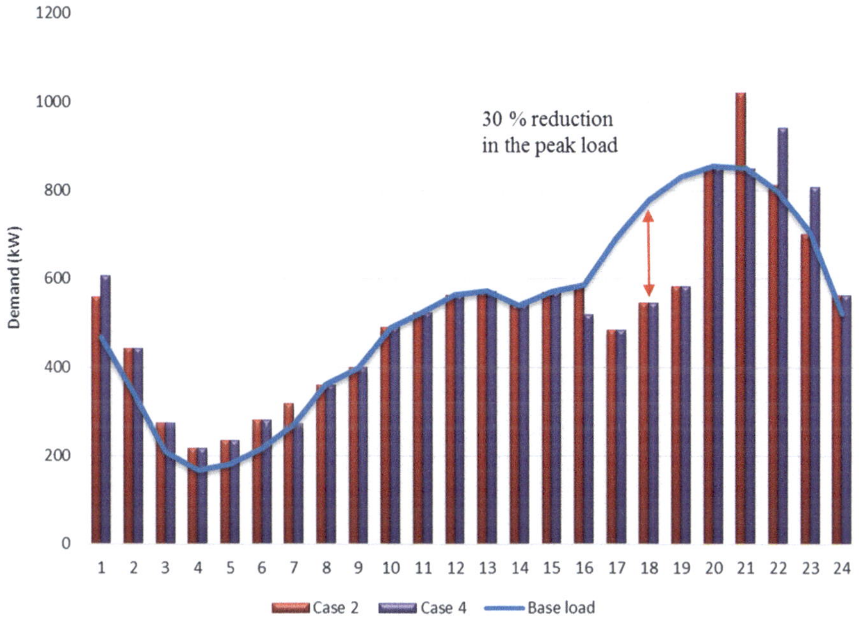

Fig. 7.12 Effect of the DR program on the electricity demand of the energy hub

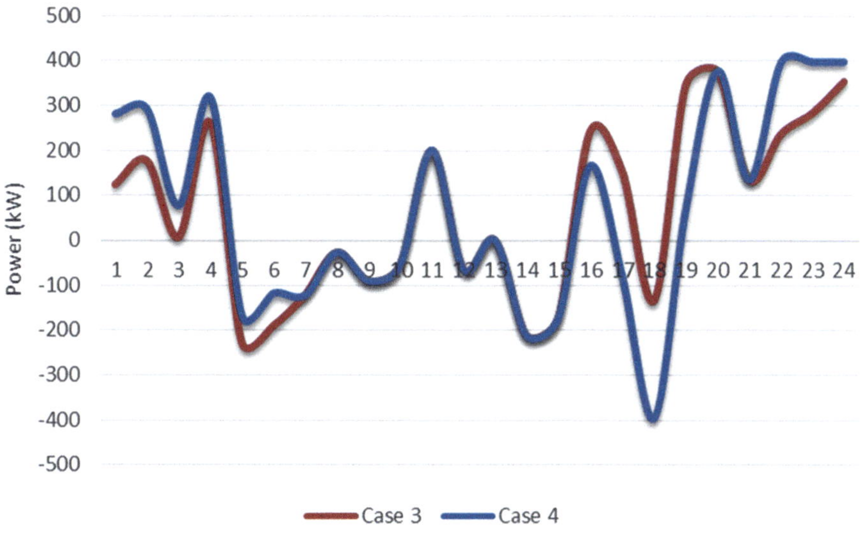

Fig. 7.13 Power exchanged with the network in cases 3 and 4

network's peak load and increase its stability. This shifted demand is provided in off-peak hours and leads to lower operational costs for the system. This amount is 61,395.5 cents for case 4, which is the smallest amount among the tested cases so far.

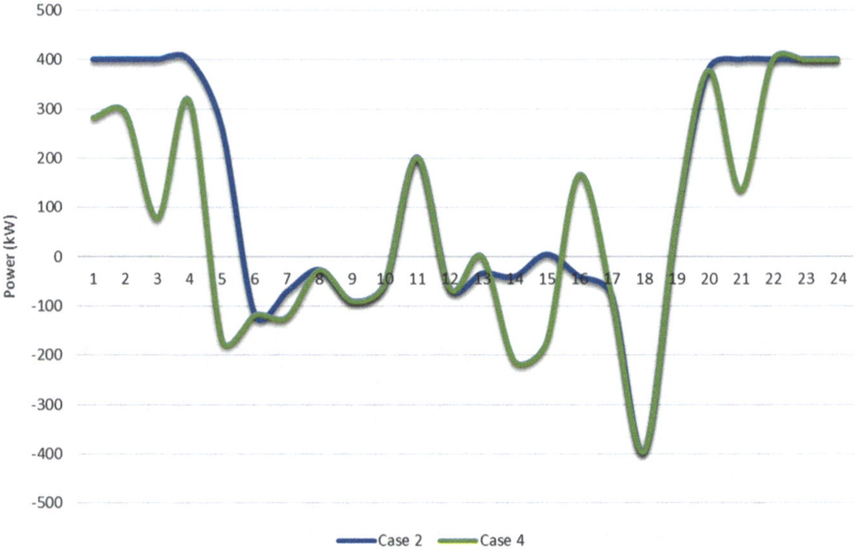

Fig. 7.14 Power exchanged with the network in cases 2 and 4

Another point can be found in comparing the power exchanged with the network in cases 2 and 4. Figure 7.14 shows this exchanged power. In both cases, energy hub participate in DR programs and demand shifts lead to increased purchases from the network at off-peak hours. But, due to lower prices for natural gas and the energy hub tendency to use cheaper energy carriers the purchase of electricity from the network in case 4 is less than case 2.

7.5.5 Numerical Results for Case 5

In this case, we assume that the studied energy hub does not have CHP and supplies its electricity through wind turbine and electricity network. Thermal energy will be supplied through the boiler. In fact, in this case, the integration between different networks through various energy carriers is lost, and shortage of power generation units for supplying different demand must be purchased directly from the relevant network. In the following, we compare this case with previous cases in two high and low prices for natural gas. Figure 7.15 shows how to exchange electrical power with the network in case 5 and in comparison with cases 1 and 2. In case 5, without the implementation of DR program, the highest electricity purchasing coincides with the peak price of the electricity (peak demand in the network) and leads to an increase in the network's peak demand and increased costumer operating costs. However, with the addition of DR capability and demand shifts to off-peak hours, electricity purchases from the network at peak hours are reduced. However, at peak

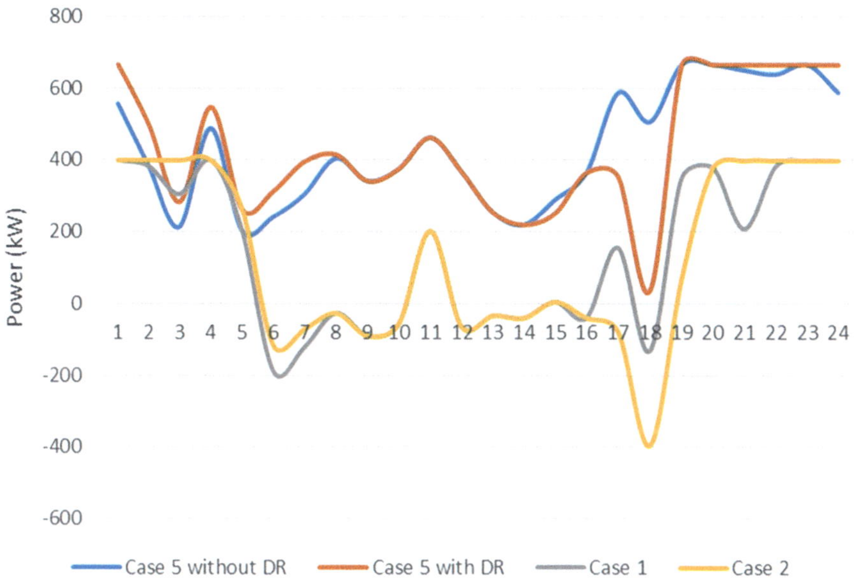

Fig. 7.15 Power exchanged with the network in cases 1, 2, and 5

hours, the system is still dependent on purchasing power from the network, and it pays a lot of money for supplying demand in expensive hours. By adding CHP to the system and the possibility of switching between different energy carriers, energy purchases from the network are significantly reduced. Even at peak hours, energy hub not only does not buy power from the network, but also sells electricity at the highest price. The energy hub participation in DR program in this case will lead to more electricity sales to the grid during high price hours and more purchases in low price hours that reduce the operating costs.

Reducing the dependence on the electricity grid due to the use of CHP leads to an increase in the purchase from the gas network, which can be seen in Fig. 7.16.

As can be seen in the figure, with the addition of CHP, gas purchases from the network are significantly increased. In the early hours of the day witch the price of natural gas is not significantly different from the electricity price, adding CHP has little effect on purchasing gas from the network. But with increasing electricity prices during the day, the desire of the energy hub increases to meet its demand from a cheaper carrier, and power generated by CHP and as a result purchasing gas from the network increases. If the electricity exchange curve for low gas price mode is considered in the form of Fig. 7.17, it can be seen that the electricity purchase from the grid in the case of using CHP is lower than the case of eliminating CHP at all hours of the day.

In general, it can be said that due to the shift of the load to low-cost periods and because of the use of natural gas for electricity production, the consumption curve in this case becomes flattened. Also, despite the increase in gas consumption in peak

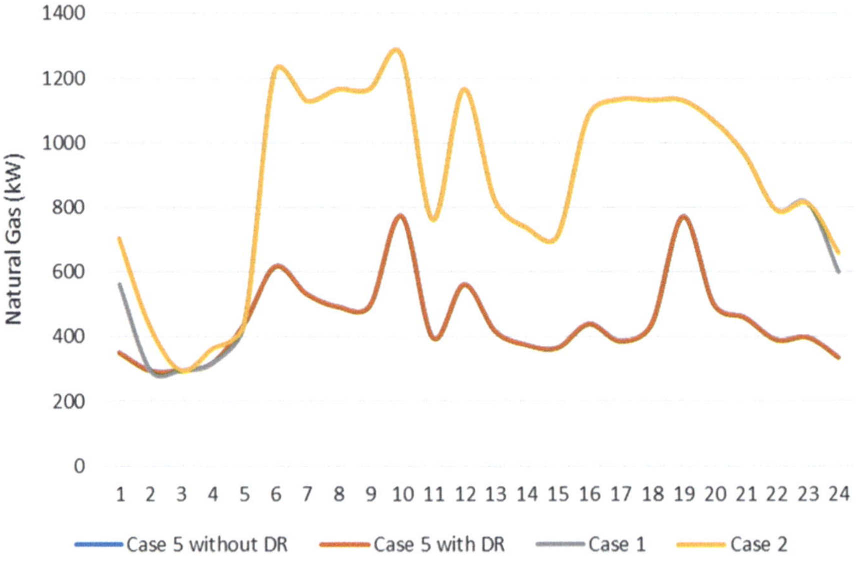

Fig. 7.16 Natural gas purchased from the network in cases 1, 2, and 5

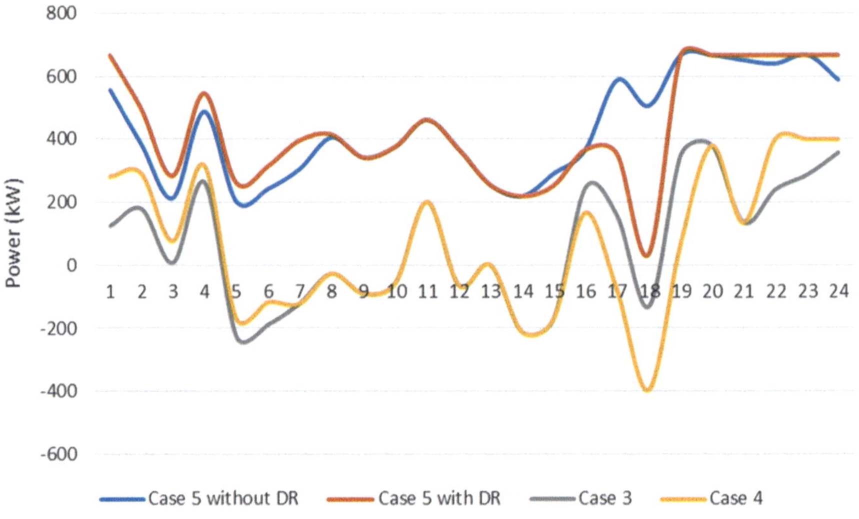

Fig. 7.17 Power exchanged with the network in cases 3, 4, and 5

hours of electricity consumption, along with the reduction of electricity purchase from the grid during these hours, the cost of customers' energy bill decreases. Also on the side of the companies, the daily costs of the electricity grid are reduced due to the flattened curve and lower production and transmission costs, while the profit of the gas company increases due to more natural gas sales. As a result, it can be said that the implementation of DR programs in the content of smart energy hubs leads to lower costs for consumers while maintaining their comfort and increasing utilities' profits, while also reducing peak demand and flattening the shape of the demand curve.

7.6 Conclusion

In this chapter, the effects of participating the renewable energy hubs in DR programs have been evaluated. For this purpose, a brief overview of the research carried out in the field of using DR programs in residential, commercial, industrial, and agricultural energy hubs has been done. In the residential sector, concepts such as smart homes using energy management and control systems as well as the use of DER have led to more responsiveness of these loads and more participation in DSM programs. Most of the research on modeling DR programs in the content of energy hubs has been done in this section. In the commercial sector, with the existence of huge loads such as HVAC and lighting systems, the optimal control of these loads and participation in DSM programs will have a huge impact on energy efficiency, reducing subscriber's energy costs. In the industrial sector, despite the more predictability of the demand for this sector than other sectors, problems such as the difficulty of change in the production plan, the existence of various energy flows and raw materials, the high initial costs for using DER have led to difficulty of implementation of DR programs in this section. In this section, more incentive and supportive policies are needed to increase energy efficiency and DER especially RES implementation and participate in DR programs. In the agricultural sector, the use of smart equipment in managing farms has a significant impact on reducing operational costs and product quality. On the other hand, the existence of agricultural production units such as greenhouses leads to a great potential for energy management and participation in DR programs. All of these require an integrated energy management system that is one of the main functions of the energy hub. Therefore, there is a high potential for using energy hub models in different consumption sectors for energy consumption management and more active participation in DSM and network management programs.

In order to clarify the discussed concepts, a complete simulation of a renewable energy hub is presented in the presence of DR. The results show that in the smart energy hubs that utilize different energy production and storage technologies, the consumer that participates in DR programs has the ability to switch between different energy carriers and technologies along with load shifting and interrupting. Therefore, it has more flexibility to operate in DR programs and it can get more benefits. Even if the customers want to receive the same level of previous service and maintains its level of comfort and simultaneously participates in DR programs, they can use from switching between different energy carriers technologies instead of shifting their demand. As a result, it can be said that the implementation of DR programs in the content of smart energy hubs leads to lower costs for consumers while maintaining their comfort and increasing utilities' profits, while also reducing peak demand and flattening the shape of the demand curve.

References

1. Mohammadi M, Noorollahi Y, Mohammadi-ivatloo B, Yousefi H, Jalilinasrabady S (2017) Optimal scheduling of energy hubs in the presence of uncertainty – a review. J Energy Manag Technol 1(1):1–17. https://doi.org/10.22109/jemt.2017.49432
2. Noorollahi Y, Itoi R, Yousefi H, Mohammadi M, Farhadi A (2017) Modeling for diversifying electricity supply by maximizing renewable energy use in Ebino City southern Japan. Sustain Cities Soc 34:371–384
3. Boshell F, Veloza O (2008) Review of developed demand side management programs including different concepts and their results. In: Transmission and distribution conference and exposition: Latin America, 2008 IEEE/PES. IEEE, pp 1–7
4. Kazemi M, Mohammadi-Ivatloo B, Ehsan M (2014) Risk-based bidding of large electric utilities using information gap decision theory considering demand response. Electr Power Syst Res 114:86–92
5. Nojavan S, Zare K, Mohammadi-Ivatloo B (2017) Optimal stochastic energy management of retailer based on selling price determination under smart grid environment in the presence of demand response program. Appl Energy 187:449–464
6. Rabiee A, Soroudi A, Mohammadi-ivatloo B, Parniani M (2014) Corrective voltage control scheme considering demand response and stochastic wind power. IEEE Trans Power Syst 29(6):2965–2973. https://doi.org/10.1109/tpwrs.2014.2316018
7. Vahid-Pakdel MJ, Nojavan S, Mohammadi-ivatloo B, Zare K (2017) Stochastic optimization of energy hub operation with consideration of thermal energy market and demand response. Energy Convers Manag 145:117–128
8. U.S. Energy Information Administration (2015) http://www.eia.gov/forecasts/aeo/er/executive_summary. Accessed 02 Feb 2016
9. Beaudin M, Zareipour H (2015) Home energy management systems: a review of modelling and complexity. Renew Sust Energ Rev 45:318–335
10. Rastegar M, Fotuhi-Firuzabad M, Lehtonen M (2015) Home load management in a residential energy hub. Electr Power Syst Res 119:322–328
11. Jabari F, Nojavan S, Mohammadi-Ivatloo B, Sharifian MB (2016) Optimal short-term scheduling of a novel tri-generation system in the presence of demand response programs and battery storage system. Energy Convers Manag 122:95–108
12. Rastegar M, Fotuhi-Firuzabad M (2015) Load management in a residential energy hub with renewable distributed energy resources. Energy Build 107:234–242
13. Brahman F, Honarmand M, Jadid S (2015) Optimal electrical and thermal energy management of a residential energy hub, integrating demand response and energy storage system. Energy Build 90:65–75
14. Bozchalui MC, Hashmi SA, Hassen H, Cañizares CA, Bhattacharya K (2012) Optimal operation of residential energy hubs in smart grids. IEEE Trans Smart Grid 3(4):1755–1766
15. Mohammadi M, Noorollahi Y, Mohammadi-ivatloo B, Yousefi H (2017) Energy hub: from a model to a concept – a review. Renew Sustain Energy Rev 80:1512–1527
16. United Nations Environment Programme (2016) http://www.unep.org/sbci/AboutSBCI/Background.asp. Accessed 02 Apr 2016
17. Steinfeld J, Bruce A, Watt M (2011) Peak load characteristics of Sydney office buildings and policy recommendations for peak load reduction. Energy Build 43(9):2179–2187

18. Bozchalui MC, Sharma R (2012) Optimal operation of commercial building microgrids using multi-objective optimization to achieve emissions and efficiency targets. In: Power and energy society general meeting, 2012 IEEE. IEEE, pp 1–8
19. U.S. Energy Information Administration (2014) http://www.eia.gov/forecasts/ieo/. Accessed 02 Jan 2016
20. Finn P, Fitzpatrick C (2014) Demand side management of industrial electricity consumption: promoting the use of renewable energy through real-time pricing. Appl Energy 113:11–21
21. Alipour M, Zare K, Mohammadi-Ivatloo B (2014) Short-term scheduling of combined heat and power generation units in the presence of demand response programs. Energy 71:289–301
22. Alipour M, Mohammadi-Ivatloo B, Zare K (2014) Stochastic risk-constrained short-term scheduling of industrial cogeneration systems in the presence of demand response programs. Appl Energy 136:393–404
23. Alipour M, Zare K, Mohammadi-Ivatloo B (2016) Optimal risk-constrained participation of industrial cogeneration systems in the day-ahead energy markets. Renew Sust Energ Rev 60:421–432
24. Alipour M, Mohammadi-Ivatloo B, Zare K (2015) Stochastic scheduling of renewable and CHP-based microgrids. IEEE Trans Ind Inf 11(5):1049–1058
25. Ding YM, Hong SH, Li XH (2014) A demand response energy management scheme for industrial facilities in smart grid. IEEE Trans Ind Inf 10(4):2257–2269
26. Xu FY, Lai LL (2015) Novel active time-based demand response for industrial consumers in smart grid. IEEE Trans Ind Inf 11(6):1564–1573
27. Ball VE, Färe R, Grosskopf S, Margaritis D (2015) The role of energy productivity in US agriculture. Energy Econ 49:460–471
28. Abbasi AZ, Islam N, Shaikh ZA (2014) A review of wireless sensors and networks' applications in agriculture. Comput Stand Interfaces 36(2):263–270
29. Gelogo YE, Park J, Kim H-K (2015) A study on U-agriculture for smart grid systems deployment. Adv Sci Tech Lett 97:64–70 http://dx.doi.org/10.14257/astl.205.97.11
30. Odara S, Khan Z (2015) Ustun TS integration of precision agriculture and smartgrid technologies for sustainable development. In: Technological innovation in ICT for agriculture and rural development (TIAR), 2015 IEEE. IEEE, pp 84–89
31. Kaviani AK, Riahy G, Kouhsari SM (2009) Optimal design of a reliable hydrogen-based stand-alone wind/PV generating system, considering component outages. Renew Energy 34(11):2380–2390
32. Haghifam MR, Manbachi M (2011) Reliability and availability modelling of combined heat and power (CHP) systems. Int J Electr Power Energy Syst 33(3):385–393. https://doi.org/10.1016/j.ijepes.2010.08.035
33. Parisio A, Del Vecchio C, Vaccaro A (2012) A robust optimization approach to energy hub management. Int J Electr Power Energy Syst 42(1):98–104
34. Dietrich K, Latorre JM, Olmos L, Ramos A (2012) Demand response in an isolated system with high wind integration. IEEE Trans Power Syst 27(1):20–29
35. Gen B (2005) Reliability and cost/worth evaluation of generating systems utilizing wind and solar energy (Ph.D. Thesis). Department of Electrical Engineering, University of Saskatchewan, Saskatoon

Chapter 8
Supply Side Management in Renewable Energy Hubs

Sayyad Nojavan, Majid Majidi, Afshin Najafi-Ghalelou, and Kazem Zare

8.1 Introduction

Energy as the most vital issue in the current century can be discussed from various viewpoints like efficiency, economy, reliability, etc. As an appropriate option for such mentioned goals, hub energy system can be used in power systems. Hub energy systems including integrated renewable [1, 2] and non-renewable generation units [3–5] can be employed to efficiently supply energy demands [6, 7] along with satisfying economic and environmental goals [6, 8].

8.1.1 Literature Review

Previously, hub energy systems have been studied and their summaries are briefly presented in the following:

In order to solve power flow problem of hub energy system in [9], heuristic based optimization algorithm called time varying acceleration coefficient-gravitational search algorithm is employed. With the aim of minimizing total operation cost of hub energy system, robust based optimization approach is used in [10]. Using energy hub concept, steady-states in microgrids have been studied in [11]. Multi-carrier energy system has been optimally planned and scheduled in the presence of renewable generation units in [12]. With the aim of improving energy efficiency, energy hub concept including various local generation units has been implemented

S. Nojavan (✉) · M. Majidi · A. Najafi-Ghalelou · K. Zare
Faculty of Electrical and Computer Engineering, University of Tabriz, Tabriz, Iran
e-mail: sayyad.nojavan@tabrizu.ac.ir; majidmajidi95@ms.tabrizu.ac.ir; afshin.najafi95@ms.tabrizu.ac.ir; kazem.zare@tabrizu.ac.ir

B. Mohammadi-Ivatloo, F. Jabari (eds.), *Operation, Planning, and Analysis of Energy Storage Systems in Smart Energy Hubs*,
https://doi.org/10.1007/978-3-319-75097-2_8

in [13]. Optimal economic operation of energy hub system has been evaluated in [14]. Energy hub concept has been implemented in [15] to create a decentralized and integrated energy system in neighborhoods. Using stochastic programming in [16], optimum performance of energy hub system under uncertainties has been evaluated. Economic dispatch problem of multi-carrier energy system has been studied using coefficient-gravitational search algorithm in [17]. Energy hub system has been investigated form viewpoint of reliability in [18, 19]. Techniques used for analyzing hub energy systems have been reviewed in [20]. Optimum operation of energy hub system has been evaluated with respect to energy balance limitation in [21]. Similar problem has been studied considering dynamic and time-of-use pricing in [22]. Optimum performance of multi-carrier energy system has been evaluated in the presence of demand response and thermal storage in [23]. Optimum operation of energy hub system embedded in a smart home has been investigated in [24, 25]. Optimum impact of heating networks on the operation of energy hub system has been investigated in [26].

Supply side management tools have been also interesting topics for various researchers. Different options can be employed as the supply side management tool in generation systems and one of them is energy storage system. Energy storage systems are various. From viewpoint of discharging time, storage systems are categorized into two groups: storage systems with short discharging time up to a few hours like batteries, flywheels, super magnetic energy storage, and super capacitors and the second group is the systems with long discharging time up to a day like compressed air energy storage system (CAES) and pumped hydro storage system. So, it can be concluded that CAES and pumped hydro are the only available mature energy storage systems with large-scale storage capacity. Using energy in off-peak time periods, CAES compresses air and later in peak time periods, stored compressed air is used to produce electric power. It should be mentioned that due to large scale size of storage capacity in CAES, this storage system is a suitable option for economic goals [27, 28].

In this chapter, a multi-objective model has been proposed for eco-emission operation of renewable-based hub energy system in the presence of CAES and DRP. Compressed air energy storage system (CAES) has been used as a supply side management tool to handle severe uncertainties created by renewable generation units in the hub energy system. In addition to CAES, demand response program has been employed to improve economic and environmental operation of renewable-based hub energy system.

8.1.2 Novelty and Contributions of This Research

Summarizing mentioned explanations above, novelty and contributions of this chapter can be expressed as follows:

- Optimum eco-emission operation of renewable-based hub energy system.
- Implementation of CAES as a supply side management for further improvement of economic and environmental performance of renewable-based hub energy system.
- Implementation of DRP for total cost and emission reduction of renewable-based hub energy system.

8.2 Problem Formulation

In this section, eco-emission performance of hub renewable-based energy system has been investigated in which compressed air storage system has been employed as a supply side management tool to handle uncertainties of renewable units. Proposed optimum eco-emission performance of renewable-based energy hub system in the presence of compressed air energy storage system and demand response program has been mathematically investigated in the following sections.

8.2.1 Objective Functions

In the proposed scheme, there are two confliction objective functions to be minimized which are total operation cost and emission of renewable-based hub energy system. Total operation cost of renewable-based energy hub system in the presence of DRP and CAES is presented through Eqs. (8.1)–(8.10).

$$\begin{aligned}\text{Min } \Phi_1 = \text{Total cost} = \; & \mathrm{C_{net}} + \mathrm{C_{Wind}} + \mathrm{C_{BS}} + \mathrm{C_{DR}} + \mathrm{C_{Ex}} + \mathrm{C_{TS}} + \mathrm{C_{Bo}} \\ & + \mathrm{C_{CHP}} + \mathrm{C_{Wa}} \end{aligned} \tag{8.1}$$

$$\mathrm{C_{net}} = \sum_{t=1}^{H} \left(\lambda_t^e \times p_t^e\right) \tag{8.2}$$

$$\mathrm{C_{Wind}} = \sum_{t=1}^{H} \left(\lambda^{\mathrm{wi}} \times p_t^{\mathrm{wi}}\right) \tag{8.3}$$

$$\mathrm{C_{BS}} = \sum_{t=1}^{H} \left(\lambda_s^b \times \left(p_t^{\mathrm{ch,BS}} + p_t^{\mathrm{dis,BS}}\right)\right) \tag{8.4}$$

$$C_{TS} = \sum_{t=1}^{H} \left(\lambda_s^h \times \left(p_t^{ch,h} + p_t^{dis,h}\right)\right) \tag{8.5}$$

$$C_{DR} = \sum_{t=1}^{H} \left(\lambda^{DR} \times \left(p_t^{e,shdo} + p_t^{e,shup}\right)\right) \tag{8.6}$$

$$C_{CHP} = \sum_{t=1}^{H} \left(\lambda^g \times g_t^{CHP}\right) \tag{8.7}$$

$$C_{Bo} = \sum_{t=1}^{H} \left(\lambda^g \times g_t^B\right) \tag{8.8}$$

$$C_{Wa} = \sum_{t=1}^{H} \left(\lambda^{wa} \times wa_t\right) \tag{8.9}$$

$$C_{Ex} = \sum_{t=1}^{H} \left(\lambda_t^e \times \left(p_t^{ch,e} - p_t^{dis,e}\right)\right) \tag{8.10}$$

Cost of purchased power from upstream network (8.2) plus the cost of wind-turbine generation (8.3) plus the operation cost of battery and thermal storage systems including charge/discharge costs (8.4) and (8.5) plus the cost of DRP (8.6) plus the operation cost of CHP and boiler (8.7) and (8.8) plus the cost of purchased water (8.9) and cost/revenue of exchanged power (8.10) result the total operation cost of renewable-based hub energy system to be minimized (8.1).

Due to utilization of CHP system and boiler in hub energy system as well as due to gas consumption in residential section and also because of burning fossil fuels in power plants which power is later transferred to the hub system, this system emits three types of pollutants, namely CO_2, SO_2, and NO_x. In order to satisfy environmental concerns, these emissions should be minimized. The objective function related to environmental operation of renewable-based energy hub system is presented in detail through Eqs. (8.11)–(8.15).

$$\text{Min } \Phi_2 = \text{Em} = \left(\text{Em}^{CHP} + \text{Em}^B + \text{Em}^L + \text{Em}^{NET}\right) \tag{8.11}$$

$$\text{Em}^{CHP} = \left(\text{EF}_{CO}^{CHP} \times g_t^{CHP}\right) + \left(\text{EF}_{SO}^{CHP} \times g_t^{CHP}\right) + \left(\text{EF}_{NO}^{CHP} \times g_t^{CHP}\right) \tag{8.12}$$

$$\text{Em}^B = \left(\text{EF}_{CO}^B \times g_t^B\right) + \left(\text{EF}_{SO}^B \times g_t^B\right) + \left(\text{EF}_{NO}^B \times g_t^B\right) \tag{8.13}$$

$$\mathrm{Em}^{L} = \left(\mathrm{EF}_{\mathrm{CO}}^{L} \times g_t^{L}\right) + \left(\mathrm{EF}_{\mathrm{SO}}^{L} \times g_t^{L}\right) + \left(\mathrm{EF}_{\mathrm{NO}}^{L} \times g_t^{L}\right) \tag{8.14}$$

$$\mathrm{Em}^{\mathrm{NET}} = \left(\mathrm{EF}_{\mathrm{CO}}^{\mathrm{NET}} \times p_t^{e}\right) + \left(\mathrm{EF}_{\mathrm{SO}}^{\mathrm{NET}} \times p_t^{e}\right) + \left(\mathrm{EF}_{\mathrm{NO}}^{\mathrm{NET}} \times p_t^{e}\right) \tag{8.15}$$

8.2.2 Electrical Model

One of the energy demands due to be supplied by renewable-based hub energy system is electrical demand. Electrical demand which is capable of participating in DRP should be satisfied through generation of wind-turbine, CHP system, purchased power from upstream network and discharged power of battery and compressed air energy storage systems (8.16).

$$\left(p_t^{l} + p_t^{\mathrm{shup}} - p_t^{\mathrm{shdo}} + P_t^{c,p}\right) = \left(A^{\mathrm{NET}} \times \eta_{\mathrm{ee}}^{T} \times p_t^{e}\right) + \left(A^{\mathrm{WIND}} \times \eta_{\mathrm{ee}}^{\mathrm{CON}} \times p_t^{\mathrm{wi}}\right) + \left(A^{\mathrm{CHP}} \times \eta_{\mathrm{ge}}^{\mathrm{CHP}} \times g_t^{\mathrm{CHP}}\right) + \left(p_t^{\mathrm{dis,BS}} - p_t^{\mathrm{ch,BS}}\right) + \left(P_t^{c,s}\right) \tag{8.16}$$

8.2.2.1 Model of Upstream Network

Each transmission system has its own components with their specific technical characteristics for power transmission which should be taken into account. Imported power from upstream network should be within the nominal capacity of transformer connecting renewable-based hub system to the upstream network (8.17).

$$\eta_{\mathrm{ee}}^{T} \times p_t^{e} \le p_c^{T} \tag{8.17}$$

8.2.2.2 Model of Renewable Energy Sources

In order to generate clean energy and reduce total cost and emission of energy hub system, wind-turbine has been used as a renewable generation unit to satisfy both economic and environmental objectives. Generation pattern according to which wind-turbine produces electrical power is presented in Eq. (8.18)

$$p_t^{\mathrm{wi}} = \begin{cases} 0 & w < w_{\mathrm{ci}} \\ p_r\left(z - y \cdot w(t) + x \cdot w^2(t)\right) & w_{\mathrm{ci}} \le w < w_r \\ p_r & w_r \le w < w_{\mathrm{co}} \\ 0 & w \ge w_{\mathrm{co}} \end{cases} \tag{8.18}$$

8.2.2.3 Model of Battery Storage

Two types of electrical storage systems with specific characteristics have been employed in renewable-based energy hub system to handle uncertainties of generation. In this section, limitations of battery storage system have been presented through (8.19)–(8.24).

$$C_t^{\text{st},e} = C_{t-1}^{\text{st},e} + p_t^{\text{ch},e} \times \eta_{\text{ch}}^e - p_t^{\text{dis},e} / \eta_{\text{dis}}^e - p_t^{\text{loss},e} \tag{8.19}$$

$$\alpha_{\min}^e \times C_c^{\text{st},e} \leq C_t^{\text{st},e} \leq \alpha_{\max}^e \times C_c^{\text{st},e} \tag{8.20}$$

$$\frac{\alpha_{\min}^e \times C_c^{\text{st},e} \times I_t^{\text{ch},e}}{\eta_{\text{ch}}^e} \leq p_t^{\text{ch},e} \leq \frac{\alpha_{\max}^e \times C_c^{\text{st},e} \times I_t^{\text{ch},e}}{\eta_{\text{ch}}^e} \tag{8.21}$$

$$\alpha_{\min}^e \times C_c^{\text{st},e} \times I_t^{\text{dis},e} \times \eta_{\text{dis}}^e \leq p_t^{\text{dis},e} \leq \alpha_{\max}^e \times C_c^{\text{st},e} \times I_t^{\text{dis},e} \times \eta_{\text{dis}}^e \tag{8.22}$$

$$p_t^{\text{loss},e} = \alpha_{\text{loss}}^e \times C_t^{\text{st},e} \tag{8.23}$$

$$I_t^{\text{ch},e} + I_t^{\text{dis},e} \leq 1 \tag{8.24}$$

Available stored energy level of battery storage is presented by (8.19). Limitations of available energy, charge and discharge power of battery storage are presented through Eqs. (8.20), (8.21), and (8.22), respectively. Battery storage energy loss is presented by Eq. (8.23). Finally, simultaneous charge and discharge of battery is restricted by Eq. (8.24).

8.2.3 Thermal Model

Heating is another type of energy demands due to be supplied by renewable-based energy hub system. Using generated heat by boiler and CHP system as well as released heat from thermal storage system, heating demand is satisfied (8.25).

$$p_t^h = \left[\eta_{\text{gh}}^B \times g_t^B\right] + \left[A^{\text{CHP}} \times \eta_{\text{gh}}^{\text{CHP}} \times g_t^{\text{CHP}}\right] + \left(p_t^{\text{dis},h} - p_t^{\text{ch},h}\right) \tag{8.25}$$

8.2.3.1 Model of Thermal Storage

Besides battery and compressed air energy storage systems which have been embedded in electrical section, thermal energy storage system has been used in thermal section to handle excess generated heat by CHP system and boiler. Limitations of employed thermal storage are presented through Eqs. (8.26)–(8.31).

$$C_t^{\mathrm{st},h} = C_{t-1}^{\mathrm{st},h} + p_t^{\mathrm{ch},h} \times \eta_{\mathrm{ch}}^h - p_t^{\mathrm{dis},h} / \eta_{\mathrm{dis}}^h - p_t^{\mathrm{loss},h} \tag{8.26}$$

$$\alpha_{\min}^h \times C_c^{\mathrm{st},h} \leq C_t^{\mathrm{st},h} \leq \alpha_{\max}^h \times C_c^{\mathrm{st},h} \tag{8.27}$$

$$\frac{\alpha_{\min}^h \times C_c^{\mathrm{st},h} \times I_t^{\mathrm{ch},h}}{\eta_{\mathrm{ch}}^h} \leq p_t^{\mathrm{ch},h} \leq \frac{\alpha_{\max}^h \times C_c^{\mathrm{st},h} \times I_t^{\mathrm{ch},h}}{\eta_{\mathrm{ch}}^h} \tag{8.28}$$

$$\alpha_{\min}^h \times C_c^{\mathrm{st},h} \times I_t^{\mathrm{dis},h} \times \eta_{\mathrm{dis}}^h \leq p_t^{\mathrm{dis},h} \leq \alpha_{\max}^h \times C_c^{\mathrm{st},h} \times I_t^{\mathrm{dis},h} \times \eta_{\mathrm{dis}}^h \tag{8.29}$$

$$p_t^{\mathrm{loss},h} = \alpha_{\mathrm{loss}}^h \times C_t^{\mathrm{st},h} \tag{8.30}$$

$$I_t^{\mathrm{ch},h} + I_t^{\mathrm{dis},h} \leq 1 \tag{8.31}$$

Stored energy level of thermal energy storage system is expressed by Eq. (8.26). Limitation of available heat and input as well as released heat of thermal storage system is expressed through Eqs. (8.27)–(8.29). Loss of heat inside the thermal energy storage system is expressed by Eq. (8.30). Heat injection and discharge to/from thermal energy storage system cannot occur at the same time which is expressed by (8.31).

8.2.3.2 Model of Gas network

The need for gas in CHP and boiler plus the gas demand in consumption side necessitates gas import from gas network. Imported gas is divided into three parts for various applications mentioned above (8.32). It should be that imported gas should be within the nominal capacity which has been set for gas network (8.33).

$$g_t^{\mathrm{net}} = g_t^B + g_t^{\mathrm{CHP}} + g_t^l \tag{8.31}$$

$$g_{\min}^{\mathrm{net}} \leq g_t^{\mathrm{net}} \leq g_{\max}^{\mathrm{net}} \tag{8.32}$$

8.2.3.3 Model of CHP system

As a common rule in each generation unit, total produced energy by each generating system should be under the systems nominal capacity. According to this definition, total generated electric power by CHP system should be less than its nominal capacity (8.33). It should be noted that since heat generation of CHP system is a function of its electrical generation, therefore by satisfying Eq. (8.33), heat generation by CHP system will be kept under nominal heat generation capacity of CHP.

$$\eta_{\text{ge}}^{\text{CHP}} \times g_t^{\text{CHP}} \leq p_c^{\text{CHP}} \tag{8.33}$$

8.2.3.4 Model of Boiler

Boiler is the only energy resource in thermal section which responsibility is only heat generation. Produced heat by this unit is constrained through Eq. (8.34).

$$\eta_{\text{gh}}^{B} \times g_t^{B} \leq p_c^{B} \tag{8.34}$$

8.2.4 Compressed Air Energy Storage System Model

In this section, model of employed CAES is presented. The air injected to the CAES is presented by Eq. (8.35). Generated electric power by CAES is presented by (8.36). Stored air in the CAES which is later pumped to the combustion chamber is mathematically modeled by (8.37) and (8.38), respectively. In order to limit operation mode of CAES which is either pumping mode or storage mode, Eq. (8.39) is employed. Available air in the CEAS is expressed and limited by Eqs. (8.40) and (8.41), respectively.

$$V_t^{\text{inj}} = \kappa^{\text{inj}} \times P_t^{c,p} \tag{8.35}$$

$$P_t^{c,s} = \kappa^{p} \times V_t^{p} \tag{8.36}$$

$$V_{\min}^{\text{inj}} \times u^{\text{inj}} \leq V_t^{\text{inj}} \leq V_{\max}^{\text{inj}} \times u^{\text{inj}} \tag{8.37}$$

$$V_{\min}^{p} \times u^{p} \leq V_t^{p} \leq V_{\max}^{p} \times u^{p} \tag{8.38}$$

$$u^{\text{inj}} + u^{p} \leq 1 \tag{8.39}$$

$$A_t = A_{t-1} + V_t^{\text{inj}} - V_t^p \tag{8.40}$$

$$A_t^{\min} \leq A_t \leq A_t^{\max} \tag{8.41}$$

8.2.5 Demand Response Program

As n new concept in energy markets, electrical consumers can participate in demand response programs to reduce their costs. By participating in these programs, consumers undertake to shift their energy demand from peak time periods to off-peak time periods. One of the common programs included in DRP is time-of-use rates (TOU) of DRP has been implemented [5, 29–31]. According to TOU, new electrical load is equal to the primary load plus the variable load. These variables can be either positive or negative meaning decrease or increase of load. The amount of increase or decrease of load which is percentage of load participation in DRP should be under a predefined limitation. Also, simultaneous increase and decrease of load is not allowed.

Summary of explanations given above is mathematically presented through Eqs. (8.42)–(8.45).

$$p_t^{\text{el,DRP}} = p_t^{\text{el}} + p_t^{\text{shup},e} - p_t^{\text{shdo},e} \tag{8.42}$$

$$0 \leq p_t^{\text{shup},e} \leq \text{LPF}^{\text{shup},e} \times p_t^l \times I_t^{\text{shup},e} \tag{8.43}$$

$$0 \leq p_t^{\text{shdo},e} \leq \text{LPF}^{\text{shdo},e} \times p_t^l \times I_t^{\text{shdo},e} \tag{8.44}$$

$$I_t^{\text{shup},e} + I_t^{\text{shdo},e} \leq 1 \tag{8.45}$$

8.2.6 Model of Water Network

As the last type of energy demand, water consumer in demand side is provided through the imported water from water network which is expressed by Eq. (8.46) and limited by Eq. (8.47).

$$\text{wa}_t^l = \text{wa}_t^{\text{net}} \tag{8.46}$$

$$\text{wa}_{\min} \leq \text{wa}_t^{\text{net}} \leq \text{wa}_{\max} \tag{8.47}$$

8.3 Case Study

8.3.1 Input Data

Studied renewable energy hub system is composed of wind generation unit, combined heat and power system, boiler and various types of energy storage systems. Schematic diagram of mentioned system is shown in Fig. 8.1.

As illustrated in this figure, four types of energy demands should be supplied by multi-carrier renewable-based energy hub system. Three types of energy storage systems, namely, batter storage system, compressed air energy storage system, and thermal storage systems have been employed to manage excess generated energy in the hub energy system. It should be noted that since battery storage system is not

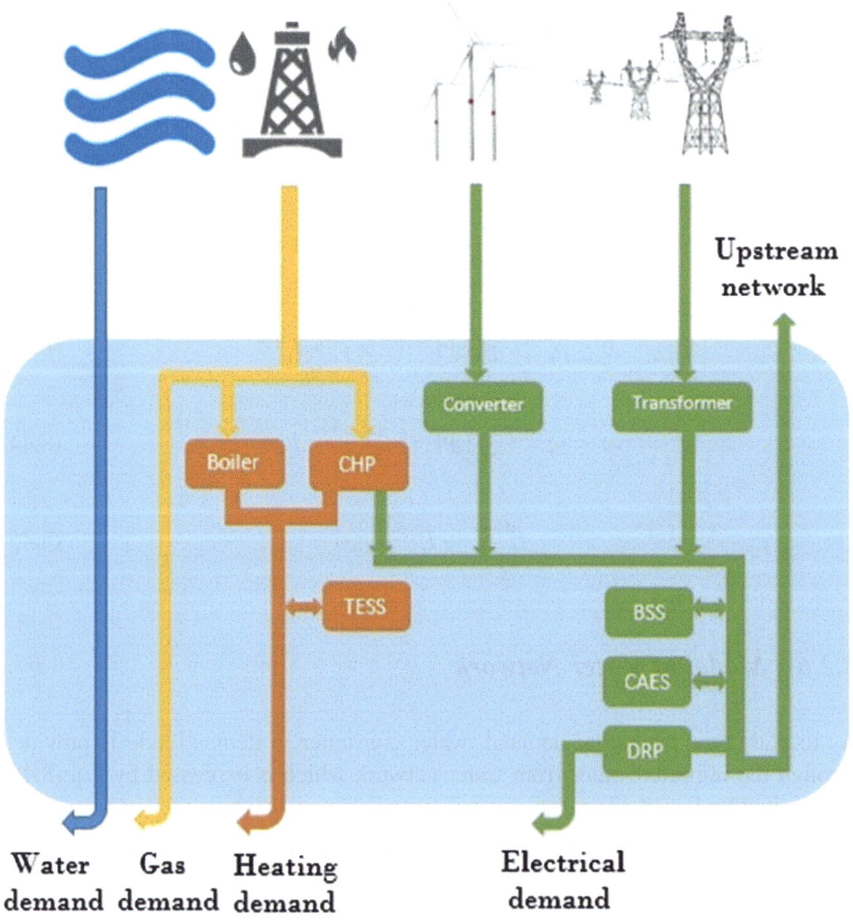

Fig. 8.1 Renewable-based hub energy system

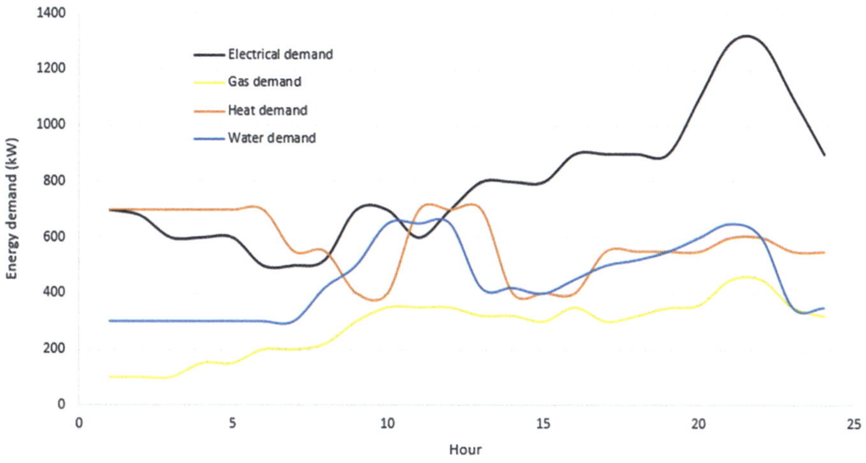

Fig. 8.2 Energy demands of renewable-based hub energy system

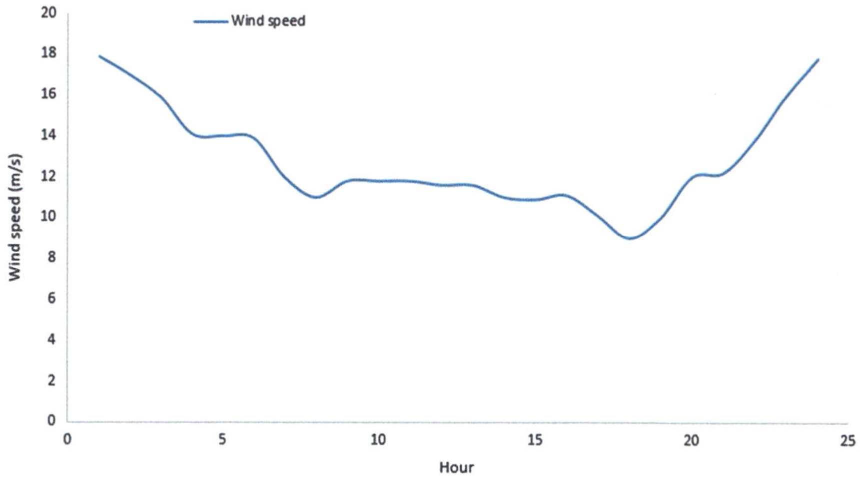

Fig. 8.3 Wind hourly speed

able to manage large quantities of uncertainty provided by wind generation and also because of that compressed air energy storage system is approximately operated at low operation cost, CAES has been used to control possible severe uncertainties caused by wind generation.

In order to model and simulate eco-emission operation of renewable-based hub energy system in the presence of CAES and DRP, the following data and info are used.

All four types of energy demands to be supplied by renewable-based hub energy system are illustrated in Fig. 8.2.

Wind speed is illustrated in Fig. 8.3.

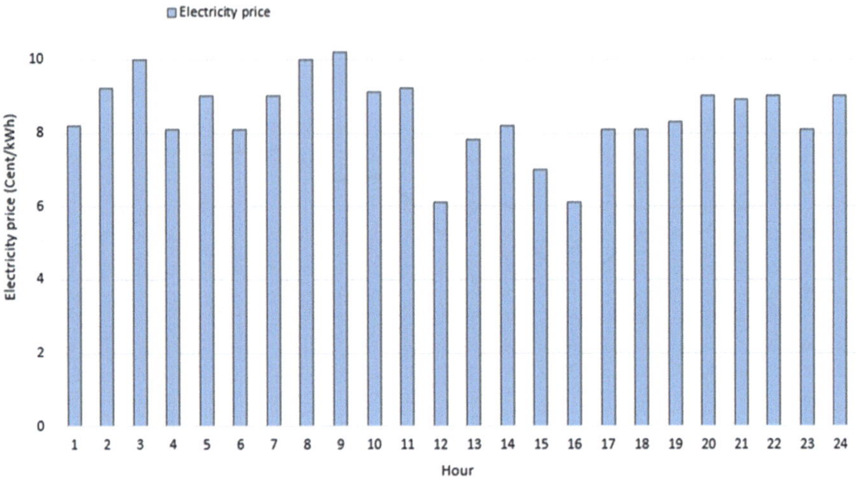

Fig. 8.4 Upper grid power price

Table 8.1 Generation unit's info

#	Unit	Value	#	Unit	Value
CHP parameter [32]			Boiler parameter [32]		
η_{ge}^{CHP}	%	40	η_{gh}^{B}	%	85
η_{gh}^{CHP}	%	35	P_{c}^{B}	kW	800
A^{CHP}	–	0.96	–	–	–
P_{c}^{CHP}	kW	800	–	–	–
Boiler emission [33]			CHP emission [33]		
#	Unit	Value	#	Unit	Value
#	Unit	Value	#	Unit	Value
EF_{CO}^{B}	kg/kWh	0.37	EF_{CO}^{CHP}	kg/kWh	0.37
EF_{SO}^{B}	kg/kWh	0.000003	EF_{SO}^{CHP}	kg/kWh	0.000003

Price of power provided by upper network is illustrated in Fig. 8.4.

Simulation data and info about storage systems are presented in Tables 8.1, 8.2, 8.3, 8.4, and 8.5.

Also, technical and environmental info of CHP system and boiler are presented in Table 8.1.

Simulation data and info about storage systems are presented in Table 8.2.

Operation cost and prices of various generation units and other sections are presented in Table 8.3.

Technical and environmental info of upper grid is presented in Table 8.4.

Finally, parameters necessary for modeling wind generation are presented in Table 8.5.

It should be mentioned that maximum capacity of gas and water networks are considered to be 1800 kW and 1000 kW, respectively. The whole simulations are carried out by CPLEX solver of GAMS under a mixed-integer linear programming [35].

Table 8.2 Storage systems data

Battery storage parameter [32]			CAES parameter [27]			Thermal storage parameter [32]		
#	Unit	Value	#	Unit	Value	#	Unit	Value
$\alpha^e_{\min}$	–	0.05	κ^{inj}	%	0.95	$\alpha^h_{\min}$	–	0.05
$\alpha^e_{\max}$	–	0.9	κ^p	%	0.95	$\alpha^h_{\max}$	–	0.9
α^e_{loss}	–	0.2	$V^{\text{inj}}_{\min}$	kWh	5	α^h_{loss}	–	0.2
η^e_{ch}	%	90	$V^{\text{inj}}_{\max}$	kWh	50	η^h_{ch}	%	90
η^e_{dis}	%	90	$V^p_{\min}$	kWh	5	η^h_{dis}	%	90
$C^{\text{st},e}_c$	kW	300	$V^p_{\max}$	kWh	50	$C^{\text{st},h}_c$	kW	200
–	–	–	$A^{\min}$	kWh	50	–	–	–
–	–	–	$A^{\max}$	kWh	500	–	–	–

Table 8.3 Operation costs and prices of various sections

Parameter [32]	Value	Unit
λ^g	6	Cent/kWh
λ^{wa}	4	Cent/kWh
λ^{wi}	0	Cent/kWh
λ^e_s	2	Cent/kWh
λ^h_s	2	Cent/kWh
λ^{DR}	2	Cent/kWh

Table 8.4 Upper grid info

Upstream network parameter [32]			Upstream network emission [34]		
#	Unit	Value	#	Unit	Value
A^{NET}	–	0.99	$\text{EF}^{\text{Net}}_{\text{CO}}$	kg/kWh	0.368
$p^e_{\max}$	kW	1000	$\text{EF}^{\text{Net}}_{\text{SO}}$	kg/kWh	0.0002
$p^e_{\min}$	kW	0	$\text{EF}^{\text{Net}}_{\text{NO}}$	kg/kWh	0.0008
p^T_c	kW	800	–	–	–

Table 8.5 Wind generation info

Parameters	Unit	Value
A^{WIND}	–	0.96
x,y,z	–	0.07, 0.01, 0.03
w_r	m/s	10
w_{ci}	m/s	4
w_{co}	m/s	22
p_r	kW	400

In order to evaluate effectiveness of CAES as the supply side management tool and also to investigate positive impacts of DRP, 4 simulation cases have been created as follows:

Case 1: Eco-environmental operation of renewable-based hub energy system without DRP and without CAES

Case 2: Eco-environmental operation of renewable-based hub energy system with DRP and without CAES
Case 3: Eco-environmental operation of renewable-based hub energy system without DRP and with CAES
Case 4: Eco-environmental operation of renewable-based hub energy system with DRP and with CAES

8.3.2 *Results*

Simulation results are presented in this section to validate effectiveness of employed techniques.

8.3.2.1 Pareto Fronts

Solving proposed eco-emission model for renewable-based hub energy system in the presence of CAES and DRP, Pareto solutions in four cases are obtained which are illustrated in Fig. 8.5.

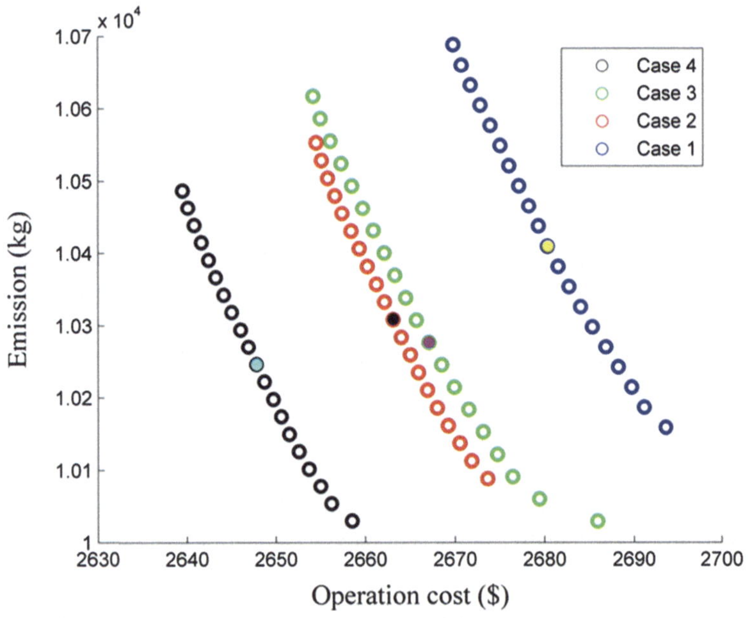

Fig. 8.5 Pareto front in four cases

It is clear from the above figure that by using CAES and DRP, Pareto front is shifted from areas with higher emission and cost to the areas with less emission and operation cost.

According to the selected solutions in each case study, total operation cost and emission of renewable-based hub energy system in case 1 are 2680.34$ and 10,410.04 kg, respectively. These values in case 2 with DRP are 2663.08$ and 10,308.98 kg, respectively. Total operation cost and emission of hub system in the presence of CAES in case 3 are 2667.09$ and 10,277.38 kg, respectively. By employing both CAES and DRP in case 4, total operation cost and emission of renewable-based hub energy system are 2647.81$ and 10,246.38 kg, respectively. Comparing the obtained results, it can be found that due to implementation of CAES and DRP in case 4, total operation cost of renewable-based hub energy system in comparison with case 1, 2, and 3 is decreased 1.24%, 0.57%, and 0.72%, respectively. Also, total generated emission of renewable-based hub energy system in case 4 is reduced 1.57%, 0.60%, and 0.30% in comparison with cases 1, 2, and 3, respectively. It can be understood from the obtained results above that both economic and environmental concerns of renewable-based hub energy system can be satisfied through utilization of CAES and DRP.

8.3.2.2 Other Results

Electrical energy demand in four cases has been illustrated in Fig. 8.6. It can be understood that in the cases 2 and 4, because of DRP implementation, electrical demand has been mostly transferred from peak periods to off-peak periods which leads to less energy procurement in peak periods and therefore more economic benefits for renewable-based hub energy system can be obtained.

As a result of DRP implementation in cases 2 and 4, total provided power by upper grid in these cases has been shifted to off-peak periods which is expressed by Fig. 8.7. Also, due to utilization of CAES, wind-turbine has been optimally used to support electrical demand which his illustrated in Fig. 8.7.

As an economic result owing to utilization of CEAS and DRP, total purchased gas has been considerably reduced in cases 2, 3, and 4. Gas import pattern is illustrated in Fig. 8.8.

By using CAES and DRP, generation of renewable units has had optimal share in supplying electrical demand. So, the role of CHP unit as one of electrical generation units has been decreased and therefore less gas has been consumed and then electrical and heat generation of this unit have been reduced. Figures 8.9, 8.10, and 8.11 illustrate the results related to CHP unit.

Since used gas by CHP unit has been changed, gas procurement pattern for boiler unit is appropriately changed and boiler has attempted to generate heat in a new pattern. Gas consumption and heat generation pattern of boiler are illustrated in Figs. 8.12 and 8.13, respectively.

Generated and consumed air by compressed air energy storage system is illustrated in Fig. 8.14. It can be seen from this figure that due to implementation of DRP,

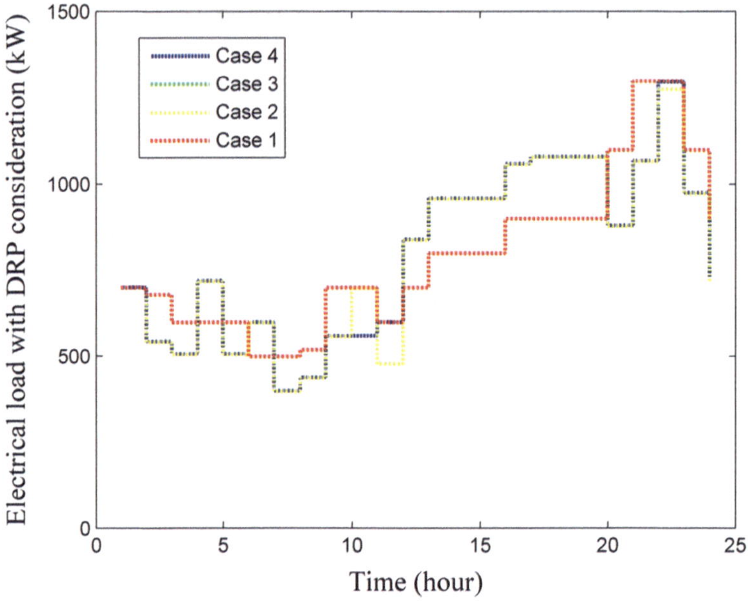

Fig. 8.6 Electrical energy demand in cases 1, 2, 3, and 4.

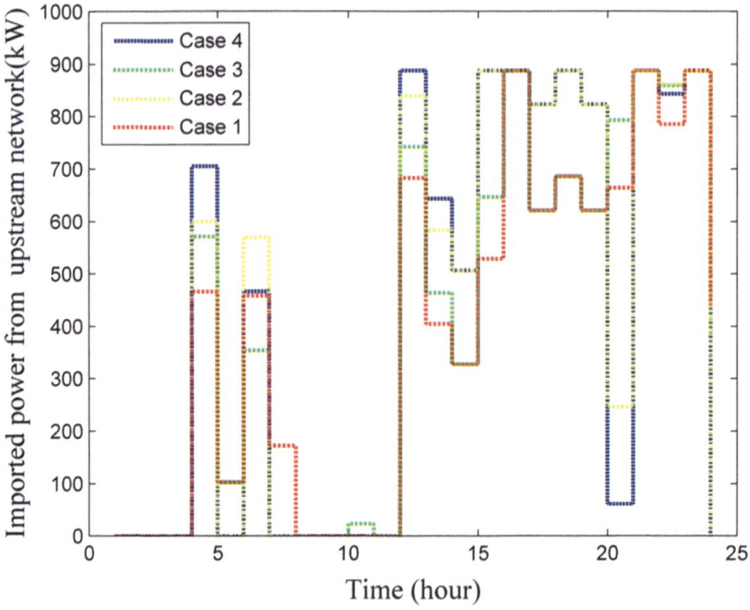

Fig. 8.7 Imported power from upper network

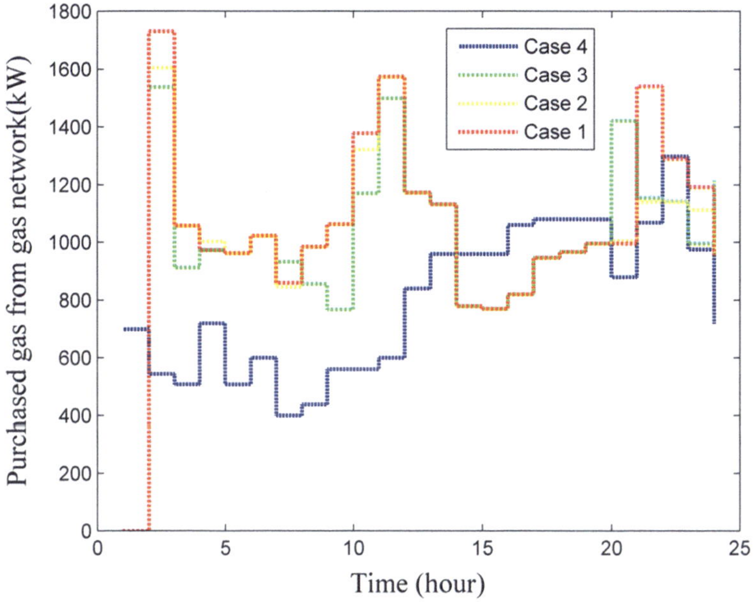

Fig. 8.8 Gas import

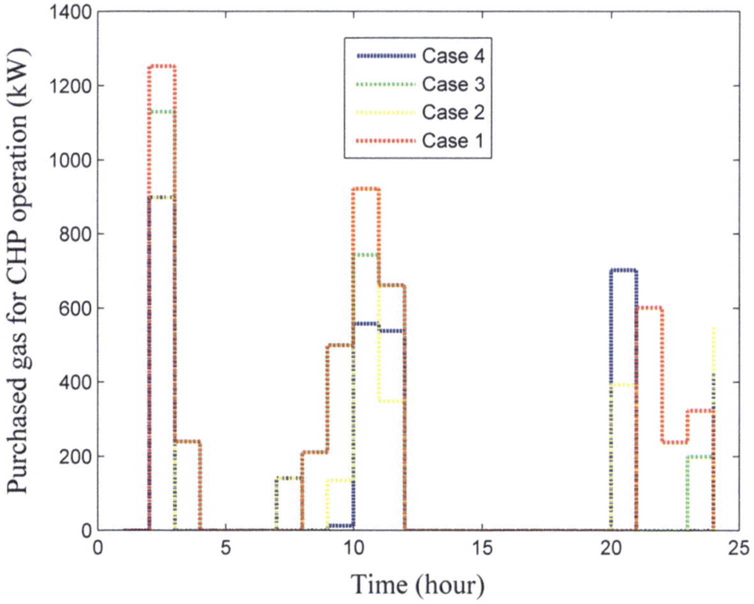

Fig. 8.9 Used gas by CHP unit

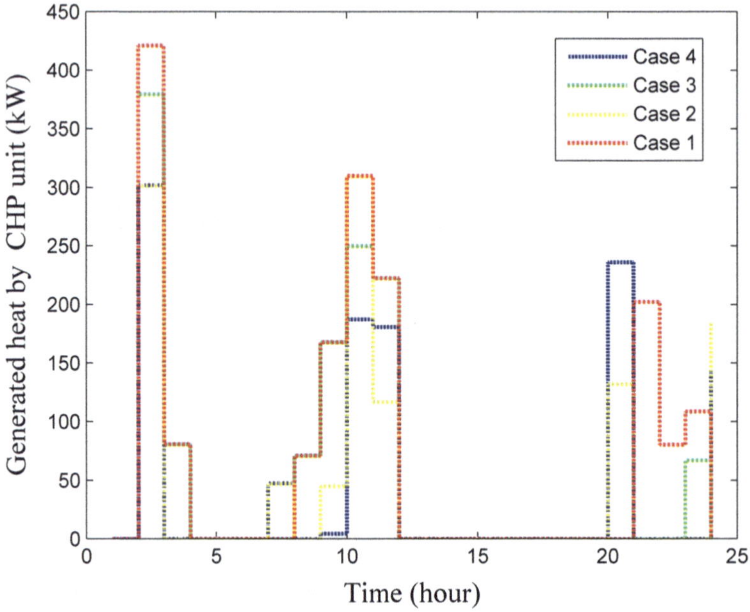

Fig. 8.10 Heat generated by CHP unit

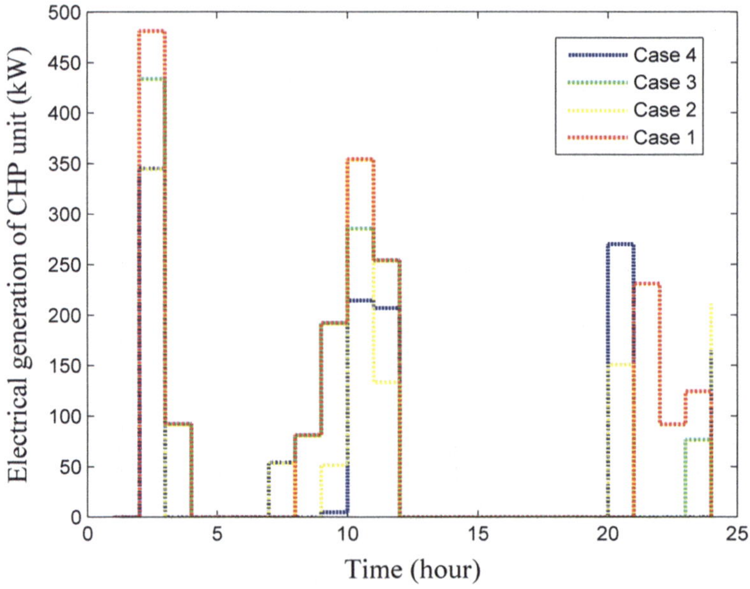

Fig. 8.11 Electrical power generated by CHP unit

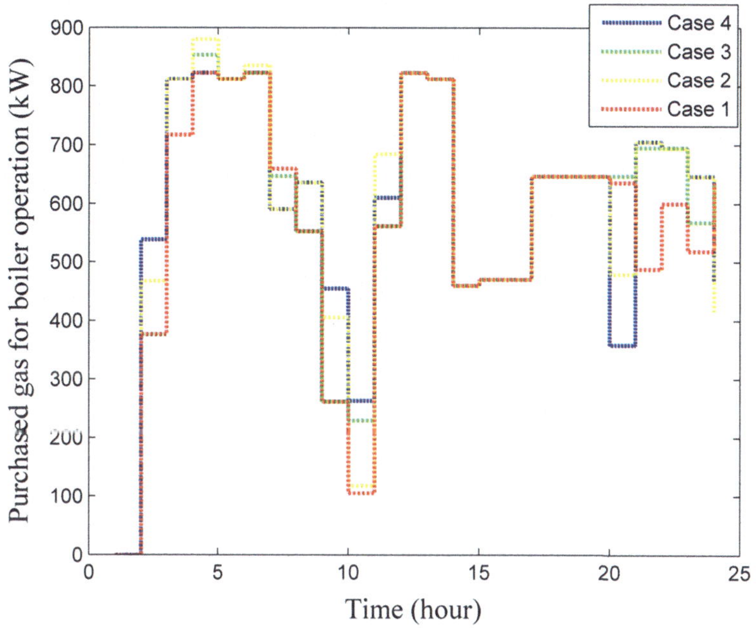

Fig. 8.12 Gas consumption of boiler

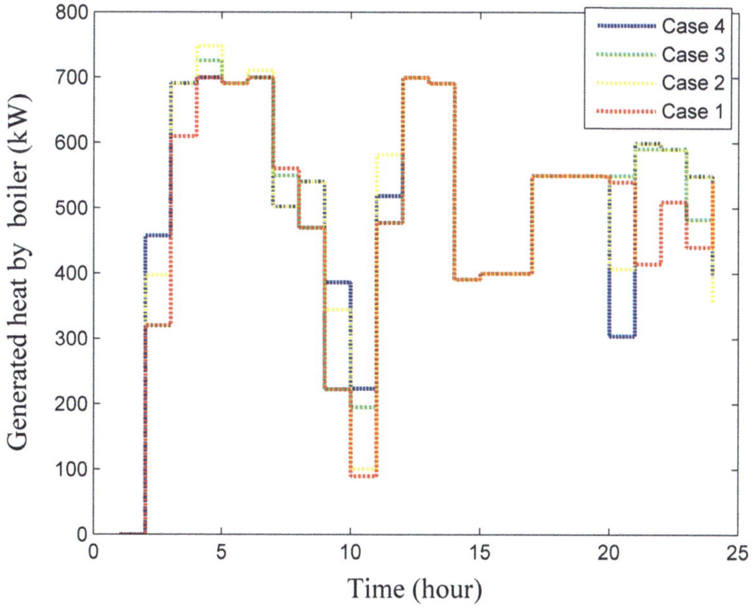

Fig. 8.13 Generated heat by boiler

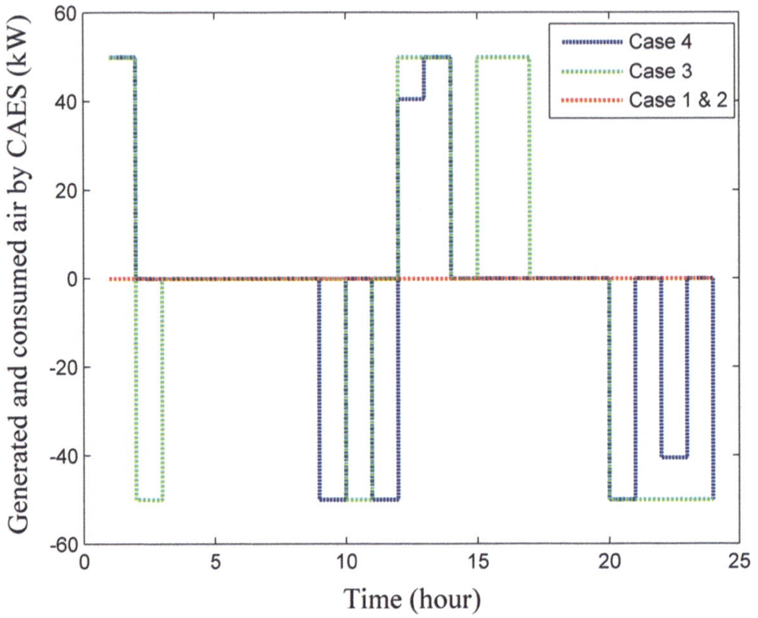

Fig. 8.14 Generated and consumed air by CAES

generation of air in CAES is increased and, on the other hand, consumption of power in CAES is reduced. Since power generation/consumption of CAES is proportional with its air generation/consumption, therefore power generation/consumption of CAES is changed with DRP which is illustrated in Fig. 8.15.

8.4 Conclusion

In this chapter, application of compressed air energy storage system as a supply side management tool has been investigated. A renewable-based hub energy system has been studied from economic and environmental viewpoints in the presence of CAES and DRP. Studied renewable-based hub system is composed of CHP system, boiler, and wind-turbine and storage systems. Since renewable generation units like wind-turbine have severe uncertainties in their outputs, CAES has been used to manage these uncertainties in the hub energy system. In simple words, CAES stores excess generation of such units and uses the saved energy in peak time periods to satisfy electrical energy demand. Optimal eco-emission operation of renewable-based hub energy system has been modeled through a mixed-integer linear programming and solved using GAMS software. Comparing the obtained results from simulations of various case studies, it can be found that due to implementation of CAES and DRP in case 4, total operation cost of renewable-based hub energy system in comparison with case 1, 2, and 3 is decreased 1.24%, 0.57%, and 0.72%, respectively. Also, total generated emission of renewable-based hub energy system in case 4 is reduced 1.57%, 0.60%, and 0.30% in comparison with cases 1, 2, and 3, respectively.

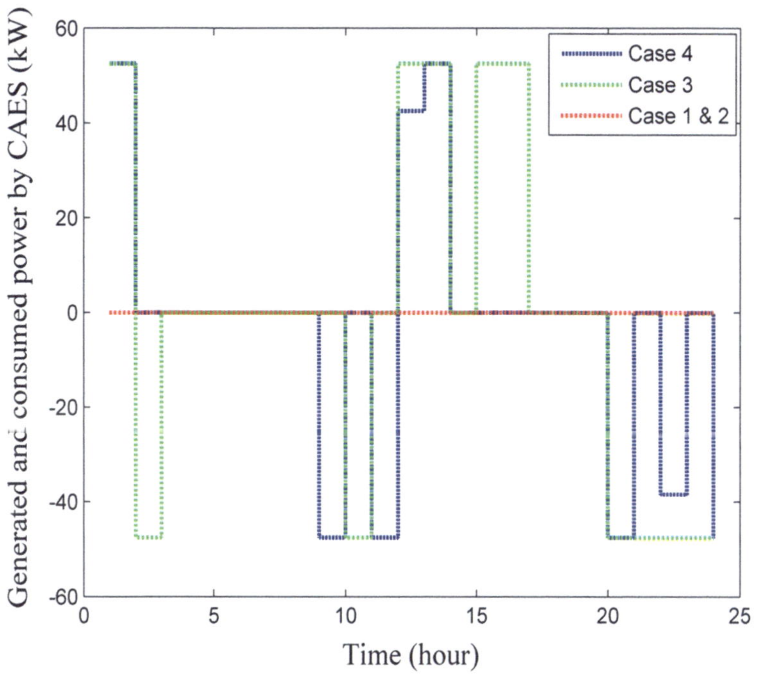

Fig. 8.15 Generated and consumed power by CAES

It can be understood from the obtained results above that both economic and environmental concerns of renewable-based hub energy system can be satisfied through utilization of CAES and DRP.

Nomenclature

Indices

t Time period index

Parameters

η_{ee}^{T} Efficiency of transformer

η_{ge}^{CHP} Gas to electricity efficiency of CHP unit

η_{ee}^{CON} Efficiency of converter unit

η_{ch}^{e} Charging efficiency of electrical storage system

η_{dis}^{e} Discharging efficiency of electrical storage system

η_{ch}^{h} Charging efficiency of heat storage system

η_{dis}^{h}	Discharging efficiency of heat storage system
$\alpha_{\min}^{e}$	Minimum limit coefficient of electrical storage system
$\alpha_{\max}^{e}$	Maximum limit coefficient of electrical storage system
$\alpha_{\mathrm{loss}}^{e}$	Loss of power coefficient for electrical storage system
$\alpha_{\min}^{h}$	Minimum limit coefficient of heat storage system
$\alpha_{\max}^{h}$	Maximum limit coefficient of heat storage system
$\alpha_{\mathrm{loss}}^{h}$	Loss of heat coefficient for electrical storage system
λ_{t}^{e}	Price of upstream network price
λ^{wi}	Wind turbine generation cost
λ^{g}	Gas price
λ^{wa}	Water price
λ_{s}^{e}	Electrical storage system operation cost
λ_{s}^{h}	Heat storage system operation cost
A^{NET}	Upstream network availability
A^{CHP}	CHP unit availability
A^{WIND}	Wind turbine availability
$C_{c}^{\mathrm{st},e}$	Rated capacity of electrical storage system
$C_{c}^{\mathrm{st},h}$	Rated capacity of heat storage system
$\mathrm{EF}_{\mathrm{CO}}^{\mathrm{CHP}}$	CO_2 emission factor for CHP unit
$\mathrm{EF}_{\mathrm{SO}}^{\mathrm{CHP}}$	SO_2 emission factor for CHP unit
$\mathrm{EF}_{\mathrm{NO}}^{\mathrm{CHP}}$	NO_x emission factor for CHP unit
$\mathrm{EF}_{\mathrm{CO}}^{B}$	CO_2 emission factor for boiler
$\mathrm{EF}_{\mathrm{SO}}^{B}$	SO_2 emission factor for boiler
$\mathrm{EF}_{\mathrm{NO}}^{B}$	NO_x emission factor for boiler
$\mathrm{EF}_{\mathrm{CO}}^{L}$	CO_2 emission factor for residential gas consumption
$\mathrm{EF}_{\mathrm{SO}}^{L}$	SO_2 emission factor for residential gas consumption
$\mathrm{EF}_{\mathrm{NO}}^{L}$	NO_x emission factor for residential gas consumption
$\mathrm{EF}_{\mathrm{CO}}^{\mathrm{Net}}$	CO_2 emission factor for upstream network
$\mathrm{EF}_{\mathrm{SO}}^{\mathrm{Net}}$	SO_2 emission factor for upstream network
$\mathrm{EF}_{\mathrm{NO}}^{\mathrm{Net}}$	NO_x emission factor for upstream network
$g_{\min}^{\mathrm{net}}$	Minimum nominal capacity of gas network
$g_{\max}^{\mathrm{net}}$	Maximum nominal capacity of gas network
g_{t}^{l}	Gas demand in residential section at time t
$p_{\min}^{e}$	Minimum nominal capacity of upstream network
$p_{\max}^{e}$	Maximum nominal capacity of upstream network

p_c^T	Nominal capacity of transformer
p_c^{CHP}	Rated capacity of CHP unit
p_c^B	Rated capacity of boiler unit
p_r	Rated power of wind turbine
p_t^{el}	Electrical demand at time t
p_t^h	Heat demand at time t
wa_t^l	Water demand at time t
$\mathrm{wa}_{\min}$	Minimum nominal capacity of water network
$\mathrm{wa}_{\max}$	Maximum nominal capacity of water network
$w_{\mathrm{ci}}, w_{\mathrm{co}}, w_r$	Cut-in, cut-out, and rated speeds of wind turbine
$w(t)$	Wind speed at time t
x,y,z	Indexes for modeling generation of wind turbine

Variables

Cost	Total operation cost of hub energy system
$\mathrm{C_{BS}}$	Operation cost of battery storage
$\mathrm{C_{DR}}$	Cost of DRP
$\mathrm{C_{Ex}}$	Cost/revenue of exchanged power
$\mathrm{C_{net}}$	Cost of purchased power from upstream network
$C_t^{\mathrm{st},e}$	Energy level of electrical storage system
$C_t^{\mathrm{st},h}$	Heat level of electrical storage system
$\mathrm{C_{TS}}$	Operation cost of thermal storage
$\mathrm{C_{Wind}}$	Operation cost of wind turbine
$\mathrm{C_{Bo}}$	Operation cost of boiler
g_t^{CHP}	Consumed gas by CHP unit
g_t^B	Consumed gas by boiler unit
g_t^{net}	Total imported gas from gas network at time t
$I_t^{\mathrm{ch},e}$	Binary variable for modeling charging state of electrical storage system
$I_t^{\mathrm{dis},e}$	Binary variable for modeling discharging state of electrical storage system
$I_t^{\mathrm{ch},h}$	Binary variable for modeling charging state of heat storage system
$I_t^{\mathrm{dis},h}$	Binary variable for modeling discharging state of heat storage system
p_t^e	Imported power from upstream network at time t
$p_t^{\mathrm{ch},e}, p_t^{\mathrm{dis},e}$	Charging and discharging power of electrical storage system
$p_t^{\mathrm{ch},h}, p_t^{\mathrm{dis},h}$	Charging and discharging heat of electrical storage system
$p_t^{\mathrm{loss},e}$	Loss of power in electrical storage system

$p_t^{\mathrm{loss},h}$	Loss of heat in heat storage system
p_t^{wi}	Electrical generation of wind turbine at time t
$\mathrm{wa}_t^{\mathrm{net}}$	Imported water at time t

References

1. Tazvinga H, Zhu B, Xia X (2015) Optimal power flow management for distributed energy resources with batteries. Energy Convers Manag 102:104–110. https://doi.org/10.1016/j.enconman.2015.01.015
2. Majidi M, Nojavan S, Esfetanaj NN, Najafi-Ghalelou A, Zare K (2017) A multi-objective model for optimal operation of a battery/PV/fuel cell/grid hybrid energy system using weighted sum technique and fuzzy satisfying approach considering responsible load management. Sol Energy 144:79–89
3. Karami H, Sanjari MJ, Gooi HB, Gharehpetian GB, Guerrero JM (2017) Stochastic analysis of residential micro combined heat and power system. Energy Convers Manag 138:190–198. https://doi.org/10.1016/j.enconman.2017.01.073
4. Li C, Gillum C, Toupin K, Park YH, Donaldson B (2016) Environmental performance assessment of utility boiler energy conversion systems. Energy Convers Manag 120:135–143. https://doi.org/10.1016/j.enconman.2016.04.099
5. Nojavan S, Majidi M, Zare K (2017) Performance improvement of a battery/PV/fuel cell/grid hybrid energy system considering load uncertainty modeling using IGDT. Energy Convers Manag 147:29–39
6. Majidi M, Nojavan S, Zare K (2017) A cost-emission framework for hub energy system under demand response program. Energy 134:157–166
7. Majidi M, Nojavan S, Zare K (2017) Optimal stochastic short-term thermal and electrical operation of fuel cell/photovoltaic/battery/grid hybrid energy system in the presence of demand response program. Energy Convers Manag 144:132–142
8. Al-Sharafi A, Yilbas BS, Sahin AZ, Ayar T (2017) Performance assessment of hybrid power generation systems: Economic and environmental impacts. Energy Convers Manag 132:418–431. https://doi.org/10.1016/j.enconman.2016.11.047
9. Derafshi Beigvand S, Abdi H, La Scala M (2016) Optimal operation of multicarrier energy systems using time varying acceleration coefficient gravitational search algorithm. Energy 114:253–265. https://doi.org/10.1016/j.energy.2016.07.155
10. Parisio A, Del Vecchio C, Vaccaro A (2012) A robust optimization approach to energy hub management. Int J Electr Power Energy Syst 42(1):98–104. https://doi.org/10.1016/j.ijepes.2012.03.015
11. Wasilewski J (2015) Integrated modeling of microgrid for steady-state analysis using modified concept of multi-carrier energy hub. Int J Electr Power Energy Syst 73:891–898. https://doi.org/10.1016/j.ijepes.2015.06.022
12. Pazouki S, Haghifam M-R (2016) Optimal planning and scheduling of energy hub in presence of wind, storage and demand response under uncertainty. Int J Electr Power Energy Syst 80:219–239. https://doi.org/10.1016/j.ijepes.2016.01.044
13. Orehounig K, Evins R, Dorer V, Carmeliet J (2014) Assessment of renewable energy integration for a village using the energy hub concept. Energy Procedia 57:940–949. https://doi.org/10.1016/j.egypro.2014.10.076
14. Ma T, Wu J, Hao L (2017) Energy flow modeling and optimal operation analysis of the micro energy grid based on energy hub. Energy Convers Manag 133:292–306. https://doi.org/10.1016/j.enconman.2016.12.011
15. Orehounig K, Evins R, Dorer V (2015) Integration of decentralized energy systems in neighbourhoods using the energy hub approach. Appl Energy 154:277–289. https://doi.org/10.1016/j.apenergy.2015.04.114

16. Najafi A, Falaghi H, Contreras J, Ramezani M (2016) Medium-term energy hub management subject to electricity price and wind uncertainty. Appl Energy 168:418–433. https://doi.org/10.1016/j.apenergy.2016.01.074
17. Beigvand SD, Abdi H, La Scala M (2017) A general model for energy hub economic dispatch. Appl Energy 190:1090–1111. https://doi.org/10.1016/j.apenergy.2016.12.126
18. Koeppel G, Andersson G (2009) Reliability modeling of multi-carrier energy systems. Energy 34(3):235–244. https://doi.org/10.1016/j.energy.2008.04.012
19. Shariatkhah M-H, Haghifam M-R, Parsa-Moghaddam M, Siano P (2015) Modeling the reliability of multi-carrier energy systems considering dynamic behavior of thermal loads. Energy Build 103:375–383. https://doi.org/10.1016/j.enbuild.2015.06.001
20. Mancarella P (2014) MES (multi-energy systems): An overview of concepts and evaluation models. Energy 65:1–17. https://doi.org/10.1016/j.energy.2013.10.041
21. Moghaddam IG, Saniei M, Mashhour E (2016) A comprehensive model for self-scheduling an energy hub to supply cooling, heating and electrical demands of a building. Energy 94:157–170. https://doi.org/10.1016/j.energy.2015.10.137
22. Kamyab F, Bahrami S (2016) Efficient operation of energy hubs in time-of-use and dynamic pricing electricity markets. Energy 106:343–355. https://doi.org/10.1016/j.energy.2016.03.074
23. Brahman F, Honarmand M, Jadid S (2015) Optimal electrical and thermal energy management of a residential energy hub, integrating demand response and energy storage system. Energy Build 90:65–75. https://doi.org/10.1016/j.enbuild.2014.12.039
24. Rastegar M, Fotuhi-Firuzabad M (2015) Load management in a residential energy hub with renewable distributed energy resources. Energy Build 107:234–242. https://doi.org/10.1016/j.enbuild.2015.07.028
25. Rastegar M, Fotuhi-Firuzabad M, Lehtonen M (2015) Home load management in a residential energy hub. Electr Power Syst Res 119:322–328. https://doi.org/10.1016/j.epsr.2014.10.011
26. Shabanpour-Haghighi A, Seifi AR (2016) Effects of district heating networks on optimal energy flow of multi-carrier systems. Renew Sust Energ Rev 59:379–387. https://doi.org/10.1016/j.rser.2015.12.349
27. Ghalelou AN, Fakhri AP, Nojavan S, Majidi M, Hatami H (2016) A stochastic self-scheduling program for compressed air energy storage (CAES) of renewable energy sources (RESs) based on a demand response mechanism. Energy Convers Manag 120:388–396. https://doi.org/10.1016/j.enconman.2016.04.082
28. Shafiee S, Zareipour H, Knight AM, Amjady N, Mohammadi-Ivatloo B (2017) Risk-constrained bidding and offering strategy for a merchant compressed air energy storage plant. IEEE Trans Power Syst 32(2):946–957
29. Nojavan S, Majidi M, Esfetanaj NN (2017) An efficient cost-reliability optimization model for optimal siting and sizing of energy storage system in a microgrid in the presence of responsible load management. Energy 139:89–97
30. Nojavan S, Majidi M, Najafi-Ghalelou A, Ghahramani M, Zare K (2017) A cost-emission model for fuel cell/PV/battery hybrid energy system in the presence of demand response program: ε-constraint method and fuzzy satisfying approach. Energy Convers Manag 138: 383–392
31. Nojavan S, Majidi M, Zare K (2017) Risk-based optimal performance of a PV/fuel cell/battery/grid hybrid energy system using information gap decision theory in the presence of demand response program. Int J Hydrog Energy 42(16):11857–11867
32. Pazouki S, Haghifam M-R, Moser A (2014) Uncertainty modeling in optimal operation of energy hub in presence of wind, storage and demand response. Int J Electr Power Energy Syst 61:335–345
33. Elsied M, Oukaour A, Gualous H, Hassan R (2015) Energy management and optimization in microgrid system based on green energy. Energy 84:139–151
34. Elsied M, Oukaour A, Gualous H, Brutto OAL (2016) Optimal economic and environment operation of micro-grid power systems. Energy Convers Manag 122:182–194
35. The GAMS Software Website (2017) [Online]. Available: http://www.gams.com/dd/docs/solvers/cplex.pdf

Chapter 9
Optimal Stochastic Short-Term Scheduling of Renewable Energy Hubs Taking into Account the Uncertainties of the Renewable Sources

Moein Moeini-Aghtaie, Amir Safdarian, Zohreh Parvini, and Fereshteh Aramoun

9.1 Introduction

Growing the number of distribution companies around the world led to the integration of various distributed generation (DG) technologies in distribution level. With the increased utilization of these new energy resources, especially co- and tri-generation technologies, energy hubs have brought into existence. These basic units of energy which serve as an interface between different infrastructures of energy can store and convert different forms of energy via an integrated framework. On the other hand, renewable-based DG units play a considerable role in future vision of energy networks. As a result, renewable energy hubs are an unavoidable part of future energy networks. Although the supply diversification in renewable energy hubs grants some degree of flexibility in feeding energy loads, it calls for new analysis tools to properly consider the effects of uncertainties associated with renewable energies. In this regard, this chapter tries to investigate the abilities of stochastic frameworks in dealing with the energy resources scheduling problem in renewable energy hubs. To reach this important goal, at first, different elements of renewable energy hubs are introduced. Then, the procedure which needs to be followed to efficiently model the input and output relations of renewable energy

---

M. Moeini-Aghtaie
Faculty of Energy Engineering, Sharif University of Technology, Tehran, Iran
e-mail: moeini@sharif.edu

A. Safdarian
Faculty of Electrical Engineering, Sharif University of Technology, Tehran, Iran
e-mail: safdarian@sharif.edu

Z. Parvini · F. Aramoun (✉)
Sharif Energy Research Institute, Sharif University of Technology, Tehran, Iran
e-mail: z.parvini@seri.sharif.edu; f.aramoun@seri.sharif.edu

B. Mohammadi-Ivatloo, F. Jabari (eds.), *Operation, Planning, and Analysis of Energy Storage Systems in Smart Energy Hubs*,
https://doi.org/10.1007/978-3-319-75097-2_9

hubs is well described. The efficient tools and algorithms which can be used to model the uncertain behavior of wind and solar energies in short-term scheduling problem are put under investigation. After modeling different uncertainties, the general framework of energy scheduling problem for a renewable energy hub will be addressed. Mathematical model of this optimization problem will be introduced and the optimization tools which can well model the features of this problem will be discussed. By briefly introducing the main features of stochastic optimization algorithms which can be used for energy scheduling problem of renewable energy hubs, the stochastic version of this optimization problem will be defined. Extracting the mathematical model of stochastic energy scheduling problem, it will be shown that how different stochastic tools can be used to deal with this problem. By following the subjects covered in this chapter, the readers learn how to model, solve, and apply the results of energy scheduling problems in renewable energy hubs.

9.2 Renewable Energy Hubs Modeling

Increasing demand of energy carriers, dwindling fossil resources, and climate change have posed new challenges to decision makers in terms of energy security and sustainability. In this regard, harnessing renewable energy sources, such as wind power, solar energy, hydropower, and biomass have considered as a promising sustainable solution to alleviate the aforementioned concerns. Integration of renewable energy facilities into power systems, especially in small distribution scales, have altered the historically centralized and bulk structure of power systems to small decentralized, and locally in-feed structures. However, intrinsic volatile and stochastic nature of renewable energy sources might raise new challenges to power system planners and operators.

On the other hand, the advent of new conversion and storage technologies, typically including micro turbines, combined heat and power (CHP), as well as thermal and electrical storage units, has brought up new paths to exploit renewable energy sources. The emergence of conversion technologies has enabled production of both electricity and heat, out of natural or biogas, biomass, and so forth. Therefore, it provides the possibility of co- and tri-generation in the system. Moreover, deployment of energy storage units, as well as prospective integration of plug-in hybrid electrical vehicles (PHEVs), evolve the prevailing passive and unidirectional structure of power system through converting the unwanted energy from uncontrollable renewable sources into a restorable form, and delivering the stored energy either to the demand side or to the supply side, in times of requirement. Implication of these conversion and storage technologies leads to complex interactions among various energy carriers, which mitigate various undesirable impacts of renewable energy sources integration to system and pave the way to the effective implication of these sources.

Introduction of promising conversion and storage technologies along with the increased penetration of renewable energy sources have changed the traditional

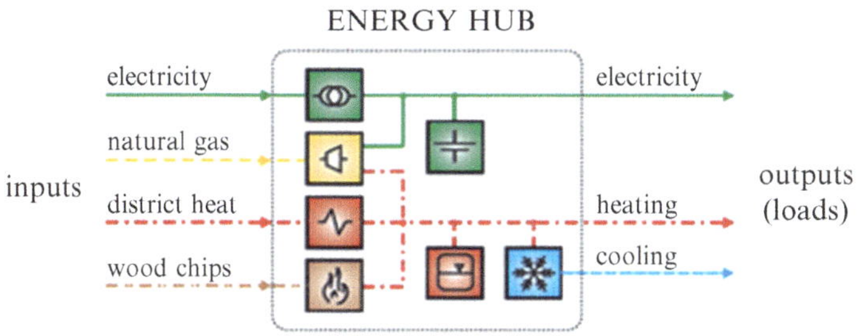

Fig. 9.1 Example of an energy hub that contains transformer, micro turbine, heat exchanger, furnace, battery, hot water storage, and absorption cooler [1]

setup of energy systems into potential, bidirectional, and active ones. Such energy systems, which provide flexibility for producing, storing, converting, and conditioning multiple energy carriers, are called renewable energy hubs. Moving toward environmentally and economically sustainable energy systems, renewable energy hubs have been considered as an inevitable part of future energy networks.

Generally, a hub represents an interface between various energy sources and demands, providing functions of input and output, conversion, and storage of multiple-energy carriers. A sample energy hub is represented in Fig. 9.1, where multiple energy carriers are received, then they are distributed to different processes of conditioning, conversion, and storage, and finally several demands are supplied from them.

As Fig. 9.1 depicts, an energy hub is typically composed of input energy carriers (electricity, gas, wood, etc.), energy converters and storage units (transformer, microturbine, gas boiler, batteries, and so on.), and output energy carriers (power, cooling, and heating). Renewable sources such as wind, solar, hydropower, and biomass are considered as inputs to the hub. The energy from these sources provides the required energy service at the output layer, either directly or indirectly, going through storage or conversion processes. Each input energy carrier, weather electricity or heat, may supply a number of loads in the output layers. For instance, in Fig. 9.1, the power originated from both grid and microturbine can meet the required electricity load. Similarly, the heat from microturbine, heat exchanger, or furnace can serve the heating demands. These redundant connections between inputs and outputs, which are provided by the hub components, are greatly beneficial from renewable energy sources point of view. The power from renewable energy sources is either adjustable like the hydropower and biogas, or stochastic such as that of wind and solar energy. By using the energy hub concept, raised concerns regarding the stochastic nature of these sources can be allayed.

Renewable energy hubs, where different dispatchable and/or stochastic energy sources are coupled and combined with storage and conversion technologies, bring three major advantages. First, the reliability and security of supplying energy to the load can be improved within a hub structure, by using multi energy complements as inputs. In this way, the load no longer depends on one single carrier to be satisfied.

For example, the electricity demand fed from renewable sources can be supplied with grid, local distribution generators, or CHP facilities, instead. Moreover, the local consumption of renewable sources will be promoted, by storing the surplus energy or converting it to other forms of energy demands. Second, from an operational perspective, cost of supplying energy can be reduced significantly, due to the great optimization potential provided by the flexibility of the hub. Redundancy of energy paths in the hub causes this flexibility by offering degrees of freedom in supplying the load. In fact, it enables the operator to choose the optimal option, in terms of, e.g., cost or emission, for supplying the demand. For example, an optimal dispatch can be run to define which facility and to what extent should provide the required heat. Third, coordinated operation of several energy carriers will enhance efficiency of the energy system, which eventually will help the energy system to improve in terms of sustainability.

In a comprehensive view, components of a renewable energy hub can be divided into five main categories (1) Energy inputs, including wind power, solar energy, natural gas, biomass, and electricity from the grid. (2) Conversion facilities, such as heat exchanger, gas turbine, boiler, and chiller. (3) Collectors and distributors which collect power, heat and cold, and distribute them through relevant networks. In these components, the output energy is equivalent to the inputs in all the time. (4) Storage units which store various energy carriers containing electricity, heat, and cold. (5) Delivering utilities which can be distribution networks of electricity, gas, and heating and cooling energy. A generic renewable energy hub is presented in Fig. 9.2.

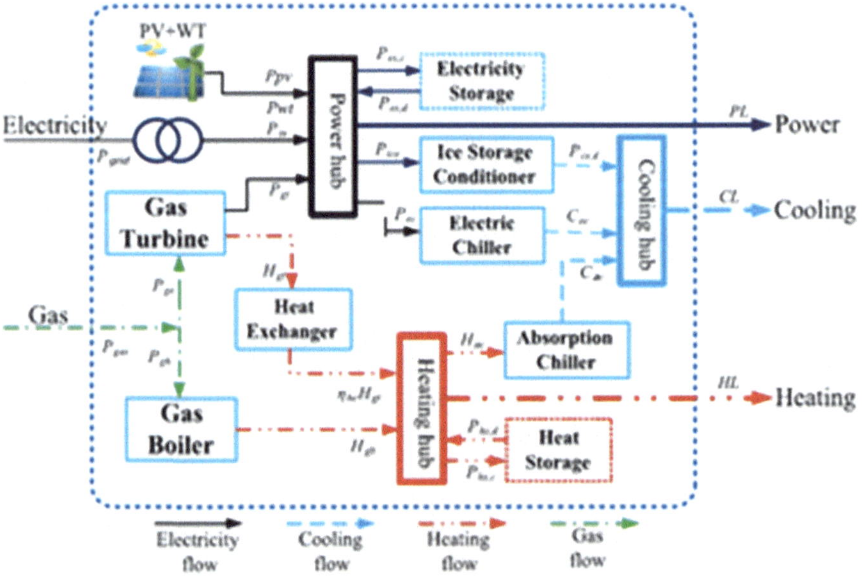

Fig. 9.2 Components within a renewable energy hub [2]

Fig. 9.3 Modeling the flow of power through an energy hub [1]

While an energy hub can be utilized just as a modeling concept, it can be attributed to a wide range of systems, from a single hybrid electric car up to an entire city as a single hub. The concept is not limited to a particular size or definite energy carriers. Examples of actual facilities, considered as a hub, include urban/rural districts, residential, industrial, and commercial complexes, as well as islanded energy systems, like a ship or aircraft.

Transportation of power through an energy hub can be modeled mathematically by input and output parameters. A set of energy carriers α, β, $\ldots \in E = \{\text{electricity, natural gas, heat}, \ldots\}$ considered as inputs or outputs depicted in Fig. 9.3 characterize the energy hub. Input sets of the hub, defined by a vector matrix $P = \{P_\alpha, P_\beta, \ldots, P_\omega\}$ are transformed to the output sets $L = \{L_\alpha, L_\beta, \ldots, L_\omega\}$ through different conversion mechanisms, represented by a coupling matrix C. The interconnections in an energy hub, with multiple inputs and outputs, can be mathematically described as follows:

$$\underbrace{\begin{bmatrix} L_\alpha \\ L_\beta \\ \vdots \\ L_\omega \end{bmatrix}}_{L} = \underbrace{\begin{bmatrix} c_{\alpha,\alpha} & c_{\beta,\alpha} & \cdots & c_{\omega,\alpha} \\ c_{\alpha,\beta} & c_{\beta,\beta} & \cdots & c_{\omega,\beta} \\ \vdots & \vdots & \ddots & \vdots \\ c_{\alpha,\omega} & c_{\beta,\omega} & \cdots & c_{\omega,\omega} \end{bmatrix}}_{C} \underbrace{\begin{bmatrix} P_\alpha \\ P_\beta \\ \vdots \\ P_\omega \end{bmatrix}}_{P} \tag{9.1}$$

Each parameter of the coupling matrix, called convertor coupling factor $c_{\alpha,\,\beta}$, links input P_α to the corresponding output L_β through considering the efficiency of conversion [3].

A renewable energy hub might consist of renewable energy conversion systems, such as wind turbines, photovoltaic arrays, hydro turbines, and fuel cells, conventional generators like diesel generators and micro turbines, as well as storage devices like batteries [4]. The components of energy hub are coordinated in order to find the optimal dispatch among input energy carriers, and satisfy demands at the output layer for harnessing the most of renewable energy sources. The first step to operate and optimize performance of a hub is to model its components. In the following, a general methodology for modeling hub components is stated in two separate sections of (1) renewable sources in energy hubs and (2) energy demands.

9.2.1 Renewable Sources in Energy Hubs

Various energy carriers, such as heat, electricity, and natural gas, might feed an energy hub. Due to recent technical advancements and reduced prices of renewable energy technologies, primarily wind and solar, they have been increasingly competitive with the other conventional energy technologies. Therefore, these sources play an inevitable role in supplying demands of an energy hub. In spite of extensive implication of wind and solar energies, the stochastic nature of these sources might cause challenges to system operation; thus, the uncertainty of these sources should be accurately considered and modeled in order to make the most of them. In this regard, modeling procedures of wind and solar energies, as the major supply sources in renewable energy hubs, are presented and the methods to address their uncertainties are discussed in the following.

9.2.1.1 Wind Energy Modeling

Electricity power produced by a wind turbine is a function of turbine's speed characteristics and wind speed at the site. The relation between mechanical energy and the electrical power produced by the turbine (Pw) can be mathematically presented by:

$$P_w = \begin{cases} \left(A + B \times V + C \times V^2\right) \times P_r & V_{\text{ci}} \leq V \leq V_r \\ P_r & V_r \leq V \leq V_{\text{co}} \\ 0 & V \leq V_{\text{ci}}, V \geq V_{\text{co}} \end{cases} \tag{9.2}$$

where V_{ci}, V_r, and V_{co} are, respectively, cut-in, rated, and cut-out speed of wind turbine, and P_r is its rated power. A, B, and C are constant parameters of wind turbine [5]. According to (9.2), when the speed of blowing wind is equal or larger than the cut-in speed, the turbine generates electricity in a nonlinear manner until it reaches its rated speed. Within the range of rated to cut-out speed, the wind turbine delivers its rated power, constantly. For wind speeds which are less (more) than the cut-in (cut-out) speed of the turbine, no power is generated by the wind turbine. Considering the wind turbine performance, according to its site characteristics, the available power from a wind turbine (P_{WT}) will be:

$$P_{\text{WT}} = P_w A_w \eta_{\text{WT}} \tag{9.3}$$

where A_w is the Total swept area; η_{WT} is the Efficiency of wind turbine system [4].

9.2.1.2 Solar Energy Modeling

A PV cell receives its energy from normal and diffuse radiations of the sun, depending on the time of the year and position of the sun in the sky [4]. The electric power generated by a PV module on an average day of *j*th month (P_j) can be defined as

$$P_j = I_{\mathrm{T}j}\eta_{\mathrm{PV}}A_{\mathrm{PV}} \tag{9.4}$$

where $I_{\mathrm{T}j}$ is the total solar radiation on the module; η_{PV} is the efficiency of PV module; A_{PV} is the area of module that is exposed to the sun [4].

A PV system in operation is usually composed of several modules connected in series and parallel to obtain higher output voltage and current. The total power generated by a PV array (P_{array}) will be equal to:

$$P_{\mathrm{array}} = N_S \times N_P \times P_{\mathrm{PV}} \tag{9.5}$$

where N_S is the Number of PV modules in series; N_P is the Number of PV modules in parallel.

The power generated by solar and wind energies are intermittent and variable, which makes them non-dispatchable, contrary to the conventional power generators. Generated power of a wind turbine is completely dependent on the stochastic speed of blowing wind. In a PV system, solar radiation incident to the arrays highly depends on atmospheric conditions. Although forecasting techniques have improved significantly, there are still unpredictable events which do not allow perfect estimation of generated power from renewable sources. Therefore, dealing with uncertain parameters becomes an inevitable part of operation and optimization of renewable energy hubs. There are several different modeling techniques for addressing uncertainty of parameters. These techniques are categorized into probabilistic approaches, possibilistic approaches, hybrid possibilistic–probabilistic approaches, information gap decision theory (IGDT), and robust optimization. By means of these methods, the effects of uncertainties can be evaluated on system performance [6]. The first category is on the basis of probability distribution functions (PDFs). In this method, the PDF of the uncertain parameter is assumed to be realized using historical data. Possibilistic approaches which use fuzzy algorithms, assign a membership function (MF) to uncertain parameters. The hybrid approaches apply a combination of probabilistic and possibilistic parameters to address uncertainties. Information gap decision theory (IGDT) studies the forecast errors of uncertain parameters. Optimization in robust techniques is run based on the worst-case scenario set. Finally, interval analysis technique considers a probable range for uncertain parameters and obtains bounds for the outputs. Among all the uncertainty modeling approaches, probabilistic methods are the most widely utilized. Probabilistic techniques are divided into two main categories: (1) numerical approaches, namely Monte Carlo Simulation techniques, and (2) analytical approaches based on linearization and PDF approximation. Further information about these techniques can be found in [6].

9.2.2 Energy Demands

The main goal of an energy hub is to satisfy the loads in the output layer. Output demands in an energy hub are typically electricity and thermal loads, which can be supplied by various input energy carriers as shown in Fig. 9.2. Thermal loads represent various heating and cooling demands, and electrical loads can range from a single PHEV and small home appliances, to power supply of an industrial complex or an entire urban district. Generally, in modeling an energy hub, demands of the same type are combined together, and they are considered as integrated output energy carriers. During operational time frames, energy demands of hubs are modeled on an hourly basis. Electricity and heat demands of a typical energy hub are presented in Fig. 9.4.

While output demands in energy hubs are often in form of electricity or thermal loads, sometimes, primary energy carriers, such as natural gas, hydrogen, and water, serve as output demand, too. Load of these demands can be modeled by hourly forecasted profiles, similar to Fig. 9.5.

Although energy demands in the output layer are not deterministic, their uncertainties are almost neglected in system scale, compared to that of renewable energy sources. In the case of very short times of operation, the uncertainty associated with loads might become more influential. The uncertainty in forecasting energy demands can be represented by a normal distribution through mean and standard deviation parameters [8].

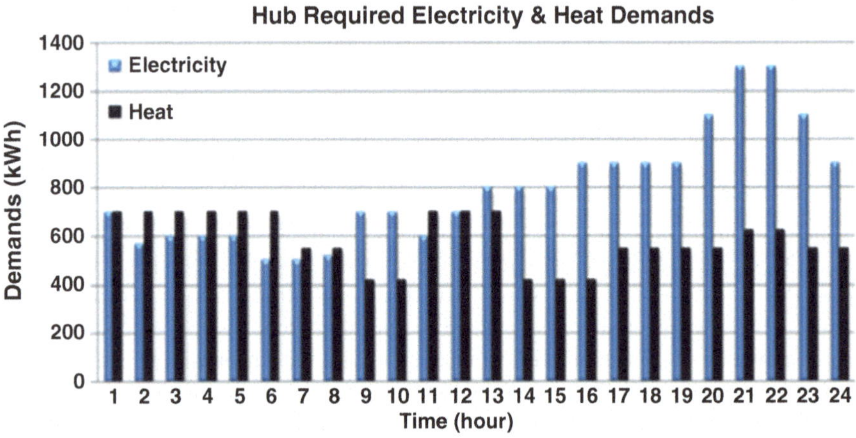

Fig. 9.4 Electricity and heat demands of a typical energy hub [7]

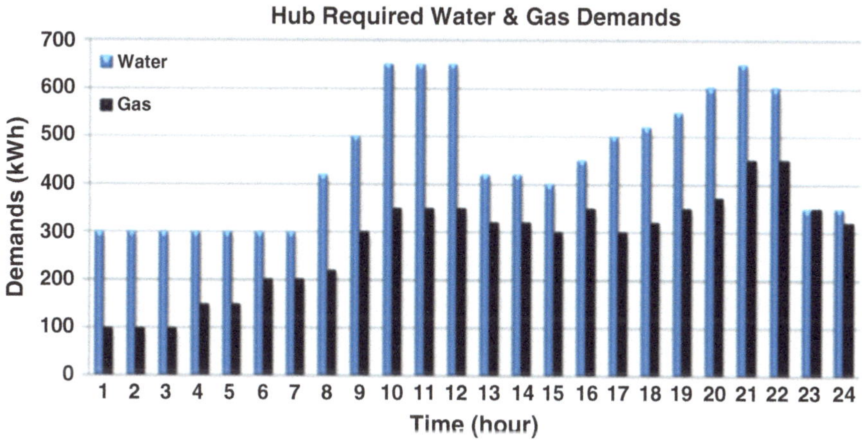

Fig. 9.5 Water and gas demands of an energy hub [7]

9.3 Energy Scheduling Problem in Renewable Energy Hubs

9.3.1 Problem Description

A renewable energy hub consists of several input energy carriers which can go through a variety of conversion and storage technologies. For example, the electricity generated by a wind turbine can directly feed the electricity demand in the output, be converted to heat by an electric heater, or be stored in a battery for a later use. On the other hand, these technologies make redundant paths for supplying different forms of energy demands in the output layer. Each energy demand can be satisfied by a variety of energy sources, redundantly. A heat exchanger, a CHP, or a furnace, for instance, can meet the heat demand.

Energy scheduling in an energy hub is intended to define how much energy should be generated within the hub or bought from the grid, based on the demands in the output, generated power by renewable sources, and energy carriers' prices. Moreover, in presence of storage units within the hub, the time interval of operation should be considered in the scheduling, since the quantity of energy in a storage unit is a matter of successive charge and discharge schedules during a period. Regarding various criteria, the operation of a renewable energy hub can be scheduled optimally. These criteria range from operating cost, energy efficiency, and expected profit, up to reliability, durability, and environmental factors.

In order to optimally schedule an energy hub, the first step is modeling of hub components, including input energy sources, connections within the hub such as transformation, conversion, and storage technologies, as well as demands in the output layer. Modeling input sources defines that whether the source is dispatchable or not, it comes from a bulk network or it is just an energy-limited source, or its generated power is constant or highly variable. Connections within the hub, categorized into four types of direct connections, transformers, converters, and storage systems, should be also comprehensively modeled. Through direct connections, such as electric cables or pipelines, input carrier is directly delivered to the output demand, without any conversion or significant change in its quality. Transformers alter the quality of inputs, e.g. voltage or pressure, in order to meet demand requirements. Conversion technologies, including combustion engines, steam/gas turbines, electrical machines, and fuel cells, change the input energy carriers into various demand forms. Storage devices are means of storing electrical or thermal energy carriers. The models of these internal connections are represented in the coupling matrix, explained in previous section. In the demand side, loads of the same type should be aggregated, thus, the demand side is modeled as integrated energy demands, i.e. electrical loads, cooling or heating demands, and natural gas.

Modeling of components leads to obtain a better understanding of the energy flows and the paths within the hub, and consequently identify the problem and manage decision-making procedure. However, it is not always possible or simple to design a perfect model, thus, there should be a trade-off between complexity and accuracy to obtain a sufficiently appropriate model [4].

The next step in optimizing performance of a renewable energy hub is modeling hub's scheduling problem mathematically according to the preferred optimization criterion. Type of the problem, which the hub operator has to deal with, is defined in this step; it can be a problem of minimizing overall costs, maximizing the hub's profit, maintaining system reliability and durability, or diminishing emissions. Based on the optimization criteria, constraints of the problem should be determined.

After defining the optimization problem and obtaining its mathematical model, an accurate solution method has to be implemented to find the best answer. There are numerous optimization methods which can be used to address various scheduling problems. Based on the problem type, the most suitable optimization technique, along with the most accurate data as the required inputs of the problem, such as cost of energy, emission, or lost load, should be utilized. In order to achieve the optimal solution, the steps depicted in Fig. 9.6 should be followed.

9.3.2 *Mathematical Model of the Problem*

In order to define the optimal schedule of the hub, which means how much of each energy carrier should supply the hub and how inputs should be dispatched through conversion and storage technologies, a mathematical model of the problem should be attained. In the following, first, a general description of the energy hub scheduling problem is provided. Then, the mathematical model of this problem is explained in detail.

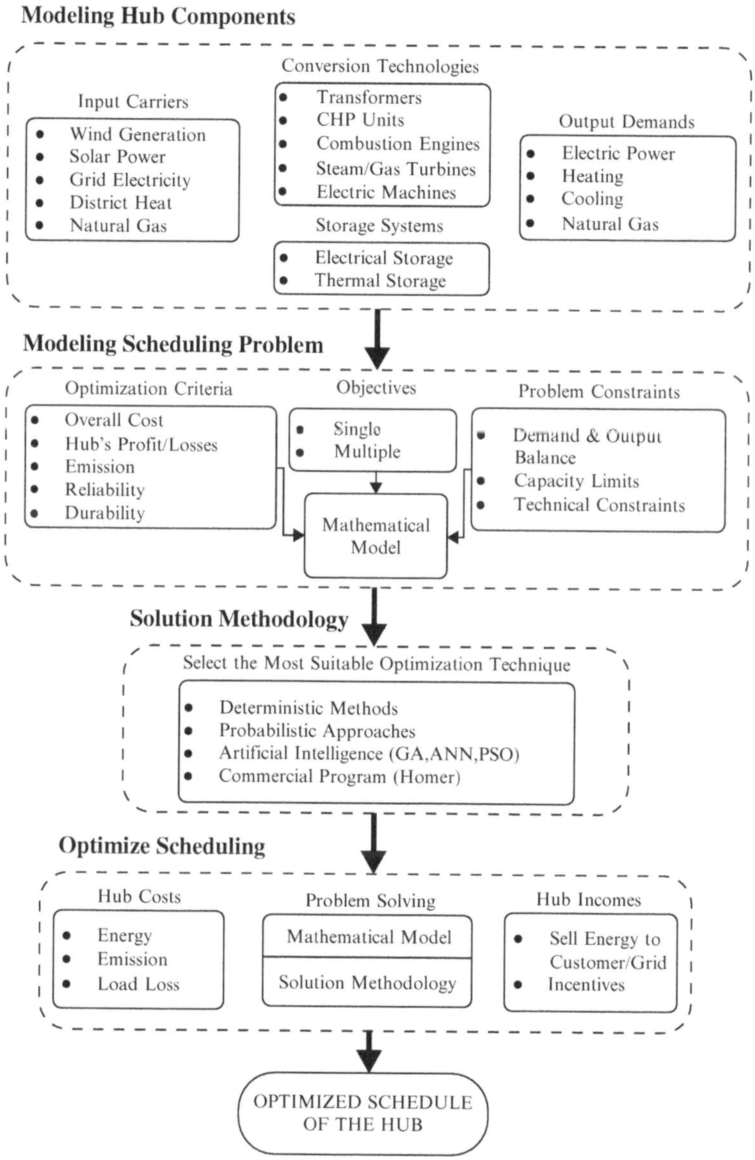

Fig. 9.6 Steps to schedule a renewable energy hub

9.3.2.1 Outline of Energy Hub Scheduling Problem

The problem of scheduling a hub is to determine the optimal operating strategy of the energy hub, subjected to the hub equations and components limits. This optimization problem, which is called multi-carrier optimal dispatch, can be generally described as follows [9]:

Subject to:
Hub equation
Input energy limitations
Physical constraints of units
Properties of dispatch factors
Storage sustainability

Various criteria, such as operating cost, energy efficiency, reliability, and environmental factors, can be considered as the objective function of a renewable energy hub. Equation of the hub, depending on its components, can vary; in presence of storage units, the stored energy should be subtracted from the energy inputs and then be represented in hub equations. Inputs of energy carriers are limited to the output generation of renewable energy sources and the available energy received from networks. Physical constraints of the hub include the power and energy limitations of converters and storage devices. Since there may be several conversion units within the hub, the share of each convertor from the input carrier is defined by a dispatch factor, which should meet requirements of power conservation. Moreover, in order to use these units sustainably, the initial and final level of energy in the storage devices should have the same amount.

9.3.2.2 Detailed Model of Energy Hub Scheduling

In order to mathematically model the aforementioned hub scheduling problem, an energy hub, including renewable energy resources and storage devices, is considered. The general structure of an energy hub is characterized by input power vector P, coupling matrix C, vector of output demands L, stored energy vector E, and storage coupling matrix S. The optimization of energy flows through a renewable energy hub can be formulated as following:

$$\text{Opt.} f = \sum_t f^{(t)}\left(P^{(t)}, \upsilon_c^{(t)}, E^{(t)}\right) \tag{9.6}$$

Subject to

$$L^{(t)} = C^{(t)}P^{(t)} - S^{(t)}\left[E^{(t)} - E^{(t-1)} + E^{ct}\right] \tag{9.7}$$

$$\underline{P} \leq P^{(t)} \leq \overline{P} \tag{9.8}$$

$$\underline{P_c} \leq \upsilon_c^{(t)} P^{(t)} \leq \overline{P_c} \tag{9.9}$$

$$\underline{E} \leq E^{(t)} \leq \overline{E} \tag{9.10}$$

$$0 \leq \upsilon_c^{(t)} \leq 1 \;\&\; \sum \upsilon_c^{(t)} = 1 \tag{9.11}$$

$$E^0 = E^T \tag{9.12}$$

Equation (9.6) is a multi-period objective function. The optimization problem can be defined as a single- or multiple-objective problem, based on the optimization criteria. In this equation, f, which represents the total objective function, is defined as a multi-period objective due to the time dependency brought by the storage devices. It aims to determine the optimal values of $P^{(t)}$, $\upsilon_c^{(t)}$, and $E^{(t)}$. The equality constraint in (9.7) corresponds to the mapping of input carriers to output demand and storage devices for each time slot in the scheduling interval. Input power flows are restricted to lower and upper bounds of available energy from resources, represented by $\underline{P}$, and $\overline{P}$ in (9.8). Equations (9.9) and (9.10) show the physical constraints in the hub structure, regarding power and energy of convertors and storage units. Constraints of power inputs to the hub components are defined in vectors of $\underline{P_c}$, and $\overline{P_c}$.

The stored energies are constrained by $\underline{E}$, and $\overline{E}$, as minimum and maximum capacity limits. The properties, which dispatch factors should display to fulfill energy conversion, are presented in (9.11). Equation (9.12) guarantees sustainable utilization of storage units during the optimization intervals, in which E^0 and E^T are the stored energies at the initial and final hours of the scheduling period. It is worth noting that the discussed model is a generic model for energy hubs; a typical hub may not have some of the aforementioned characteristics or technologies, such as storage devices or multi in- and output converters.

9.3.3 Solution Methodology

The solution algorithm for scheduling a hub is closely related to the hub's mathematical model. Thus, in order to solve the optimization problem of hub scheduling, it is vital to identify the type and characteristics of the problem. In hub scheduling problem, the optimization objective is the first thing to be defined, in terms of its optimization criteria, number of variables, and type of the objective function. As discussed before, there can be various objectives for optimizing operation of an energy hub. Depending on the considered optimization criteria, type of the objective function may change; and it will be a convex or non-convex function. If the energy cost is employed as the optimization criterion, similar to classical optimal power flow (OPF), the cost may be a quadratic function of the power provided by the sources [10]. Minimizing gaseous emissions can also be considered as an objective. These emissions are commonly modeled as polynomial or exponential functions of the source's power output [10]. Moreover, multi-criteria optimizations can be performed through a weighted linear combination of individual objectives, where multi-objective optimization approaches are applied to solve them.

Energy hub optimization equations, as a whole problem, can be divided into different classes, depending on the considered objective function, as well as hub structure and components. Based on the problem structure, a suitable optimization approach should be selected to find the optimal solution. In continuous nonlinear problems, where optimization variables are continuous and the problem is subjected to equality and inequality constraints, nonlinear programming (NLP) approaches are implemented to obtain optimums. Objective functions, including integer variables along with continuous optimization parameters, form mixed-integer problems. Integer variables emerge, when on and off status of components should be involved in the optimization problem. Whether optimization constraints are linear or nonlinear, the problems can be categorized into mixed-integer linear programing (MILP) or mixed-integer nonlinear program (MINLP).

On the other hand, scheduling a hub for multiple time periods rather than a single snapshot results in multi-period (MP) optimization problems. These MP problems can be characterized by continuous or mixed integer variables, and linear or nonlinear constraints. Scheduling of energy hubs including storage units or units with ramping limitations should be optimized for coupling hours during a period, therefore it is a MP optimization.

The general multi-carrier optimal dispatch, presented by (9.6)–(9.12), is a nonlinear, inequality constrained, multivariable, and time-coupled objective function optimization problem. In this problem, despite the convex objective function, the solution space is nonconvex, due to the nonlinearity in constraints. Therefore, numerical methods cannot be utilized to find the global optimum solution for the above problem; they can just provide feasible region in this situation [9]. In order to obtain the global optimums of this nonconvex problem, either heuristic methods should be implemented or nonlinear constraints should be linearized [10].

For an energy hub with a constant coupling matrix, which results in linear transformation of power, optimization problem is subject to linear constraints. Under this condition, dispatch factors are not involved in the objective function as an optimization variable, since they are set to be predefined values. Therefore, the nonlinear constraints of (9.7)–(9.11) in the considered hub scheduling problem are dropped out. In this simple presentation, the convex objective function along with linear constraints forms an optimization problem with convex solution space which can be solved by implementing numerical methods [10]. Constant coupling factor is considered as the simplest structure of hubs, however, in reality each energy carrier is dispatched to several components via different internal dispatch factors. In hubs with multiple input and output components and internal dispatch factors, coupling factors are no longer constant and they should be optimized. These dispatch factors result in nonlinear constraints, which should be either linearized or solved by heuristic methods [11].

9.4 Stochastic Optimization scheduling

9.4.1 Stochastic Renewable Energy Hubs

The problem of scheduling a renewable energy hub and obtaining the optimal values for energy inputs, conversion flows, and stored energies, which was defined in the previous part, has been based on the assumption that hub parameters are definite or perfectly forecasted, and none of the hub inputs, outputs, or external variables were considered as uncertain and stochastic parameters. However, in reality, the quantities for most of these parameters are not completely known and despite the advanced forecasting techniques, there are errors in predictions and they are never fully reliable. Available input energy, output demands, and energy prices are typical uncertain parameters in an energy hub. Generally, minor uncertainties do not cause considerable problems in scheduling of energy hubs. In a renewable-based energy hub, however, the intermittent and variable generation of renewable sources is a serious challenge. As it was mentioned in previous parts, one of the main advantages of implementing hub concept is effectively harnessing intermittent and stochastic renewable energy sources, such as wind and solar energy. An energy hub not only offers various options for using and storing generated power by renewables, but also provides high flexibility for compensating its associated uncertainties.

In order to make the best use of an energy hub, the uncertainties, raised from renewable integration, should be considered in operation scheduling, so that it can be able to manage the deviations in predicted and realized values. In this regard, applying stochastic optimization techniques to the energy hub scheduling problem is necessary for taking into account uncertain nature of renewable energies, and finding the most beneficial solution. There are several methods for addressing uncertainty of parameters, including confidence intervals in forecasts of neural networks, and time series, interval analysis, and fuzzy numbers [12], as well as probability distribution functions (PDFs). Among various uncertainty modeling techniques, PDFs are the most commonly adopted approach in uncertainty studies of renewable energy hubs. Simulation or approximate sampling methods, such as point estimate method (PEM) [11, 13], Monte Carlo Simulation (MCS) [14, 15], and scenario trees [7], are utilized along with these PDFs to enable analyzing the hub optimization problem probabilistically. In order to obtain optimums in the stochastic problem of hub scheduling, various optimization techniques exist, dealing with stochastic parameters. These techniques include stochastic methods [8, 14, 15], as well as robust techniques [13, 16]. Based on the modeling process of uncertainty and the defined scheduling problem, analytical or heuristic optimization methods can be implemented for solving the problem and obtaining the optimal results. In the following, different approaches for addressing uncertainties, as well as solution algorithms are discussed.

9.4.2 Uncertainty Modeling

Stochastic programming has been considered as one of the strongest methods for addressing probabilistic problem [14]. In this method, uncertain parameters are represented by several realizations, in form of scenarios, and associated probabilities. The MCS technique can be used to obtain scenarios based on the probability density functions (PDFs) corresponding to uncertain parameters. Considering the obtained scenarios, the stochastic optimization aims to find the optimal solution for each of the scenarios. These uncertainties are characterized using the PDFs whose statistics are obtained from historical data.

9.4.2.1 Uncertainty in Renewable Energy Resources

The unpredicted behavior regarding the intermittent and volatile characteristic of renewable energy sources, i.e. wind and solar energy, can be realized from the historical data records of these sources. Statistical methods are employed to predict uncertain behavior of renewable sources. Based on the spatial and temporal characteristics of energy source under study, various probability distribution functions can be utilized to fit into the historical data. The performance of models and the best distribution fitted for each data set is selected based on goodness of fit criteria, such as errors between the actual and predicted data of different PDFs [17]. The stochastic models of wind and solar energies are described hereunder, respectively.

(a) Wind speed uncertainty modeling

The intermittent characteristics of wind speed data are usually modeled by PDFs including Weibull, Rayleigh, and Gamma [17]. Among these distribution functions, Rayleigh distribution is the most widely used PDF in modeling statistical characteristics of wind speed data in time scales of energy hub scheduling [8, 15]. The Rayleigh PDF, representing fluctuation of wind speed v, can be expressed by:

$$\mathrm{PDF}(v) = \left(\frac{v}{c^2}\right) \exp\left[-\left(\frac{v^2}{2c^2}\right)\right] \tag{9.13}$$

where c is the scale parameter of the distribution

(a) Solar radiation uncertainty modeling

There are several PDFs which can be utilized in modeling solar irradiance, including Lognormal, Beta, and Logistic distributions [17]. It is shown solar radiation follows Beta distribution the best [17]. The PDF of Beta distribution for solar irradiance r is characterized as follows [18]:

$$F(r) = \frac{\Gamma(\alpha+\beta)}{\Gamma(\alpha)\Gamma(\beta)}\left(\frac{r}{r_{\max}}\right)^{\alpha-1}\left(1-\frac{r}{r_{\max}}\right)^{\beta-1} \tag{9.14}$$

where $r_{\max}$ is the Maximum solar irradiance; α is the Shape parameter; β is the Shape parameter; Γ is the Gamma function [18].

The generated electric power from solar panels can be obtained by (9.4) and (9.5), as described in Sect. 9.3.2.2.

9.4.2.2 Other Sources of Uncertainty

Apart from uncertainties raised by stochastic nature of wind and solar energy sources, an energy hub may encounter several other uncertainties depending on its structure and the setting in which it is operated. Load, energy price, and demand response are other sources of uncertainty that may exit in a typical energy hub [7, 14]. The general procedure for modeling uncertainties corresponding to these parameters is to implement widely implemented normal (Gaussian) PDF, which is characterized by two factors of mean μ, and standard deviation σ as the following [15]:

$$\mathrm{PDF}(x) = \frac{1}{\sigma_x\sqrt{2\pi}}\exp\left[-\frac{(x-\mu_x)^2}{2{\sigma_x}^2}\right] \tag{9.15}$$

9.4.2.3 Scenario Generation and Reduction

In order to enable stochastic evaluation of hub scheduling problem, a finite number of realizations should be obtained from uncertain parameters' PDFs, in form of scenarios. MCS technique [14] is broadly implemented to generate sufficient number of scenarios for uncertain parameters, including wind speed, solar radiation, load, and energy price, by using their assigned PDFs. By combining the obtained scenario sets for each of the uncertain parameters, the final scenario vector can be obtained for the general optimization problem of hub scheduling. Despite comprehensive realization of the stochastic scheduling problem, considering all combinations of the scenarios applies a great computational burden to the solving process of hub scheduling problem. Therefore, it is essential to reduce large number of generated

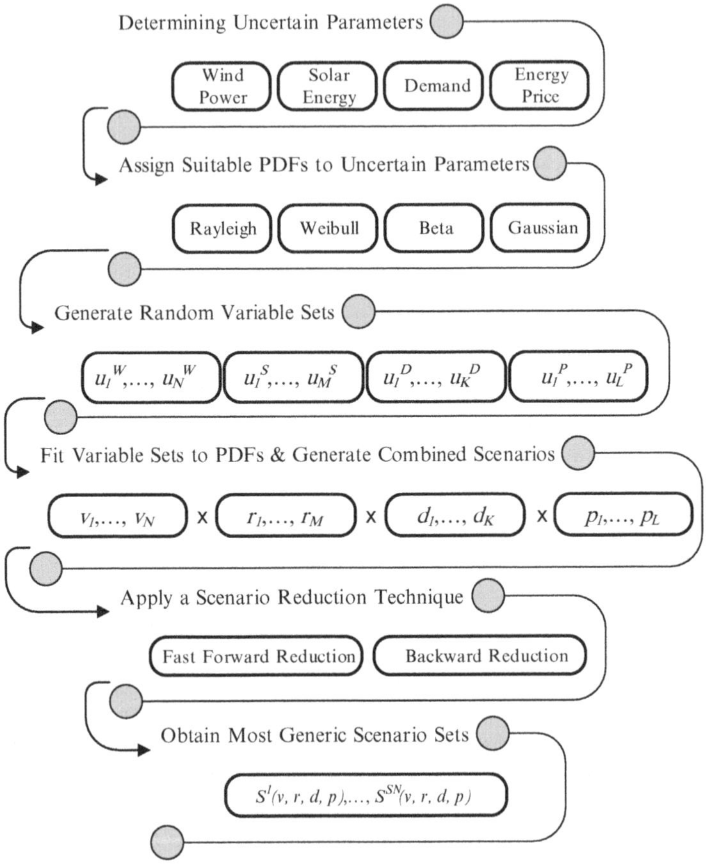

Fig. 9.7 Flowchart of uncertainty modeling for hub stochastic optimization

scenarios to obtain the most generic ones, featuring the prevailing characteristics of uncertain parameters. Different scenario reduction techniques, such as fast forward reduction and backward reduction, are provided by the SCENRED tool of GAMS environment [15]. The best scenario reduction technique can be chosen according to its reduction time and performance accuracy. The whole process of uncertainty modeling for the stochastic optimization of hub scheduling is presented in Fig. 9.7.

9.4.3 Stochastic Scheduling Formulation of Energy Hub

The uncertainty of parameters in a renewable energy hub can be properly incorporated to its scheduling problem by the use of stochastic programming model. In this model, contrary to the deterministic method presented in Sect. 9.3, several realizations of the stochastic parameter are considered in form of scenarios with different probabilities. The scenario-based optimization problem results in the optimal strategy in facing each of the scenarios [15].

9.4.3.1 Objective of Scheduling Energy Hub

Similar to the deterministic form of hub scheduling, the objective function of a stochastic problem can be based on different criteria of operating cost, reliability, and environmental impacts. The main difference in a stochastic problem is that the objective function is a weighted sum of the realized value of optimization criteria for all scenarios. The stochastic objective function of hub scheduling can be generally formulated as:

$$\text{Opt.}\, f = \sum_{t}\sum_{s}\rho_s\left\{f^{(t,s)}\left(P^{(t,s)}, \upsilon_c^{(t,s)}, E^{(t,s)}\right)\right\} \tag{9.16}$$

where ρ_s is the probability of sth scenario; $f^{\,(s,t)}$ is the realized objective function for scenario s at time t.

The uncertainty in hub parameters makes all variables to be scenario-based [14].

9.4.3.2 Constraints

The previously discussed operation constraints in Sect. 9.3 are still valid for the stochastic scheduling. However, constraints should be satisfied for each realization of scenarios. Different operation constraints can be modeled stochastically as the following.

- Energy Balance

The optimal solution in hub scheduling should meet the balance constraints for all types of energy carriers in each scenario s at time t as the following:

$$L^{(t,s)} = C^{(t,s)}P^{(t,s)} - S^{(t,s)}\left[E^{(t,s)} - E^{(t-1,s)} + E^{ct}\right] \tag{9.17}$$

Equation (9.17), in its matrix form, can represent balance of electrical energy flow from input carriers to the demand loads and electrical storage devices, as well as balance of natural gas received from the grid to its converted and stored forms. The storage units cause the scheduling problem to be time-coupled.

- Technical limitations

The internal components of the energy hub and input energy carriers are restricted to technical and practical constraints, which should be considered in all scenarios.

$$P_{\min} \leq P^{(t,s)} \leq P^{\max} \tag{9.18}$$

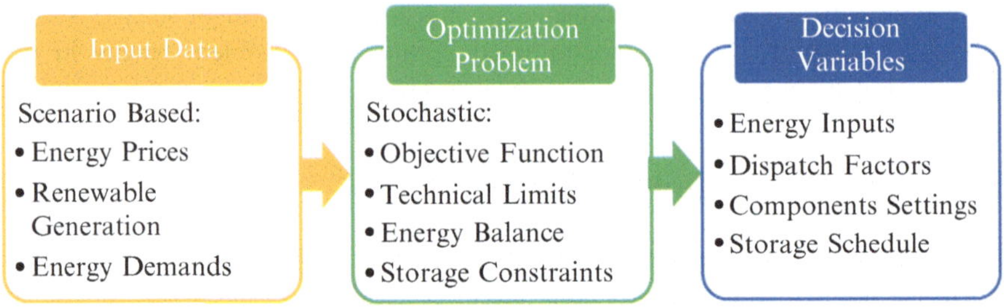

Fig. 9.8 Overview of the stochastic optimization problem for hub scheduling

Equation (9.18) limits the energy flow in the hub to the maximum and minimum allowed value, $P_{\min}$ and $P_{\max}$. This constraint accounts for limitations of input electrical energy, natural gas, and heat, and also power limitations of conversion and storage units.

- Storage Constraints

Inclusion of storage units in the energy hub employs new constraints to the problem:

$$E_{\min} \le E(t, s) \le E^{\max} \tag{9.19}$$

This constraint defines the minimum and maximum allowable limits of the stored energy in storage units, either electrical or thermal.

The stochastic optimization problem of scheduling a renewable energy hub is summarized in Fig. 9.8. As this figure presents, the inputs of the optimization problem are composed of prices of energy carriers, generation of renewable resources, and energy demands, which are scenario based parameters. The scheduling optimization problem consists of a stochastic objective function subjected to technical limitations of hub components, energy balance, and storage constraints. The stochastic optimization will define the amount of each energy carrier input, the dispatch factors, the optimal settings of hub components, and charge and discharge schedule of storage units.

9.4.4 Probabilistic Optimization Methods

In order to address the uncertainty in operation parameters and solve the stochastic optimization problem, various probabilistic methods can be used with different principals. Probabilistic approaches can be categorized into two general groups of (1) numerical and (2) analytical approaches [6]. Each of these groups includes several types of optimization techniques which are depicted in Fig. 9.9.

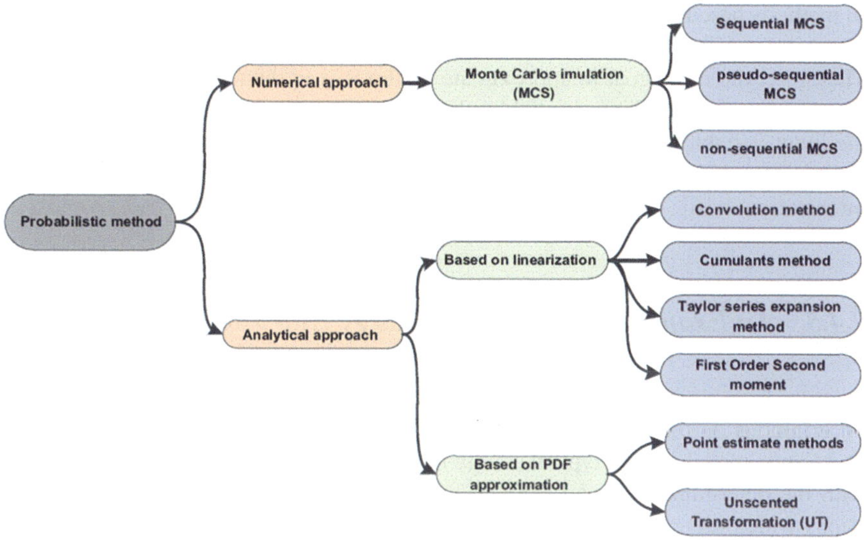

Fig. 9.9 Different categories of probabilistic methods [6]

9.4.4.1 Numerical Approaches

Numerical optimization approaches are applied when a problem is highly nonlinear, complicated, and stochastic [6]. Monte Carlo simulation (MCS) is the basis of numerical optimization methods, and it is utilized in three different types of sequential MCS, pseudo-sequential MCS, and non-sequential MCS [6].

In order to preserve temporal characteristics of time dependent variables, sequential Monte Carlo (SMC) methods are utilized. SMC methods are easy to implement, parallelizable, flexible, and applicable in general settings [6]. However, the large number of random numbers and iterations required in SMC methods results in a high computational burden to the problem. Therefore, Pseudo-sequential Monte Carlo simulation technique is developed, which is much faster than SMC [6]. Non-sequential Monte Carlo simulation method, which is simply a state sampling approach, is widely implemented optimizing probabilistic and stochastic problems.

9.4.4.2 Analytical Approaches

In analytical approaches, both inputs variables and output results are expressed through mathematical equations. In other words, the analytical approaches relate the stochastic inputs variables to the outputs by doing arithmetic with PDFs [6]. In these approaches, the analytical relationship between inputs and outputs is established either by linearization or by approximation. Linearization includes methods of convolution, cumulant, Taylor series expansion, and first order speed moment [6].

In methods based on approximation, PDF of stochastic variables are approximated, since it is easier than approximating a nonlinear transformation function [6]. Therefore, the main goal in approximate methods is to define and obtain the most appropriate samples of input variables that can maintain sufficient information about the input variable's PDF [6]. There are two techniques for this purpose: (1) point estimation method, (2) Unscented transformation. Interested readers can find further information in [6].

9.4.5 An Example of Stochastic Energy Hub Modeling

In order to evaluate the effects of renewable energy sources and their associated uncertainties on the optimal operation of energy hubs, the stochastic optimization problem is modeled for a typical energy hub, and the obtained results are compared with the deterministic model. This example is based on the modeling and evaluations which are conducted in [7]. In this reference, an energy hub, composed of wind turbine, transformer, convertor, CHP unit, boiler, electrical and thermal storage units, as well as demand response is considered as shown in Fig. 9.10. The energy carriers of this hub include electricity, gas, water, and heat.

The objective of optimization problem is to minimize the cost functions of energy from the grid, charge and discharge of electric and thermal storage units, wind

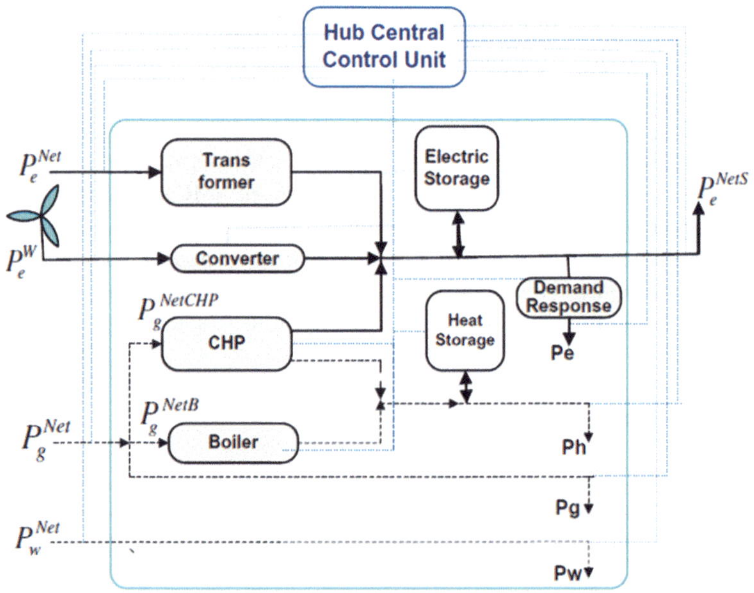

Fig. 9.10 The energy hub under study [7]

zero cost, demand response, reliability, and emission costs [7]. The revenue from selling energy to the grid is modeled as a negative cost. Wind, electricity price, and demand are considered as the uncertain parameters in hub scheduling and the optimization is based on two-stage stochastic programming. Energy carriers are considered as decision variables of stage one, which do not correspond to different wind scenarios of the second stage. On the other hand, variables of energy storage, demand response, and energy not supplied (ENS) take different values in different wind scenarios [7]. It is assumed that minimizing the operation costs of energy hub is the main objective of the scheduling problem. The stochastic formulation of this objective function can be described as follows [7]. The nomenclature for the utilized parameters is presented in Appendix.

$$\begin{aligned}
\text{Min} \sum_{s=1}^{10}\sum_{t=1}^{24} & \left(\pi_e^{\text{Net}}(t,s)\,\text{Pro}_{\text{RTP}}(s) P_e^{\text{Net}}(t) + \sum_{s=1}^{10}\sum_{t=1}^{24} \pi_e^{w} P_e^{w}\left(t,s\right)\text{Pro}_w(s)\right) \\
&+ \sum_{t=1}^{24}\left(\left[\pi_g^{\text{Net}} P_g^{\text{Net}}(t)\right] + \left[\pi_g^{\text{Net}} P_g^{\text{Net}B}(t)\right] + \left[\pi_w^{\text{Net}} P_w^{\text{Net}}(t)\right]\right) \\
&+ \left[\pi_h^{s}\left(P_h^{\text{ch}}(t) + P_h^{\text{dis}}(t)\right)\right] + \left[\sum_{\text{em}=1}^{3} \pi_{\text{em}} \text{EF}_{\text{em}}^{\text{Net}} P_{\text{em}}^{\text{Net}}(t)\right] \\
&+ \left[\sum_{\text{em}=1}^{3} \pi_{\text{em}} \text{EF}_{\text{em}}^{\text{CHP}} P_{\text{em}}^{\text{NetCHP}}(t)\right] + \left[\left[\sum_{\text{em}=1}^{3} \pi_{\text{em}} \text{EF}_{\text{em}}^{B} P_{\text{em}}^{\text{Net}B}(t)\right]\right] \\
&+ \sum_{s=1}^{10}\sum_{t=1}^{24}\left(\left[\pi_e^{\text{Net}}(t)\left(P_e^{\text{ch}}(t,s) - P_e^{\text{dis}}\left(t,s\right)\right)\right] + \left[\pi_e^{s}\left(P_e^{\text{ch}}(t,s) + P_e^{\text{dis}}\left(t,s\right)\right)\right]\right. \\
&\left. + \left[\pi_e^{\text{DR}}\left(P_e^{\text{shdo}}(t,s) + P_e^{\text{shup}}\left(t,s\right)\right)\right] + \left[\pi_e^{\text{ENS}} P_e^{\text{ENS}}(t,s)\right]\right)
\end{aligned} \tag{9.20}$$

The optimization is subject to various constraints as follows:

- Wind Power

Wind power is limited to the mechanical power generated by the wind speed. Thus, for each wind speed scenario with probability of $\text{Pro}_w(t,s)$, there is a different limit for wind power as:

$$P_e^{W}(t,s) = \begin{cases} 0 & w(t,s) < w_{\text{ci}} \\ P_r\left(z - y \times w(t,s) + x \times w^2(t,s)\right) & w_{\text{ci}} \le w(t,s) < w_r \\ P_r & w_r \le w(t,s) < w_{\text{co}} \\ 0 & w(t,s) \ge w_{\text{co}} \end{cases} \tag{9.21}$$

- Energy Balance

There should be always a balance between the consumed energy in the demand side, the received energy to the input layer, and the stored energy in the storage units. Stochastic electricity demand Pe(t,s) should be equal to the electricity received from the network, CHP, wind turbine, electrical storage units, minus the demand response, stored electricity, and energy not supplied. Heat, gas, and water demands are considered certain parameter and should be supplied accordingly. The formulations for balancing energy carriers is as following:

$$\begin{aligned}\mathrm{Pro}_d(s)P_e(t,s) &= \left[A^{\mathrm{Net}}\eta_{\mathrm{ee}}^{T}P_e^{\mathrm{Net}}(t)\right] + \left[A^{\mathrm{CHP}}\eta_{\mathrm{ge}}^{\mathrm{CHP}}P_g^{\mathrm{NetCHP}}(t)\right] \\ &+ \left[A^{\mathrm{Wind}}\eta_{\mathrm{ee}}^{\mathrm{con}}P_g^{W}(t,s)\right] + \left[P_e^{\mathrm{dis}}(t,s) - P_e^{\mathrm{ch}}(t,s)\right] \\ &+ \left[P_e^{\mathrm{shdo}}(t,s) - P_e^{\mathrm{shup}}(t,s)\right] + \left[P_e^{\mathrm{ENS}}(t,s)\right]\end{aligned} \tag{9.22}$$

$$P_h(t) = \left[A^{\mathrm{CHP}}\eta_{\mathrm{gh}}^{\mathrm{CHP}}P_g^{\mathrm{NetCHP}}(t)\right] + \left[\eta_{\mathrm{gh}}^{B}P_g^{\mathrm{Net}B}(t)\right] + \left[P_h^{\mathrm{dis}}(t) - P_h^{\mathrm{ch}}(t)\right] \tag{9.23}$$

$$P_g(t) = \left[P_g^{\mathrm{Net}}(t)\right] - \left[P_g^{\mathrm{NetCHP}}(t)\right] - \left[P_g^{\mathrm{Net}B}(t)\right] \tag{9.24}$$

$$P_w = \left[P_w^{\mathrm{Net}}(t)\right] \tag{9.25}$$

- Network Constraints

Energy from electricity, gas, and water networks is limited, and this limitation should be considered in scheduling problem:

$$P_e^{\mathrm{Net\ min}} \le P_e^{\mathrm{Net}}(t) \le P_e^{\mathrm{Net\ max}} \tag{9.26}$$

$$P_g^{\mathrm{Net\ min}} \le P_g^{\mathrm{Net}}(t) \le P_g^{\mathrm{Net\ max}} \tag{9.27}$$

$$P_w^{\mathrm{Net\ min}} \le P_w^{\mathrm{Net}}(t) \le P_w^{\mathrm{Net\ max}} \tag{9.28}$$

- Converter Constraints

Physical constraints of converters, including transformer, CHP, and boiler, should also be taken into account:

$$\eta_{\mathrm{ee}}^{T}P_e^{\mathrm{Net}}(t) \le P^{T} \tag{9.29}$$

$$\eta_{\text{ge}}^{\text{CHP}} P_g^{\text{NetCHP}}(t) \leq P^{\text{CHP}} \tag{9.30}$$

$$\eta_{\text{gh}}^{B} P_g^{\text{Net}B}(t) \leq P^{B} \tag{9.31}$$

- Storage Constraints

Constraints regarding electrical and thermal storage units can be formulated as the following.

Electrical Storage:

$$P_e^S(t,s) = P_e^S(t-1,s) + P_e^{\text{ch}}(t,s) - P_e^{\text{dis}}(t,s) - P_e^{\text{loss}}(t,s) \tag{9.32}$$

$$P_e^{\text{loss}}(t,s) = \alpha_e^{\text{loss}} P_e^{\text{loss}}(t,s) \tag{9.33}$$

$$\alpha_e^{\min} P_e^M \leq P_e^S(t,s) \leq \alpha_e^{\max} P_e^M \tag{9.34}$$

$$\alpha_e^{\min} \frac{1}{\eta_e^{\text{ch}}} P_e^M I_e^{\text{ch}}(t,s) \leq P_e^{\text{ch}}(t,s) \leq \alpha_e^{\max} P_e^M \frac{1}{\eta_e^{\text{ch}}} P_e^M I_e^{\text{ch}}(t,s) \tag{9.35}$$

$$0 \leq I_e^{\text{ch}}(t,s) + I_e^{\text{dis}}(t,s) \leq 1 \tag{9.36}$$

Thermal Storage:

$$P_h^S(t) = P_h^S(t-1) + P_h^{\text{ch}}(t) - P_h^{\text{dis}}(t) - P_h^{\text{loss}}(t) \tag{9.37}$$

$$P_h^{\text{loss}}(t) = \alpha_h^{\text{loss}} P_h^{\text{loss}}(t) \tag{9.38}$$

$$\alpha_h^{\min} P_h^M \leq P_h^S(t) \leq \alpha_h^{\max} P_h^M \tag{9.39}$$

$$\alpha_h^{\min} \frac{1}{\eta_h^{\text{ch}}} P_h^M I_h^{\text{ch}}(t) \leq P_h^{\text{ch}}(t) \leq \alpha_h^{\max} P_h^M \frac{1}{\eta_h^{\text{ch}}} P_h^M I_h^{\text{ch}}(t) \tag{9.40}$$

$$0 \leq I_h^{\text{ch}}(t) + I_h^{\text{dis}}(t) \leq 1 \tag{9.41}$$

- Demand Response Constraints

Implementing demand response programs is restricted to the limits on up and down demand shifts:

$$\sum_{t=1}^{24} P_e^{\text{shup}}(t,s) = \sum_{t=1}^{24} P_e^{\text{shdo}}(t,s) \tag{9.42}$$

$$0 \le P_e^{\text{shup}}(t,s) \le \text{LPF}^{\text{shup}} P_e(t,s)\, I_e^{\text{shup}}(t,s) \tag{9.43}$$

$$0 \le P_e^{\text{shdo}}(t,s) \le \text{LPF}^{\text{shdo}} P_e(t,s)\, I_e^{\text{shdo}}(t,s) \tag{9.44}$$

$$0 \le I_e^{\text{shup}}(t,s) + I_e^{\text{shdo}}(t,s) \le 1 \tag{9.45}$$

Various indices can be utilized in order to assess reliability of the scheduled operation. Loss of load expectation (LOLE), loss of load probability (LOLP), and energy not supplied (ENS) are the most important reliability indices that are used in reliability assessments. These indices are calculated by the following equations:

$$\text{LOLE} = \sum_{t=1}^{24} t_{P_e^{\text{ENS}}(t)} \tag{9.46}$$

$$\text{LOLP} = \sum_{t=1}^{24} \frac{t_{P_e^{\text{ENS}}(t)}}{T} \tag{9.47}$$

$$\text{ENS} = \sum_{t=1}^{24} P_e^{\text{ENS}}(t) \tag{9.48}$$

MCS technique is used to generate scenarios based on the forecasted wind power, energy price, and energy demand. The backward/forward technique in GAMS SCENRED tool is employed to obtain 10 most prevailing scenarios. The optimal schedule for the energy hub in Fig. 9.11 is obtained for two cases of certain and uncertain parameter. Different factors are investigated when the result of scheduling of the energy hub is obtained, including importing or selling grid electricity, charging and discharging power of electrical and thermal storage, shiftable power to up and down, importing gas power for boiler and CHP, and the ENS [7].

In case of wind, price, and electricity demand uncertainty, as the results in Fig. 9.11 show, more electricity is purchased from grid and less is sold to the grid. Therefore, the operation cost has increased which is the actually cost of enhanced hub reliability.

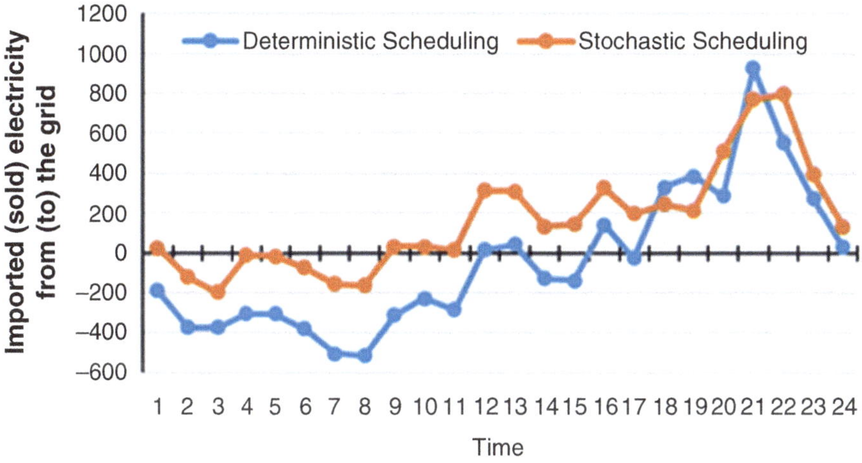

Fig. 9.11 Imported (sold) electricity from (to) the grid in 24 hours

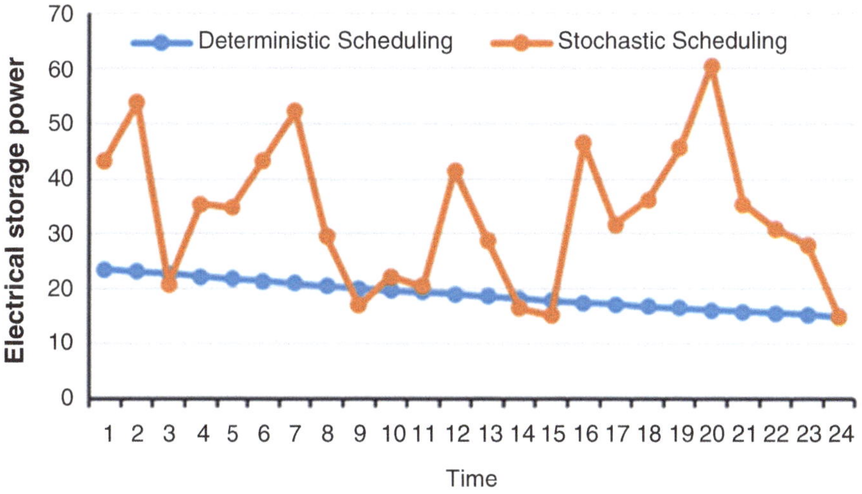

Fig. 9.12 Charge and discharge schedule of electrical storage

In addition, as Fig. 9.12 depicts, electrical storage is charged when electricity price and demand are low and it is discharged in times of high electricity price and demand. The charge and discharge schedule of electrical storage is more sensible in the case with uncertainty compared to the deterministic case, which shows the significant role of electrical storage in uncertain environments.

As Fig. 9.13 illustrates, electricity power shifts up and down according to the low and high prices of electricity when optimization is done deterministically. In the stochastic scheduling, power shifts up with high wind, low price and demand;

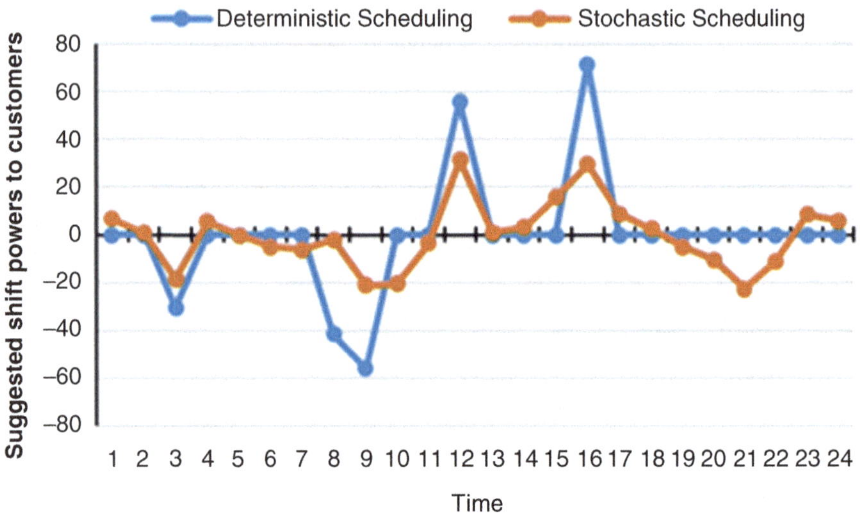

Fig. 9.13 Suggested shift powers to customers

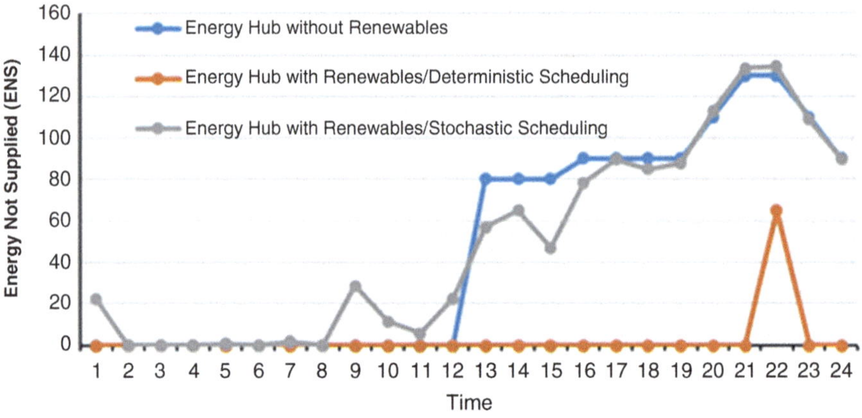

Fig. 9.14 Evaluation of energy not supplied for three cases

and it shifts down when demand and price are high and wind is low. Among different uncertainties of wind, price, and demand, demand shifting is more sensitive to price uncertainty.

Investigating

ENS for three cases of energy hub without renewable energy, renewable energy hub with perfect forecast, and renewable energy hub with uncertainty (Fig. 9.14) reveals that integration of renewable energy resources to the energy hub reduces ENS and enhances system reliability. By taking into account the uncertainties

Table 9.1 Total operation costs and reliability assessments specification of thermal units

Energy hub characteristics	Costs	ENS	LOLE	LOLP
Without renewables	953,633.8	1170	12	0.5
With renewables	190,300	65	1	0.042
Deterministic scheduling	749,196.6	1181.3	13.1	0.546

of parameters and scheduling the energy hub stochastically, the resulting ENS increases in comparison to the renewable integrated case with perfect forecast. Despite this, the results of stochastic scheduling of renewable based hub shows that integrating uncertain renewable resources is sometimes beneficial to system reliability compared to the case of energy hub without renewables.

Total operation cost and reliability assessments, presented in Table 9.1 indicate that operation cost of energy hub without renewables is higher than the other cases and hub reliability is lower. In the case of renewable energy hub with perfect forecast, a significant positive change happens in operation costs and system reliability. Nevertheless, considering parameters' uncertainties in hub scheduling brings about higher operation costs and lower reliability level. This investigation shows that scheduling renewable energy hubs with the assumption of certain parameters does not lead to realistic results and they might be misleading. Therefore, it is necessary to take into account uncertainty of parameters when optimizing operation of a hub in order to find optimal and realistic schedules.

9.5 Conclusion

In this chapter, the energy scheduling of renewable-based energy hub was put under investigation. Introducing the main elements of a renewable-based energy hub, it was discussed how the concept of energy hub can help the system operators to properly model the intermittent nature of renewable resources. The mathematical model of energy scheduling problem in renewable-based energy hubs is extracted based on input and output relations of energy hubs. Also, procedures need to be followed to model the stochastic behavior of renewable energy sources are explained. To reach an efficient solution algorithm for such probabilistic optimization problems, i.e. energy scheduling of renewable-based energy hubs, the stochastic algorithm is classified and it is shown how these algorithms should be applied to deal with uncertain parameters of this problem. Step by step implementation of energy scheduling in an energy hub is explained by an example and it is discussed how the presence of renewable-based DG units can affect the operation of an energy hub.

Nomenclature

Indices

em	emission of
s	scenario
t	time

Variables

$I_e^{\mathrm{ch}}(t)$	binary for charge of electrical storage
$I_e^{\mathrm{dis}}(t)$	binary for discharge of electrical storage
$I_e^{\mathrm{shdo}}(t)$	binary for shift down of DR
$I_e^{\mathrm{shup}}(t)$	binary for shift up of DR
$I_h^{\mathrm{ch}}(t)$	binary for charge of thermal storage
$I_h^{\mathrm{dis}}(t)$	binary for discharge of thermal storage
$p_e^{\mathrm{ch}}(t)$	charge power of electrical storage
$P_e^{\mathrm{dis}}(t)$	discharge power of electrical storage
$P_e^{\mathrm{ENS}}(t)$	energy not supplied
$P_e^{\mathrm{Net}}(t)$	purchased or sold power from network
$P_e^{\mathrm{loss}}(t)$	loss power of electrical storage
$P_e^{s}(t)$	electrical storage contain
$P_e^{\mathrm{shdo}}(t)$	shifted down power via DR
$P_e^{\mathrm{shup}}(t)$	shifted up power via DR
$P_g^{\mathrm{Net}}(t)$	imported gas power from network
$P_g^{\mathrm{Net}B}(t)$	purchased gas for boiler from network
$P_g^{\mathrm{NetCHP}}(t)$	purchased gas for CHP from network
$P_h^{\mathrm{ch}}(t)$	charge power of thermal storage
$P_h^{\mathrm{dis}}(t)$	discharge power of thermal storage
$P_h^{\mathrm{loss}}(t)$	loss power of thermal storage
$P_h^{S}(t)$	thermal storage contain
$P_w^{\mathrm{Net}}(t)$	purchased water from network

Constants

α_e^{loss}	loss efficiency of electrical storage
α_h^{loss}	loss efficiency of thermal storage
η_e^{ch}	electrical storage charge efficiency
η_e^{dis}	electrical storage discharge efficiency
η_h^{ch}	thermal storage charge efficiency

${\eta_h}^{\mathrm{dis}}$	thermal storage discharge efficiency
${\eta_{\mathrm{ee}}}^{\mathrm{con}}$	electricity efficiency of ac/ac converter
${\eta_{\mathrm{ee}}}^{T}$	electricity efficiency of transformer
${\eta_{\mathrm{ge}}}^{\mathrm{CHP}}$	gas to electricity efficiency of CHP
${\eta_{\mathrm{gh}}}^{\mathrm{CHP}}$	gas to heat efficiency of CHP
${\eta_{\mathrm{gh}}}^{B}$	gas to heat efficiency of boiler
${\pi_e}^{\mathrm{DR}}$	DR operation cost
${\pi_e}^{\mathrm{ENS}}$	energy not supplied cost
${\pi_e}^{\mathrm{Net}}(t)$	hourly electricity price
${\pi_e}^{S}$	electrical storage operation cost
${\pi_e}^{w}$	produced wind power cost
π_{em}	emission cost for
${\pi_g}^{\mathrm{Net}}$	gas price
${\pi_h}^{s}$	thermal storage operation cost
${\pi_w}^{\mathrm{Net}}$	water price
A^{CHP}	CHP availability
A^{Net}	electricity network availability
A^{wind}	wind turbine availability
$\mathrm{EF}_{\mathrm{em}}$	emission factor for CHP, boiler and grid
EFL	equivalent loss factor
$\mathrm{EFL}^{\max}$	maximum equivalent loss factor
$\mathrm{LPF}^{\mathrm{shdo}}$	load participation factor for shift down of DR
$\mathrm{LPF}^{\mathrm{shup}}$	load participation factor for shift up of DR
P^{B}	boiler capacity
P^{CHP}	CHP capacity
P^{T}	transformer capacity
$P_e(t)$	hourly electricity demand
${P_e}^{M}$	electrical storage capacity
${P_e}^{\mathrm{Netmax}}$	maximum capacity of electricity network
${P_e}^{w}(t)$	hourly wind power
$P_g(t)$	hourly gas demand
${P_g}^{\mathrm{Netmax}}$	maximum capacity of gas network
$p_h(t)$	hourly heat demand
P_r	rated power of wind turbine
$P_w(t)$	hourly water demand
${P_w}^{\mathrm{Netmax}}$	maximum capacity of water network
$\mathrm{Pro}_d(s)$	reduced demand scenarios
$\mathrm{Pro}_{\mathrm{RTP}}(s)$	reduced RTP scenarios
$\mathrm{Pro}_w(s)$	reduced wind power scenarios
$w(t)$	hourly wind speed
w_{ci}	cut-in state of wind turbine
w_{co}	cut-out state of wind turbine

References

1. Geidl M et al (2007) The energy hub–a powerful concept for future energy systems. In: Third annual Carnegie mellon conference on the electricity industry, vol 13
2. Ma T, Wu J, Hao L (2017) Energy flow modeling and optimal operation analysis of the micro energy grid based on energy hub. Energy Convers Manag 133:292–306
3. Krause T et al (2011) Modeling interconnected national energy systems using an energy hub approach. In: PowerTech, 2011 IEEE Trondheim. IEEE
4. Bhandari B et al (2015) Optimization of hybrid renewable energy power systems: A review. Int J Precis Eng Manuf Green Technol 2(1):99–112
5. Parvini Z et al (2016) An analytical framework for operational reliability studies of highly wind integrated power systems. In: 2016 International Conference on probabilistic methods applied to power systems (PMAPS), IEEE
6. Aien M, Hajebrahimi A, Fotuhi-Firuzabad M (2016) A comprehensive review on uncertainty modeling techniques in power system studies. Renew Sustain Energ Rev 57:1077–1089
7. Pazouki S, Haghifam M-R, Moser A (2014) Uncertainty modeling in optimal operation of energy hub in presence of wind, storage and demand response. Int J Electr Power Energy Syst 61:335–345
8. Dolatabadi A et al (2017) Optimal stochastic design of wind integrated energy hub. IEEE Trans Ind Inf 99:1
9. Geidl M, Andersson G (2007) Optimal coupling of energy infrastructures. In: Power Tech, 2007 IEEE Lausanne. IEEE
10. Geidl M (2007) Integrated modeling and optimization of multi-carrier energy systems. Dissertation
11. Geidl M, Andersson G (2007) Optimal power flow of multiple energy carriers. IEEE Trans Power Syst 22(1):145–155
12. Vaccaro A, Pisani C, Zobaa AF (2015) Affine arithmetic-based methodology for energy hub operation-scheduling in the presence of data uncertainty. IET Gener Transm Distrib 9(13):1544–1552
13. Moeini-Aghtaie M et al (2014) Multiagent genetic algorithm: an online probabilistic view on economic dispatch of energy hubs constrained by wind availability. IEEE Trans Sustain Energy 5(2):699–708
14. Vahid-Pakdel MJ et al (2017) Stochastic optimization of energy hub operation with consideration of thermal energy market and demand response. Energy Convers Manag 145:117–128
15. Dolatabadi A, Mohammadi-Ivatloo B (2017) Stochastic risk-constrained scheduling of smart energy hub in the presence of wind power and demand response. Appl Therm Eng 123:40–49
16. Parisio A, Del Vecchio C, Vaccaro A (2012) A robust optimization approach to energy hub management. Int J Electr Power Energy Syst 42(1):98–104
17. Abdulkarim A, Abdelkader SM, John Morrow D (2015) Statistical analyses of wind and solar energy resources for the development of hybrid microgrid. In: 2nd International congress on energy efficiency and energy related materials (ENEFM2014), Springer, Cham
18. Karaki SH, Chedid RB, Ramadan R (1999) Probabilistic performance assessment of autonomous solar-wind energy conversion systems. IEEE Trans Energy Convers 14(3):766–772

Chapter 10
Risk-Constraint Scheduling of Storage and Renewable Energy Integrated Energy Hubs

Parinaz Aliasghari, Manijeh Alipour, Mehdi Jalali, Behnam Mohammadi-Ivatloo, and Kazem Zare

10.1 Introduction

Energy security and environment concerns related to the energy efficiency, exhaustible fossil resources and greenhouse gas emissions have been increased in recent years. In order to deal with these concerns, two ways are taken into account: firstly, improving energy efficiency and secondly, utilizing renewable energy resources as a promising way to deal with these concerns. Residential, commercial, and industrial customers have various kinds of demands such as gas, heat, and electricity. On the other hand, energy service companies have separate and independent infrastructures to meet different demands. In order to improve energy management, researches and practical attempts were focused on the optimal planning and scheduling of energy distribution. In the last two decades, energy hub has been raised as a promising and powerful concept to increase the overall reliability and efficiency, reduce the consumption of energy and the emission of greenhouse gases [1, 2]. This concept has been used to describe an integrated system with input and output ports. It consists of various kinds of units such as distributed energy resources, connecting equipment to upstream, converters, energy storages, and transformers. Additionally, the energy hub could fulfill the output demands both connecting with upstream networks and with own generators and storage units. A considerable amount of literature has been published on energy hub

P. Aliasghari · M. Alipour (✉) · M. Jalali · K. Zare
Faculty of Electrical and Computer Engineering, University of Tabriz, Tabriz, Iran
e-mail: p.aliasghari@tabrizu.ac.ir; alipour@tabrizu.ac.ir; mehdi.jalali@tabrizu.ac.ir; kazem.zare@tabrizu.ac.ir

B. Mohammadi-Ivatloo
Department of Electrical and Computer Engineering, University of Tabriz, Tabriz, Iran
e-mail: bmohammadi@tabrizu.ac.ir

B. Mohammadi-Ivatloo, F. Jabari (eds.), *Operation, Planning, and Analysis of Energy Storage Systems in Smart Energy Hubs*,
https://doi.org/10.1007/978-3-319-75097-2_10

concept [3–5]. These studies have been focused on planning, operating, scheduling, and reliability issues of energy hubs. The authors in [3] have utilized a robust method to optimize an energy hub operation consisting distributed generator, storage and converter units. The aim of this paper is to minimize the cost function of energy hub. The importance of connecting several energy hubs together has been analyzed in [4]. The case studies reported in [5] indicate that the more connection among the energy hubs causes the improvement of both economical and environment performances. Environmental effects of energy hubs have been investigated in [6]. In this paper, the multi-objective function consists of both economic and environmental performances of energy hub. An iterative optimization algorithm has been presented in [7] to solve the optimization problem of energy hub. Different scenarios have been used to evaluate the various conditions in terms of the existence or absence of energy storage systems, renewable energy sources, and various electricity tariffs. In [5], a goal attainment based technique has designed to solve the multi-objective problem of optimal energy management in energy hubs with nonlinear constraints. A model of residential energy hubs has been utilized in [8] for the energy management of a smart home. The main goal of this study was to determine that simultaneous planning and operation of gas and electrical infrastructures are more efficient than separate planning and operation. In [9] the classical model of energy hubs has been upgraded to the smart energy hub model in the smart environment. Throughout the study game theory methods have been used to simulate the demand side management to decrease the ratio of peak to average. The rendered model in [9] has been considered in [10]. The authors have extended demand response (DR) programs for both natural gas and electricity to reduce the energy consumption of the customer side. The authors in [11] have implemented demand response programs for both electrical and thermal demands to improve the efficacy of the proposed energy hub in the presence of wind farm. The model based on energy flow has been introduced in [12] to schedule an energy hub system with energy storage units in both input and output sides to meet daily electricity, cooling, and heating demands. A mixed-integer nonlinear programming (MINP) model was utilized to solve the scheduling problem while maximizing the profit. In [13], the reliability metrics of the proposed energy hub have been evaluated by considering the dynamic behavior of loads, which has been modeled by Markov-chains and Monte-Carlo simulations. To answer the social concerns about air polluting and global warming, new forbidden lows have been passed to limit fossil fuel usage. Moreover, various kinds of incentives are considered for investing on renewable energy resources. So, recent trends in reducing fossil fuel consumption have led to a proliferation of studies about renewable energy resources such as wind and solar. A technique to integrate decentralized energy units at neighborhood scale has been stated in [14]. In order to determine the size and value of energy units based on energy-autonomy and ecological factors, the energy demands and local energy storage units of building are taken into account in the urban area. Reference [15] has rendered an incentive framework for installing the units of distributed generation for optimal designing of an energy hub. The output demands of the proposed energy hub, concluding electricity and gas, have been met by taking into account the constraint of carbon dioxide emission. One major challenge of utilizing renewable energy resources is their uncertainty behavior. It is essential that an appropriate method be used as the uncertainty model of generated power of RES units. The stochastic method has been rendered in [16] to design an energy hub equipped with integrated wind power plants. The presented stochastic method has been formalized as MILP problem for optimal designing of energy hub in time horizon. For investigating the effectiveness of the proposed method, the indexes of reliability such as EENS (expected energy not supply) and LOLE (loss of load expectation) have been considered. The authors in [17] have used multi-agent genetic algorithm (MAGA) as heuristic technique to optimize an online probabilistic economic dispatch (ED) model of systems equipped with energy hubs and WTs. PV units based residential energy hub have been introduced in [18]. The aim of this study was minimizing the consumption of electricity and gas energy to reduce the peak demand and consumer's bill. Implementing of the model under the smart grids concluding auto decision technology has increased the efficiency and punctuality of the proposed method. A linear formulation of the optimal designed energy hub taking into account the adequate reliability indexes as constraint was rendered in [19, 20]. Power to gas technology has been installed in the proposed renewable energy hub in [21] to produce hydrogen from surplus power generating of PV and WT. The hydrogen can provide the fuel of transportation fleet or enter into the natural gas piles. The main aim of this paper is to extend the energy hub model by implementing various technology of energy generation while decreasing the cost of energy producing and air pollution.

In the current chapter, a renewable-based energy hub (REH) which contains WT, PV, energy storages, boiler, and combined heat and power (CHP) units is considered. One important challenge of implementing RES is the existing uncertainties in their power generations. In this regard, an optimal stochastic short-term scheduling considering the uncertainties of the renewable source generations is presented. To model the uncertainties a scenario-based technique is utilized while applying a proper scenario generation and reduction method. Additionally, risk management problem has been employed considering conditional value-at-risk (CVaR) as a risk measurement in the short-term scheduling of the REH.

10.2 Risk Controlling Model

Scheduling of the energy hub in the stochastic environment has increased the value of the risk. It is necessary that the risk be controlled in the satisfying range. Various approaches based on measuring risk have been applied in the literature [22] in order to schedule the operation of conventional players including generators, retailers, and large consumers. Value-at-risk (VAR) has been introduced as a useful meter to estimate the monetary losses of large consumers or generators caused by uncertain behavior of effective players such as fluctuating of generation in RES with a specified confidence degree. The extension version of VAR meter which named CVaR is utilized in this study. It is known as one of the most practical and impressive risk meters in stochastic programing. This index provides all four specifications that describe a coherent risk measure, i.e., subadditivity, monotonicity, translation invariance, and positive homogeneity.

To obtain the CVaR at α degree of confidence, α − CVaR is equal with the expected cost of the $(1-\alpha)$ of 100% of scenarios taking into account the highest cost [22]. The CVaR can be formulated as equation while the uncertainties are presented in discrete form:

$$< \text{Min VaR} + \frac{1}{1-\alpha} \times \left\{ \max \sum_{\omega \in N_\omega} \pi_\omega \times \xi_\omega \right\} \tag{10.1}$$

Subject to:

$$\text{TC}_\omega - \text{VaR} \leq \xi_\omega \tag{10.2}$$

$$\xi_\omega \geq 0 \tag{10.3}$$

In the above equations, if the value of total cost, TC_ω, at ωth scenario is lower than VaR, ξ_ω is set to zero. Otherwise, the value of ξ_ω is equal to the difference of TC_ω and VaR. In other words, the optimal value of cost is equal to VaR under the confidence degree α [23].

10.3 Configuration of REH

Currently, different kinds of consumers can provide their various kinds of energy demands, independently. A system based on energy hub concept can provide the energy demands from combined energy infrastructures which behave more reliable and economical than separate energy carriers. The suggested REH is connected to the gas and electricity grids in the input side. It could also receive wind and solar radiation as input element to utilize the WT and PV cells. Additionally, it delivers electricity and thermal energy to meet the output side demands. The structure of the energy hub and its components including CHP, WT, PV, boiler, electrical and thermal storage units as well as input/output multi carriers are depicted in Fig. 10.1.

10.3.1 REH Scheduling Objective Function

The aim of this chapter is to schedule the proposed REH taking into account the optimal operation in time horizon. As it can be seen in Table 10.1, different paths demonstrate the potential capability of the REH's components to meet the output demands. Even though some of these paths are not cost-effective under normal conditions, they can be utilized in emergency conditions such as deficit of natural gas, electricity, and congestion of electrical and natural gas connection lines. According to this description, minimization of purchased energy from the upstream

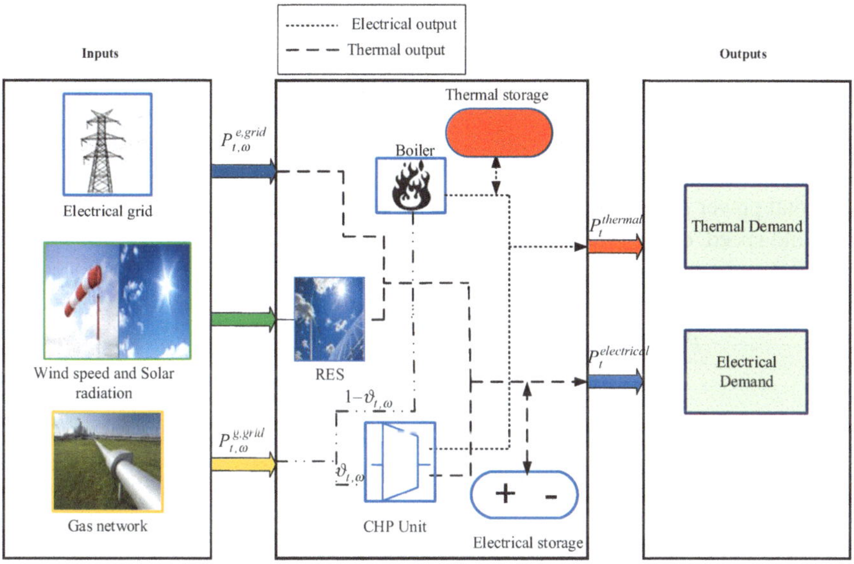

Fig. 10.1 Configuration of the REH model

Table 10.1 Potential capabilities of REH components in supplying demands

Demand	WT	PV	CHP	Boiler	TSU	ESU
Electrical	✓	✓	✓	–	–	✓
Thermal	–	–	✓	✓	✓	–

grid and the start-up/shut-down cost of boiler and CHP units are formulated as objective function in (10.4).

$$\text{Min} \sum_{t=1}^{N_t} \left\{ \begin{array}{l} \sum_{\omega=1}^{N_\omega} \pi_\omega \times \left(P_{\omega,t}^{e,\text{grid}} * \lambda_t^e + P_{\omega,t}^{g,\text{grid}} * \lambda_t^g \right) \\ + \text{SUC} \times u_{\text{SU},t}^{\text{CHP}} + \text{SDC} \times u_{\text{SD},t}^{\text{CHP}} \end{array} \right\} \tag{10.4}$$

where $P_{\omega,t}^{e,\text{grid}}$ and $P_{\omega,t}^{g,\text{grid}}$ are the amount of purchased electricity and gas from upstream grids, respectively. In addition, SUC and SDC are the cost of the startup/shutdown of CHP unit, respectively.

10.3.2 Components of REH

10.3.2.1 Wind Turbine

Despite intermittent and uncertain generating of wind power, plants based on wind power have been penetrated in high level recently. To deal with this uncertainty, variety of approaches have been utilized to estimate wind speed with respect to historical data. ARMA model is adopted to forecast the wind speed in the current study. According to the difference between estimated and real values, the scenarios are updated. Afterwards, the function of Weibull distribution is utilized to produce the random numbers of wind speed. Eventually, to reduce the number of scenarios SCENRED tool under GAMS is used. More details have been presented in [24]. The total power generated by a WT, which is a function of WT's specifications and the wind speed, can be modeled by the following equations [25]:

$$P_{\omega,t}^{A,\mathrm{WT}} = \begin{cases} 0 & S_{\omega,t} < S^{\mathrm{CI}}, S_{\omega,t} > S^{\mathrm{CO}} \\ P_{\max}^{\mathrm{WT}} \times \left(\frac{S_{\omega,t}-S^{\mathrm{CI}}}{S^{R}-S^{\mathrm{CI}}}\right) & S^{\mathrm{CI}} \le S_{\omega,t} \le S^{R} \\ P_{\max}^{\mathrm{WT}} & S^{R} \le S_{\omega,t} \le S^{\mathrm{CO}} \end{cases} \tag{10.5}$$

Generated power of WT can be calculated with respect to $S_{\omega,t}$, hourly speed, and $P_{\max}^{\mathrm{WT}}$, maximum wind power, by implementing technical constraints including cut-in, cut-out, and rated speeds. Furthermore, the power generated by the WT at time interval t is limited by the available and maximum wind power which are represented by $P_{\omega,t}^{A,\mathrm{WT}}$ and $P_{\max}^{\mathrm{WT}}$, respectively. Wind power spillage is also allowed. The algorithm is responsible for deciding about the amount of utilizing and spillage wind power taking the operational constraints and the total cost into consideration. To force this restriction (10.6) is used.

$$P_{\omega,t}^{\mathrm{WT}} < P_{\omega,t}^{A,\mathrm{WT}} \tag{10.6}$$

10.3.2.2 Photovoltaic (PV) System

The other renewable energy resource, which has exploded in the recent years is power plant based on solar radiation. PV system has produced power by absorbing solar radiation. The amount of producing power is related to various factors including the array of utilizing panel, solar radiation, and temperature. Producing power can be calculated by the following formula:

$$P_{\omega,t}^{\mathrm{PV}} = A \times I_{\omega,t}^{\mathrm{PV}} \left(1 - 0.005\left(T_t^{\mathrm{Out}} - 25\right)\right) \times \eta_{\mathrm{PV}} \tag{10.7}$$

In the above equation the efficiency of the solar cell is denoted by η_{PV}; A is represented the array area in m^2. Such intermittency creates the necessity of an appropriate and accurate forecasting method. In this study, ARMA model is utilized as time series forecasting model to estimate solar radiation based on historical data. More details about applying the ARMA model in order to forecast the solar radiation are available in [26].

10.3.2.3 Combined Heat and Power (CHP) Unit

Utilization of CHP units has been expanded to meet both heat and electricity demands much more efficient. Gas and steam turbine cycles are usually utilized in the CHP units. Despite their different operation strategies, characterizing feasible generation of the thermal and electricity powers is available. Several attempts have been made to design an appropriate strategy for optimal operation of CHPs with respect to some factors such as maximum and minimum values of fuel, electrical, and heat power. In the current study, the introduced method in [16, 27] is utilized to determine the feasible region of studying CHP unit with respect to the limit of ramp-up/ramp-down rates and the generation capacity.

$$P_{\omega,t}^{\mathrm{CHP}} = \vartheta_{\omega,t}{}^{*}\eta_e^{\mathrm{CHP}*}\left(P_{\omega,t}^{g,\mathrm{grid}}\right) \tag{10.8}$$

$$P_{\min}^{\mathrm{CHP}} \times \upsilon_t^{\mathrm{CHP}} \leq P_{\omega,t}^{\mathrm{CHP}} \leq P_{\max}^{\mathrm{CHP}} \times \upsilon_t^{\mathrm{CHP}} \tag{10.9}$$

$$P_{\omega,t}^{\mathrm{CHP}} - P_{\omega,t-1}^{\mathrm{CHP}} \leq R^U \tag{10.10}$$

$$P_{\omega,t-1}^{\mathrm{CHP}} - P_{\omega,t}^{\mathrm{CHP}} \leq R^D \tag{10.11}$$

where $\vartheta_{\omega,\,t} \in [0, 1]$ is a variable which determines the amount of dispatched $P_{\mathrm{Gas}}^{\mathrm{grid}}$ between CHP and boiler units. To simulate the startup/shutdown cost of CHP, the startup and shutdown binary variable $u_{\mathrm{SU},t}^{\mathrm{CHP}}$ and $u_{\mathrm{SD},t}^{\mathrm{CHP}}$ are calculated according to:

$$u_{\mathrm{SU},t}^{\mathrm{CHP}} = \upsilon_t^{\mathrm{CHP}} \times \left(1 - \upsilon_{t-1}^{\mathrm{CHP}}\right) \tag{10.12}$$

$$u_{\mathrm{SD},t}^{\mathrm{CHP}} = \upsilon_{t-1}^{\mathrm{CHP}} \times \left(1 - \upsilon_t^{\mathrm{CHP}}\right) \tag{10.13}$$

The amount of heat generation is also restricted between the minimum and maximum heat capacity. Note that the limit of ramp-up/down rates for power generation of CHP unit also imposes for generating of heat.

$$H_{\omega,t}^{\mathrm{CHP}} = \vartheta_{\omega,t}{}^{*}\eta_h^{\mathrm{CHP}*}\left(P_{\omega,t}^{g,\mathrm{grid}}\right) \tag{10.14}$$

$$H_{\min}^{\mathrm{CHP}} \times \upsilon_t^{\mathrm{CHP}} \leq H_{\omega,t}^{\mathrm{CHP}} \leq H_{\max}^{\mathrm{CHP}} \times \upsilon_t^{\mathrm{CHP}} \tag{10.15}$$

10.3.2.4 Boiler Unit

Boiler, as an important backup unit, is utilized into the proposed REH structure to provide thermal demand. The producing heat power of boiler unit is restricted with the maximum and minimum generation capacity. Moreover, the model of the startup/shutdown state of boiler is depicted by (10.17) and (10.18).

$$H^B_{\omega,t} = (1 - \vartheta_{\omega,t}) * \eta^{B*}\left(P^{g,\text{grid}}_{\omega,t}\right) \tag{10.16}$$

$$H^B_{\min,t} \times \upsilon^B_t \le H^B_{\omega,t} \le H^B_{\max,t} \times \upsilon^B_t \tag{10.17}$$

$$u^B_{\text{SU},t} = \upsilon^B_t \times \left(1 - \upsilon^B_{t-1}\right) \tag{10.18}$$

$$u^B_{\text{SD},t} = \upsilon^B_t \times \left(1 - \upsilon^B_{t-1}\right) \tag{10.19}$$

10.3.2.5 Thermal and Electrical Storages

Thermal and electrical energy storages are key components to store exceed and low price thermal and electrical powers to consume later. Specially, the role of the electrical storage unit has been highlighted by increasing the penetration of renewable energy resources such as solar radiation and wind speed. With respect to uncertain behavior of these energy resources, it is essential that the electrical storage be used to store power in high producing power and consume it in low producing power. The storage units should also store energy under their operational restriction, which are presented in the following:

$$E^s_{\min} \le E^s_{\omega,t} \le E^s_{\max} \quad e, h \in s \tag{10.20}$$

in which h and e stand for the thermal and electrical storage units, respectively. The charging/discharging efficiency of the storage systems can be expressed by (10.21):

$$\eta = \begin{cases} \eta_s, & \text{charging state} \\ 1/\eta_s, & \text{discharging state} \end{cases} \tag{10.21}$$

To clarify the state of energy in the storage systems, (10.22) is implemented. Moreover, the amount of charged/discharged energy is bounded by maximum char rates as shown in (10.23) and (10.24).

$$E^s_{\omega,t} = E^s_{\omega,t-1} + \eta_s \times E^{s,\text{ch}}_{\omega,t} - \left(1/\eta_s \times E^{s,\text{dis}}_{\omega,t}\right) \tag{10.22}$$

$$0 \leq E_{\omega,t}^{s,\text{ch}} \leq E_{\max}^{s,\text{ch}} \tag{10.23}$$

$$0 \leq E_{\omega,t}^{s,\text{dis}} \leq E_{\max}^{s,\text{dis}} \tag{10.24}$$

10.3.3 Electrical and Thermal Powers Balance

During the scheduling of the REH, matching between the various demands and generated and purchased gas and power from the upstream grid are important issues. There are two kinds of demands in the presented REH. So, the balance should be established for both of the electrical and thermal demands in every scenario at each time for the consideration time horizon scheduling. To satisfy these constraints, the following equation are expressed:

$$P_{\omega,t}^{\text{PV}} + P_{\omega,t}^{\text{wind}} + P_{\omega,t}^{\text{CHP}} + P_{\omega,t}^{e,\text{grid}} + E_{\omega,t}^{e,\text{dis}} - E_{\omega,t}^{e,\text{ch}} = P_t^{\text{electrical}} \tag{10.25}$$

$$H_{\omega,t}^{\text{CHP}} + H_{\omega,t}^{B} + E_{\omega,t}^{h,\text{dis}} - E_{\omega,t}^{h,\text{ch}} = P_t^{\text{thermal}} \tag{10.26}$$

10.3.4 REH Risk-Averse Scheduling Model

Finally, the objective function corresponding to REH scheduling program is adjusted to the CVaR formula as following equation:

$$\begin{aligned} \text{Min } (1-\beta) \times & \left\{ \sum_{t=1}^{N_t} \left\{ \sum_{\omega=1}^{N_\omega} \begin{matrix} \pi_\omega \times \left(P_{\omega,t}^{e,\text{grid}*} \lambda_t^e + P_{\omega,t}^{g,\text{grid}*} \lambda_t^g \right) \\ + \text{SUC} \times u_{\text{SU},t}^{\text{CHP}} + \text{SDC} \times u_{\text{SD},t}^{\text{CHP}} \end{matrix} \right\} \right\} \\ & + \beta \times \left\{ \text{VaR} + \frac{1}{1-\alpha} \times \sum_{\omega \in N_\omega} \pi_\omega \times \xi_\omega \right\} \end{aligned} \tag{10.27}$$

in which the CVaR of the cost is multiplied with the weighting parameter $\beta \in [0, 1]$. The weighting parameter is applied to the cost function in order to evaluate the tradeoff between the expected cost and risk aversion. According to (10.27), when the value of the parameter β is set to zero, a risk neutral scheduling problem will be resolved, whereas the energy hub behaves more risk averse by increasing the value of β.

10.4 Simulations

In this section, in order to evaluate the efficiency and accuracy of the proposed energy hub model, the method is applied on optimal scheduling problem of the REH for the 24-hour time horizon. In the test system, the inputs of the REH consist of the electricity and natural gas delivered from upstream grids. Moreover, it receives wind speed and solar radiation as input sources of WT and PV, respectively. It is also worth noting that these energy resources should provide the electrical and thermal demands. The optimal REH scheduling problem is formed as a mixed-integer nonlinear problem (MINLP) formulation and solved using SBB/CONOPT solver [28] under GAMS [29].

The operational restrictions and economic information of boiler and CHP units are presented in Table 10.2. The value of η^B, η_e^{CHP}, and η_h^{CHP} are set to 75%, 35%, and 45%, respectively [30]. The characteristics of the thermal and electrical storages are depicted in Table 10.3. Additionally, the characteristics of PV and WT are adapted from [24, 31] and presented in Tables 10.4 and 10.5, respectively. The total capacity of usage PV and WT are considered to be 3 MW and 2 MW, respectively. The uncertain behavior of the REH scheduling program caused by fluctuating nature of the renewable resources is simulated with multistage stochastic method. The uncertainties are the speed of wind and the solar radiation. The stochastic program is based on scenario generation. The ARMA model is utilized to produce 1000 scenarios for each uncertain element through implementing the hourly historical data. Afterwards, in order to decrease the number of the scenarios and participate the scenarios with high probability, fast backward scenario reduction method has been applied. Eventually, ten scenarios of both wind speed and solar radiation have been remained. In other words, total 100 scenarios are implemented in the stochastic scheduling of the REH problem. The hourly electricity and thermal demands are adopted from [24], which shown in Fig. 10.2. Moreover, the price of the electricity shown in Fig. 10.3 is extracted from [32] and natural gas price is chosen 30 ($/MWh) [32].

Table 10.2 Characteristics of boiler and CHP units [27]

Unit	R^u	R^d	CSU	CSD	$P_{\max}$	$P_{\min}$
Boiler	–	–	9	9	8	0
CHP	6	6	20	20	9	9

Table 10.3 Characteristics of energy storage units [30]

Unit	p^{disch} (MW)	p^{char} (MW)	$E_{\min}$ (MWh)	$E_{\max}$ (MWh)	η
Thermal storage	5	5	0	8	0.9
Electrical storage	3	3	0	6	0.9

Table 10.4 Characteristics of each PV panel [31]

Unit	A (m^2)	T_{Out} (°C)	η_{PV}	Maximum output power (MW)
PV	1500	25	15.7	0.2

Table 10.5 Characteristics of each WT [24]

Unit	S^{CI} (m/s)	S^{CO} (m/s)	S^{R} (m/s)	Maximum output power (MW)
WT	3.5	25	11.9	0.9

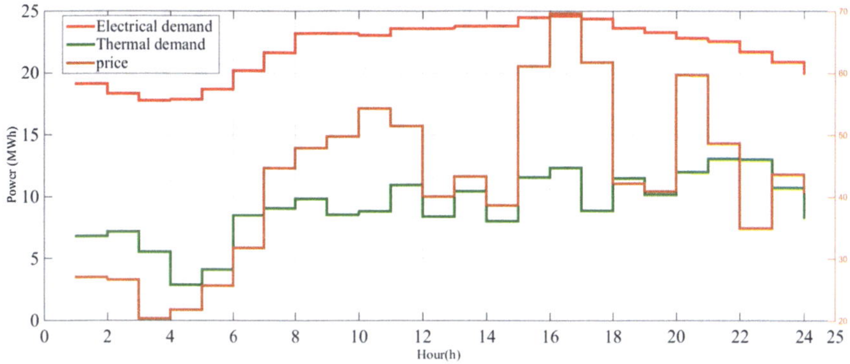

Fig. 10.2 Thermal and electrical demands profile and the price of electricity

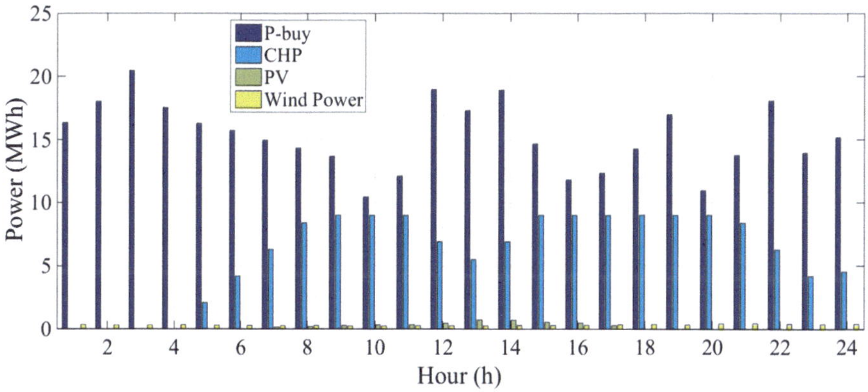

Fig. 10.3 Generated and purchased electrical power

Two case studies including the optimal scheduling of REH with and without consideration of the CVaR term are analyzed in this chapter. At the first part of the simulations, the objective function is solved without implementing the CVaR term (case study 1). The produced power of the REH components as well as purchased power from the market during the short-term scheduling problem is portrayed in Fig. 10.3. According to the figure, the zero cost power producing by the PV and WT is consumed whole of the scheduling time horizon. The PV is able to produce power during the day between hours 7 and 17. At the early hours, a great part of the electrical demand is met by purchasing power from the market, whereas CHP unit is not operated. It means that the cost of purchased electricity is lower than the CHP's operation cost during these hours. After hour 5, the CHP unit starts up to produce

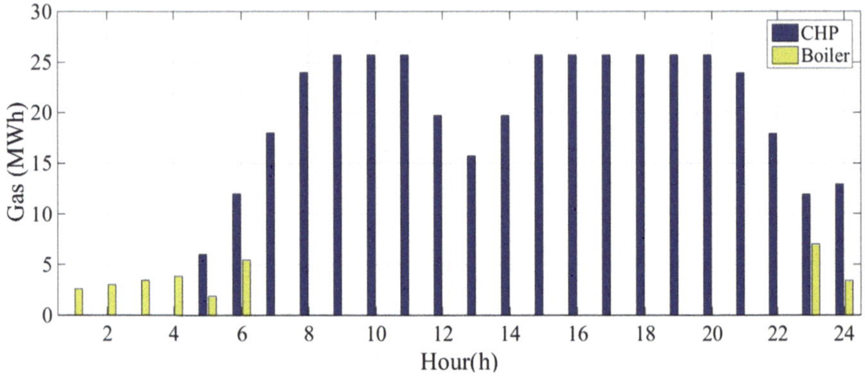

Fig. 10.4 Gas distribution between CHP and boiler units

power by taking into account the ramp constraints. It is continuously operated up to hour 24. In these hours, the purchased power is more expensive than the produced power through the CHP unit. During the hours 12–14, the CHP unit reduces its production because of the reduced thermal demand and low market price.

The amount of the gas distribution between CHP and boiler units is depicted in Fig. 10.4. As it seen in the figure, from hour 1 to 5 only the boiler unit is fed by the natural gas. It means that at this time interval the thermal demand is provided by the boiler unit, which acts more economical than the CHP unit. In the rest of hours the boiler unit is turned out except hours 23 and 24. CHP unit decreases its production at hours 12–14, as thermal demand has been reduced. Moreover, the value of expected cost without considering the risk term in the objective function formula is equal to 31,582.795$.

In the second part of the simulations, the objective function is solved by taking into account the CVaR term (case study 2). The confidence parameter α is taken to be 0.9. The impacts of risk factor on the REH scheduling problem are analyzed by increasing the value of the weighted parameter, β, step by step from 0 to 1. In terms of the risk level, if the CvaR is increased, the value of the expected cost must be decreased. It means that the expected cost become more risk averse. And conversely, the expected cost of the REH must be increased while the CVaR is decreased. This is obviously confirmed in the results of the proposed method shown in Fig. 10.5. The comparison of the various values of the expected cost regarding the level of β is illustrated. As it can be seen in the figure, the highest value of expected cost is achieved when β is equal to 0 (completely risk-averse) and the lowest value is carried out for $\beta = 1$ (risk neutral). As expected, the value of the expected cost both without and with employing the CVaR term is equal for $\beta = 1$.

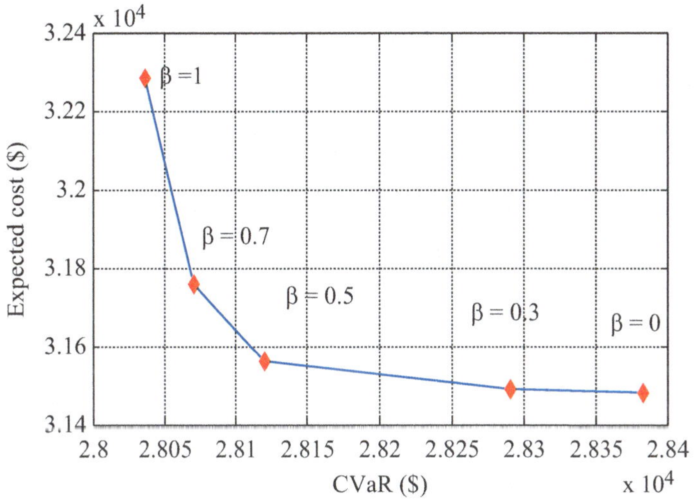

Fig. 10.5 Risk term impact on the optimal solution of the REH

10.5 Conclusion

This chapter addresses the optimal scheduling of renewable-based energy hub considering a risk-constrained two-stage stochastic programing model. This object would be achieved through the least operating cost of the energy hub components while satisfying electrical and thermal demands. The proposed REH model contains the renewable generators including PV and WT as components. Therefore, the solar radiation and wind speed uncertainties affect the REH scheduling in time horizon. The uncertain behavior of the solar radiation and wind speed is simulated by employing two-stage stochastic programming. The ARMA model is utilized as an appropriate scenario generation method. Afterwards, in order to minimize the scale of the problem and computation burden, the proper scenario reduction method is used, without reducing the problem-solving accuracy. Additionally, the proposed model incorporates the tradeoff between minimizing the expected total cost of the REH without and with considering a risk measure. This study focuses on a widely applied and practical risk measure CVaR. The comparison between the risk neutral and risk averse problems confirm the increment of the value of expected cost related to risk averse problem.

Nomenclature

Indices

ω	Scenario index $[1 : N_{\omega}]$
e	Index of electrical storage unit
h	Index of thermal storage unit
s	Index of energy storage units
t	Time index $[1 : N_t]$

Parameters

α	Confidence level
η_h^B	The efficiency of boiler unit
$\eta_e^{\text{CHP}}/\eta_h^{\text{CHP}}$	The efficiency of electrical/thermal generation of CHP unit
η^{PV}	The efficiency of PV panel
λ_t^e/λ_t^g	Electricity/gas price of the grid at tth hour
$E_{\min}^s/E_{\max}^s$	Minimum/maximum stored energy of storage unit
$H_{\min}^{\text{CHP}}/H_{\max}^{\text{CHP}}$	Minimum/maximum heat production of CHP unit
$H_{\min}^B/H_{\max}^B$	Minimum/maximum heat production of boiler unit
$P_{\max}^{\text{WT}}$	Maximum output power of WT
$P_{\min}^{\text{CHP}}/P_{\max}^{\text{CHP}}$	Minimum/maximum electrical power production of CHP unit
$P_t^{\text{electrical}}/P_t^{\text{thermal}}$	Electrical/thermal demand of REH
R^U/R^D	Ramp-up/down power CHP unit
S^R	The value of rated wind speed
$S^{\text{CI}}/S^{\text{CO}}$	The value of cut-in/cut-out wind speed

Variables

ξ_{ω}	Auxiliary variable used for CVaR computing
υ_t^{CHP}	Binary variable depicted on/off state of CHP unit
π_{ω}	Probability of ωth scenario
E_t^s	Amount of stored energy in energy storage at tth hour and ωth scenario
$E_{t,\omega}^{s,\text{ch}}/E_{t,\omega}^{s,\text{dis}}$	Charging/discharging of energy storage
$H_{\omega,t}^B$	Thermal generation of boiler unit at tth hour and ωth scenario
$H_{\omega,t}^{\text{CHP}}$	Thermal generation of CHP unit at tth hour and ωth scenario
$I_{\omega,t}^{\text{PV}}$	Solar radiation
$P_{\omega,t}^{\text{PV}}$	Utilizing solar power
$P_{\omega,t}^{\text{WT}}$	Utilizing wind power

$P_{\omega,t}^{\mathrm{CHP}}$	Power generation of CHP unit at tth hour and ωth scenario
$P_{\omega,t}^{A,\mathrm{WT}}$	Available wind power at tth hour and ωth scenario
$P_{\omega,t}^{g,\mathrm{grid}}/P_{\omega,t}^{e,\mathrm{grid}}$	Purchased gas/power from the grid at tth hour and ωth scenario
$S_{\omega,t}$	Wind speed at tth hour and ωth scenario
T_t^{Out}	Environment temperature
$u_{\mathrm{SU},t}^{\mathrm{CHP}}/u_{\mathrm{SD},t}^{\mathrm{CHP}}$	Binary variable depicting start-up/shutdown status of CHP unit at tth hour
VaR	Value-at-risk (VaR)

References

1. Geidl M, Koeppel G, Favre-Perrod P, Klockl B, Andersson G, Frohlich K (2007) Energy hubs for the future. IEEE Power Energy Mag 5(1):24–30
2. Hemmes K, Zachariah-Wolf J, Geidl M, Andersson G (2007) Towards multi-source multi-product energy systems. Int J Hydrogen Energy 32(10):1332–1338
3. Parisio A, Del Vecchio C, Vaccaro A (2012) A robust optimization approach to energy hub management. Int J Electr Power Energy Syst 42(1):98–104
4. Maroufmashat A, Elkamel A, Fowler M, Sattari S, Roshandel R, Hajimiragha A, Walker S, Entchev E (2015) Modeling and optimization of a network of energy hubs to improve economic and emission considerations. Energy 93:2546–2558
5. La Scala M, Vaccaro A, Zobaa A (2014) A goal programming methodology for multiobjective optimization of distributed energy hubs operation. Appl Therm Eng 71(2):658–666
6. Majidi M, Nojavan S, Zare K (2017) A cost-emission framework for hub energy system under demand response program. Energy 134: 157–166
7. Roldán-Blay C, Escrivá-Escrivá G, Roldán-Porta C, Álvarez-Bel C (2017) An optimisation algorithm for distributed energy resources management in micro-scale energy hubs. Energy 132:126–135
8. Rastegar M, Fotuhi-Firuzabad M, Lehtonen M (2015) Home load management in a residential energy hub. Electr Power Syst Res 119:322–328
9. Sheikhi A, Rayati M, Bahrami S, Ranjbar AM, Sattari S (2015) A cloud computing framework on demand side management game in smart energy hubs. Int J Electr Power Energy Syst 64:1007–1016
10. Sheikhi A, Bahrami S, Ranjbar AM (2015) An autonomous demand response program for electricity and natural gas networks in smart energy hubs. Energy 89:490–499
11. Vahid-Pakdel M, Nojavan S, Mohammadi-ivatloo B, Zare K (2017) Stochastic optimization of energy hub operation with consideration of thermal energy market and demand response. Energy Convers Manag 145:117–128
12. Moghaddam IG, Saniei M, Mashhour E (2016) A comprehensive model for self-scheduling an energy hub to supply cooling, heating and electrical demands of a building. Energy 94:157–170
13. Shariatkhah M-H, Haghifam M-R, Parsa-Moghaddam M, Siano P (2015) Modeling the reliability of multi-carrier energy systems considering dynamic behavior of thermal loads. Energy Build 103:375–383
14. Orehounig K, Evins R, Dorer V (2015) Integration of decentralized energy systems in neighbourhoods using the energy hub approach. Appl Energy 154:277–289
15. Salimi M, Ghasemi H, Adelpour M, Vaez-ZAdeh S (2015) Optimal planning of energy hubs in interconnected energy systems: a case study for natural gas and electricity. IET Gener Transm Distrib 9(8):695–707

16. Dolatabadi A, Mohammadi-ivatloo B, Abapour M, Tohidi S (2017) Optimal stochastic design of wind integrated energy hub. IEEE Trans Ind Inf 99:1
17. Moeini-Aghtaie M, Dehghanian P, Fotuhi-Firuzabad M, Abbaspour A (2014) Multiagent genetic algorithm: an online probabilistic view on economic dispatch of energy hubs constrained by wind availability. IEEE Trans Sustain Energy 5(2):699–708
18. Bozchalui MC, Hashmi SA, Hassen H, Cañizares CA, Bhattacharya K (2012) Optimal operation of residential energy hubs in smart grids. IEEE Trans Smart Grid 3(4):1755–1766
19. Shahmohammadi A, Moradi-Dalvand M, Ghasemi H, Ghazizadeh M (2015) Optimal design of multicarrier energy systems considering reliability constraints. IEEE Trans Power Deliv 30(2):878–886
20. Moeini-Aghtaie M, Farzin H, Fotuhi-Firuzabad M, Amrollahi R (2017) Generalized analytical approach to assess reliability of renewable-based energy hubs. IEEE Trans Power Syst 32(1):368–377
21. Sharif A, Almansoori A, Fowler M, Elkamel A, Alrafea K (2014) Design of an energy hub based on natural gas and renewable energy sources. Int J Energy Res 38(3):363–373
22. Conejo AJ, Carrión M, Morales JM (2010) Decision making under uncertainty in electricity markets, vol 153. Springer Science & Business Media, New York
23. Hosseini-Firouz M (2013) Optimal offering strategy considering the risk management for wind power producers in electricity market. Int J Electr Power Energy Syst 49:359–368
24. Alipour M, Mohammadi-Ivatloo B, Zare K (2015) Stochastic scheduling of renewable and CHP-based microgrids. IEEE Trans Ind Inf 11(5):1049–1058
25. Abbaspour M, Satkin M, Mohammadi-Ivatloo B, Lotfi FH, Noorollahi Y (2013) Optimal operation scheduling of wind power integrated with compressed air energy storage (CAES). Renew Energy 51:53–59
26. Ji W, Chee KC (2011) Prediction of hourly solar radiation using a novel hybrid model of ARMA and TDNN. Sol Energy 85(5):808–817
27. Dolatabadi A, Mohammadi-Ivatloo B (2017) Stochastic risk-constrained scheduling of smart energy hub in the presence of wind power and demand response. Appl Therm Eng 123:40–49
28. The GAMS software website (2012)
29. Brooke A, David K, Meeraus A (1996) GAMS release 2.25; a user's guide. GAMS Development Corporation, Washington, DC (EUA)
30. Alipour M, Zare K, Abapour M (2017) MINLP probabilistic scheduling model for demand response programs integrated energy hubs. IEEE Trans Ind Inf. https://doi.org/10.1109/TII.2017.2730440
31. Honarmand M, Zakariazadeh A, Jadid S (2014) Integrated scheduling of renewable generation and electric vehicles parking lot in a smart microgrid. Energy Convers Manag 86:745–755
32. Soroudi A, Keane A (2015) Risk averse energy hub management considering plug-in electric vehicles using information gap decision theory. In: Plug in electric vehicles in smart grids. Springer, Singapore, pp 107–127

Chapter 11
Grid Integration of Large-Scale Electric Vehicles: Enabling Support Through Power Storage

Prateek Jain and Trapti Jain

11.1 Comparing the Greenhouse Gas (GHG) Emissions of Electric Vehicles (EV) and Conventional Internal Combustion Engine (ICE) Vehicles: A Large-Scale Perspective

Various energy losses occur at every single stage of fuel life cycle, i.e., in delivering fuel from primary (ultimate) source to final conversion into vehicular motion. For example, energy is expended, and emissions take place in the extraction of crude oil, combustion of fossil fuels, etc. in the whole operation of internal combustion engine (ICE) vehicles, whereas losses occur in generating electricity from various sources, its transmission and distribution, utilization in charging the battery, etc. during the whole operation of an electric vehicle (EV). The life cycle energy and greenhouse gas (GHG) assessment, popularly known as well-to-wheel analysis, is carried out to assess the environmental impact of the above two vehicular technologies. It consists of two stages: (1) well-to-tank—an upstream stage, and (2) tank-to-wheel—the downstream stage. The well-to-tank stage involves evaluating the energy dissipated and the associated GHG emissions in delivering the refined fuel from the primary source into onboard (tank) the vehicle. The tank-to-wheel refers to evaluating the exhausted energy and associated GHG emissions from the fuel onboard the vehicle in achieving a particular driving range. The addition of the estimates of the two stages will give the total well-to-wheel energy expenditure and associated GHG emissions.

P. Jain (✉) · T. Jain
Discipline of Electrical Engineering, Indian Institute of Technology Indore, Simrol, Indore, 453552, MP, India

B. Mohammadi-Ivatloo, F. Jabari (eds.), *Operation, Planning, and Analysis of Energy Storage Systems in Smart Energy Hubs*,
https://doi.org/10.1007/978-3-319-75097-2_11

Table 11.1 Vehicle data

Electric vehicle (EV) Tesla (2017)	
Electric car model	Tesla Model S
Battery capacity (kWh)	75
Average speed (miles per hour)	45
Distance possible with the battery capacity at average speed (miles)	393
ICE vehicle parameters equivalent of EV	
Average fuel economy (miles per gallon (MPG))	25
Gallons of gasoline required for 393 miles	15.72

In this section, a comparison of fuel life cycle GHG emissions from the battery electric vehicles (BEV) and that of equivalent ICE vehicles during the whole operation has been made. A Tesla Model S with a typical battery capacity of 75 kWh has been selected as a representative BEV. Based on the selected BEV, the equivalent ICE vehicle parameters were devised. A total of 0.2 million representatives BEVs and hence the equivalent ICE vehicles are assumed considering a mid-size city for the comparison. The assumed scenario on vehicle data has been listed in Table 11.1.

The well-to-wheel energy usage and GHG emissions for the above two categories of vehicles are discussed below.

11.1.1 Battery Electric Vehicle (BEV)

Table 11.2 summarizes well-to-wheel energy expenditure and GHG emission analysis for the considered scenario of EVs. The two comprising stages in the assessment are as follows:

11.1.1.1 Well-to-Tank Assessment

While charging a battery, some of the power is utilized in pushing the electrons through the battery, decreasing the actual energy being stored and available for driving. The typical number for this loss in the battery is 10% Markowitz (2013). The average electricity transmission and distribution (T&D) losses as estimated by the US Energy Information Administration (EIA) EIA (2017a) is about 5% of the electricity that is transmitted and distributed annually in the USA. Based on this information, the T&D losses while supplying the charging energy to the battery of an EV are taken as 5%. Adding the above two gives a total 15% losses. Thus, for a given EV capacity of 75 kWh, the energy corresponding to 86.25 kWh has to be supplied from the sources mix to meet these losses. The major sources of electricity generation in the USA at utility-scale facilities in 2016 EIA (2017b) as well as the life cycle GHG emissions by each source (in g CO_2/kWh) Edenhofer et al (2012) are

Table 11.2 Well-to-wheel analysis of electric vehicles (EV)

Stage 1: Well-to-tank				
Sources/ technology	Percentage generation	Gram CO_2/kWh	kWh from each sources	Gram CO_2 emission
Coal	30.4	1001	26.22	26,246.22
Natural gas	33.8	469	29.1525	13,672.5225
Nuclear	19.7	16	16.99125	271.86
Hydroelectric	6.5	4	5.60625	22.425
Wind	5.6	12	4.83	57.96
Solar	0.9	46	0.77625	35.7075
Biomass	1.5	18	1.29375	23.2875
Geothermal	0.4	45	0.345	15.525
Stage 2: Tank-to-wheel				
Electric vehicles emit no gasses at all at the point of operation, i.e., CO_2 emissions = 0				
Total GHG emissions under the assumed scenario				
GHG emissions per vehicle (g)				4.0345×10^4
GHG emissions of 0.2 million EVs (kg)				8.0691×10^6

summarized in Table 11.2. It can be correlated that the above energy of 86.25 kWh per EV is supplied via these sources as per their percentage shares in the generation mix. From this, the gram CO_2 emission per vehicle from these sources for the total energy supplied can be evaluated as recorded in Table 11.2. The total well-to-tank GHG emissions per vehicle are found to be 40.345 kg.

11.1.1.2 Tank-to-Wheel Assessment

The BEVs are zero emission vehicles as no gases are generated at the point of operation. The batteries are sealed, having a gel with no harmful fumes produced Sachen (2015). Therefore, the tank-to-wheel GHG emissions from the BEVs can be treated as zero. Hence, the overall GHG emissions of 0.2 million BEVs as considered in the scenario is estimated as 8.0691×10^6 kg.

11.1.2 Internal Combustion Engine (ICE) Vehicle

Table 11.3 summarizes well-to-wheel energy dissipation analysis for the ICE vehicles scenario equivalent of considered EVs. Again, the two comprising stages in the assessment are as follows:

Table 11.3 Well-to-wheel analysis of ICE equivalent of EV

Stage 1: Well-to-tank					
GHG emissions from crude oil production			GHG emissions from petroleum refining		
Source	Emission (g/gallon crude)	Emission per vehicle (g/gallon crude)	Source	Emission (g/gallon crude)	Emission per vehicle (g/gallon crude)
ROG	0.7	11.004	ROG	0.2	3.144
CO	0.3	4.716	CO	0.5	7.86
NO_x	0.3	4.716	NO_x	0.4	6.288
			SO_x	0.7	11.004
Stage 2: Tank-to-wheel					
CO_2 emissions per gallon of gasoline (g)					8887
CO_2 emission per vehicle (g)					139,703.64
Total GHG emissions under the assumed scenario					
Well-to-tank GHG emissions per vehicle (g)					48.732
Tank-to-wheel GHG emissions per vehicle (g)					1.3970×10^5
GHG emissions of 0.2 million ICE vehicles (kg)					2.79504×10^7

11.1.2.1 Well-to-Tank Assessment

An equivalent ICE vehicle having the same driving range (393 miles) as of the considered EV above would require 15.72 gallons of gasoline with an average fuel economy of 25 miles per gallon (MPG) Naughton (2015). Now, there are GHG emissions accompanied with crude oil production and then from petroleum refining to feed these gallons of gasoline onboard tank of the vehicle. The various emissions per ICE vehicle along with their sources considering crude oil production TIAX LLC (2007) and petroleum refining TIAX LLC (2007) in this stage are quantified in Table 11.3.

11.1.2.2 Tank-to-Wheel Assessment

The grams of CO_2 dissipated per gallon of gasoline combustion is evaluated by multiplying the heat content of the gasoline per gallon with the kg CO_2 per heat content of the fuel. The conversion factor of 8887 g of CO_2 Federal Register (2010) emissions per gallon of gasoline consumed have been taken as the standard assuming all the carbon in the gasoline is converted to CO_2 Eggleston et al (2006). Using this factor, the CO_2 emissions per ICE vehicle which is consuming 15.72 gallons of gasoline can be evaluated as shown in Table 11.3. The sum of well-to-tank and tank-to-wheel GHG emissions would yield total life cycle GHG emissions, which is found to be 2.79504×10^7 kg with 0.2 million ICE vehicles equivalent of the considered scenario of EVs.

From the above assessment of life-cycle GHG emissions for the two categories of vehicular technology, it can be concluded that an internal combustion engine vehicle emits about 3.5 times the emissions with the equivalent driving range battery electric vehicle. As per International Energy Agency (IEA) IEA (2011), transportation sector alone accounts for 30% of global energy consumption, being the second largest source of CO_2 emissions contributing to 20% of global GHG emissions. Also, it is anticipated that there will be a tremendous increase in energy consumption in the transportation with growing demand for personal vehicles EIA (2013). Hence, transportation electrification with growing use of EVs presents excellent prospects for reducing the discharge of CO_2 and other toxic GHG, apart from saving the depleted stock of fossil fuels. Further, these benefits will increase manifold if renewable energy sources are being exploited to the fullest to charge the batteries of this energy efficient breed of vehicles.

11.2 Development of Charging Load Profiles of Electric Vehicles[1]

In order to ascertain whether the existing grid capacity will be able to support additional EV load with random charging, the assessment of charging load profiles based on the driving pattern of the owners is integral. The selection of charging power magnitude among the existent charging standards as well as the charging physics plays a crucial role in shaping the load profiles generated by the EVs. In this regard, this part analyzes the charging load profiles of the large-scale EVs employing the possible combinations of charging physics of constant time (CT) charging and constant power (CP) charging Darabi and Ferdowsi (2011) along with two distinct charging rates of 3.3 and 6.6 kW.

11.2.1 Process of Developing the Charging Load Profiles

11.2.1.1 Electric Vehicle Characteristics

Three types of EVs are considered. Their relevant characteristics and composition percentages RWTH (2010) in the system are detailed in Table 11.4. The Battery Electric Vehicle (BEV) and City-BEV are fully electric vehicles powered solely by the onboard battery. The PHEV 90, carrying an electric range of 90 km is a hybrid electric vehicle having ICE as a range extender unit. A total of 0.17 million vehicles is assumed in the system for the case study. Based on the vehicles' characteristics and composition percentages in the system, the weighted average values for the assumed scenario are also summarized in Table 11.4.

[1]Section adapted from work published by the authors in reference Jain and Jain (2014a).

Table 11.4 Characteristics of electric vehicles

Type of vehicle	Battery capacity (kWh)	Consumption (kWh/km)	All-electric range (km)	Composition (%)
BEV	35	0.20	175	37
City-BEV	16	0.12	133	10
PHEV 90	18	0.20	90	53
Weighted average values				
Battery capacity (kWh)				24
Consumption (kWh/km)				0.192
All-electric range (km)				125

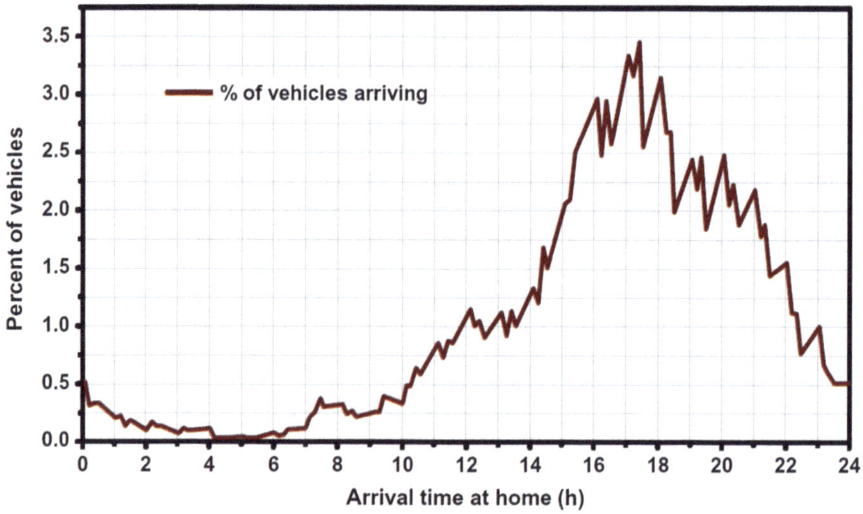

Fig. 11.1 Final arrival times of vehicles at home

11.2.1.2 Arrival Pattern

Figure 11.1 shows the percentage of vehicles arriving against their final arrival times at home. The arrival pattern has been developed taking the data inputs from Darabi and Ferdowsi (2011); NHTS (2001). The final arrival time of the vehicles has been treated as the charging start time because it is inferential that the commuters would plug their vehicles for charging soon after arriving at home. It can be observed that higher percentages of vehicles are arriving at home in the evening and late evening hours characterizing the routine driving behavior of commuters, returning to home from work or other related trips.

Table 11.5 Electric vehicle charging standards

SAE J1772 standard			
Charging type	Voltage level	Power level	Phase
Level 1	120 V AC	1.2–2.0 kW	Single-phase
Level 2 (low)	208–240 V AC	2.8–3.8 kW	Single-phase
Level 2 (high)	208–240 V AC	6.0–19.2 kW	Single-phase
Level 3	208–240 V AC	15–96 kW	3-phase
DC charging (level 1, 2 and 3)	200–600 V DC	>15–240 kW	DC
EPRI charging standard			
Charging type	Electrical ratings		
AC level 1	120 V AC, 12–16 A, 1.44–1.92 kW, single-phase		
AC level 2	208–240 V AC, 12–80 A, 2.5–19.2 kW, single-phase		
DC level 1, 2 and 3	200–600 V DC, $\leq$80–400 A, $\leq$19.2–$\leq$240 kW		

11.2.1.3 Charging Standards

The two EV charging standards namely SAE J1772 Kalhammer et al (2009) and EPRI-NEC Duvall and et al (2011) are summarized in Table 11.5. However, both the standards are proportionate seeing the electrical ratings of voltage, current, and power. Most of the contemporary charging infrastructure are suited for domestic AC low charging, as well as the worldwide top selling model of electric cars, supports charging with SAE J1772 AC Level 1 or 2 connectors up to 6.6 kW. Installation of DC fast charging (DCFC) station for typical residential applications is debatable because, first, its setup is very expensive, and second, there will be a huge burden of utility-scale distribution capacity upgradation in order to allow such a huge amount of power to flow through the distribution end power equipment. Considering this, the charging power levels of 3.3 and 6.6 kW, considering AC level 2 (low) and level 2 (high) of both the standards, are taken to develop the load profile of the EVs. With every hour of charging, these power levels add an electric range of approximately 17 and 34 km, respectively.

11.2.1.4 Charging Physics

Constant Time (CT) Charging Approach In constant time charging approach Darabi and Ferdowsi (2011), the total charging time is a fixed duration and is decided by the charging power standard for a given battery. This results in variation of charging power as per the SOC of the battery. For example, a battery with capacity 24 kWh has fixed charging time of 7.3 and 3.6 h, respectively with a given charging power standards of 6.6 and 3.3 kW.

Constant Power (CP) Charging Approach In this approach Darabi and Ferdowsi (2011), the charging power is fixed at the level specified. So, the charging time varies depending upon the SOC of the battery. Thus, the charging power levels of 3.3 and 6.6 kW results in a maximum charging time of 7.3 and 3.6 h, respectively, for an average battery capacity of 24 kWh.

11.2.1.5 Energy Required from the Grid

The arrival times of the vehicles are discretized into four arrival times per hour, and hence a total 96 arrival times throughout the day. Within the average all-electric range of 125 km (Table 11.4), the vehicles were classified into various driven distance groups (n). Finally, the driven distance groups are dispersed into the considered arrival times of the vehicles throughout the day. Electrical energy is consumed by the vehicles in driving, causing depleted energy state of the battery. This energy state is specified by the term state of charge (SOC). The SOC of a battery is expressed as the percentage of the energy state of a fully charged battery. For example, a vehicle driven completely to its capacity (up to AER) would carry 0% SOC. Likewise, a vehicle driven half of its AER would carry a SOC of 50%. The charging energy required to bring the battery back to the full is the complement of this SOC.

The charging energy required by the EVs from the grid at various arrival times of the day will be:

$$E^t = \sum_{m=1}^{n} \left(d_m^t \times n_m^t \times E_{\text{avg}}\right) \qquad \forall\, t \in (1, 2, \ldots, 96) \tag{11.1}$$

where E^t is the charging demand of EV aggregation arriving at time t, d_m^t is the driven distance by the mth distance group of vehicles arriving at time t, n_m^t is the number of mth distance group of vehicles arriving at time t, and E_{avg} is the average energy consumed by the vehicle.

11.2.2 The Charging Load Profiles

The charging load profiles of EVs as realized with the possible combination of charging power levels and charging physics are shown in Fig. 11.2. It has been considered that the vehicles start charging as soon as they arrive at home after finishing the trip(s). Figure 11.2 contains all the charging curves, i.e., the profiles obtained by employing the two charging powers of 6.6 and 3.3 kW individually with the constant time (CT) and constant power (CP) approaches. It can be observed that the load curves with CT approach are less peaking as compared to the load curves with CP approach for both the power levels. Similarly, for both the charging approaches, the load curves of 3.3 kW power level has lesser peak value when compared with the load curves of 6.6 kW charging power level. In addition to this, the peaks with CT scheme are shifted toward the right in comparison to the peaks with CP scheme for both the charging powers. Likewise, for both the charging schemes, the peaks with a lesser charging power of 3.3 kW is shifted toward the right in contrast to the charging load peaks caused by the 6.6 kW power level. The above two features are summarized quantitatively in Table 11.6. This is so,

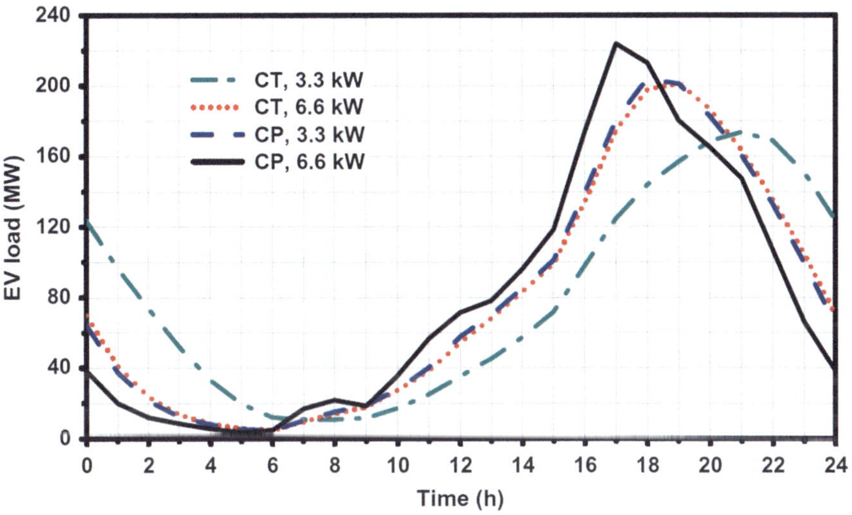

Fig. 11.2 Charging load profiles of EVs under various approaches

Table 11.6 Features of charging load curves of EVs

	Charging approach			
	Constant power (CP)		Constant time (CT)	
Charging level (kW)	Peak load (MW)	Peak time	Peak load (MW)	Peak time
3.3	204	18:00	174	21:00
6.6	224	17:00	202	19:00

as, for a given amount of charging energy required from the grid as per the SOCs of the vehicles arriving, use of high charge power level would supply the energy fast (in a lesser time), resulting in an increased peak that too near the arrival time of the vehicles. Also, in constant power charging approach, the charging power is constant whereas the charging time is being scaled as per the SOC of the vehicle, causing fast charging of vehicles in opposition to constant time charging approach, where the charging power is being scaled down to lower values in order to keep the total charging time a fixed duration. Thus, it can be concluded that the load is more peaking as well as drifted toward early hours with a combination of high charge power level with constant power charging approach, increasing the degree of fluctuation. In opposition to this, the load profiles originating from a combination of low charge power level with constant time charging approach are the flatter ones, as the peak is less as well as shifted toward late hours.

11.3 V2G and G2V Profiles Under Varying Equilibrium of EV Aggregation[2]

In transportation, the average car is parked almost 90% of the time leaving enormous time margin during the day to exploit the storage potential of the battery for grid support services. This led the researchers to propose the vehicle-to-grid (V2G) mode of operation of EVs in which a proportion of energy stored in the battery (after accounting for driving consumption) can be injected back into the grid as an aggregated storage device. In view of this, in this section V2G profiles are developed with various discharge power levels characterizing the mobility behavior. The heterogeneity in the vehicles as well as in the mobility behavior is also incorporated to determine the grid-to-vehicle (G2V) and V2G power capabilities of the aggregation at different moments of parking under varying penetrations of the electric vehicles. The quantification of the effects of the simultaneous combination of resulting G2V and V2G profiles with the conventional load on hourly loading and electricity market price is presented taking IEEE Test Bus system as an example.

11.3.1 Mobility Attributes

Four specimens EVs, namely BEV, City-BEV, PHEV 90, and PHEV 30, for representing the small, medium, and large version of electric cars in the market are considered as shown in Table 11.7. Speed dependent energy consumption of EVs considering four different phases of driving Pasaoglu et al (2012); RWTH (2010) viz. road, downtown, highway, and traffic for each of the four vehicles have been modeled. Three penetration percentages 25, 50, and 100% for the presence of EVs in the customer segment are created, and the proportion of various electric cars at these penetration levels was varied. Again, the total number of EVs are assumed to be 0.17 million (at 100% penetration). The percentage proportions in EV adoption at various penetration levels are influenced by several factors like socioeconomic capability, charging infrastructure availability, the cost of EVs, etc. Based on the above factors, RWTH (2010) presented a trend of adoption figures of EVs in various proportions which formed the basis for the selection of above composition percentages of various EVs at these penetration levels. In this case, 120 distance groups of vehicles, from 1 to 196 km, are considered. Also, it is supposed that, distance groups of EVs up to 67 km complete 40% of their trip on the road, 30% from downtown, 20% on the expressway, and remaining 10% in moving through traffic. The remaining distance groups, from 67 to 196 km, are assumed to perform 40% of their travel on the expressway, 30% from downtown, 20% on the road, and the remaining 10%

[2]Section adapted from work published by the authors in references Jain and Jain (2016) and Jain and Jain (2014b).

Table 11.7 Mobility attributes

Vehicle type	Battery capacity (kWh)	Energy consumption (kWh/km)				All-electric range (km) (based on 80% DoD)				Percentage composition			Driving course	Speed range (kmph)
		Road	Downtown	Expressway	Traffic	Road	Downtown	Expressway	Traffic	25%	50%	100%		
BEV	35	0.130	0.181	0.209	0.212	215	154	134	132	32	45	37	Road	55–67
City-BEV	16	0.120	0.167	0.193	0.195	107	76	66	65	20	12	10	Downtown	35–43
PHEV 90	18	0.150	0.209	0.241	0.244	96	69	60	59	24	21.5	26.5	Expressway	80–98
PHEV 30	12	0.150	0.209	0.241	0.244	64	46	40	39	24	21.5	26.5	Traffic	21–27

in traffic driving. This is to signify that the trips with short distances are mainly taking place in the urban zone while the trips with large distances include a high proportion of transit through expressways. The depth-of-discharge of the battery in driving as well as V2G supply is limited up to 80% in the analysis with a purpose of EV owners' obligation of maintaining a reasonable battery lifetime, as deep charge-discharge cycles shorten the battery life. Based on the premise, the weighted average parameters of the aggregation at the three penetration scenarios of 25, 50, and 100% are summarized in Table 11.8.

11.3.2 Development of V2G and G2V Profiles

11.3.2.1 Energy Consumption in Driving

The energy consumed in driving by the EV aggregation of 120 distance groups arriving at various (96) arrival times through the day is given by:

$$E^{\tau} = \sum_{m=1}^{120} (\alpha + \beta + \gamma + \delta) \qquad \forall\, \tau \in (1, 2, \ldots, 96) \tag{11.2}$$

where,

$$\alpha = 0.4 \mid 0.2 \begin{cases} pc_m^{\tau}\,(k_m^{\tau R} \cdot E_{\text{avg}}^{R}) & \forall\ k_m^{\tau R} \le \text{AER}_{\text{avg}}^{R} \\ pc_m^{\tau}\,(\text{AER}_{\text{avg}}^{R} \cdot E_{\text{avg}}^{R}) & \forall\ k_m^{\tau R} > \text{AER}_{\text{avg}}^{R} \\ 0 & \forall\ \text{BEVs and City} - \text{BEVs} \end{cases} \tag{11.2.1}$$

$$\beta = 0.3 \begin{cases} pc_m^{\tau}\,(k_m^{\tau D} \cdot E_{\text{avg}}^{D}) & \forall\ k_m^{\tau D} \le \text{AER}_{\text{avg}}^{D} \\ pc_m^{\tau}\,(\text{AER}_{\text{avg}}^{D} \cdot E_{\text{avg}}^{D}) & \forall\ k_m^{\tau D} > \text{AER}_{\text{avg}}^{D} \\ 0 & \forall\ \text{BEVs and City} - \text{BEVs} \end{cases} \tag{11.2.2}$$

$$\gamma = 0.2 \mid 0.4 \begin{cases} pc_m^{\tau}\,(k_m^{\tau E} \cdot E_{\text{avg}}^{E}) & \forall\ k_m^{\tau E} \le \text{AER}_{\text{avg}}^{E} \\ pc_m^{\tau}\,(\text{AER}_{\text{avg}}^{E} \cdot E_{\text{avg}}^{E}) & \forall\ k_m^{\tau E} > \text{AER}_{\text{avg}}^{E} \\ 0 & \forall\ \text{BEVs and City} - \text{BEVs} \end{cases} \tag{11.2.3}$$

Table 11.8 Scenario description

	Average battery capacity (kWh)		Average energy consumption (kWh/km)				Average all-electric range (km)			
Penetration ratio	80% DoD	100% DoD	Road	Downtown	Expressway	Traffic	Road	Downtown	Expressway	Traffic
25%	17.697	22.113	0.1376	0.1916	0.2212	0.2240	129	92	80	79
50%	19.771	24.728	0.1374	0.1914	0.2208	0.2237	144	103	90	88
100%	18.543	23.196	0.1396	0.1944	0.2244	0.2273	133	95	83	81

$$\delta = 0.1 \begin{cases} pc_m^{\tau}\left(k_m^{\tau \mathrm{Tr}} \cdot E_{\mathrm{avg}}^{\mathrm{Tr}}\right) & \forall\ k_m^{\tau \mathrm{Tr}} \le \mathrm{AER}_{\mathrm{avg}}^{\mathrm{Tr}} \\ \begin{cases} pc_m^{\tau}\left(\mathrm{AER}_{\mathrm{avg}}^{\mathrm{Tr}} \cdot E_{\mathrm{avg}}^{\mathrm{Tr}}\right) & \forall\ k_m^{\tau \mathrm{Tr}} > \mathrm{AER}_{\mathrm{avg}}^{\mathrm{Tr}} \\ \begin{cases} 0 \end{cases} & \forall\ \text{BEVs and City} - \text{BEVs} \end{cases} \end{cases} \tag{11.2.4}$$

Here, E^{τ} is the energy consumed in driving by the EVs arriving at time τ and pc_m^{τ} is the percentage of mth mileage group of vehicles arriving at time τ. Further, $k_m^{\tau R}$, $k_m^{\tau D}$, $k_m^{\tau E}$, and $k_m^{\tau \mathrm{Tr}}$ are the km traveled by m^{th} mileage group of vehicles arriving at time τ, respectively, while moving through road, downtown, expressway, and traffic driving periods; $\mathrm{AER}_{\mathrm{avg}}^{R}$, $\mathrm{AER}_{\mathrm{avg}}^{D}$, $\mathrm{AER}_{\mathrm{avg}}^{E}$, and $\mathrm{AER}_{\mathrm{avg}}^{\mathrm{Tr}}$ are the average values of all-electric range (AER) given by vehicles; and E_{avg}^{R}, E_{avg}^{D}, E_{avg}^{E}, and $E_{\mathrm{avg}}^{\mathrm{Tr}}$ are the average values of energy consumed in driving per km by the vehicles, respectively, when they move through road, downtown, expressway, and traffic driving periods. The figures 0.4, 0.3, 0.2, and 0.1 signify the travel percentage of vehicles, respectively for the driving periods road, downtown, expressway, and traffic for mileage groups with short trips (up to 67 km). Though, these figures are 0.2, 0.3, 0.4, and 0.1, correspondingly for these driving courses for mileage groups with long trips (above 67 km). The computation yields energy required by the EVs in driving along the number of vehicles arriving at various arrival times. Given the total storage capacity of the aggregation, the complement of the energy required for the driving is the net available energy for V2G supply.

11.3.2.2 V2G and G2V Moments

The mobility pattern of vehicles is defined by considering only work purpose trips in which vehicles commute between home and workplace. Thus, the G2V and V2G moments can be ascertained, once the arrival and departure times, travel and parking duration, as well as the commute circuit are fixed. This analysis accounts the average workplace parking duration to be 7 h Pasaoglu et al (2012) and average commuting duration 1.3 h, resulting in 15.7 h of average home parking time. It is hypothesized that the vehicles are plugged into the grid at workplace only soon after their arrivals to supply V2G power, while they are connected to the grid for charging (G2V) as soon as they finally arrive at home to bring the battery back to the full. Figure 11.1 shows the pattern of the final arrival time of vehicles at home. By employing the workplace parking and commuting duration Pasaoglu et al (2012) as considered above, the pattern of arrival of vehicles at the workplace can be obtained, which is shown in Fig. 11.3. A greater concentration of final arrivals of vehicles at home exists in the evening hours, though the concentration shifts into morning hours for the arrivals at the workplace, characterizing the regular office/work timings. Each distance group of vehicles was split into the considered 96 arrival times in the same proportions as derived from the arrival patterns.

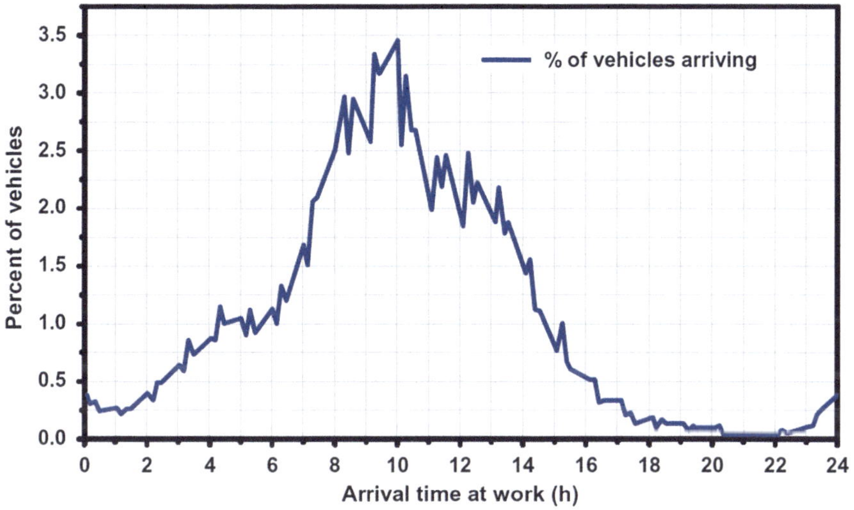

Fig. 11.3 Arrival times of vehicles at work

Table 11.9 Charging time duration and electric range added

Charging power (kW)	Charging time duration (h)						Electric range added per hour of charging (km)		
	Constant power (CP)			Constant time (CT)					
	25%	50%	100%	25%	50%	100%	25%	50%	100%
2.5	4.8	5.5	5.0	8.8	9.9	9.3	13.9	14.0	13.7
3.3	3.6	4.2	3.8	6.7	7.5	7.0	17.4	18.4	18.1
6.6	1.8	2.1	1.9	3.3	3.7	3.5	36.8	36.9	36.3

11.3.2.3 Charging and Discharging Power Levels

The V2G profiles have been realized with the discharge power levels of 1.44, 1.64, 1.92, 2.5, 3.3, and 6.6 kW, which covers both AC Level 1 and AC Level 2 range of SAE J1772 Kalhammer et al (2009) and EPRI Duvall and et al (2011) charging standards. However, the G2V profiles have been developed with charging power levels of 2.5, 3.3, and 6.6 kW only, which also comprise AC Level 1 and 2 of the two charging standards. The charging power cannot be selected below 2.5 kW in this study because of the constraint of 15.7 h available maximum charging time at home to bring the battery back to the full. The electric range added per hour of charging with these charging powers is shown in Table 11.9.

11.3.2.4 Charging and Discharging Approach

The nonlinear charging characteristics of a typical Li-ion battery consist of two stages of charging. The first stage is the constant current (CC) stage Simpson

(2011); Young et al (2013) which is analogous to constant power (CP) charging Darabi and Ferdowsi (2011) and persists till the battery is about 70% charged. In this stage, charging current remains constant, while the battery voltage rises to the reference voltage limit. This results in variable charging time depending upon the SOC of the battery as discussed in Sect. 11.2.1.4. The second stage takes over after it and lasts till the battery is fully charged. This stage is called constant voltage (CV) stage Simpson (2011); Young et al (2013) and is analogous to constant time (CT) charging Darabi and Ferdowsi (2011) approach. During this stage, the charging current decays exponentially (power scaling) resulting in a high charging time in comparison to the CC stage. Considering this, the G2V (charging) profiles of the aggregation have been developed considering charging from 0 to 70% battery capacity through constant power (CP) approach, while the next 70 to 100% capacity through constant time (CT) approach. The charging times with the two approaches at various charge power levels are listed in Table 11.9.

11.3.3 V2G and G2V Profiles of the Aggregation

11.3.3.1 V2G Profiles

Figure 11.4 shows the V2G profiles of the aggregation at the two terminal discharge power levels of 1.44 and 6.6 kW of the considered range under the three penetration scenarios. A range buffer corresponding to 20 km Pasaoglu et al (2012) as well as vehicle-grid interfacing converter efficiency of 93% has been assumed while evaluating the actual V2G power being supplied through these profiles. Thus, after accounting for above two deductions, the V2G profiles shown contains energy corresponding to 24.3, 30, and 26.2% of the average battery capacities, respectively, at the three penetration ratios of 25, 50, and 100%. The relative G2V and V2G MW values of the aggregation under the various scenarios are summarized in Table 11.10. It can be observed that the V2G profiles at the three penetration ratios are not proportionately modified. For example, neither the V2G peak is proportionately altered with the penetration ratios nor the shifting of V2G peak times with the increase in V2G power from 1.44 to 6.6 kW is proportionate with the variation in penetration ratios. This is due to the presence of heterogeneity in the mobility attributes, resulting in the changed G2V/V2G energy equilibrium of the aggregation at these penetration levels. The important characteristics of these profiles due to changed equilibrium at these penetrations are shown in Table 11.11.

11.3.3.2 G2V Profiles

Figure 11.5 shows the G2V profiles of the aggregation at the two terminal charge power levels of 2.5 and 6.6 kW of the considered range under the three penetration ratios. This G2V power is the sum of energy required by the EVs for driving as well as power consumed from the batteries in V2G supply. It can be observed from the profiles that, as a result of increased charging rate, the G2V peak increases as well

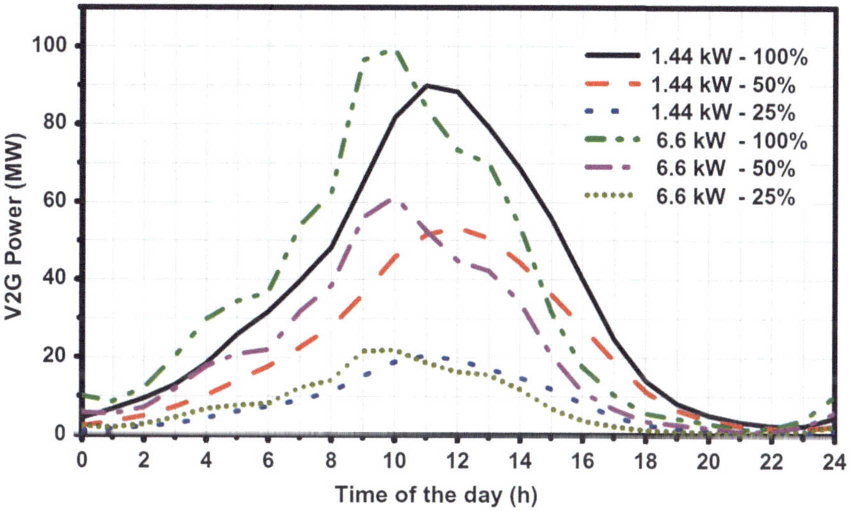

Fig. 11.4 V2G profiles of aggregation

Table 11.10 V2G and G2V energy balance of the aggregation

Particular	MWs 25%	50%	100%	% of total battery capacity 25%	50%	100%
V2G power by aggregation	182.53	503.08	824.70	24.27	29.93	26.17
Aggregated inverter loss	13.74	37.87	62.07	1.83	2.25	1.97
Driving consumption of aggregation	403.57	835.45	1645.89	53.66	49.71	52.24
Total G2V demand of aggregation	599.85	1376.40	2532.67	79.75	81.90	80.38

Table 11.11 Characteristics of V2G profiles

V2G power level (kW)	V2G peak power (MW) 25%	50%	100%	Peak time 25%	50%	100%	V2G duration (h) 25%	50%	100%	Total V2G power (MW) 25%	50%	100%
1.44	20.48	53.32	89.88	11:00	12:00	11:00	2.98	4.11	3.37			
1.64	20.96	54.23	94.55	11:00	12:00	11:00	2.62	3.61	2.96			
1.92	21.82	56.23	94.80	10:00	11:00	11:00	2.24	3.08	2.53	182.535	503.084	824.699
2.5	21.93	57.60	95.59	10:00	11:00	10:00	1.72	2.37	1.94			
3.3	22.60	59.63	98.99	10:00	10:00	10:00	1.30	1.79	1.47			
6.6	23.24	61.39	99.55	10:00	10:00	10:00	0.65	0.90	0.73			

as shifts toward left with the increase in charging power level from 2.5 to 6.6 kW at all the penetration ratios. Conversely, the minimum G2V load decreases and shifts toward early hours with the increase of charging power. The characteristic thing to be noted that the average increase in G2V peak or decrease in minimum G2V load is not proportionate at the three penetration ratios at various charge power levels.

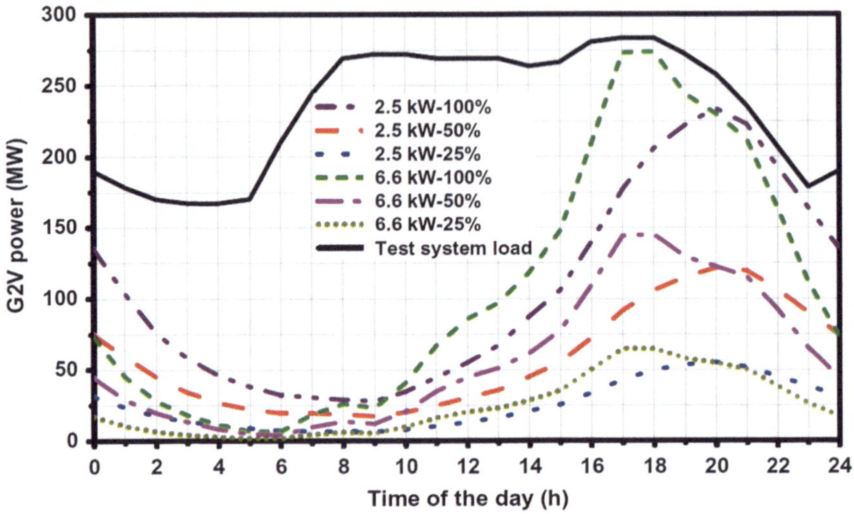

Fig. 11.5 G2V profiles of aggregation

Table 11.12 Characteristics of G2V profiles

G2V power level (kW)	Peak G2V power (MW)			Peak time			Least G2V power (MW)			Time of occurrence		
	25%	50%	100%	25%	50%	100%	25%	50%	100%	25%	50%	100%
2.5	55.15	121.61	233.74	20:00	20:00	20:00	6.64	17.67	28.23	09:00	09:00	09:00
3.3	57.97	129.76	245.03	19:00	19:00	19:00	4.62	14.15	20.36	06:00	09:00	06:00
6.6	64.90	145.08	273.91	17:00	18:00	18:00	1.58	4.30	06.82	05:00	06:00	06:00

Again, this is on account of heterogeneity in the vehicles' attributes at the three penetration ratios, varying the energy equilibrium. The variation in the concentration of different capacity (types) EVs alters the G2V/V2G capacities of the aggregation under the various penetration scenarios. Table 11.12 summarizes the characteristics of resulting G2V profiles of the EV aggregation.

11.3.3.3 Effect of V2G and G2V Profiles Integration on Grid Load and Electricity Pricing

Integration of the resulting V2G and G2V profiles with the system will modify the daily load pattern. The variations in market clearing volume (MW) due to this can alter the electricity market clearing price (MCP) due to the tweaks in unit commitment. A single-sided auction mechanism has been employed to determine the hourly market clearing price (MCP) Gutierrez et al (2005). Modified IEEE 30-Bus system Shahidehpour et al (2002) composed of nine generating units is taken as the test system to demonstrate the effect on electricity market price. The combined

Table 11.13 Generator data

Generating unit	P_{min} (MW)	P_{max} (MW)	Marginal cost ($/MWh)
G1	7	28	6.010
G2	14	56	8.005
G3	20	84	10.004
G4	25	100	13.345
G5	20	130	16.504
G6	15.2	76	18.012
G7	10	55	25.928
G8	4	20	37.575
G9	2.4	12	39.922

Table 11.14 Characteristics of net load on the system

Charge/ discharge	Highest slump in load (MW)			Time of highest slump			Maximum hike in load (MW)			Time of maximum hike		
power (kW)	25%	50%	100%	25%	50%	100%	25%	50%	100%	25%	50%	100%
2.5	13.56	34.88	60.78	10:00	10:00	10:00	54.36	119.16	230.00	20:00	20:00	20:00
3.3	14.13	40.74	65.42	10:00	10:00	10:00	56.95	126.72	240.31	19:00	19:00	19:00
6.6	16.03	43.82	73.24	09:00	09:00	09:00	63.41	141.50	268.32	18:00	18:00	18:00

generating capacity of the system is 561 MW. Generator limits, as well as their marginal cost of generating electricity, are presented in Table 11.13. The generator data are obtained from Djurovic et al (2012); Qiaozhu Zhai et al (2009). Hourly conventional load, expected to be fixed, on the system is computed for a regular winter weekday according to IEEE reliability test system Wong et al (1999) taking daily peak loads value from Shahidehpour et al (2002). Generators are assumed to bid their true marginal cost of generating power. The conventional daily load curve thus obtained for the selected modified IEEE 30-Bus test has been shown in Fig. 11.5 via solid black line.

The effect of integration of resulting V2G and G2V profiles with the selected test system on hourly loading and electricity market price is quantified in Tables 11.14 and 11.15, respectively. The two distinct attributes of the resulting system load and hourly MCP are: (1) reduction in net load and hence the MCP in the morning hours due to V2G supply of the aggregation with arrival of vehicles at workplace, and (2) rise in net load and hence the MCP due to G2V demand of the aggregation with arrival of vehicles at home. Table 11.14 summarizes the relative MW values of maximum hike and reduction in the original test load as well as their timings due to the integration of V2G and G2V profiles, under the three penetration scenarios.

Table 11.15 Characteristics of electricity market pricing

G2V and V2G power level (kW)	MCP reduction intervals			New MCP ($/MWh)	Old MCP ($/MWh)	MCP hike intervals			New MCP ($/MWh)	Old MCP ($/MWh)
	25%	50%	100%			25%	50%	100%		
2.5						03:00–05:00	03:00–05:00	03:00–05:00	13.345	10.004
						14:00–16:00	14:00–16:00	24:00–02:00	16.504	13.345
						20:00–22:00	20:00–23:00	14:00–16:00	16.504	13.345
	08:00–13:00	08:00–14:00	08:00–14:00	13.345	16.504			23:00–24:00	16.504	13.345
								21:00–23:00	18.012	13.345
								20:00–21:00	25.928	13.345
								17:00–18:00	18.012	16.504
								18:00–20:00	25.928	16.504
3.3						03:00–05:00	03:00–05:00	03:00–05:00	13.345	10.004
						14:00–16:00	14:00–16:00	24:00–01:00	16.504	13.345
						20:00–22:00	20:00–23:00	14:00–16:00	16.504	13.345
			05:00–06:00	10.004	13.345			22:00–24:00	16.504	13.345
	08:00–13:00	08:00–14:00	08:00–13:00	13.345	16.504			21:00–22:00	18.012	13.345
								20:00–21:00	25.928	13.345
							18:00–20:00	16:00–17:00	18.012	16.504
								17:00–20:00	25.928	16.504
6.6							03:00–04:00		13.345	10.004
						14:00–16:00	14:00–16:00	14:00–16:00	16.504	13.345
						20:00–22:00	20:00–23:00	22:00–24:00	16.504	13.345
	05:00–06:00	05:00–06:00	05:00–06:00	10.004	13.345			22:00–23:00	18.012	13.345
								20:00–21:00	25.928	13.345
	08:00–12:00	08:00–12:00	08:00–12:00	13.345	16.504		17:00–20:00	16:00–17:00	18.012	16.504
								19:00–20:00	25.928	16.504
								17:00–18:00	37.575	16.504
								18:00–19:00	39.922	16.504

The lower charge and discharge power level results in a flatter G2V and V2G profiles, respectively, resulting in a lesser increase of electricity market price with G2V as scheduling of costly generators are avoided. The two critical mobility attributes namely the driven distance and the base case arrival time at homes are independent of each other. Thus, the equilibrium of various EVs regulates the amount of V2G support and hence the net load on the system and hourly market prices with the integration of G2V and V2G profiles at various charge/discharge power levels. The above analysis concludes that the V2G support is not only dependent on the number of vehicles available to support the grid but is also dependent on the heterogeneity of the aggregation where battery electric vehicles (BEVs) may contribute more to V2G than plug-in hybrid electric vehicles (PHEVs), an important factor necessary to be incorporated to create any future robust model of EV dominated transportation system in order to accurately predict the fleet level effects on the grid.

11.4 Electric Vehicles Charging and Discharging Coordination for Reserve Capacity Commitment

This section presents the coordination of the EV aggregation during the charging and discharging phases to obtain the MW capacity that can be contracted as the capacity commitment (energy and reserve) in the volatile ancillary services market. After accounting for the driving consumption in transportation, the available battery capacity is the storage that can be supplied to the grid through V2G as a coordinated aggregation to produce MW level effect on the grid. Based on the mobility pattern defined in the previous section, the two parking places of home and work can be considered as the two operational places to simulate the G2V and V2G activities. The changeable locations of vehicles also suggest that the grid services function can be segregated based on the zone of operation (control area) of vehicles. In view of this, the aggregation of vehicles can either charge (G2V) at home and discharge (V2G) at work, or it can charge (G2V) at work while discharge (V2G) at home.

As discussed in the previous section, the vehicles arrive at the two places—home and work as per the pattern shown in Figs. 11.1 and 11.3 respectively, all over the day. Let there is n work as well as home arrival times and the number of vehicles arriving at either of the two places during these times are:

$$N_1,\ N_2,\ N_3,\ \ldots\ldots\ldots,\ N_n$$

The Charging (G2V) Phase Let the selected charging and discharging power levels (kW) for the aggregation of vehicles are P_c and P_d. Then, the CP phase charging power of the aggregation in MW arriving at a particular time n is given by,

$$\mathrm{MW}^{c}_{\mathrm{CP}} = \left(\frac{P_c}{1000}\right) \cdot N_n \tag{11.3}$$

and the CT phase charging power of the aggregation in MW arriving at a particular time n

$$\mathrm{MW}^{c}_{\mathrm{CT}} = \left(\frac{P_c}{1000}\right) \cdot \left(\frac{\mathrm{MWh}^{c}_{\mathrm{CT}}}{B}\right) \tag{11.4}$$

where B is the average battery capacity of the vehicles considered in the aggregation.

Total charging duration of the aggregation is given by,

$$T^c = \left(\frac{\mathrm{MWh}^{c}_{\mathrm{CP}}}{\mathrm{MW}^{c}_{\mathrm{CP}}}\right) + \left(\frac{\mathrm{MWh}^{c}_{\mathrm{CT}}}{\mathrm{MW}^{c}_{\mathrm{CT}}}\right) = T^c_{\mathrm{CP}} + T^c_{\mathrm{CT}} \tag{11.5}$$

where $\mathrm{MWh}^{c}_{\mathrm{CP}}$ and $\mathrm{MWh}^{c}_{\mathrm{CT}}$ are the total energy required by the aggregation arriving at particular time in CP and CT phase of charging, respectively.

The Discharging (V2G) Phase The discharging power of the aggregation in MW arriving at a particular time n is,

$$\mathrm{MW}_{\mathrm{dis}} = \frac{P_d}{1000} \cdot N_n \tag{11.6}$$

and, the total discharging duration of the aggregation,

$$T^d = \frac{\mathrm{MWh}^{\mathrm{dis}}_{n}}{\mathrm{MW}_{\mathrm{dis}}} \tag{11.7}$$

where $\mathrm{MWh}^{\mathrm{dis}}_{n}$ is the total V2G energy made available by the aggregation arriving at time n.

Determination of Capacity Reserve Let, there are n variables (α) along the n arrival times for each of the two CP and CT phases of charging, i.e., $\alpha_{\mathrm{CP}}(n)$ and $\alpha_{\mathrm{CT}}(n)$, respectively. There are total 1440 min (denoted by t) in a day's timeline starting from 00:00 till 23.59. For a given total charging duration of the aggregation arriving at various times of the day, the α variables take the value unity or zero as per the following conditions:

$$\alpha_{\mathrm{CP}} = \begin{cases} 1 & \forall\ n \leq t \leq (n + T^c_{\mathrm{CP}}) \\ 0 & \text{otherwise} \end{cases} \tag{11.8}$$

$$\forall\ t = 1, 2, 3, \ldots\ldots\ldots, 1440$$

$$\alpha_{\mathrm{CT}} = \begin{cases} 1 \ \forall \ (n + T^c_{\mathrm{CP}}) \ \leq t \leq (n + T^c) \\ 0 \quad \text{otherwise} \end{cases} \tag{11.9}$$

$$\forall \ t = 1, 2, 3, \ldots\ldots\ldots, 1440$$

Here, n is the nth arrival time of the vehicles. The total charging (G2V) power drawn by the EVs at any minute of the day is obtained as,

$$\mathrm{MW}^c = \sum_{i=1}^{n} \left[\left(\alpha_{\mathrm{CP}}(i) \cdot \mathrm{MW}^c_{\mathrm{CP}} \right) + \left(\alpha_{\mathrm{CT}}(i) \cdot \mathrm{MW}^c_{\mathrm{CT}} \right) \right] \tag{11.10}$$

and the total discharged power (V2G) of the EVs at any minute of the day,

$$\mathrm{MW}^d = \sum_{i=1}^{n} \left[\alpha_{\mathrm{CP}}(i) \cdot \mathrm{MW}_{\mathrm{dis}} \right] \tag{11.11}$$

Hence, the net power supplied or drawn from the grid at any minute of the day is given by,

$$\mathrm{MW}^{\mathrm{Net}} = \mathrm{MW}^c - \mathrm{MW}^d \tag{11.12}$$

Nonetheless, if only unidirectional flow of power during charging (G2V) is possible due to infrastructure limitations, the $\mathrm{MW}^d = 0$ and,

$$\mathrm{MW}^{\mathrm{Net}} = \mathrm{MW}^c \tag{11.13}$$

The capacity reserve (generation/demand) for any m minutes time-interval in a day's timeline can be obtained by taking the average value of net power supplied/drawn from the grid over that m minutes, i.e.,

$$\text{Capacity reserve (generation/demand)} = \overline{\mathrm{MW}}^{\mathrm{Net}} \quad \forall \, t \in (1, 2, 3, \ldots, m) \tag{11.14}$$

The reserve capacity of the aggregation at any moment is dependent upon vehicles' arrival patterns at home and work as shown in Figs. 11.1 and 11.3, respectively. Consequently, when the aggregation chooses to charge at work and discharge at home, the demand capacity (G2V) of the reserve will be dominating the generation capacity (V2G) of reserve at the morning hours while vice versa in the evening. Conversely, when the aggregation is selected

(continued)

to charge at home and discharge at work, the demand capacity (G2V) of the reserve will be dominating the generation capacity (V2G) of reserve at the evening hours while vice versa in the morning. The ancillary services market for the capacity reserve is a volatile high-value market. Thus, for a defined mobility pattern a capacity commitment (energy and reserve) in these competitive services market on a long-term basis could yield a notable revenue stream, in addition to increasing the grid reliability.

11.5 Load Leveling Through Charging and Discharging Coordination

The load levelization simplifies the load forecasting and dispatch exercises in the system operation by reducing the complexities associated with the oscillating load, and thereby the regulation services requirements. With a defined mobility pattern of home–work commute with work and home as the two parking slots available for G2V and V2G activities, the charging and discharging modes of EV aggregation can be coordinated to fill the valley(s) and shave the peak(s) of a fluctuating load with a purpose of its levelization. This can be realized by enacting the G2V (charging) mode and V2G (discharging) mode of the aggregation respectively, during the valley periods and peaking times of the original load.

Let us construe a case of coordinating the charging of the vehicles during their availability at home (G2V) and discharging during their availability at the workplace (V2G), with a purpose of valley filling and peak shaving, respectively. The assumptions on vehicles and its parameters are shown in Table 11.16.

The pattern of arrival of vehicles at home and the workplace is previously shown in Figs. 11.1 and 11.3, respectively. In this case, a total 24 arrival times are considered for both home and workplace arrival of vehicles. With the total number

Table 11.16 Assumptions on vehicle parameters

Parameter	Value
Type of vehicle	Nissan LEAF electric car (BEV)
Total number of vehicles	0.25 million
Battery capacity	24 kWh
Battery capacity with 80% DoD	19.2 kWh
Capacity required for 20 km range buffer	3.632 kW
Net available capacity for driving	15.568 kWh
Average home–work commute distance	27.5 km
Energy consumed in home–work commute	5 kWh
Net available battery capacity for V2G	10.567 kWh

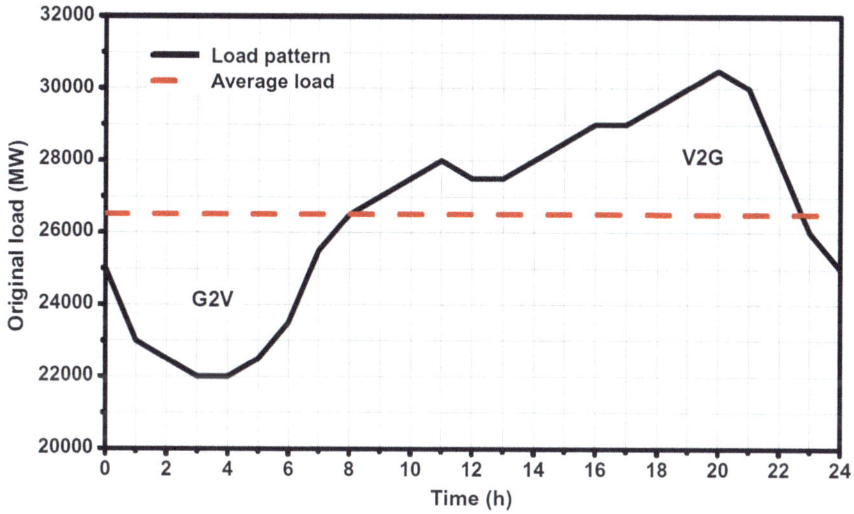

Fig. 11.6 Hourly load pattern of CAISO load (Monday, June 12, 2017)

of vehicles assumed to be 0.25 million in the system, the actual number of vehicles arriving at these 24 arrival times can be obtained from the above patterns. Here, the home parking, workplace parking, and travel duration constraints are assumed to be 15, 7, and 2 h, respectively. Figure 11.6 shows the demand curve of a typical day (Monday, June 12, 2017) of California Independent System Operator (CAISO) system CAISO (2017), with the dashed line showing the average load of 26,520 MW of the day. The maximum demand is 30,500 MW occurring at hour 20:00 while the minimum demand being 22,000 MW occurring at hours 03:00 and 04:00. Between the two demands, the load pulsates requiring ramping up and down of the generation sources in order to follow the load pattern.

In order to levelize the load around the average value, the load points (MWs) above the average load has to be curtailed through V2G (peak shaving), while the load points (MWs) below the average values are to be lifted through G2V (valley filling) by the aggregation. Thus, from Fig. 11.6, the V2G supply is required between hours 08:00 and 23:00, while the G2V is required between the hours 00:00–08:00 and 23:00–24:00, in order to levelize the load around the average. For simplification of G2V and V2G coordination, the charging and discharging power per vehicle are set to 10.60 and 15.57 kW, respectively, so that the aggregation is able to discharge and charge completely to the limits considered within an hour. These charging powers fall in the gamut of AC Level 2 gamut of EV charging standards (Table 11.5). The vehicles are charged and discharged with constant power charging approach.

Let the various hours of the day is denoted by h_n, where $n \in (1, 2, 3, \ldots, 24)$, then MW drawn, i.e., G2V by the aggregation in an hour h_n to h_{n+1},

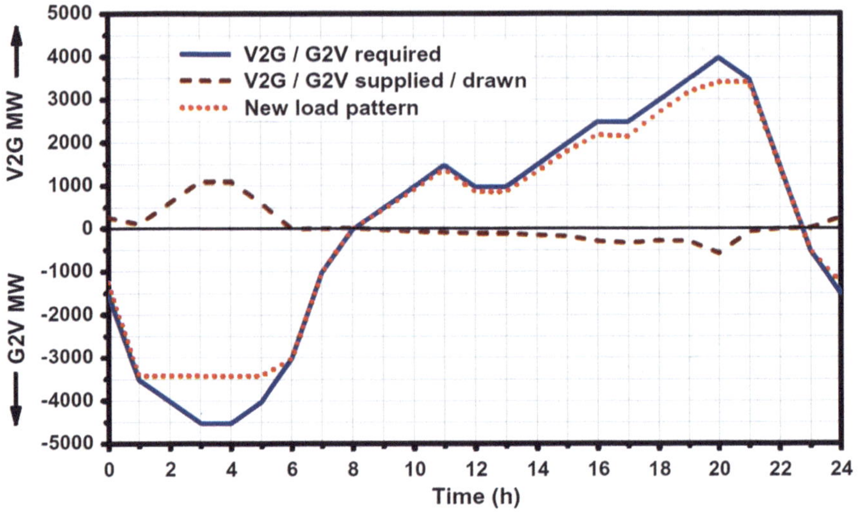

Fig. 11.7 V2G and G2V MWs required as well as supplied by the aggregation at various times

$$MW_{G2V} = \left(\frac{10.60}{1000}\right) N_{h_n} \tag{11.15}$$

and the MW supplied, i.e., V2G by the aggregation in an hour h_n to h_{n+1},

$$MW_{V2G} = \left(\frac{15.57}{1000}\right) N_{h_n} \tag{11.16}$$

Here, N_{h_n} is the number of vehicles arriving (hence available) at hour n of the day.

The vehicles are scheduled V2G and G2V modes with the set charging and discharging power levels, as per the constraints work and home parking duration specified. It should be noted that as the vehicles are primarily accompanied by transportation (driving) energy consumption, the V2G support MWs by the aggregation would inherently be lesser than the charging demand (G2V support). In other words, the total G2V demand from the grid is the sum of driving energy consumption and the energy supplied through V2G. The accessibility of net V2G and G2V MWs for peak shaving and valley filling respectively is limited by the total number of vehicles in the system as well as the pattern of arrival of vehicles which governs the availability of number vehicles at the two locations for charging and discharging. Therefore, here, the net V2G/G2V supplied/drawn by the scheduling the vehicles for peak shaving and valley filling respectively during the various hours is lesser than the V2G/G2V required to completely levelize the load around the average, as shown in Fig. 11.7. In this figure, for depiction, the average load is shown by zero and pattern of G2V and V2G required are plotted below and above the zero average, respectively. Nonetheless, a significant amount of peak shaving and valley filling is achieved thereby leveling the load. Table 11.17 summarizes the relative

Table 11.17 MW proportions in V2G/G2V coordination

Particular	Value
V2G MWs supplied by the EV aggregation	2572.37
Travel demand of the EV aggregation (MW)	1248.87
G2V MWs drawn by the EV aggregation	3821.16

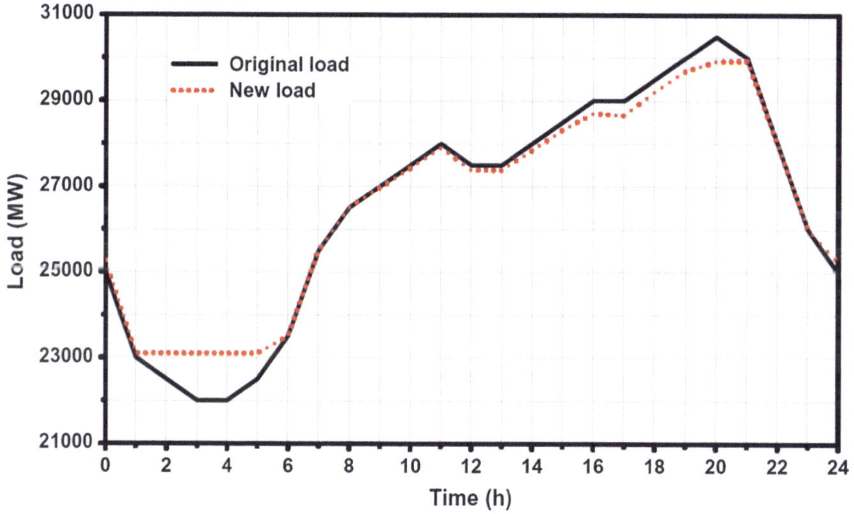

Fig. 11.8 New equalized load pattern

MW proportions of G2V and V2G energy transfer in this scheme for load leveling. Figure 11.8 shows the new hourly load curve of the day after the levelization through V2G and G2V coordination.

Conventionally, the ramp up and ramp down energy and reserve requirements to match the fluctuating load throughout the day are provided by expensive and slow response coal, gas, or fossil fuel based power sources. The oscillating and unpredictable nature of load necessitates procurement of these costly ancillary services (energy and reserve capacity) by the system operator to maintain the stable and reliable operation of the interconnected system. The increased cost of electricity is ultimately ended up being passed on to the final consumers. The load leveling mechanism through V2G and G2V coordination by a large pool of EVs as demonstrated can be an effective measure to reduce the dependability of ramping commitments on traditional sources, thereby reducing the ultimate cost of electricity to the consumers. In addition, the G2V and V2G energy storage and transactions take place on the distribution side (receiving end) avoiding the transmission line congestion, mostly at the peaking times.

11.6 Electric Vehicle Grid Interfacing to Enable Support Through Power Storage

Figure 11.9 shows the representative schematic of the electric vehicle and electric utility interfacing to facilitate the grid support services through aggregated EV storage. The EVs offer the advantage of high ramp up and ramp down speed capabilities but at the same time possess the limitations of changeable availability affecting the contract sizes and absence of stabilizing inertia as of the large generators. This makes them appropriate for short-term high-value ancillary services markets Kempton and Tomić (2005a) like regulation and spinning reserves instead for the base load sources, as shown in Fig. 11.9.

The charging points having the ability to enable two-way communication between the charging station and the EV to limit the charging current to the safe limits are also termed as the electric vehicle supply equipment (EVSE). In order to have control over the charge as well as discharge rates of the vehicle battery, the EVSE must be designed to have the bidirectional power and communication flow capabilities. Also, to support V2G, the EVSE should be designed to have the capability to handle different charge/discharge power levels and support AC as well as DC power transfer to/from the vehicle as per the infrastructure requirements.

The standard battery capacity of an electric vehicle is only in the range of few kilowatt hours, creating negligible impact at the grid level operations. The V2G support services would require a controllable capacity of MWs to have a substantial impact on the system. This is possible only with the aggregated battery storage necessitating the grouping of a large number of available EVs at the moment. Also, it is almost impractical for the system operator to interact with each individual vehicle. Thus, an interfacing entity called vehicle aggregator Guille and Gross (2009); Lopes et al (2011) is proposed, for managing the groups of battery storage to provide overall load (G2V) and generation (V2G) services to the electric utility (system operator). To the system operator, the aggregator provides a single point of contact—managing a resource of rapidly controllable electric reserve and its

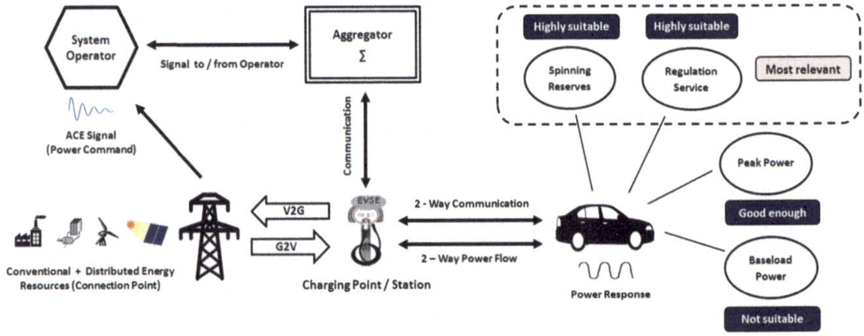

Fig. 11.9 Schematic of EV and electric utility interactions for grid support through V2G

participation in grid support services. Principally, the vehicle aggregator would be in control of (1) location monitoring of vehicles, (2) tracking their grid connectivity, (3) integrating participants to ensure sufficient capacity, (4) ensuring their participation, (5) communication/control (command) signals from/to the system operator, (6) establishing contracts with the operator, and (7) coordinating the payment streams down to the connected vehicles for the grid services. The system requirement for this include added communication and controls with the electric utility to ensure the energy transfer between the vehicle owner and system operator in an optimal way. An unregulated utility transacting electricity or an independent third party entity like an automobile manufacturer, a battery manufacturer, or a mobile network provider having expertise in communication functions and automated customer transactions may serve as a vehicle aggregator in the future scenario Briones et al (2012); Kempton and Tomić (2005b).

Justification of the economic feasibility of V2G is contingent upon numerous factors. High battery cost, long charging time, range anxiety, costly charging infrastructure, etc. are the first few hurdles in the greater EV adoption. However, in V2G, the electric utilities or in turn the vehicle aggregators will have control and access to charging/discharging of vehicle batteries for the purpose of improving the system reliability through grid support. Thus, the bidirectional power flow in V2G allows the commitment of energy and capacity services via the grid-connected vehicles for which the aggregator and hence the vehicle owners will be compensated. The value created on the part of these services would be a vital motivation toward consumers' willingness to participate in V2G. Not to mention, the V2G support services should be in addition to the primary function of the vehicles, i.e., transportation, in order to not to jeopardize the customers' comfort of vehicle utilization in travel.

Nomenclature

Abbreviations, Acronyms, & Symbols

α_{CP}	Variable along CP phase of charging
α_{CT}	Variable along CT phase of charging
$\overline{\mathrm{MW}}^{\mathrm{Net}}$	Capacity reserve (generation/demand)
$\mathrm{AER}_{\mathrm{avg}}^{D}$	Average value of AER achievable by vehicle aggregation in downtown driving
$\mathrm{AER}_{\mathrm{avg}}^{E}$	Average value of AER achievable by vehicle aggregation in express-way driving
$\mathrm{AER}_{\mathrm{avg}}^{R}$	Average value of AER achievable by vehicle aggregation in road driving
$\mathrm{AER}_{\mathrm{avg}}^{\mathrm{Tr}}$	Average value of AER achievable by vehicle aggregation in traffic driving

B	Average battery capacity of the vehicles considered in the aggregation
d_m^t	Driven distance by the mth distance group of vehicles arriving at time t
E^τ	Energy consumed in driving by the EV aggregation arriving at time τ
E^t	Charging demand of EV aggregation arriving at time t
E_{avg}	Average energy consumed by the vehicle
E_{avg}^D	Average value of energy consumed per km by the vehicle aggregation in downtown driving
E_{avg}^E	Average value of energy consumed per km by the vehicle aggregation in expressway driving
E_{avg}^R	Average value of energy consumed per km by the vehicle aggregation in road driving
$E_{\text{avg}}^{\text{Tr}}$	Average value of energy consumed per km by the vehicle aggregation in traffic driving
h_n	n^{th} hour of the day
$k_m^{\tau D}$	km traveled by m^{th} mileage group of vehicle aggregation arriving at time τ in downtown driving
$k_m^{\tau E}$	km traveled by m^{th} mileage group of vehicle aggregation arriving at time τ in expressway driving
$k_m^{\tau R}$	km traveled by m^{th} mileage group of vehicle aggregation arriving at time τ in road driving
$k_m^{\tau \text{Tr}}$	km traveled by m^{th} mileage group of vehicle aggregation arriving at time τ in traffic driving
MW^c	Total charging (G2V) power drawn by the EVs at any minute of the day
MW^d	Total discharged power (V2G) of the EVs at any minute of the day
MW^{Net}	Net power supplied or drawn from the grid at any minute of the day
MW_{CP}^c	CP phase charging power of the aggregation in MW
MW_{CT}^c	CT phase charging power of the aggregation in MW
MW_{dis}	Discharging power of the aggregation in MW
MW_{G2V}	MW drawn by the aggregation
MW_{V2G}	MW supplied by the aggregation
MWh_{CP}^c	Energy required by the aggregation in CP phase of charging
MWh_{CT}^c	Energy required by the aggregation in CT phase of charging
$\text{MWh}_n^{\text{dis}}$	V2G energy made available by the aggregation arriving at time n
n	n^{th} arrival time of the vehicles
N_{h_n}	number of vehicles arriving at hour n of the day
n_m^t	Number of mth distance group of vehicles arriving at time t
N_n	Number of vehicles arriving at a particular time n
P_c	Selected charging power level (kW) for the aggregation of vehicles
P_d	Selected discharging power level (kW) for the aggregation of vehicles
pc_m^τ	Percentage of mth mileage group of vehicle aggregation arriving at time τ
T^c	Total charging duration of the aggregation
T^d	Total discharging duration of the aggregation
T_{CP}^c	CP phase charging duration of the aggregation

T^{c}_{CT}	CT phase charging duration of the aggregation
AER	All-electric range
BEV	Battery electric vehicle
CAISO	California Independent System Operator
CC	Constant current
CP	Constant power
CT	Constant time
CV	Constant voltage
DCFC	Direct current fast charging
DoD	Depth of discharge
EIA	Energy Information Administration
EPRI	Electric Power Research Institute
EV	Electric vehicle
EVSE	Electric vehicle supply equipment
G2V	Grid-to-vehicle
GHG	Greenhouse gas
ICE	Internal combustion engine
IEA	International Energy Agency
MCP	Market clearing price
MPG	Miles per gallon
PHEV	Plug-in hybrid electric vehicle
SAE	Society of Automotive Engineers
SOC	State-of-charge
T&D	Transmission and distribution
V2G	Vehicle-to-grid

Acknowledgements The authors thank Rishil Lakhe, third year B.Tech. Electrical Engineering student from Sardar Vallabhbhai National Institute of Technology Surat, Surat, India, for technical assistance in preparing Sects. 11.1 and 11.5 of the chapter.

References

1. Briones A, Francfort J, Heitmann P, Schey M, Schey S, Smart J (2012) Vehicle-to-grid (V2G) power flow regulations and building codes review by the Avta. Technical report INL/EXT-12-26853, Idaho National Laboratory, U.S. Department of Energy National Laboratory, Idaho Falls, Idaho [Online]. Available https://energy.gov/eere/vehicles/downloads/avta-vehicle-grid-power-flow-regulations-and-building-codes-review
2. CAISO (2017) California Independent System Operator [Online]. http://www.caiso.com/Pages/default.aspx
3. Darabi Z, Ferdowsi M (2011) Aggregated impact of plug-in hybrid electric vehicles on electricity demand profile. IEEE Trans. Sustainable Energy 2(4):501–508
4. Djurovic MZ, Milacic A, Krsulja M (2012) A simplified model of quadratic cost function for thermal generators. In: Annals and proceedings of DAAAM international, Vienna, vol 23, No 1
5. Duvall M et al (2011) Transportation electrification: a technology overview. Technical report, CA: 2011.1021334, Electrical Power Research Institute, Palo Alto, CA, pp 3.1–3.2, 5.10

6. Edenhofer O, Pichs-Madruga R, Sokona Y et al (2012) Renewable energy sources and climate change mitigation. Summary for policymakers and technical summary, Potsdam Institute for Climate Impact Research, The Intergovernmental Panel on Climate Change
7. Eggleston S, Buendia L, Miwa K et al (2006) 2006 IPCC guidelines for national greenhouse gas inventories, prepared by the national greenhouse gas inventories programme. IPCC report, The Intergovernmental Panel on Climate Change, IGES, Japan
8. EIA (2013) Annual Energy Outlook 2013 - with projections to 2040. Report DOE/EIA-0383(2013), U.S. Energy Information Administration, Washington, DC
9. EIA (2017) How much electricity is lost in transmission and distribution in the United States? U.S. Energy Information Administration. https://www.eia.gov/tools/faqs/faq.php?id=105&t=3 [Online]. Accessed May 2017
10. EIA (2017) What is U.S. electricity generation by energy source? U.S. Energy Information Administration. https://www.eia.gov/tools/faqs/faq.php?id=427&t=3 [Online]. Accessed May 2017
11. Federal Register (2010) Light-duty vehicle greenhouse gas emission standards and corporate average fuel economy standards; Final Rule, p 25330. Federal Register/Vol 75, No. 88, Part - II: Environmental Protection Agency, Department of Transportation
12. Guille C, Gross G (2009) A conceptual framework for the vehicle-to-grid (V2G) implementation. Energy Policy 37(11):4379–4390
13. Gutierrez G, Quinonez J, Sheble GB (2005) Market clearing price discovery in a single and double-side auction market mechanisms: Linear programming solution. In: Proceedings of 2005 IEEE Russia power technologies, St. Petersburg
14. IEA (2011) Technology roadmap: electric and plug-in hybrid electric vehicles. Technical report, International Energy Agency, France
15. Jain P, Jain T (2014) Assessment of electric vehicle charging load and its impact on electricity market price. In: 2014 international conference on connected vehicles and expo (ICCVE), Vienna, pp 74–79
16. Jain P, Jain T (2014) Impacts of G2V and V2G power on electricity demand profile. In: 2014 IEEE international electric vehicle conference (IEVC), Florence, pp 1–8
17. Jain P, Jain T (2016) Development of V2G and G2V power profiles and their implications on grid under varying equilibrium of aggregated electric vehicles. Int J Emerg Electr Power Syst 17(2):101–115
18. Kalhammer FR, Kamath H, Duvall M, Alexander M, Jungers B (2009) Plug-in hybrid electric vehicles: promise, issues and prospects. In: Proceedings of EVS24 international battery, hybrid and fuel cell electric vehicle symposium, Stavanger, pp 1–11
19. Kempton W, Tomić J (2005) Vehicle-to-grid power fundamentals: calculating capacity and net revenue. J Power Sources 144(1):268–279
20. Kempton W, Tomić J (2005) Vehicle-to-grid power implementation: from stabilizing the grid to supporting large-scale renewable energy. J Power Sources 144(1):280–294
21. Lopes JAP, Soares FJ, Almeida PMR (2011) Integration of electric vehicles in the electric power system. Proc IEEE 99(1):168–183
22. Markowitz M (2013) Wells to wheels: electric car efficiency. https://matter2energy.wordpress.com/2013/02/22/wells-to-wheels-electric-car-efficiency/ [Online]. Accessed May 2017
23. Naughton N (2015) Average U.S. mpg edges up to 25.5 in May. Automotive News. http://www.autonews.com/article/20150604/OEM05/150609925/average-u.s.-mpg-edges-up-to-25.5-in-may [Online]. Accessed May 2017
24. NHTS (2001) National Household Travel Survey (NHTS), U.S. Department of Transportation [Online]. Available http://nhts.ornl.gov
25. Pasaoglu GK, Fiorello D, Martino A, Scarcella G, Alemanno A, Zubaryeva A, Theil C (2012) Driving and parking patterns of European car drivers - a mobility survey. Technical report JRC77079, EUR - scientific and technical research reports, Institute for Energy and Transport, European Commission, Joint Research Centre (2012) [Online]. Available http://publications.jrc.ec.europa.eu/repository/handle/JRC77079

26. Qiaozhu Zhai Q, Guan X, Yang J (2009) Fast unit commitment based on optimal linear approximation to nonlinear fuel cost: error analysis and applications. Electr Power Syst Res 79(11):1604–1613
27. RWTH (2010) WP:1.3 Parameter manual. V05 ed., Grid for vehicles (G4V), RWTH Aachen [Online]. Available http://www.g4v.eu/
28. Sachen R (2015) EV vs gas - part 2 emissions. Sunspeed Enterprises. http://sunspeedenterprises.com/ev-vs-gas-part-2-emissions/ [Online]. Accessed May 2017
29. Shahidehpour M, Yamin H, Li Z (2002) Example systems data. In: Market operations in electric power systems: forecasting, scheduling, and risk management. Wiley, New York. IEEE, appx. D, sec. D.4, pp 477
30. Simpson C (2011) Battery charging. Litrature No. SNVA557, 2011, National Semiconductor, Texas Instrument [Online]. Available http://www.ti.com/lit/an/snva557/snva557.pdf
31. Tesla (2017) https://www.tesla.com/models [Online]. Accessed May 2017
32. TIAX LLC (2007) Full fuel cycle assessment, well to tank energy inputs, emissions, and water impacts. Consultant report (draft), California Energy Commission, Cupertino, CA
33. Wong P, Albrecht P, Allan R, Billinton R, Chen Q, Fong C, Haddad S, Li W, Mukerji R, Patton D, Schneider A, Shahidehpour M, Singh C (1999) The IEEE reliability test system-1996. IEEE Trans Power Syst 14(3):1010–1020
34. Young K, Wang C, Wang LI, Strunz K (2013) Electric vehicle battery technologies. In: Electric vehicle integration into modern power networks. Power electronics and power systems, 1st edn. Springer, New York

Chapter 12
Optimal Operation of Renewable-Based Residential Energy Hubs for Minimizing PV Curtailment

Soroush Senemar, Alireza Seifi, and Mohammad Rastegar

12.1 Introduction

Increasing rate of energy consumption, lack of fossil fuels, reliability, and environmental concerns are the main challenges of the energy systems. These reasons motivate us to use energy sources more efficiently. As a promising solution, cogeneration technologies have been proposed recently to make the energy generation and consumption more efficient. This technology uses a kind of energy carrier such as natural gas, biofuels, or solar energy to produce electricity, heating and cooling energies. Combined heat and power (CHP) units, as the most popular cogeneration technologies, is penetrating drastically in the worldwide energy system. For example, it is forecasted to have more than two million micro-CHP units in Japan by 2030, especially beside the residential consumers [1]. The CHP units are fed with natural gas and generate electricity and heat in the output.

An energy system including cogeneration technologies is named multi-carrier energy system. Energy hub is a usual structure to model the energy transfer, storage, and conversion in a multi-carrier energy system [2]. Because of interactions between different energy carriers, the operation of energy hub is more complex than a single energy carrier system. The main operational questions in a residential energy hub are "how much the energy carrier is consumed at the input port of the energy hub" and "how energy carriers are flowing in the different ways from inputs to outputs of the energy hub." These questions are usually answered by solving an optimization-based problem.

Since almost 40% of total energy is consumed in residential sectors [3], it seems essential to study residential energy hubs, using CHP units [4]. A boiler is usually

S. Senemar · A. Seifi · M. Rastegar (✉)
Electrical and Computer Engineering Department, Shiraz University, Shiraz, Iran
e-mail: soroush_senemar@shirazu.ac.ir; seifi@shirazu.ac.ir; mohammadrastegar@shirazu.ac.ir

B. Mohammadi-Ivatloo, F. Jabari (eds.), *Operation, Planning, and Analysis of Energy Storage Systems in Smart Energy Hubs*,
https://doi.org/10.1007/978-3-319-75097-2_12

used next to the CHP unit to guarantee providing the heat demand. In other words, the demanded heat can be provided by the CHP or the boiler. In addition, a share of electrical demand is fed by the CHP unit and remained can be supplied by electrical grid and other electrical resources. Ren et al. [4] developed a mixed integer nonlinear programing (MINLP) optimization and determined the optimal overall energy cost for a test year and the hourly operation schedules of CHP unit.

Recently, many studies focused on the optimal operation of multi-carrier energy systems especially residential energy hubs [5–10]. The main objective functions of proposed optimization problems in the residential energy hub are minimizing customers' costs, such as the energy cost or the emission cost. For example in [2], a general modeling of multi-carrier energy systems is developed and the concept of energy hubs is explained. Then, power flow equations for each energy carrier are used to model the energy hub. Finally, total energy cost is considered as the objective function and is minimized. In [10], a residential energy hub is modeled and a multi-objective optimization is proposed. The authors used weighted sum technique to converting a multi-objective function to a single objective one. The single objective function elements are containing energy costs, energy consumption, CO_2 emission costs, and the peak load of electricity demand. Then, based on the weights set by customers, the optimization is solved and the amount of each energy carrier at inputs of the energy hub and the interaction between different energy carriers are determined.

In addition, renewable resources particularly photovoltaic (PV) panels although have a high installation cost are becoming more widespread in the residential buildings because of low operational and emission costs. In [11], a residential energy hub is proposed and solar panels are providing power for the energy hub. Other components like CHP unit, plug-in hybrid electric vehicle (PHEV), and heat storage as parts of future smart homes are incorporated. The objective function is customer's energy cost. These resources such as PV panels, which have low operational costs, should be utilized maximally to become profitable for their owners [10, 12]. In renewable residential energy hubs that contain a CHP generator and PV panels, high heat demand may have impact on the PV utilization. Because, at hours with high heat demand, CHP units are utilized to meet heat demands, which causes supplying a large portion of the power demand. Thus, the PV generation power is less than what it could potentially produce, that can affect the revenue of PV installation. In other words, when a large portion of power demand is supplied by CHP, a part of the PV generation is left unused, which is called PV curtailment. High heat demand can increase the PV curtailment, especially in sunny hours of the day [13]. In addition, heat and electrical storage systems can be considered in the energy hub [11, 14]. Reference [14] has investigated the effect of energy storage on optimal operation of residential energy hub. The investment payback and benefits of thermal and electrical storage on other component of energy hub is analyzed in [14]. The objective function contains customer's energy costs and penalty for carbon emission. Plug-in Hybrid Electrical Vehicle (PHEV) as one of the new emerging technologies is recently considered in the optimal operation study of residential energy hubs [11, 12]. These vehicles consume electrical energy at charging periods and can bring the charged energy back to the grid by discharging at the time of availability. In summary, although a number of works have studied the operation of residential energy hubs, a few of them have considered all the emerging components, i.e., PV panels, CHP units, boiler, and storage systems, in the smart home. Among these few works, to the best of authors' knowledge, PV curtailment concern has not been considered in the optimal operation of the renewable-based residential energy hub.

This chapter optimizes the day-ahead operation of a renewable residential energy hub. The energy hub model includes a CHP unit, a PHEV, PV panels, and a gas boiler (GB) to provide the electrical and heat demand. Electricity and natural gas are assumed as the energy hub inputs. In addition, opposing to the most of previous works, the gas to heat and gas to power efficiencies of CHP are considered dependent on the output power. According to the proposed model, an optimization problem is designed for optimal operation of the proposed framework. The objective functions are minimizing customer energy cost and PV curtailment simultaneously. The problem is subjected to different operational constraints of the components and energy and heat balance in the energy hub. Solving the proposed problem determines the amount of each energy carrier and interaction between different types of energy. In addition, it determines the time scheduling of PHEV. The proposed method is applied to a home as a renewable residential energy hub to demonstrate the effectiveness of the proposed method. In addition, the impact of incorporating a heat storage system on the PV curtailment is studied in this chapter. Based on recent contents, the main contributions of this chapter are:

- Integrating new technologies such as PHEV, PV panels, and storage systems in the energy hub model,
- Accurate modeling of CHP units especially in part load operation, and
- Considering PV curtailment in the optimal operation of energy hub.

12.2 Methodology and Problem Formulation

12.2.1 Proposed Renewable-Based Residential Energy Hub Structure

Energy hub concept can be used for every multi-carrier framework that includes different kinds of demands and resources. This chapter models a home as an energy hub with electrical and heat demand, PHEV, PV panels, storage systems, a CHP unit, and a boiler. The structure of the proposed residential energy hub is shown in Fig. 12.1. As shown in Fig. 12.1 the electrical demand can be provided directly from the grid, generation power of PV, and output power of CHP unit. In addition, the PHEV can be discharged to provide a part of electrical demand at the time of availability.

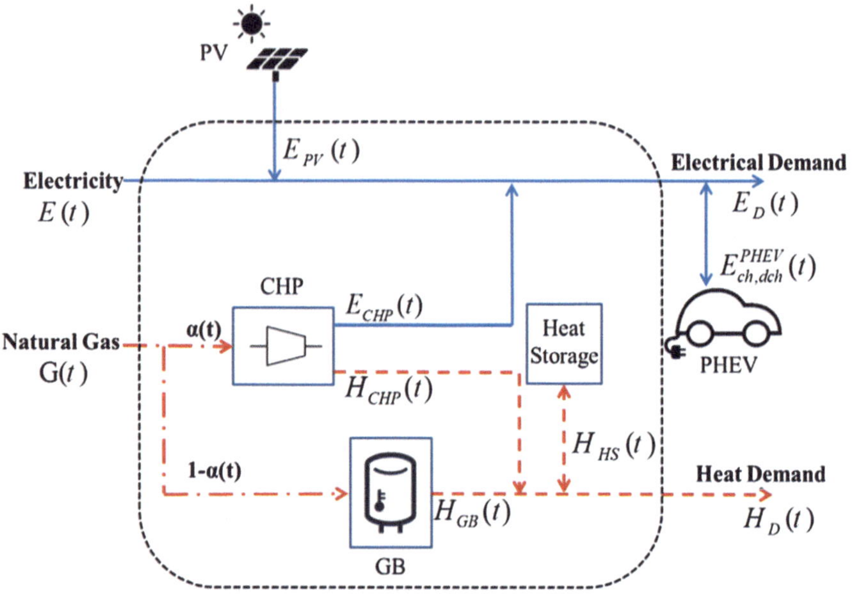

Fig. 12.1 Structure of the renewable residential energy hub

Output heat from CHP unit and gas boiler provides the heat demand. The heat storage can provide a part of the heat demand. In the proposed energy hub structure, the input gas can flow in two paths, i.e., to CHP unit or to gas boiler. Distribution of input gas between components is determined by dispatch factor ($\alpha(t)$). The dispatch factor is variable between 0 and 1. This factor determines the amount of natural gas flowing into the CHP unit and gas boiler.

12.2.2 *Problem Formulation*

Here, the optimization-based formulations for the operation of the residential energy hub are presented. The goal is determining the optimal value of dispatch factors, i.e., $\alpha(t)$, and the optimal flow of heat and electricity in different paths in Fig. 12.1. The optimization is studied for one-day period with 1-hour time-step. The optimization problem, i.e., objectives and constraints, is presented as follows:

12.2.2.1 Objective Function

The main objective for households is minimizing the energy cost. Hence, the energy cost is assumed as the first criterion in the optimization problem. It is mathematically formulated as follows:

$$\mathrm{EC} = \sum_{t=1}^{24} \left[\lambda_e(t) * E(t) + \lambda_g(t) * G(t)\right] \tag{12.1}$$

$\lambda_e(t)$ and $\lambda_g(t)$ are the electrical and gas tariffs in hour t. $E(t)$ and $G(t)$ are, respectively, the electrical energy and the natural gas received from the grid in hour t.

In addition, the curtailment of PV panels can be considered as another important criterion in the operation of the energy system. This criterion clearly provides the customer's welfare. PV curtailment is presented as follows:

$$\mathrm{PC} = \sum_{t=1}^{24} \left[E_{\mathrm{PV}}^{G}(t) - E_{\mathrm{PV}}(t)\right] \tag{12.2}$$

where $E_{\mathrm{PV}}^{G}(t)$ is the capability of PV panels for generation and $E_{\mathrm{PV}}(t)$ is the energy drawn from the PV panels in hour t.

As a conclusion, we meet a multi-objective optimization problem. The problem can be solved by weighted sum method. This method adds multiple objectives with different weights to make a single objective problem. Here, per unit values of objective functions are added together to develop the objective function, as follows:

$$\min\ \mathrm{CU} = w_1 * \mathrm{EC}_{\mathrm{pu}} + w_2 * \mathrm{PC}_{\mathrm{pu}} \tag{12.3}$$

where CU is the customer utilization, $\mathrm{EC}_{\mathrm{pu}}$ is equal to $\frac{\mathrm{EC}}{\mathrm{EC}_{\mathrm{base}}}$ and, $\mathrm{PC}_{\mathrm{pu}}$ is $\frac{\mathrm{PC}}{\mathrm{PC}_{\mathrm{base}}}$. The base values of EC and PC, i.e., $\mathrm{EC}_{\mathrm{base}}$ and $\mathrm{PC}_{\mathrm{base}}$, are the energy cost and PV curtailment in a non-optimal case, in which CHP units are used to provide the electrical demand at its rated value.

w_1 and w_2 are importance factors of objectives i.e., weights for $\mathrm{EC}_{\mathrm{pu}}$ and $\mathrm{PC}_{\mathrm{pu}}$. The procedure of determining w_1 and w_2 to consider the preferences of the customer is performed by a fuzzy decision making (FDM) method. The detailed descriptions of this method are presented in [15]. In this chapter, based on the importance of the objectives from the customer's viewpoint the w_1 and w_2 are determined in the Numerical Studies Section, using the FDM method.

12.2.2.2 Optimization Constraints

The proposed objective function is subjected to different constraints, including power and heat flow equations and different technical constraint of the different components in the proposed structure of Fig. 12.1.

- *Power and heat flow equations*

According to Fig. 12.1, power and heat flow equations can be presented as follows:

$$E(t) + E_{\mathrm{PV}}(t) + E_{\mathrm{CHP}}(t) + \eta_{\mathrm{PHEV}}^{\mathrm{dch}} * E_{\mathrm{PHEV}}^{\mathrm{dch}}(t) - \frac{E_{\mathrm{PHEV}}^{\mathrm{ch}}(t)}{\eta_{\mathrm{PHEV}}^{\mathrm{ch}}} = E_D(t) \tag{12.4}$$

$$H_{\mathrm{GB}}(t) + H_{\mathrm{CHP}}(t) + \eta_{\mathrm{HS}}^{\mathrm{dch}} * H_{\mathrm{HS}}^{\mathrm{dch}}(t) - \frac{H_{\mathrm{HS}}^{\mathrm{ch}}(t)}{\eta_{\mathrm{HS}}^{\mathrm{ch}}} = H_D(t) \tag{12.5}$$

$$E_{\mathrm{CHP}}(t) = \alpha(t) * G(t) * \eta_{\mathrm{CHP}}^{g-e}(t) \tag{12.6}$$

$$H_{\mathrm{GB}}(t) = (1 - \alpha(t)) * G(t) * \eta_{\mathrm{GB}} \tag{12.7}$$

$$H_{\mathrm{CHP}}(t) = \frac{E_{\mathrm{CHP}}(t)}{\beta_{\mathrm{CHP}}^{e-h}(t)} \tag{12.8}$$

where $E_{\mathrm{CHP}}(t)$ is the output power of CHP in hour t and $\eta_{\mathrm{CHP}}^{g-e}(t)$ is gas to power efficiency in hour t, $E_{\mathrm{PHEV}}^{\mathrm{ch}}(t)$ and $E_{\mathrm{PHEV}}^{\mathrm{dch}}(t)$ are the PHEV charging and discharging power of PHEV, respectively, in hour t, $\eta_{\mathrm{PHEV}}^{\mathrm{ch}}$ and $\eta_{\mathrm{PHEV}}^{\mathrm{dch}}$ are the charging and discharging efficiency of PHEV, respectively, and $E_D(t)$ is the demanded power of electrical appliances in hour t.

In addition, $H_{\mathrm{CHP}}(t)$ is the heat generation of CHP unit in hour t and $\beta_{\mathrm{CHP}}^{e-h}(t)$ is the power to heat ratio of CHP unit in hour t, $H_{\mathrm{GB}}(t)$ is the output heat of gas boiler in hour t and η_{GB} is the efficiency of gas boiler. $H_D(t)$ is the heat demand of energy hub in hour t. $H_{\mathrm{HS}}^{\mathrm{ch}}(t)$ and $H_{\mathrm{HS}}^{\mathrm{dch}}(t)$ are the charging and discharging heat of heat storage, respectively, in hour t. $\eta_{\mathrm{HS}}^{\mathrm{ch}}$ and $\eta_{\mathrm{HS}}^{\mathrm{dch}}$ are the charging and discharging efficiency of heat storage.

Equations (12.4) and (12.5) show that the demanded power and heat is provided by grid, CHP unit, and storage. Note that selling the energy to the grid is not possible. Equation (12.6) presents the electrical generation from CHP unit considering the gas to power efficiency, i.e., $\eta_{\mathrm{CHP}}^{g-e}(t)$. The gas to power efficiency of CHP depends on the electricity generation of the CHP, i.e., $E_{\mathrm{CHP}}(t)$. It is mathematically modeled later. Equation (12.7) calculates the share of heat demand that is provided by the gas boiler. In (12.8), the heat generation of CHP is calculated based on the output power of CHP and power to heat ratio of CHP unit. This efficiency is also a function of output power of CHP.

- *CHP*

Cogeneration system is one of the essential parts of future homes. CHP units are the most popular kinds of cogeneration technologies because of its high efficiency. As the main constraint of CHP unit, output power of CHP cannot go beyond from CHP capacity that presented as follows:

$$E_{\mathrm{CHP}}(t) \le E_{\mathrm{CHP}}^{\max} \tag{12.9}$$

where $E_{\text{CHP}}^{\max}$ is the nominal output power of CHP. The output power and heat of CHP units can be either calculated by gas to heat and gas to power efficiencies or gas to power efficiency and power to heat ratio. These factors are changing versus the electrical load of CHP, which is called CHP part load ratio [16]. Thus, Eqs. (12.10) and (12.11) can mathematically present $\eta_{\text{CHP}}^{g-e}(t)$ and $\beta_{\text{CHP}}^{e-h}(t)$, respectively. This is in line with the presented models in the literature [16].

$$\begin{aligned}\eta_{\text{CHP}}^{g-e}(t) = 0.9033\text{PL(t)}^5 - 2.9996\text{PL(t)}^4 + 3.6503\text{PL(t)}^3 \\ - 2.0704\text{PL(t)}^2 + 0.4623\text{PL(t)} + 0.3747\end{aligned} \tag{12.10}$$

$$\beta_{\text{CHP}}^{e-h}(t) = 1.0785\text{PL(t)}^4 - 1.9739\text{PL(t)}^3 + 1.5005\text{PL(t)}^2 - 0.2817\text{PL(t)} + 0.6838 \tag{12.11}$$

where PL is the part load ratio of CHP calculated as follows:

$$\text{PL}(t) = \frac{E_{\text{CHP}}(t)}{E_{\text{CHP}}^{\max}} \tag{12.12}$$

Equations (12.10) and (12.11) are valid for $\text{PL}(t) \geq 0.05$. For $\text{PL}(t) \leq 0.05$, mentioned factors are $\eta_{\text{CHP}}^{g-e}(t) = 0.2716$ and $\beta_{\text{CHP}}^{e-h}(t) = 0.6816$.

- *PHEV*

PHEVs, as a part of future smart homes, have a storage capability. Charging and discharging rates are limited to a definite value as follows:

$$E_{\text{PHEV}}^{\text{ch}}(t) \leq E_{\max}^{\text{ch}} \tag{12.13}$$

$$E_{\text{PHEV}}^{\text{dch}}(t) \leq E_{\max}^{\text{dch}} \tag{12.14}$$

where $E_{\max}^{\text{ch}}$ and $E_{\max}^{\text{dch}}$ are the maximum charging and discharging rate, respectively.

It is assumed that PHEV is going out in hour g and coming back home at hour c. During $[g,c]$, PHEV is not available at home and charging and discharging is impossible. Therefore, charge level of PHEV battery in hour t ($E_{\text{PHEV}}(t)$) is determined as follows:

$$E_{\text{PHEV}}(t) = \begin{cases} E_{\text{PHEV}}^{0} + \sum\limits_{h=1}^{t} E_{\text{PHEV}}^{\text{ch}}(h) - E_{\text{PHEV}}^{\text{dch}}(h) & \forall t \leq g-1 \\ E_{\text{PHEV}}^{0} + \left[\sum\limits_{h=1}^{t} E_{\text{PHEV}}^{\text{ch}}(h) - E_{\text{PHEV}}^{\text{dch}}(h)\right] - E_{\text{PHEV}}^{\text{out}} & \forall t \geq c+1 \end{cases} \tag{12.15}$$

Equation (12.15) defines the charge level of PHEV in hour t, where E_{PHEV}^{0} is the charge level of the PHEV at the beginning of the day called initial charge value and $E_{\mathrm{PHEV}}^{\mathrm{out}}$ is the consumed energy of PHEV in out of home hours, i.e., $[g,c]$.

PHEV battery package has a capacity and cannot be charged more than nominal battery capacity. This constraint is modeled as follows:

$$\begin{cases} E_{\mathrm{PHEV}}(t) \geq 0 \\ E_{\mathrm{PHEV}}(t) \leq \mathrm{cap}_{\mathrm{PHEV}} \end{cases} \tag{12.16}$$

where $\mathrm{cap}_{\mathrm{PHEV}}$ is the PHEV battery capacity.

In addition, customer tends to go out with fully charged battery, as follows:

$$E_{\mathrm{PHEV}}\,(g-1) = \mathrm{cap}_{\mathrm{PHEV}} \tag{12.17}$$

PHEV charge level at the end of the day should be more than or equal to its initial charge value. This constraint is presented as follows:

$$E_{\mathrm{PHEV}}(24) \geq E_{\mathrm{PHEV}}^{0} \tag{12.18}$$

- *PV panels*

PV generation is dependent on daily sunlight, thus the output power depends on the radiation. Equation (12.19) shows the relation between the output power of PV and the daily radiation as follows:

$$E_{\mathrm{PV}}^{G} = \begin{cases} (\eta_s/K_s) * R^2 & R \leq K_s \\ \eta_s {}^* R & K_s \leq R \leq R_n \\ P_n & R \geq R_n \end{cases} \tag{12.19}$$

where R is the solar radiation, η_s is the constant efficiency, K_s is the defined knee point, and P_n is nominal output power of PV.

- *Gas boiler*

Beside cogeneration systems, a gas boiler usually exists to guarantee providing the heat demand. The input natural gas of boiler should be capped to nominal capacity as follows:

$$H_{\mathrm{GB}}(t) \leq H_{\mathrm{GB}}^{\max} \tag{12.20}$$

$H_{\mathrm{GB}}^{\max}$ is the nominal capacity of gas boiler.

- *Heat storage*

Previous studies [11, 14] show that the heat storage devices can have an effective role in improving system operation. Thus, a heat storage unit is considered in the proposed structure for better operation of the energy hub. In addition, the heat storage may have impacts on decreasing the PV curtailment. The constraints of heat storage unit are mathematically modeled as follows:

$$\mathrm{SOH}_{\mathrm{HS}}(t) = H_{\mathrm{HS}}^{0} + \sum_{h=1}^{t} H_{\mathrm{HS}}^{\mathrm{ch}}(h) - H_{\mathrm{HS}}^{\mathrm{dch}}(h) \tag{12.21}$$

The state of the stored heat in the heat storage in hour t is calculated in Eq. (12.21). H_{HS}^{0} is the stored heat in the storage at the beginning of the day.

Charging and discharging rates of heat storage are limited to their predetermined values, as follows:

$$H_{\mathrm{HS}}^{\mathrm{ch}}(t) \le H_{\mathrm{HS}}^{\mathrm{max,ch}} \tag{12.22}$$

$$H_{\mathrm{HS}}^{\mathrm{dch}}(t) \le H_{\mathrm{HS}}^{\mathrm{max,dch}} \tag{12.23}$$

Heat storage charge level at the end of the day is assumed to be more than or equal to its initial value at the beginning of the day. This is shown in Eq. (12.24):

$$\mathrm{SOH}_{\mathrm{HS}}(24) \ge H_{\mathrm{HS}}^{0} \tag{12.24}$$

The state of heat in heat storage cannot be negative. On the other hand, the state of heat is always less than heat storage capacity. Equation (12.25) shows this as follows:

$$\begin{cases} \mathrm{SOH}(t) \ge 0 \\ \mathrm{SOH}(t) \le \mathrm{CAP}_{\mathrm{HS}} \end{cases} \tag{12.25}$$

where $\mathrm{CAP}_{\mathrm{HS}}$ is the capacity of heat storage.

12.3 Numerical Studies

In this section, a sample home is assumed as the proposed renewable-based residential energy hub and the proposed optimization problem is applied to the home. The purpose of this study is to investigate the effect of emerging components in future smart homes on the proposed objective function and interactions between these components. Four cases are defined here.

- *Case 1*: PHEV and heat storage are omitted from the presented energy hub in Fig. 12.1. In addition, the gas to power efficiency and the power to heat ratio of CHP unit are considered constant. Results reported in this case are the amount of objective function and the dispatch factor of CHP unit at each hour.
- *Case 2*: the gas to power efficiency and power to heat ratio of CHP unit are considered dependent on the output power of CHP. The gas to power efficiency and power to heat ratio of CHP unit at each hour are also reported in case two.
- *Case 3*: a PHEV is added to the proposed energy hub and the effect of PHEV on the operation of energy hub is taken into account, in Case 3. In this case, charging and discharging schedule of PHEV is also presented at each hour. Although PHEV is shown at the output of energy hub, the PHEV battery can be discharge at some hours and provide a part of the household electrical load. This result is deeply presented in the following.
- *Case 4*: the heat storage is added to the proposed energy hub. The impact of heat storage presence on the objective function, especially, the PV curtailment is presented in the next subsections.

Application of the proposed optimization procedure in each case and the results are reported in the following. The proposed optimization problems are solved using a Nonlinear Programing (NLP) solver in the GAMS environment. Computations are performed on a PC with a 2.53 GHz processor and 4 GB RAM, in less than 1 min.

12.3.1 Assumptions

Since a part of objective function is energy cost, energy tariff has an effective role in determining the value of objective function. The gas tariff is a fixed rate in a day and it is assumed to be 0.05 $/kWh in this study. It is also assumed that the consumer accepted to participate in a price-based demand response program. One of the most commonly used time-varying pricing for residential consumers is time of use (TOU). Based on this pricing method, different price levels, usually two or three levels, are considered for a day and electricity prices are fixed for a specific period of the day. Here, a two-level TOU tariff is assumed for a day as shown in Fig. 12.2. Figure 12.3 shows the variation of gas to power efficiency of CHP unit based on the CHP part load. The nominal output power of CHP is assumed to be 1.8 kW. In addition, capacity of gas boiler is 0.6 kW with the efficiency of 0.9. Table 12.1 shows PHEV parameters, i.e. PHEV battery capacity, maximum charging and discharging rate, initial charge value, the departure and arrival time, out of home energy consumption of PHEV and charging and discharging efficiencies. In Fig. 12.4, the capability of PV panels for power generation is shown. These values are obtained by inserting radiation values for a sample city in the proposed model for PV panels in Eq. (12.19). Table 12.2 summarizes heat storage parameters, i.e.

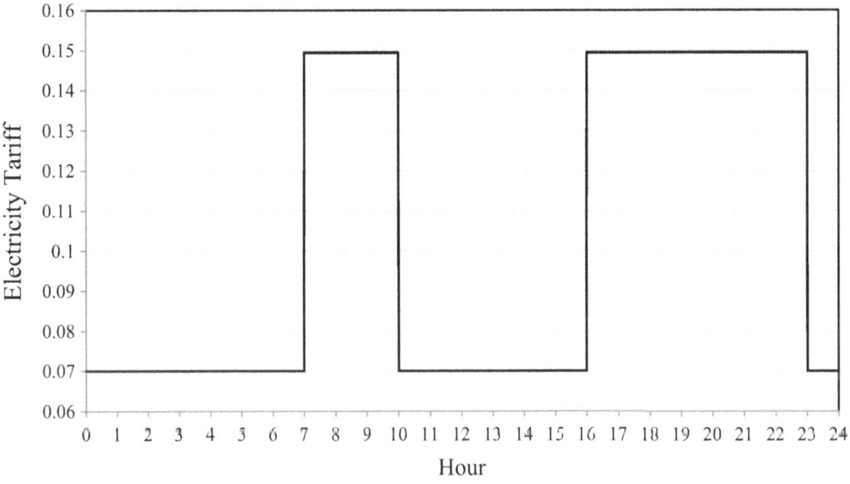

Fig. 12.2 Two-level TOU for a day

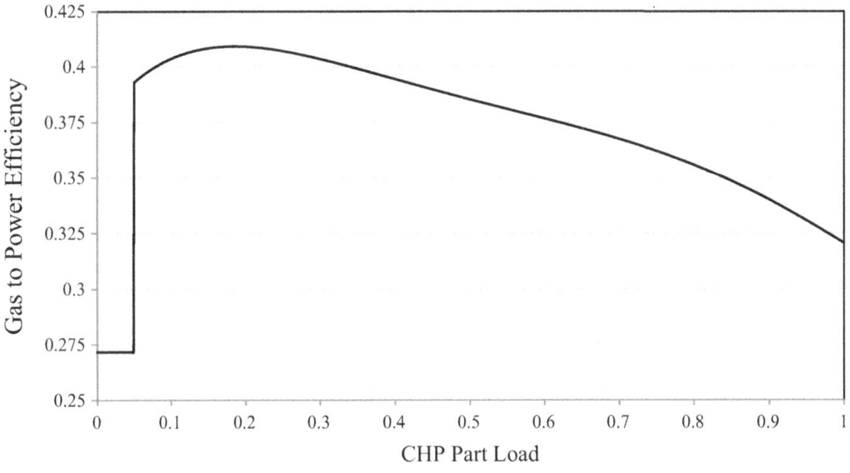

Fig. 12.3 Gas to power efficiency based on the Part Load

Table 12.1 PHEV parameters

cap_{PHEV}	E_{max}^{ch}	E_{max}^{dch}	E_{PHEV}^{0}	g	c	E_{PHEV}^{out}	η_{PHEV}^{ch}	η_{PHEV}^{dch}
7.8 kW	1.4 kW	1.4 kW	2 kW	8	17	3.9 kW	0.88	0.88

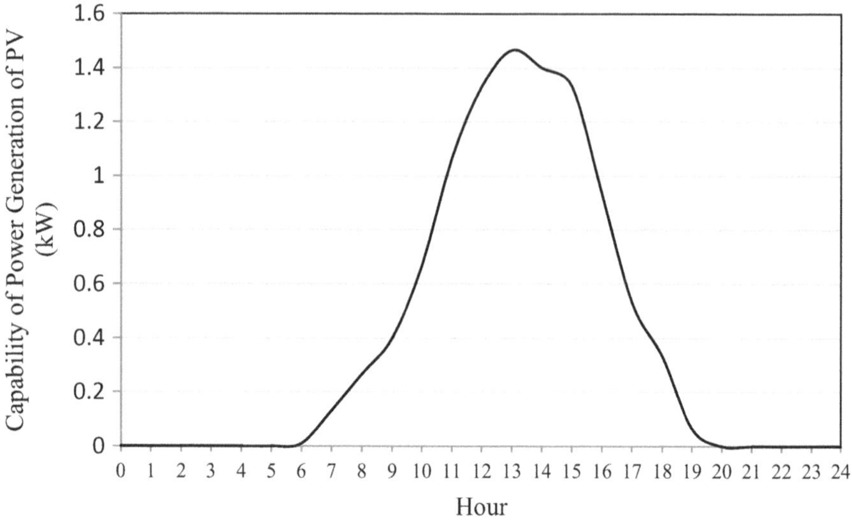

Fig. 12.4 Capability of power generation of PV

Table 12.2 Heat storage parameters

CAP_{HS}	$H_{HS}^{max,ch}$	$H_{HS}^{max,dch}$	H_{HS}^{0}	η_{HS}^{ch}	η_{HS}^{dch}
0.6 kW	0.2 kW	0.2 kW	0.1 kW	0.9	0.9

heat storage capacity, maximum charging and discharging rate of heat storage, initial charge value and charging and discharging efficiencies. Electrical and heat demands for a sample home is shown in Fig. 12.5 for a day. w_1 and w_2, which are the weights of the objectives in (12.3), are determined according to the customer's preferences through the proposed FDM method in [15]. Here, it is assumed that the importance of both objective functions for the customer is similar. In this case, FDM method is applied and the weights, i.e. w_1 and w_2 are also achieved equal to 0.5.

12.3.2 *Case 1*

This case is simple enough to investigate the impact of the proposed optimization procedure on the operation of the residential energy hub. In this case, according to Fig. 12.1, CHP unit, gas boiler, and PV panels are components of energy hub and electricity grid and natural gas as the inputs of energy hub supply the electrical and heat demands in the output. The gas to power efficiency and power to heat ratio of CHP are assumed to be 0.37 and 0.77, respectively.

Energy hub operation is optimized by Eq. (12.3) subjected to (12.4)–(12.25). The results show that the value of objective function is 3.024. Natural gas and electricity purchased from the grid are determined as the outcome of the problem and are shown in Figs. 12.6 and 12.7, respectively. Total input gas for whole day

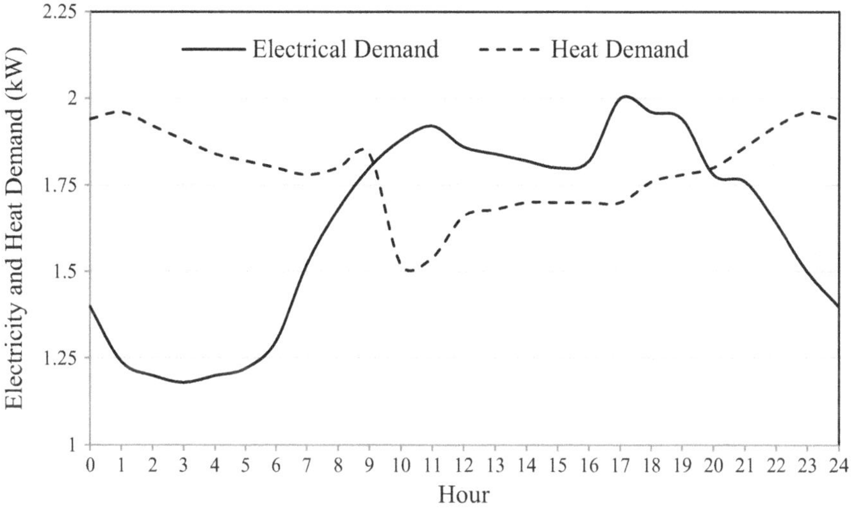

Fig. 12.5 Electrical and heat demand in a sample day

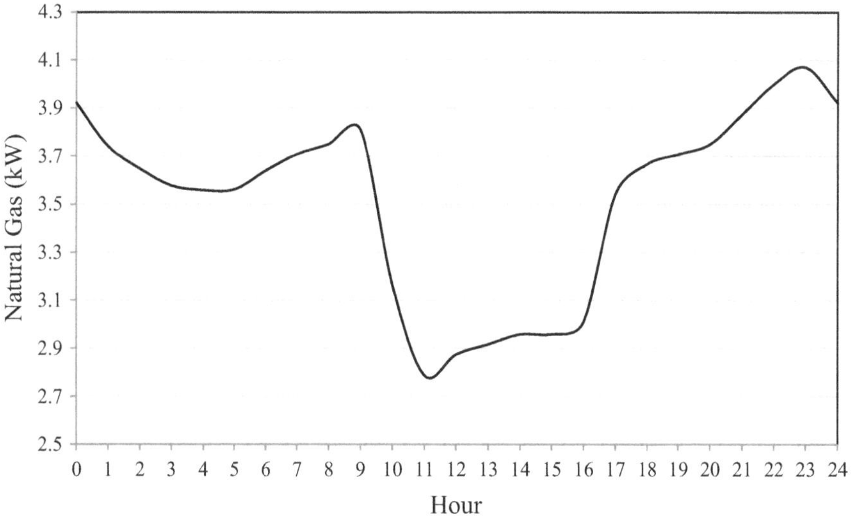

Fig. 12.6 Natural gas purchased from the grid in case 1

is 84.19 kWh and total purchased electricity from the grid is 1.89 KWh. Dispatch factor $\alpha(t)$ is another outcome of the problem and its value at each hour is shown in Table 12.3. At hours (17–23), dispatch factors are equal to one and all the input gas enters the CHP unit. The reason for this is that these hours are high-peak tariff hours and electrical demand is high. Therefore, it is profitable to use CHP unit with maximum output power and supply the remained electrical demand through the grid.

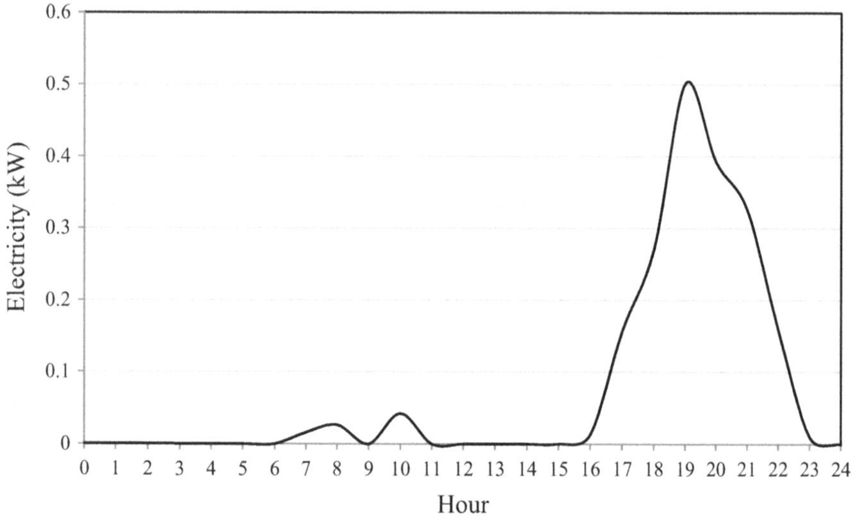

Fig. 12.7 Electricity purchased from the grid in case 1

Table 12.3 Dispatch factor of CHP in case 1

Hour	*1*	*2*	*3*	*4*	*5*	*6*	*7*	*8*
Value	0.895	0.889	0.891	0.911	0.925	0.965	1	1
Hour	*9*	*10*	*11*	*12*	*13*	*14*	*15*	*16*
Value	0.993	1	0.827	0.768	0.771	0.774	0.774	0.797
Hour	*17*	*18*	*19*	*20*	*21*	*22*	*23*	*24*
Value	1	1	1	1	1	1	0.996	0.964

Between hours (11–16), PV panels supply a part of electrical demand. Therefore, the electrical demand at the output of CHP is reduced and CHP can be fed with less natural gas. In the early hours of the day, i.e., hours (1–6), the heat demand is high and the gas boiler cannot provide whole the heat demand. Thus, the CHP supplies the rest of the heat demand. Generated power of CHP at these hours is enough for low electrical demand during these hours and purchased electricity from the grid is zero. As previously discussed, the ratio of output power to rated power is called part load ratio. Figure 12.8 shows the part load ratio of CHP for each hour. This ratio depends on the amount of input natural gas and the dispatch factor $\alpha(t)$. According to Figs. 12.6 and 12.8, the pattern of the input natural gas and part load ratio is similar. Table 12.4 shows the hours in which PV is curtailed. The amount of PV curtailment is also presented in the table. The results show that during hours

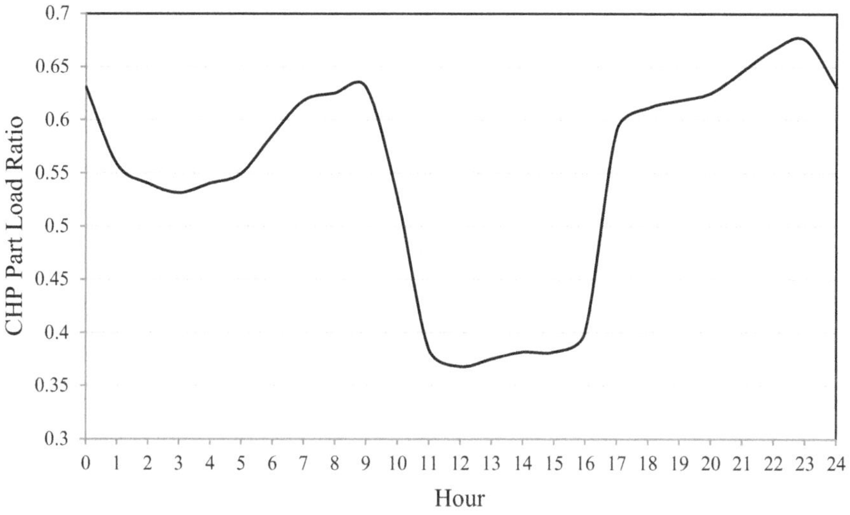

Fig. 12.8 CHP part load ratio in case 1

Table 12.4 PV curtailment in case 1

Hours	*12*	*13*	*14*	*15*
Value (kW)	0.289	0.458	0.428	0.380

(12–15) a part of capability of PV panels for generating power is unused. Since, the PV generation and the power generation of CHP at these hours are high. Thus, most of the electrical demand is provided by the CHP and the PV is curtailed. In this case, total PV curtailment is 1.555 kWh for whole the day.

12.3.3 *Case 2*

The gas to power efficiency and power to heat ratio of CHP are varying based on the CHP load according to Eqs. (12.10) and (12.11). As presented in Fig. 12.3, gas to power efficiency of CHP can vary between 0.271 and 0.409. In addition, the power to heat ratio can be between 0.668 and 1.007. In this case, the proposed optimization procedure also determines the optimum values for the CHP efficiency.

The objective function, i.e., the weighted summation of energy cost and the PV curtailment, is minimized subjected to equations (12.4)–(12.25). The resulted value of objective function is 2.925 in this case, which 3.44% is lower than Case 1. Inputted natural gas and electricity are shown in Figs. 12.9 and 12.10, respectively. The daily natural gas and electrical energy purchased from the grid are 81.62 kWh and 4.38 kWh, respectively. However, daily energy of input natural gas is decreased. Although the total received energy is increased a little, decreased PV curtailment

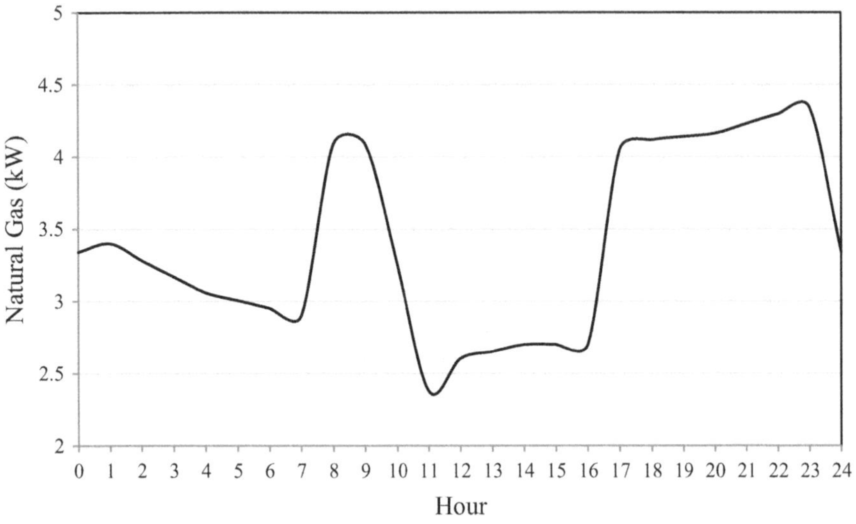

Fig. 12.9 Natural gas purchased from the grid in case 2

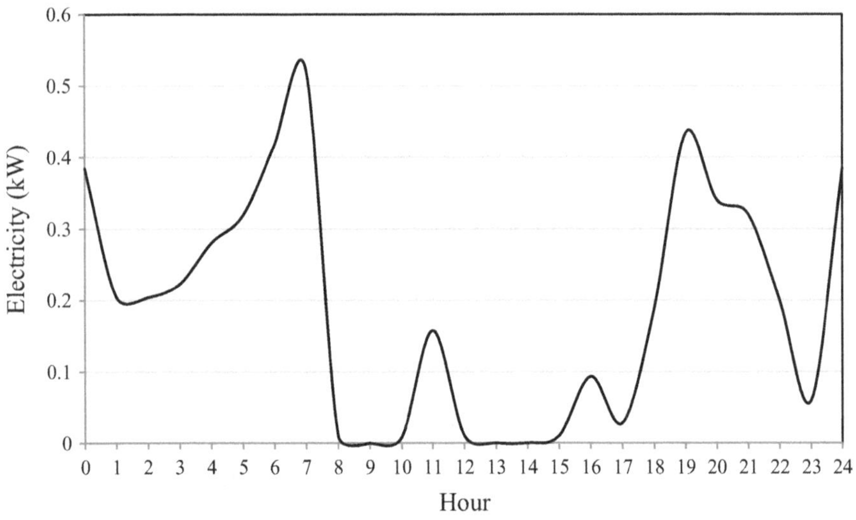

Fig. 12.10 Electricity purchased from the grid in case 2

reduces the objective function. In addition, the peak of natural gas received from the grid is increased in Case 2 in comparison with that of Case 1. In Table 12.5, resulted dispatch factor are shown. According to Fig. 12.9, in Case 2, inputted natural gas is decreased in comparison with case 1 at hours 1–6 and dispatch factor is reduced at these hours. This result causes that the electrical demand at hours 1–6 is not

Table 12.5 Dispatch factor of CHP in case 2

Hour	*1*	*2*	*3*	*4*	*5*	*6*	*7*	*8*
Value	0.804	0.797	0.790	0.782	0.778	0.774	0.770	0.967
Hour	*9*	*10*	*11*	*12*	*13*	*14*	*15*	*16*
Value	0.954	1	0.740	0.744	0.749	0.753	0.753	0.753
Hour	*17*	*18*	*19*	*20*	*21*	*22*	*23*	*24*
Value	0.999	0.982	0.977	0.972	0.957	0.942	0.932	0.800

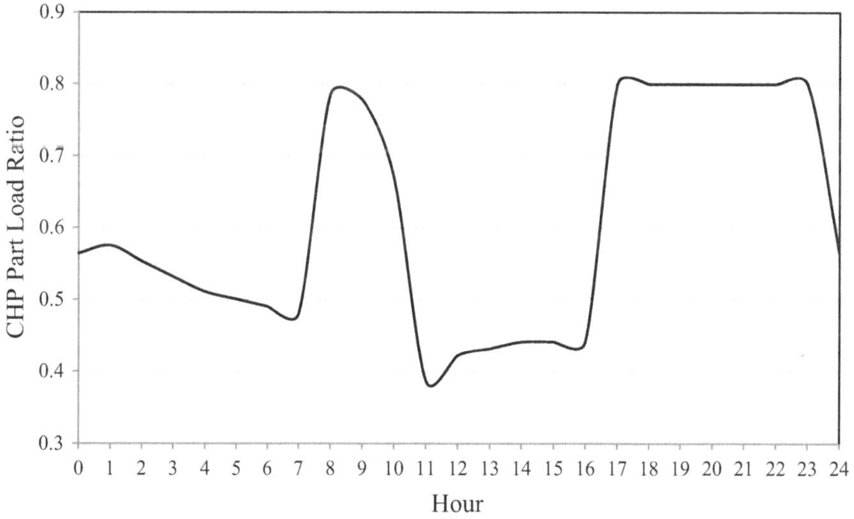

Fig. 12.11 CHP part load ratio in case 2

completely provided by CHP unit. Therefore, the remained electrical demand is supplied from the grid. Figures 12.11 and 12.12 show the CHP part load ratio and gas to power efficiency of CHP, respectively. The pattern of input natural gas and CHP part load is similar. Comparing Figs. 12.9 and 12.12 verifies that operation of CHP at low efficiency leads to high consumption of natural gas at the input to meet the household heat demand. Table 12.6 shows PV curtailment in case 2. Total PV curtailment is 1.337 kWh for whole the day, which is 14% lower than that of Case 1. The reason for this is considering the part load efficiency in Case 2, which is more realistic.

12.3.4 *Case 3*

In this case, a PHEV is added to the proposed residential energy hub in Case 2. The PHEV is out of home between hours 8 and 17, and its outdoor consumption

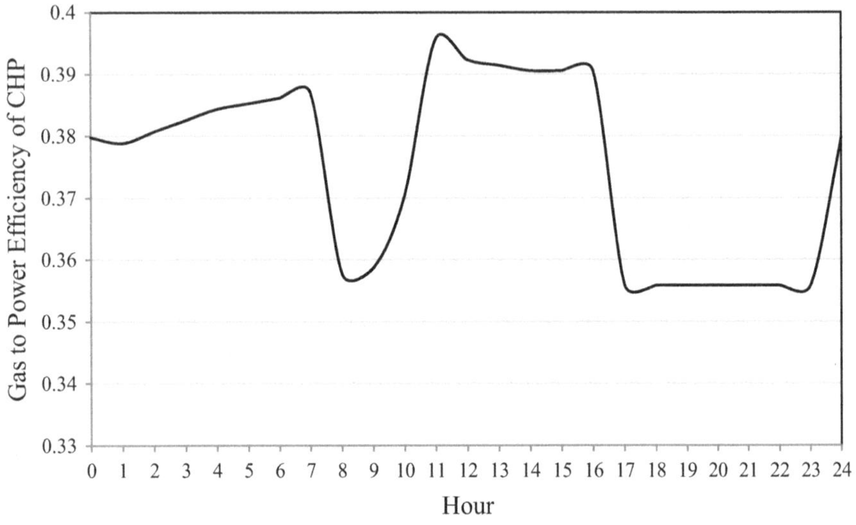

Fig. 12.12 Gas to power efficiency of CHP in case 2

Table 12.6 PV curtailment in case 2

Hours	*12*	*13*	*14*	*15*
Value (kW)	0.232	0.403	0.375	0.327

is 3.9 kWh. It is assumed that the PHEV goes out with fully charged battery. At available hours at home, PHEV battery package can be charged or discharged with the aim of reducing the objective function according to its constraints. Although the PHEV outdoor consumption is added as an electrical demand, it is worthwhile to note that PHEV battery package acts as electrical energy storage and can improve the objective function at available hours.

The PHEV needs a definite amount of energy to consume out of home. This increases the household electrical demand. Thus, the resulted objective function is 3.009 in this case, which is more than Case 2. In Figs. 12.13 and 12.14, purchased natural gas and electricity is shown, respectively. CHP dispatch factor as another resulted output is shown in Table 12.7 for each hour. CHP part load ratio is represented in Fig. 12.15. Charging and discharging schedule of PHEV is shown in Fig. 12.16. In Fig. 12.16, positive and negative values show, respectively, charging and discharging of PHEV battery.

Total input gas and electricity are, respectively, 78.14 kWh and 11.08 kWh for whole the day. According to Fig. 12.16, during hours (17–23), the PHEV is discharged to supply a part of electrical demand. In addition, PHEV is charged at hours 23, 24, and (1–7). Discharging hours are high-peak tariff hours while PHEV is charged during low-tariff hours. The variation pattern of inputted natural gas in Cases 2 and 3 is similar. Since, the heat demand is the same in both cases. However, between hours 17 and 23, in which the PHEV is discharged, electrical demand of

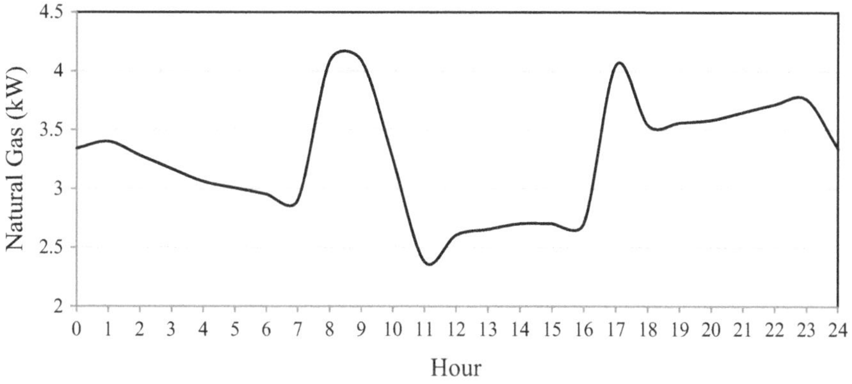

Fig. 12.13 Purchased Natural Gas in Case 3

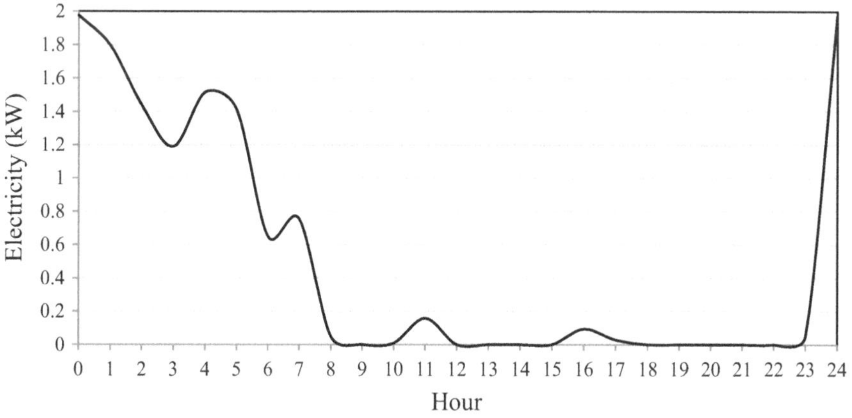

Fig. 12.14 Purchased Electricity in Case 3

Table 12.7 Dispatch factor of CHP in case 3

Hour	*1*	*2*	*3*	*4*	*5*	*6*	*7*	*8*
Value	0.804	0.797	0.790	0.782	0.778	0.774	0.770	0.967
Hour	*9*	*10*	*11*	*12*	*13*	*14*	*15*	*16*
Value	0.954	1	0.740	0.744	0.749	0.753	0.753	0.753
Hour	*17*	*18*	*19*	*20*	*21*	*22*	*23*	*24*
Value	0.999	0.926	0.921	0.915	0.898	0.882	0.872	0.800

CHP unit and, in turn, CHP part load ratio is reduced. Part load reduction causes a higher CHP efficiency, presented in Fig. 12.3, which lowers input gas for providing the heat demand. According to Fig. 12.14, purchased electricity during hours (17–23) becomes zero because PHEV is discharged and beside the output power of the

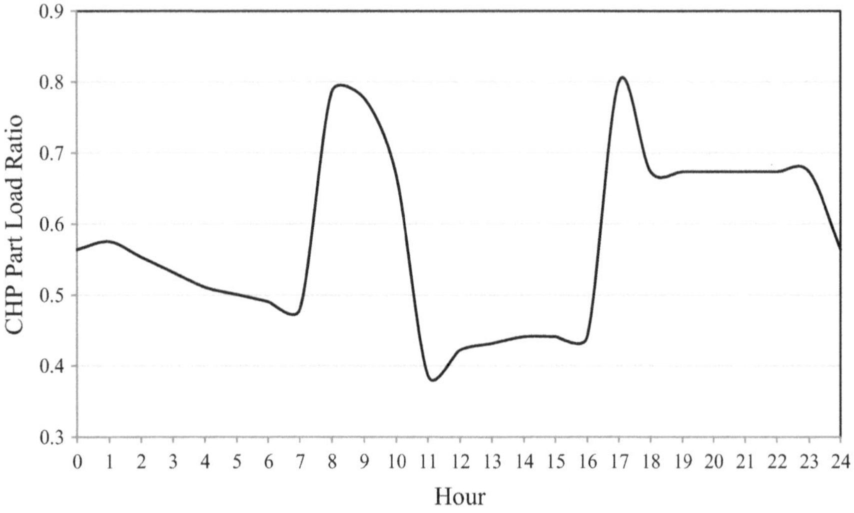

Fig. 12.15 CHP Part Load Ratio in Case 3

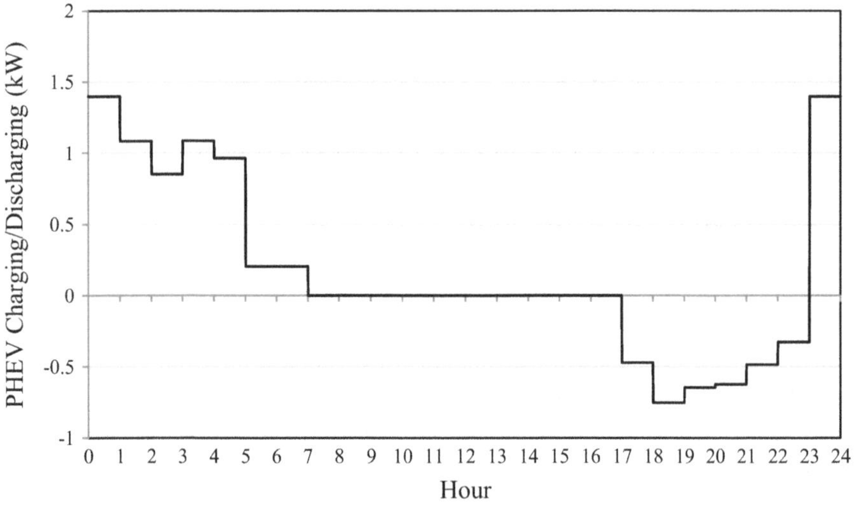

Fig. 12.16 PHEV charging and discharging schedule in Case 3

CHP, it provides the electrical demand. On the other hand, at hours 23, 24, and (1–7), in which PHEV is charged, purchased electricity is increased in comparison with Case 2, and its peak reached 1.975 kW, which is more than that of Case 2. In Case 3, during hours (12–15), PHEV is out of home and there is no change in electrical and heat demand in comparison with Case 2. Therefore, the PV curtailment in this case is the same as Case 2.

12.3.5 Case 4

It was previously discussed that in the proposed energy hub, high heat demand of CHP unit leads to more CHP output power and less using output power of PV panels, which leads to PV curtailment. In this case, the heat storage unit is added to the proposed energy hub. The heat storage can be discharged when needed and provide a part of heat demand. Therefore, the heat drawn from CHP unit is reduced. As a result, output power of CHP is reduced and PV panels can provide more part of electrical demand. The heat storage specifications were presented in Table 12.2.

The objective function is minimized subjected to Equations (12.4)–(12.25). The amount of objective function is 2.772 and it is 7.87% lower than Case 3. In Figs. 12.17 and 12.18, purchased natural gas and electricity from the grid are shown, respectively. Total consumed natural gas is 7.35 kWh and total purchased electricity is 11.21 kWh for the day. CHP dispatch factors are shown in Table 12.8 for each hour. CHP part load ratio for each hour is shown in Fig. 12.19. Charging and discharging schedule of PHEV and heat storage are concluded in Figs. 12.20 and 12.21, respectively.

According to Fig. 12.21, during hours (11–15), the heat storage is discharged to provide a part of the heat demand. Therefore, inputted natural gas, CHP dispatch factors, and CHP part load ratio are decreased at these hours. Thus, more PV generation is used to provide the electrical load and the PV curtailment is decreased in comparison with previous cases.

The presented results in Fig. 12.21 show that discharging/charging the heat storage is occurred at low-tariff/high-tariff hours of electricity. Since, at high-tariff hours of electricity, it is profitable to provide more electricity demand by CHP. Hence, the heat demand of CHP is forced to be increased by charging the storage.

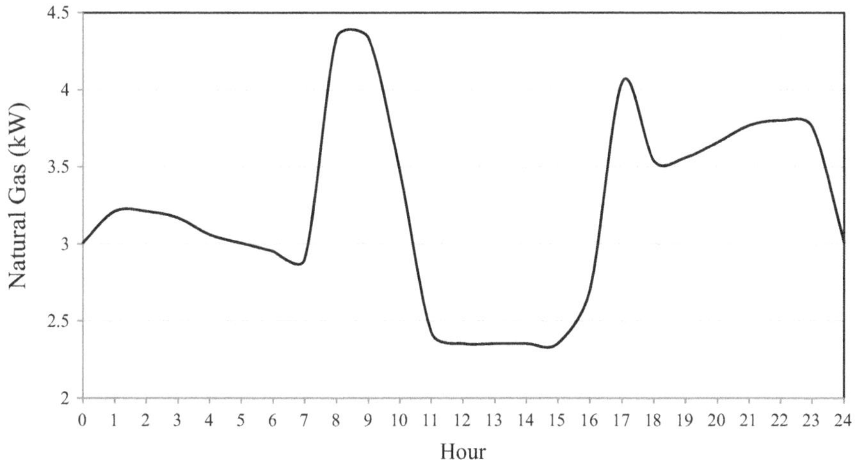

Fig. 12.17 Purchased natural gas in Case 4

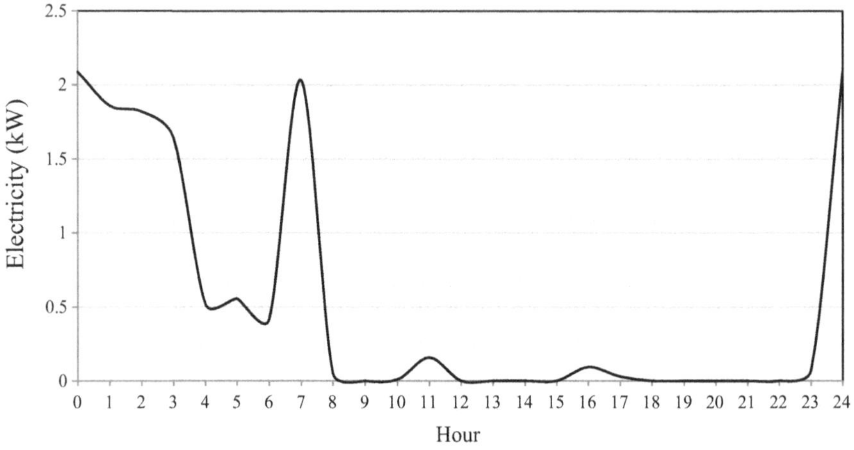

Fig. 12.18 Purchased electricity in Case 4

Table 12.8 Dispatch factor of CHP in case 4

Hour	*1*	*2*	*3*	*4*	*5*	*6*	*7*	*8*
Value	0.792	0.792	0.790	0.782	0.778	0.774	0.770	0.912
Hour	*9*	*10*	*11*	*12*	*13*	*14*	*15*	*16*
Value	0.900	0.945	0.725	0.717	0.717	0.717	0.717	0.753
Hour	*17*	*18*	*19*	*20*	*21*	*22*	*23*	*24*
Value	0.999	0.926	0.921	0.895	0.869	0.862	0.872	0.778

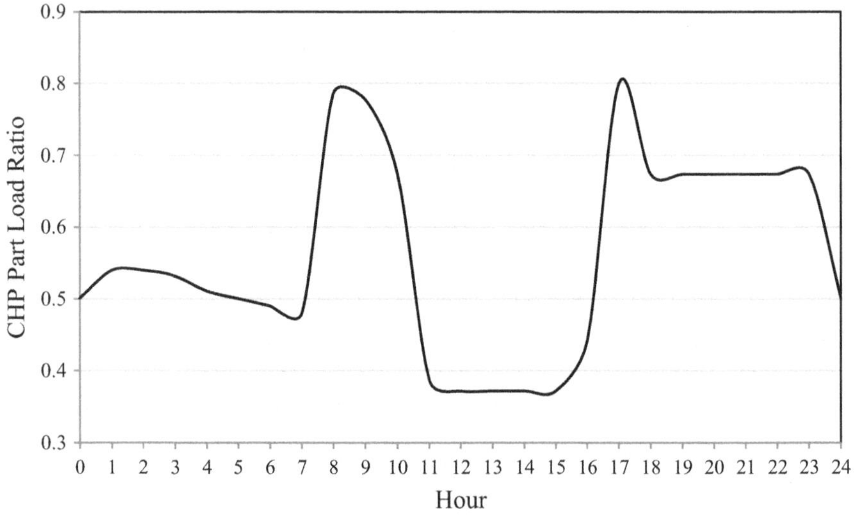

Fig. 12.19 CHP Part load in Case 4

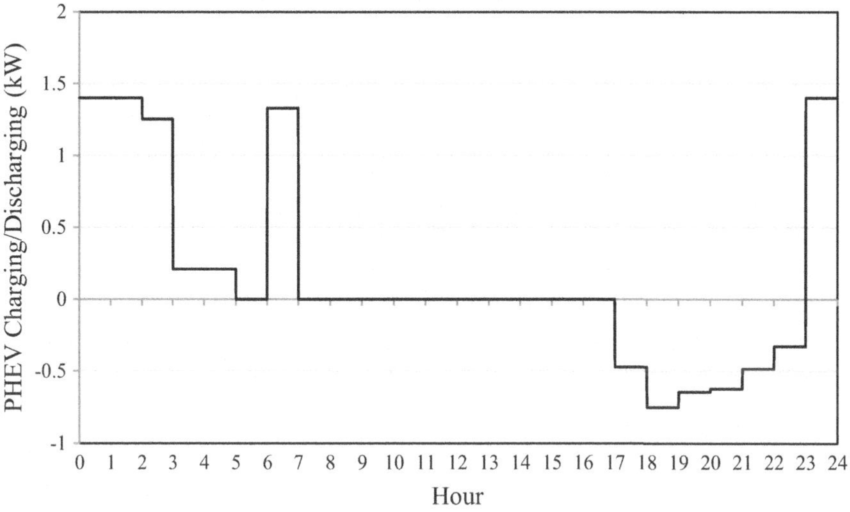

Fig. 12.20 PHEV charging and discharging schedule in Case 4

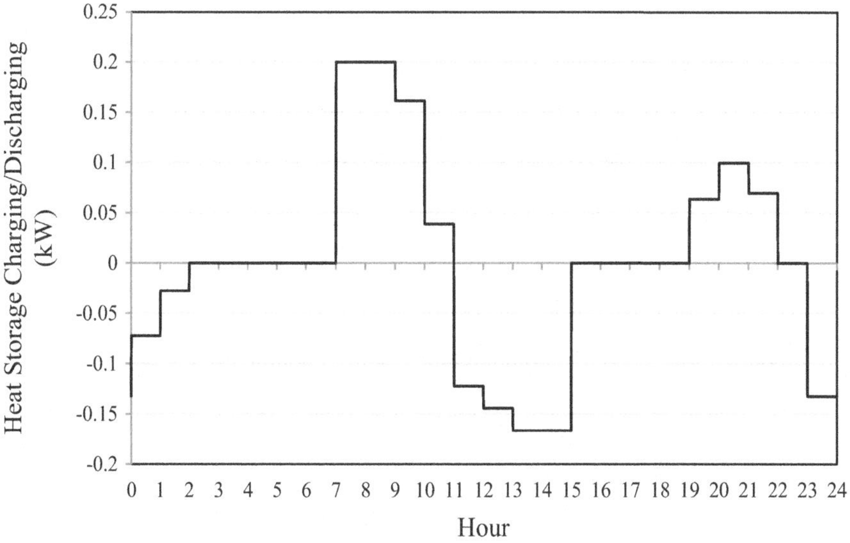

Fig. 12.21 Heat storage charging and discharging schedule in Case 4

This raises the CHP output power, subsequently. This leads to lower need for purchasing the electricity from the grid at high-tariff hours.

According to Fig. 12.20, PHEV provides a part of electrical demand during hours (17–23). This makes the purchased electrical from the grid zero. PHEV battery is charged in low-tariff hours, i.e., (23–24) and (1–7), which causes an increment

Table 12.9 PV curtailment in Case 4

Hours	*12*	*13*	*14*	*15*
Value (kW)	0.141	0.295	0.249	0.201

in the required electricity from the grid. This leads to 2.08 kW peak load at hour 24. In addition, between hours 12 and 15, output power of CHP along with PV panels can provide electrical demand completely and the purchased electricity from the grid becomes zero. Because of PHEV discharging between hours 17 and 23, output power and part load ratio of CHP are reduced. In Case 4, the amount of PV curtailment is shown in Table 12.9. Total PV curtailment is 0.886 kWh for whole the day. The heat storage unit descends the PV curtailment. As previously discussed, heat storage is discharged between hours (12–15) and reduces the heat demand of CHP unit. Therefore, the inputted natural gas and output power of CHP are reduced. Thus, PV panels can provide more part of electrical demand and PV curtailment is reduced. In Case 4, PV curtailment is reduced about 33% in comparison with Case 2.

12.4 Conclusion

This chapter optimizes the operation of a residential energy hub, including a CHP, a boiler, a PHEV, PV panels, and a heat storage to provide the electrical and heat demand of a home. The objective function is the weighted summation of minimizing the energy cost and PV curtailment. The outputs of the objective function were the dispatch factor of the energy hub at different hours, the amount of purchased electricity and natural gas from the grid, the charge/discharge scheduling of PHEV battery and the heat storage, and the manner of energy flow in the energy hub. Different cases were designed to investigate the impact of PHEV and the heat storage on the results. In addition, the impact of CHP part load efficiency was studied on the results. This makes the results more realistic.

The results show that PV curtailment is occurred at midday, which are the high-generation hours of the PV panels. Thus, a PHEV, which is not usually available at these hours, cannot reduce the PV curtailment significantly. Because of the charge/discharge efficiency of PHEV battery, the average electricity demand of PHEV is positive. Thus, PHEV presence may increase the energy cost in comparison with the case, in which there is no PHEV. However, the heat storage presence not only reduces the energy cost but also decreases the PV curtailment. The results show a 33% decrement in the PV curtailment by using the heat storage system.

References

1. Pales AF (2013) The IEA CHP and DHC collaborative (CHP/DHC country scorecard: Japan), International Energy Agency (IEA) Insights Series, Paris, France, Technical Report
2. Geidl M, Andersson G (2007) Optimal power flow of multiple energy carriers. IEEE Trans Power Syst 22(1):145–155
3. Energy Information Administration (2016) http://www.eia.gov/electricity/annual. Accessed 30 July 2016
4. Ren H, Gao W, Ruan Y (2008) Optimal sizing for residential CHP system. Appl Therm Eng 28:514–523
5. Martinez-Mares A, Fuerte-Esquivel CR (2012) A unified gas and power flow analysis in natural gas and electricity coupled networks. IEEE Trans Power Syst 27(4):2156–2166
6. Geidl M, Andersson G (2006) Operational and structural optimization of multi-carrier energy systems. Eur Trans Electr Power 16:463–477
7. Ruan Y, Gao W (2005) Optimization of co-generation system for housing complex. J Environ Eng (Trans AIJ) 592:15–22
8. Du P, Lu N (2011) Appliance commitment for household load scheduling. IEEE Trans Smart Grid 2(2):411–419
9. Brahman F, Honarmand M, Jadid S (2014) Optimal electrical and thermal energy management of a residential energy hub, integrating demand response and energy storage system. Energy Build 90:65–75
10. Bozchalui MC, Hashmi SA, Hassen H (2012) Optimal operation of residential energy hubs in smart grids. IEEE Trans Smart Grid 3(4):1755–1766
11. Rastegar M, Fotuhi-Firuzabad M, Zareipoor H, Moeini-aghtaie M (2016) A probabilistic energy management scheme for renewable-based residential energy hubs. IEEE Trans Smart Grid 8:2217–2227. https://doi.org/10.1109/TSG.2016.2518920
12. Rastegar M, Fotuhi-Firuzabad M, Lehtonen M (2015) Home load management in a residential energy hub. J Electr Power Syst Res 119:322–328
13. Fujimoto Y et al (2016) Distributed energy management for comprehensive utilization of residential photovoltaic outputs. IEEE Trans Smart Grid. https://doi.org/10.1109/TSG.2016.2581882
14. Barmayoon MH, Fotuhi-Firuzabad M, Rajabi-Ghahnavieh A, Moeini-Aghtaie M (2016) Energy storage in renewable-based residential energy hubs. IET Gener Transm Distrib 10(13):3127–3134
15. Moeini-aghtaie M, Abbaspour A, Fotuhi-Firuzabad M (2012) Incorporating large-scale distant wind farms in probabilistic transmission expansion planning – part I: theory and algorithm. IEEE Trans Power Syst 27(3):1585–1593
16. Karami H, Sanjari MJ, Hosseinian SH, Gharehpetian GB (2014) An optimal dispatch algorithm for managing residential distributed energy resources. IEEE Trans Smart Grid 5(5):2360–2367

Chapter 13
Long-Term Smart Grid Planning Under Uncertainty Considering Reliability Indexes

Bruno Canizes, João Soares, Mohammad Ali Fotouhi Ghazvini, Cátia Silva, Zita Vale, and Juan M. Corchado

13.1 Introduction

The electricity sector is extremely important to the society. The increasing energy needs are mostly satisfied by nonrenewable energy sources like coal or natural gas. However, this energy resources are scarce and can bring negative consequences to environment. In this way, there is a necessity to find new alternatives to, at least, reduce their use. In fact, environmental and techno-economic factors have motivated the widespread adoption of Distributed Generation (DG) technologies in distribution networks [1]. Therefore, the portion of DG based generated electricity is increasing as a consequence and will play an important role in distribution network systems. Nevertheless, DG is based on renewable sources such as solar and wind and therefore carry an inherent variability [2].

Stochastic expansion model for the transmission problem has been proposed in [3–5], suggesting superior results compared with deterministic approaches when likely realizations are considered. Cao et al. [5] propose a multiple resource expansion planning in smart grids. The two-stochastic model minimizes the expected

B. Canizes (✉) · J. Soares · M. A. F. Ghazvini · C. Silva · Z. Vale
GECAD–Research Group on Intelligent Engineering and Computing for Advanced Innovation and Development, Institute of Engineering, Polytechnic of Porto (ISEP/IPP), Porto, Portugal
e-mail: brmrc@isep.ipp.pt; joaps@isep.ipp.pt; mafgh@isep.ipp.pt; cvcds@isep.ipp.pt; zav@isep.ipp.pt

J. M. Corchado
University of Salamanca, Salamanca, Spain

Osaka Institute of Technology, Osaka, Japan

University of Technology Malaysia, Pusat Pentadbiran Universiti Teknologi Malaysia, Skudai, Malaysia
e-mail: corchado@usal.es

B. Mohammadi-Ivatloo, F. Jabari (eds.), *Operation, Planning, and Analysis of Energy Storage Systems in Smart Energy Hubs*,
https://doi.org/10.1007/978-3-319-75097-2_13

cost in the entire planning horizon and in the second stage the realization of the load and wind generation are found. The results reveal that the expansion plans depend on the uncertainty level of prospective wind generation, existing capacity, and transmission capacity. A stochastic planning approach of distribution lines is presented in [6]. The work is based on Monte Carlo and optimization procedure to minimize the conductor profile of a power line and the transformer capacity. The net present value of the total average annual costs is evaluated for the planning period of 30 years. The stochastic approach is compared with the deterministic one, and the results reveal that the former can increase the net present value by 13–25%. The work presented in [1] concerns a multi-year distributed generation investment planning. The stochastic model considers uncertainty on emission price, demand growth, and renewable generation. The results in the real network suggest that compared to the naive decisions, the stochastic model yields better and more robust decisions, namely amounting to more than 7%.

Combined heat and power (CHP) planning has demonstrated value in previous works [7, 8]. By definition, CHP plants can produce heat and power simultaneously, saving the heat that would be wasted in electricity production while saving up to 30% compared to conventional condensing power plants. Rong and Lahdelma [7] refer that when steam or hot water is produced for an industrial plant or a residential area, power can be produced as a by-product. Excess heat from an electric power plant can be used for industrial purposes, or for heating space and water. CHP is applied in the district heating concept. A district heating scheme comprises a network of insulated pipes used to deliver heat, in the form of hot water or steam, from the generation point to the final user. A district heating plant is often a CHP plant but renewables sources, for example biomass or solar energy, can be applied in district heating utilities, either completely or as a complement to traditional fossil fuels.

Rong and Lahdelma [7] propose efficient algorithms for combined heat and power production planning in the electricity markets. Authors propose algorithms up to 1860 times faster than CPLEX. Fast solutions of hourly CHP models are important, because long-term planning model requires solving several hourly models, and a large number of scenarios in stochastic approaches. In [8] multiple energy infrastructures are addressed, namely for supplying electricity and gas loads. The planning model determines a least-cost network of transmission lines for both infrastructures. The authors demonstrate that the coupling multiple energy hubs offer advantages and more flexible options between the interconnected systems.

Considering current literature, in this work we propose to include heat and power demand in the grid expansion problem (new lines construction) to improve reliability indexes ensuring the radial topology of the distribution power network at minimum costs. Results indicate that it influences the grid planning and a joint planning is more indicated.

This chapter is organized as follows: After the brief introductory part, Sect. 13.2 presents the modeling of system uncertainties; Sect. 13.3, the problem formulation; Sect. 13.4, the adopted case study; Sect. 13.5, the results and its discussion; and Sect. 13.6, the conclusions.

13.2 System Uncertainties

Grid expansion and planning problems can be modeled as deterministic or stochastic problems. Usually, power system planners have considered this problem as a deterministic model, i.e., they considered parameters and inputs based on the assumption that the data for the problem is known accurately. Nonetheless, the inputs of the expansion model must be estimated, such as the load demand and the renewables penetration in the project lifespan, which is usually a decade at least. However, the projections are done with a large anticipation process depending on many factors and as a consequence they are not 100% accurate. The high deviations in the projections can have a relevant impact on the economic and technical aspects of daily grid operation. Therefore, the recent advances in expansion planning models are moving from deterministic to stochastic approaches in order to incorporate the uncertainty in projections for future in the planning models [9, 10]. In practice, it is possible to feed a deterministic model with several likely scenarios and run each optimization independently. However, advanced stochastic models can provide better alternatives [11–14]. To capture the underlying uncertainty in the problem data, a sophisticated energy planning model is developed here. The goal is to find a solution that is feasible for all the supplied scenarios while minimizing or maximizing the objective function, e.g., the expected investment cost [10].

The steps involving stochastic programming are typically developing the possible scenarios that represent the underlying uncertainty. This step is usually a cumbersome task where lot of possible scenarios might be generated. Therefore, a second step is generally applied using scenario reduction techniques. The objective is to obtain a reduced set of likely scenarios that is feasible to be solved [15]. The third step involves developing a multi-scenario stochastic model to accommodate for the set of scenarios. In the proposed model, the distribution system operator (DSO) faces several sources of uncertainty for the projections in 30 years, namely the forecast errors of load demand, number of consumers, and the potential production of renewable units. These parameters are considered as potential uncertainties in this model [16]. In stochastic models, the optimal decisions are taken on the basis of future adaptability against a set of predicted scenarios [9]. The uncertainties related to these inputs are considered in the model and the planning problem is developed as a stochastic scenario-based optimization model [10].

In stochastic problems, where a set of scenarios needs to be handled, the main issue is to construct a set of realizations for the random variable. These scenarios should adequately represent the probabilistic characteristics of the data [17]. In this stochastic planning model, the initial set of scenarios is a large data set generated by the Monte Carlo Simulation (MCS) technique for representing the uncertainties which the DSO faces while solving the problem. The MCS parameters are the probability distribution functions of the forecast errors [18]. To include the forecast error, an additional term which can be positive or negative is added to the forecasted profile ($x^{\text{forecasted}}$) [10].

$$x(s) = x^{\text{forecasted}} + x^{\text{error}}(s), \qquad \forall s \tag{13.1}$$

The error term (x^{error}) is a zero-mean noise with standard deviation σ [17, 19]. Scenarios, which are projections for a specific date in future, are represented with $x(s)x^s$. The uncertainties of the forecast errors are modeled with the probability distribution functions, which are usually obtained from the historical data [17]. In this model, the forecast errors for the uncertain inputs are all represented by normal distribution functions.

Including all the generated scenarios in the planning problem results in a large-scale optimization problem [17]. Generally, there should be a tradeoff between model accuracy and computation speed [16, 20]. In order to handle the computational tractability of the problem, the standard scenario reduction techniques developed in [21] are used. These scenario reduction algorithms exclude the scenarios with low probabilities and combines those that are close to each other in terms of statistic metrics [21]. They determine a scenario subset of the prescribed cardinality and probability which is closest to the initial distribution in terms of a probability metric [18]. The key purpose of scenario reduction is to decrease the dimensions of the problem. The number of variables and equations are reduced after applying these algorithms. Accordingly, the solutions can be found more efficiently, without losing the main statistical characteristics of the initial dataset [22]. However, the potential cost of applying these approaches is introducing imprecision in the final plans [20].

The reduction algorithms proposed in [21] consists of algorithms with different computational performance and accuracy, namely fast backward method, fast backward/forward method, and fast backward/backward method. The selection of the algorithms depends on the problem size and the expected solution accuracy [18, 21]. For example, the best computational performance with the worst accuracy can be provided by the fast backward method for large scenario tress. Furthermore, the forward method provides best accuracy and highest computational time. Thus, it is usually used where the size of reduced subset is small [18]. These algorithms are also incorporated in a General Algebraic Modeling System (GAMS) tool called SCENRED. SCENRED can be used to reduce the randomly generated scenarios [23].

13.3 Problem Formulation

The growing trend of electricity demand prompts an expansion of the distribution network. Thus, one of the proposals will be the construction of new lines, as it may influence the values of energy losses and energy not supplied. Costs related to the investment, network operation, and satisfaction of all operational, physical, and financial constraints lead to a planning problem.

A distribution network planning problem can be of two types [24, 25]: static and multi-step. The first one considers that the construction/expansion of medium voltage (MV) distribution network can be carried out in a single step, usually associated with small interventions.

The multi-stage planning problem is related to a long term where the investments are carried out at different stages of planning. One of the way to solve this problem is considering only a single step with several static problems, where the next step starts with the solution of the previous step as input.

The distribution network is spited into two subsystems: a primary one, supplied by MV, and a secondary one, supplied by low voltage LV. Carrying out the planning of these two subsystems simultaneous is very complex, so one of the solutions is to make the planning for the different subsystems separated. Thus, there is a decrease in complexity since the method no longer involves a high number of decision variables and also different voltage levels.

The problem considered in this chapter is related to a MV primary network with several objectives. The objective function considers the energy loss cost, the expected energy not supplied cost and the cost related to the investments, which in this case will be in the construction of new lines.

The main goal of this problem is to minimize the costs referred above subject to all technical network constraints. Indirectly the methods also minimize the number of switches to be operated, since there are constraints to deal with the network radiality. With this, the problem must consider the following constraints:

- Power balance—Kirchhoff's first law;
- Generation limits;
- Lines/cable thermal limits;
- Only one direction of power flow can exist;
- Radiality condition.

13.3.1 Economic Evaluation

The uncertainty associated with any project that involves a large amount of investment requires careful and detailed economic analysis. One of the difficulties faced during the economic evaluation of projects is that the cash flows (entry and exit of money) are staggered over time. Gallo [26] says that it's a common sense that the money owned today is more valuable than the same amount after a few years (inflation rate decreases purchasing power). Thus, using a discount rate and converting the financial amounts between different time periods it is possible to solve the above-mentioned difficulty.

Bruni et al. [27] mention that an economic evaluation of projects usually involves a set of parameters to establish the viability of the project. Thus, the author refers to three commonly used tools:

- Net Present Value—NPV: is the difference between the money flows, duly updated during the project analysis period. This value should be positive indicating that the results achieved allow to cover the initial investment and still make a profit. If it is null, there was only recovery of the initial investment;
- Internal Rate of Return—IRR: is the rate that nullify the NPV. Obtaining an IRR above the discount rate indicates that the project is economically feasible. In other words, the project manages to generate a rate of return greater than the cost of capital;
- Payback: is the number of years required to recover the investment, assuming that the investment was done all in year zero.

The planning method proposed in this chapter considers the acquisition and connection of new power lines as the investment to be applied to the distribution network. Thus, the economic evaluation considers, in addition to this Investment (INV), the profits achieved with the application of this new solution—through the reduction of power losses (PL) and expected energy not supplied (EENS).

For the investment economic evaluation, the lifetime project and the discount rate must be defined by the investor. The typical duration for planning distribution networks is approximately 25 years [28].

All necessary investments and all obtained benefits to improve the reliability indexes are considered in the economic evaluation. The investment is considered profitable when the present value (PV) of the incoming related to the improvement of reliability indexes and losses reduction is greater than the investment made in new power lines construction. This means that the net present value (NPV) is positive (Eq. (13.2)).

The benefit (BNF) corresponds to the savings related to the reliability indexes improvement and losses savings. Investment is the total investment for the planning project.

$$\begin{aligned} &\text{NPV} = \text{BNF} - \text{Investment} \\ &\text{NPV} > 0 \end{aligned} \tag{13.2}$$

The present value of the savings that are related to the reliability indexes improvement and losses savings can be calculated by the capital recovery factor (CRF). CRF, presented in Eq. (13.3), is the ratio of a constant annuity to the present value of receiving that annuity for a given project lifetime. Thus, for t periods bnf1 = bnf2 = … = bnft = bnf Eq. (13.4).

$$\text{CRF} = \frac{\text{dr}}{1 - e^{-\text{dr}\cdot t}} \cong \frac{\text{dr} \times (1 + \text{dr})^t}{(1 + \text{dr})^t - 1} \tag{13.3}$$

$$\text{BNF} = \frac{\text{bnf}}{\text{CRF}} = \text{bnf} \times \frac{(1 + \text{dr})^t - 1}{\text{dr} \times (1 + \text{dr})^t} \tag{13.4}$$

where dr is the discount rate, and t the project lifetime.

13.3.2 Target Reliability Values

Distribution system reliability is one of the most important issues in system planning and operation [29]. Institute of Electrical and Electronics Engineers (IEEE) [30] as well as other authors like Canizes et al. [31] use the basic reliability indexes:

- Failure rate (λ)—is the number of faults of a given equipment in a given period of time. The failure rate represents the probability of an equipment failure;
- Repair time (r)—is the failure average duration;
- Unavailability (U)—is the annual outage duration.

In energy distribution systems, these indices are mathematically related according to the equation:

$$U \cong \lambda \times r \tag{13.5}$$

With this, it will be possible to determine the Forced Outage Rate (FOR), another relevant index in the reliability analysis. FOR represents the probability of an unavailability network equipment when it is requested. This index is defined as the number of hours that the equipment is unavailable dividing by the difference between the number of total hours of a year (T), 8760 hours, and the repair time of equipment i.

$$\mathrm{FOR}_i = \frac{U_i}{T - r_i} \tag{13.6}$$

The FOR is used to determine the power not supplied in each distribution network line by the following equation:

$$\mathrm{PNS}_{ij} = \mathrm{FOR}_{ij} \times S_{ij} \qquad \mathrm{kW} \tag{13.7}$$

Thus, the expected energy not supplied is:

$$\mathrm{EENS} = \sum_{ij=1}^{\mathrm{NL}} \mathrm{PNS}_{ij} \times 8760 \qquad \mathrm{kWh/year} \tag{13.8}$$

where ij is the line between bus i and bus j, and NL is the number of distribution network lines.

The reliability indexes such as System Average Interruption Duration Index (SAIDI), System Average Interruption Frequency Index (SAIFI), and Expected Energy Not Supplied (EENS), adopted by the IEEE standard [32], are used to evaluate reliability of the system.

The network operator defines target values for the reliability indexes. To achieve the new reliability values, the system operator should improve the repair times and the failure rates.

The following reliability indexes (13.9)–(13.12) are considered in the proposed method:

- System Average Interruption Duration Index

$$\text{SAIDI} = \frac{\text{Total customer interruption durations}}{\text{Total number of costumers in the system}} \tag{13.9}$$

$$\text{SAIDI} = \frac{\sum_{i=1 \in L} U_i \times N_i}{\sum_{i=1 \in L} N_i} \quad \text{hour/customer year} \tag{13.10}$$

- System Average Interruption Frequency Index

$$\text{SAIFI} = \frac{\text{Total number of customer interruptions}}{\text{Total number of costumers in the system}} \tag{13.11}$$

$$\text{SAIFI} = \frac{\sum_{i=1 \in L} \lambda_i \times N_i}{\sum_{i=1 \in L} N_i} \quad \text{interruptions/customer year} \tag{13.12}$$

13.3.3 Stochastic Planning Model

The planner in the decision making under uncertainty should make optimal decisions throughout a decision horizon with incomplete information. A number of stages can be defined for the considered decision horizon, representing a point in time where decisions are made or where uncertainty partially or totally vanishes [33].

In this chapter is considered a two-stage planning method with a stochastic process represented by a set of scenarios. Thus, two types of decisions can be used in the planning process:

First stage: The decision is made before stochastic process execution. Thus, the variables that represent the first stage do not depend on each stochastic process execution. These variables are known as "here and now" variables.

Second stage: The decision is made after knowing the stochastic process. Thus, the decision depends on each vector of stochastic process execution. When the stochastic process is represented by a set of scenarios, the second stage decision variables are defined for each considered scenario.

The two-stage stochastic programming is an effective approach to include the impacts of the decision in stochastic optimization problems. More theoretical background on stochastic programming models can be found in [32, 33].

Usually the distribution network planning is treated as a multiobjective optimization problem with nonlinear programming. This is because in the formulation of the problem there are nonlinear constraints related to the power flow, binary variables among others. Thus, the planning model can be formulated as a mixed integer nonlinear programming (MINLP); however, this problem is complex and difficult to solve. Thus, it is important to find a simple method to avoid this complexity. The DC power flow constraints are considered in the optimization model (13.23). The usage of a DC model is justified because in many countries, like in Portugal, the distribution networks have voltage regulators and capacitors banks carefully positioned along the grid to keep the voltage and reactive power between the desire limits. Usually, the voltage stability is placed at the HV/MV substation level. However, in the Portuguese case the MV/LV transformers also have voltage regulators. Therefore, the problem will be formulated as a mixed integer linear programming (MILP).

13.3.3.1 Power Losses Linearization

To make the problem linear it is necessary to linearize the objective function. In this case, the only nonlinear term in the objective function is the power losses. The linearization of power losses is done according to the Venikov method [34]. This approach considers that the lines and cables in the system work close to the nominal current, i.e., the economic current density (J_{eco}).

$$I = J_{eco} \times S_{cc} \tag{13.13}$$

where J_{eco} is the economic current density (A/mm^2); S_{cc} is the line section (mm^2).

Thus, the power losses can follow the Eq. (13.14):

$$\Delta P = k' \times R \times I^2 = k' \times R \times I \times I \tag{13.14}$$

Replacing in (13.14) the Eq. (13.13):

$$\Delta P = k' \times R \times I \times J_{eco} \times S_{ec} \tag{13.15}$$

where:

$$I = k' \times \frac{S}{U_l} \tag{13.16}$$

$$R = \frac{\rho \times L}{S_{cc}} \tag{13.17}$$

in which k' and k'' are constants that depend on the type of service (one or three phases); S is the load (kVA); R is the line resistance (Ω/km); I is the current that flow in the line (A); ρ is the line resistivity at operating temperature (Ω mm^2/km); L is the line length (km).

Replacing (13.16) and (13.17) into Eq. (13.15) the linear equation of power losses is:

$$\Delta P = \frac{k \times \rho \times L \times J_{eco}}{U_l} \times S \tag{13.18}$$

The current density value is calculated by Eq. (13.19):

$$J_{eco} = \sqrt{\frac{q \times 10^{-3}}{n \times \rho \times h \times p \times \text{CRF}}} \tag{13.19}$$

where n is the number of active conductors; h is the number of service hours for the electric conduits per year; q is the constant value dependent of the line/cable type; p is the energy price €/kWh.

13.3.3.2 Proposed Methodology

Figure 13.1 presents the scheme of the proposed methodology. The proposed methodology has five main steps, which are presented in more detail as follows.

Input Data

The first step is to prepare all the input data to be considered in the model, such as generation and load points, lines and new lines option characteristics, and reliability data. The data regarding the predicted values for solar power and wind power, load and heat demand, and the number of consumers as well as their standard deviation values are also considered.

Scenarios Generation

In this step a set of scenarios is generated using Monte Carlo Simulation (MCS) following a normal distribution. The predicted and standard deviation values referred above are used as inputs for the MCS, which is implemented in MATLAB software.

Scenarios Reduction

A set of thousand of scenarios is generated, and scenarios reduction becomes imperative to handle with the computational tractability of the problem. Thus, the standard scenario reduction techniques developed in [21] are used. These scenario reduction algorithms exclude the scenarios with low probabilities and combine those

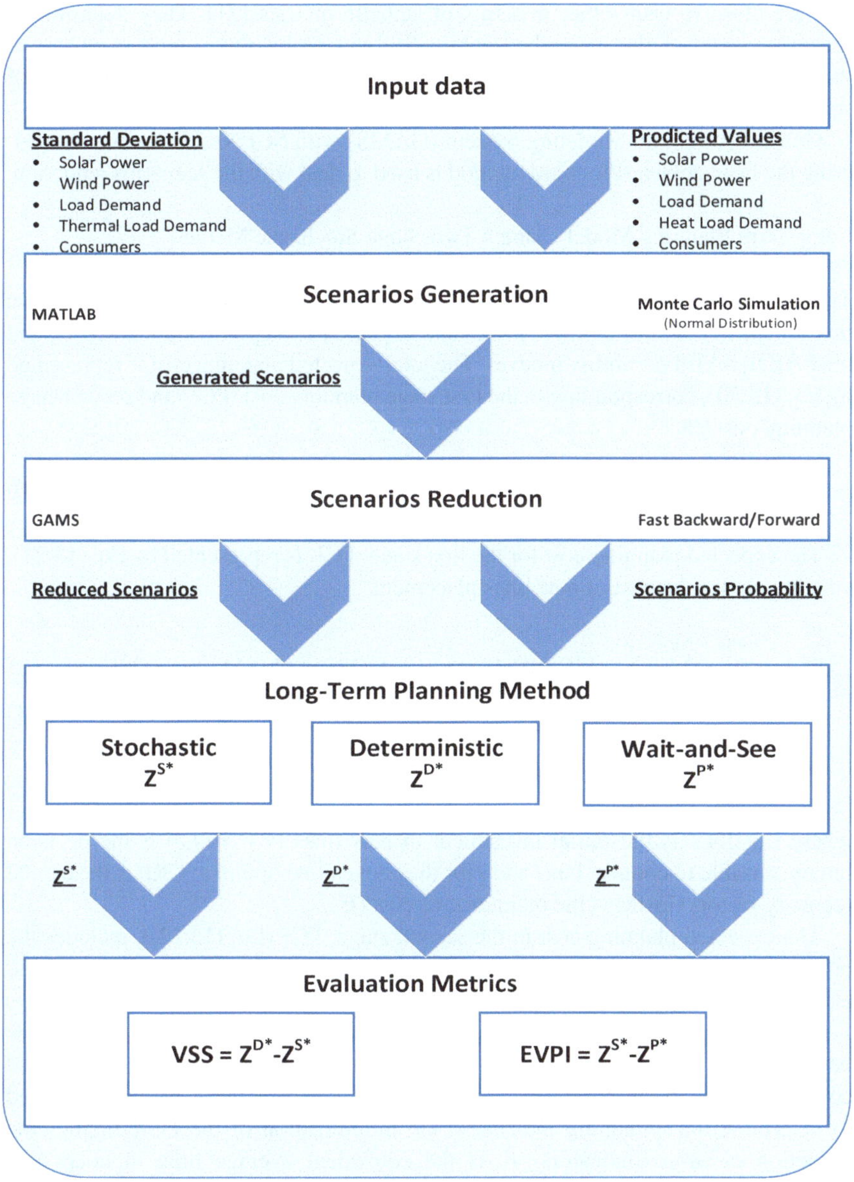

Fig. 13.1 Methodology diagram

that are close to each other in terms of statistic metrics [21]. They determine a scenario subset of the prescribed cardinality and probability which is closest to the initial distribution in terms of a probability metric [32]. The main purpose of scenario reduction is to reduce the size of the problem.

General Algebraic Modeling System (GAMS) with SCENRED toolbox considering the fast backward/forward method is used to deal with the scenarios reduction.

Long-Term Planning Model Using a Two-Stage Stochastic Method

This optimization model has as outputs the decision variables regarding the investment in new lines, power losses and expected energy not supplied costs, and the SAIDI, SAIFI reliability indexes. The total expected planning cost is represented by Eq. (13.20), corresponding to the first stage planning cost (PC^1) and second stage planning cost (PC^2).

$$\text{Minimize} \quad E\left(\text{PC}_{\text{Total}}\right) = \text{PC}^1 + E\left(\text{PC}^2\right) \tag{13.20}$$

The expected planning cost for the first stage, PC^1, is represented by Eq. (13.21), which includes the cost of new lines placement.

$$\text{PC}^1 = \sum_{i=1}^{\text{NB}} \sum_{\substack{j=1 \\ j \neq i}}^{\text{NB}} \sum_{c=1}^{\text{NO}} \left\{\left[\left(\text{CostINV} \times y_{(i,j,c)} + \text{CRF} \times \text{CostM} \times y_{(i,j,c)}\right)\right]\right\} \tag{13.21}$$

$$\forall y \in \{0,1\}, \quad \forall\ (i,j,c) \in \Omega_l$$

where CostINV is the initial investment in new lines (€); $y(i,j,c)$ is the decision binary variable to connect bus i and j for the chosen line option c; CRF is the capital recovery factor; CostM is the maintenance cost (€).

The expected planning cost in the second stage, PC^2 (Eq. (13.22)), includes the power losses cost (first term), expected energy not supplied costs (second term), and excess of power supply costs (third term).

$\text{FOR}_{(i,j,c)}$ and $P_{(i,j,c)}$ are respectively the forced outage rate and the power flow between bus i and bus j according to the chosen line option c. FOR is calculated considering the basis reliability indexes r and λ. Since these indexes are used to determine the remaining indicators, the minimization of the FOR implies the reduction of those indicators. T_e is the equivalent average time in hours and according to Gustafson [35] is the average number of hours during which it would be necessary for the peak load to be carried to give the same energy loss as that given by the actual load throughout the year. To obtain more reliable results, it is necessary to subtract to the T_e the number of probable hours in which the lines may be out of service in the 8760 hours of the year.

$$E\left(\mathrm{PC}^2\right)=\sum_{i=1}^{\mathrm{NB}}\sum_{\substack{j=1\\ j\neq i}}^{\mathrm{NB}}\sum_{c=1}^{\mathrm{NO}}\sum_{s=1}^{\mathrm{NS}}\left\{\begin{array}{l}\left[\left(\mathrm{FOR}_{(i,j,c,s)}\times \mathrm{CostEENS}\times \mathrm{CRF}\times P_{(i,j,c,s)}\right)\right]+\\ \left[\left(\begin{array}{l}\mathrm{CRF}\times (k'\times R\times I\times J_{\mathrm{eco}}\times S_{\mathrm{ec}})\times\\ \mathrm{CostPL}\times\left(T_e-8760\times \mathrm{FOR}_{(i,j,c,s)}\right)\times P_{(i,j,c,s)}\end{array}\right)\right]\end{array}\right\}+$$
$$\sum_{i=1}^{\mathrm{NB}}\sum_{s=1}^{\mathrm{NS}}\left[\left(\mathrm{CostGCP}\times p_{\mathrm{GCP}(i,s)}\right)\right]\quad \forall\ (i,j,c)\in \Omega_l \tag{13.22}$$

The objective function Eq. (13.20) is subjected to several constraints. Below it is possible to find all the model constraints (Eqs. (13.23)–(13.45)).

Network grid constraints:

- Power balance (first Kirchhoff law)

$$\begin{array}{l}\sum\limits_{g\in\Omega_{\mathrm{DG}}^{\mathrm{nd}}}\left(p_{\mathrm{DG}(g,s)}-p_{\mathrm{GCP}(g,s)}\right)+\sum\limits_{g\in\Omega_{\mathrm{DG}}^{d}}\left(p_{\mathrm{DG}(g)}\right)+\sum\limits_{\mathrm{sp}\in\Omega_{\mathrm{SP}}^{b}}p_{\mathrm{Supplier}\,(\mathrm{sp})}+\\ \sum\limits_{e\in\Omega_E^b}\left(p_{\mathrm{Discharge}\,(e,s)}-p_{\mathrm{Charge}\,(e,s)}\right)-\sum\limits_{l\in\Omega_L^b}p_{\mathrm{Load}\,(l,s)}-\sum\limits_{v\in\Omega_V^b}p_{\mathrm{Charge}\,(v,s)}+\\ \sum\limits_{i=1}^{\mathrm{NB}}\sum\limits_{c=1}^{\mathrm{NO}}P_{(i,j,c,s)}-\sum\limits_{j=1}^{\mathrm{NB}}\sum\limits_{c=1}^{\mathrm{NO}}P_{(j,i,c,s)}=0\ \forall(i,s)\end{array} \tag{13.23}$$

- Maximum admissible line flow

$$p_{(i,j,c,s)}\le p_{(i,j,c)}^{\max}\times y_{(i,j,c)}\quad \forall s,\ \ \forall y\in\{0,1\},\ \ \forall\ (i,j,c)\in\Omega_l \tag{13.24}$$

- Radiality condition

  This constraint ensures the radial topology of the distribution network.

$$\sum_{(i,j,c)=1}^{\mathrm{NL}} y_{(i,j,c)}=\mathrm{NB}-\mathrm{NBS}\ \ \forall y\in\{0,1\},\ \ \forall\ (i,j,c)\in\Omega_l \tag{13.25}$$

- Unidirectional power flow

  This constraint ensures the power unidirectionality between bus i and bus j and also the choice of only one line option c in that direction.

$$y_{(i,j,c)}+y_{(j,i,c)}\le 1\ \ \forall y\in\{0,1\},\ \forall\ (i,j,c)\in\Omega_l \tag{13.26}$$

- Transfer buses

  A bus with no generation or demand is referred as a transfer bus. This kind of buses are used to connect a load bus to other load bus and is not a terminal bus (main condition to use the transfer buses), i.e., there are at least two more circuits "leaving" the transfer bus.

To model the use of a transfer bus, first a binary variable must be defined such that is equal to 1 if the transfer bus is used; otherwise, is equal to 0. To consider transfer buses the Eq. (13.25) is replaced by Eq. (13.27):

$$\sum_{(i,j,c)=1}^{NL} y_{(i,j,c)} = \mathrm{NB} - \mathrm{NBS} - \sum_{w=1}^{NW} \left(1 - z_{(w)}\right) \quad z_{(w)} \in \{0,1\} \quad \forall w \in \Omega_{\mathrm{BT}} \quad \forall\,(i,j,c) \in \Omega_l \tag{13.27}$$

$$y_{(i,j,c)} \le z_{(w)} \quad z_{(w)} \in \{0,1\} \quad \forall w \in \Omega_{\mathrm{BT}} \tag{13.28}$$

$$y_{(j,i,c)} \le z_{(w)} \quad z_{(w)} \in \{0,1\}, \quad \forall w \in \Omega_{\mathrm{BT}}, \quad \forall\,(i,j,c) \in \Omega_l \tag{13.29}$$

$$\sum_{(i,j,c)=1}^{NL} y_{(i,j,c)} + \sum_{(j,i,c)=1}^{NL} y_{(j,i,c)} \ge 2 \times z_{(w)} \quad z_{(w)} \in \{0,1\}, \quad \forall w \in \Omega_{\mathrm{BT}}, \quad \forall\,(i,j,c) \in \Omega_l \tag{13.30}$$

where z_j is the binary variable related to the transfer buses.

Constraints (13.27)–(13.30) avoid the loop generation due to the presence of transfer buses and also prevent the appearance of a terminal transfer bus (with only one connected circuit).

- Avoid distributed generator isolation from substation

$$\sum_{(i,j,c)=1}^{NL} d_{(i,j,c)} - \sum_{(i,j,c)=1}^{NL} d_{(j,i,c)} - D_{(g)} = 0 \quad \forall g \in \Omega_B \quad \forall\,(i,j,c) \in \Omega_l \tag{13.31}$$

$$D_{(g)} = 1 \quad \forall g \in \Omega_{\mathrm{DG}} \tag{13.32}$$

$$D_{(g)} = 0 \quad \forall g \notin \Omega_{\mathrm{DG}} \cup \Omega_{\mathrm{BS}} \tag{13.33}$$

$$\left|d_{(i,j,c)}\right| \le \mathrm{nDG} \times y_{(i,j,c)} \quad \forall\,(i,j,c) \in \Omega_l \tag{13.34}$$

where $D_{(g)}$ is a fictitious load of each distributed generator that only can be fed by the substation. $d_{(i,j,c)}$ is the fictitious flow associated with branch i,j for c line option. If it is allowed the distributed generators supply some loads independently, then (13.31)–(13.34) are not considered in the model.

Controllable DG units and external suppliers:

- Maximum and minimum limits for active generated power

$$p_{\mathrm{DG}(g)} \geq P_{\mathrm{DGMinLimit}(g)} \quad \forall g \in \Omega_{\mathrm{DG}}^{d} \tag{13.35}$$

$$p_{\mathrm{DG}(g)} \leq P_{\mathrm{DGMaxLimit}(g)} \quad \forall g \in \Omega_{\mathrm{DG}}^{d} \tag{13.36}$$

- The upstream supplier limits

$$p_{\mathrm{Supplier(sp)}} \geq P_{\mathrm{SMinLimit(sp)}} \quad \forall \mathrm{sp} \tag{13.37}$$

$$p_{\mathrm{Supplier(sp)}} \leq P_{\mathrm{SMaxLimit(sp)}} \quad \forall \mathrm{sp} \tag{13.38}$$

Reliability indexes limits:

- System Average Interruption Frequency Index

$$\mathrm{SAIFI} \leq \mathrm{SAIFI}^{\max} \tag{13.39}$$

- System Average Interruption Duration Index

$$\mathrm{SAIDI} \leq \mathrm{SAIDI}^{\max} \tag{13.40}$$

Energy storage systems constraints:

- The charging and discharging status of the ESSs are respectively represented by x_{ESS} and a_{ESS}. Charging and discharging cannot occur simultaneously.

$$x_{\mathrm{ESS}(e,s)} + a_{\mathrm{ESS}(e,s)} \leq 1 \qquad \forall e, \forall s \tag{13.41}$$

- The maximum discharge limit for each ESS

$$p_{\mathrm{Discharge}(e,s)} \leq P_{\mathrm{DischargeLimit}(e)} \times x_{\mathrm{ESS}(e,s)} \qquad \forall e, \forall s \tag{13.42}$$

- The maximum charge limit for each ESS

$$p_{\mathrm{Charge}(e,s)} \leq P_{\mathrm{ChargeLimit}(e)} \times a_{\mathrm{ESS}(e,s)} \quad \forall e, \forall s \tag{13.43}$$

Parking lot constraints:

The EVs are treated as virtual batteries in the proposed model. A virtual battery can represent a parking lot or a set of EVs located in the network. In this model the EV charge is equal to charge limit multiplied by simultaneity factor (sf). sf is considered equal to 1.

- The charge limit for each virtual battery v is represented by (13.44):

$$p_{\text{Charge}(v,s)} \leq P_{\text{ChargeLimit}(v,s)} \times \text{sf}_v \quad \forall v, \forall s \tag{13.44}$$

Generation curtailment power:

- The generation curtailment power of non-dispatchable DG units cannot be higher than the predicted amount of generation

$$p_{\text{GCP}(g,s)} \leq P_{\text{DGScenario}(g,s)} \quad \forall t, \forall g \in \Omega_{\text{DG}}^{\text{nd}}, \forall s \tag{13.45}$$

District heating:

The use and development of district heating (DH) are increasing in several countries, namely in north of Europe. The generating heat plants in DH send out the heat to the households as water or steam. Thus, the constraints (13.46) and (13.47) could be considered to incorporate the heat demand in the proposed model. The considered heat plants in this model are the CHP and boiler plants.

- Power balance considering CHP (first Kirchhoff law)

$$\begin{aligned}
&\sum_{g\in\Omega_{\text{DG}}^{\text{nd}}} \left(p_{\text{DG}(g,s)} - p_{\text{GCP}(g,s)}\right) + \sum_{g\in\Omega_{DG}^{d}} \left(p_{\text{DG}(g)} + p_{\text{CHP}(g)}\right) + \sum_{\text{sp}\in\Omega_{\text{SP}}^{b}} p_{\text{Supplier}(\text{sp})} + \\
&\sum_{e\in\Omega_{E}^{b}} \left(p_{\text{Discharge}(e,s)} - p_{\text{Charge}(e,s)}\right) - \sum_{l\in\Omega_{L}^{b}} p_{\text{Load}(l,s)} - \sum_{v\in\Omega_{V}^{b}} p_{\text{Charge}(v,s)} + \\
&\sum_{i=1}^{\text{NB}} \sum_{c=1}^{\text{NO}} P_{(i,j,c,s)} - \sum_{j=1}^{\text{NB}} \sum_{c=1}^{\text{NO}} P_{(j,i,c,s)} = 0 \ \forall i, s
\end{aligned} \tag{13.46}$$

- Heat balance

$$\sum_{h\in\Omega_{\text{heatboiler}}} \left(\text{hb}_{(h,s)}\right) + \sum_{\text{hp}\in\Omega_{\text{chp}}} \left(\text{hchp}_{(\text{hp},s)}\right) - \sum_{\text{hl}\in\Omega_{\text{heatload}}} \left(\text{hload}_{(\text{hl},s)}\right) = 0 \tag{13.47}$$

- CHP constraints

CHP plants in this model have the following operation region (Fig. 13.2).

Each line equation (linear equation–algebraic equation) of this region are the constraints for these units, i.e., lines equation 1, 2, 3, 4.

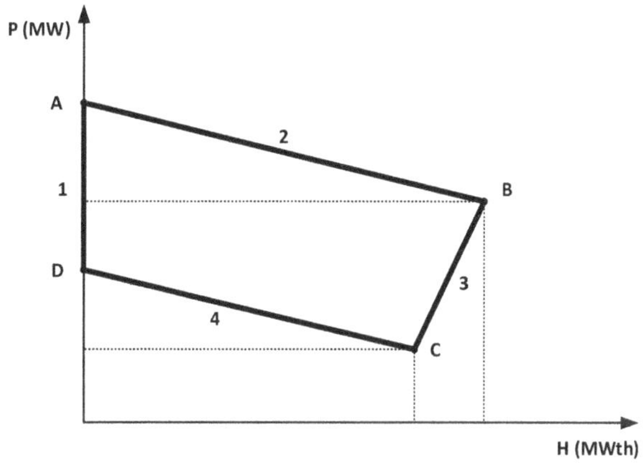

Fig. 13.2 CHP operation region

Evaluation Metrics

The well-known indices, such as the expected value of perfect information (EVPI) and the value of stochastic solution (VSS), are used to evaluate the benefits of the stochastic programming [33]. The EVPI represents the amount that the decision maker is not able to gain due to the presence of imperfect information, e.g., forecasts. It is useful to evaluate how the uncertainty factors affect the evaluated optimal problem. Regarding VSS, it represents the advantage of using stochastic programming over a deterministic approach [33].

EVPI for minimization problems can be represented by (13.48). The stochastic solution represented by Z^{S^*} is calculated by the stochastic programming approach and represents the total expected cost (S). Z^{P^*} represents the wait-and-see solution (WSS). The WSS can be obtained by using the deterministic approach for each scenario. Then, WSS is computed by multiplying the individually obtained cost by each scenario probability.

$$\text{EVPI} = z^{S*} - z^{P*} \tag{13.48}$$

The VSS equation for minimization problems is represented through Eq. (13.49):

$$\text{VSS} = z^{D*} - z^{S*} \tag{13.49}$$

where Z^{D^*} is the optimal value of the modified stochastic problem. It is a deterministic version of the original problem with an average scenario. The optimal decision variables of the original stochastic problem are considered as input in the modified problem.

13.4 Case Study

This section presents a case study to demonstrate how the proposed method is applied. A distribution network (Fig. 13.3) with 13 buses, 30 kV, and one substation (located in bus 1) is used in this chapter. Connections between buses are made by AA 90 overhead lines type. The dashes lines presented in Fig. 13.3 are new connection

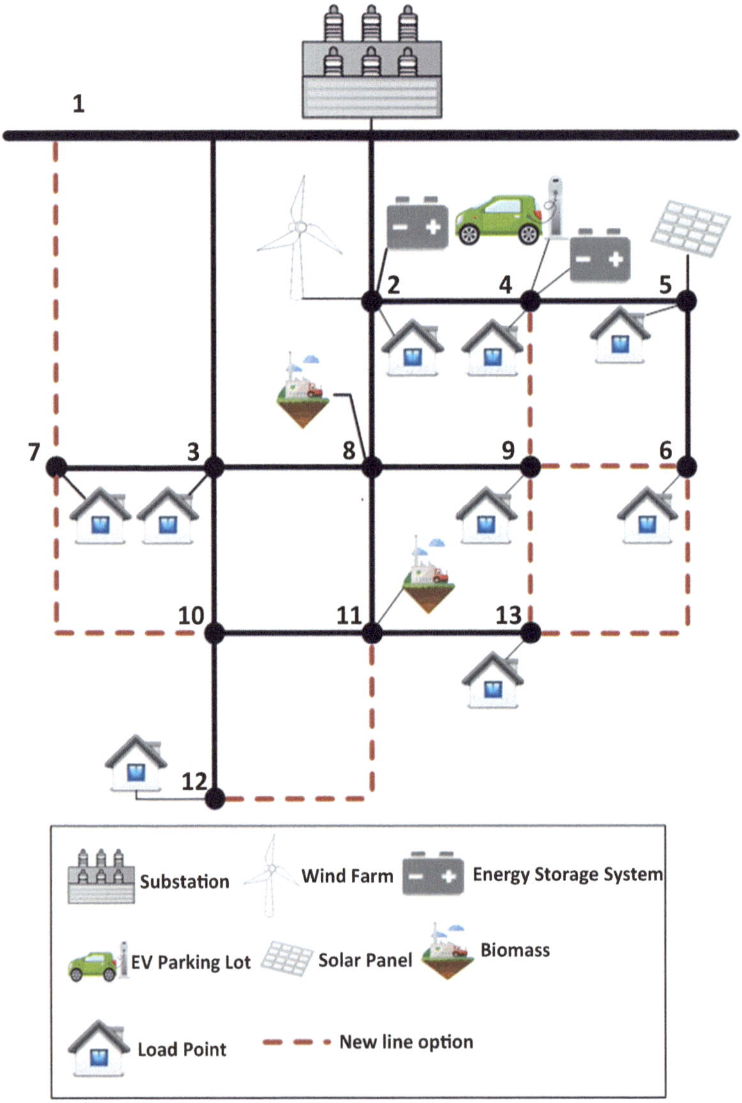

Fig. 13.3 13-bus distribution network

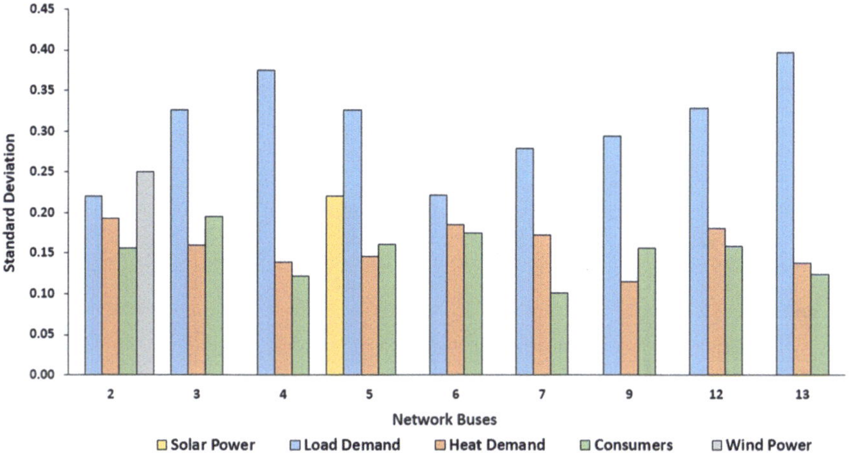

Fig. 13.4 Standard deviation for each resource

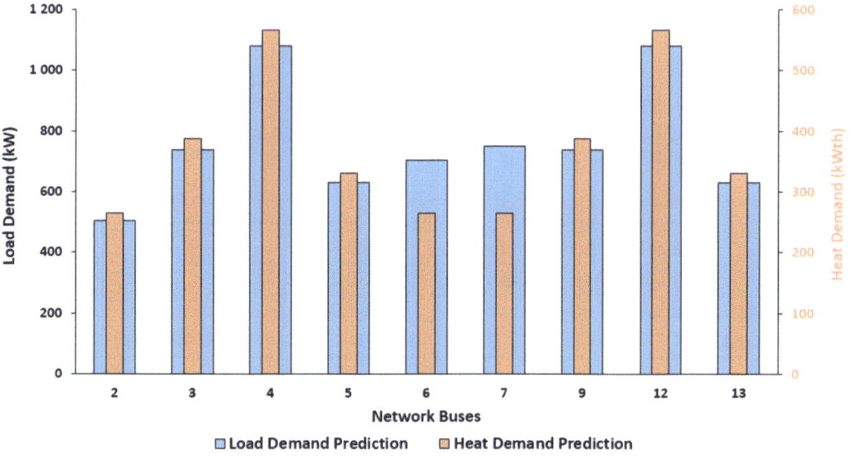

Fig. 13.5 Load and heat demand predictions

options and do not exist in the actual network. Figures 13.4, 13.5, and 13.6 present respectively the standard deviation for each resource, the load and heat demand predictions, and the intermittent energy resources (solar and wind) and consumers predictions for the year 2050.

This network has nine load points, one parking lot for EVs, four DG units (one wind generator, one solar generator, and two biomass units). This distribution network has also two storage systems located at buses 2 and 4. The energy resources data as well as the prediction for the number of consumers are shown in Tables 13.1 and 13.2, respectively.

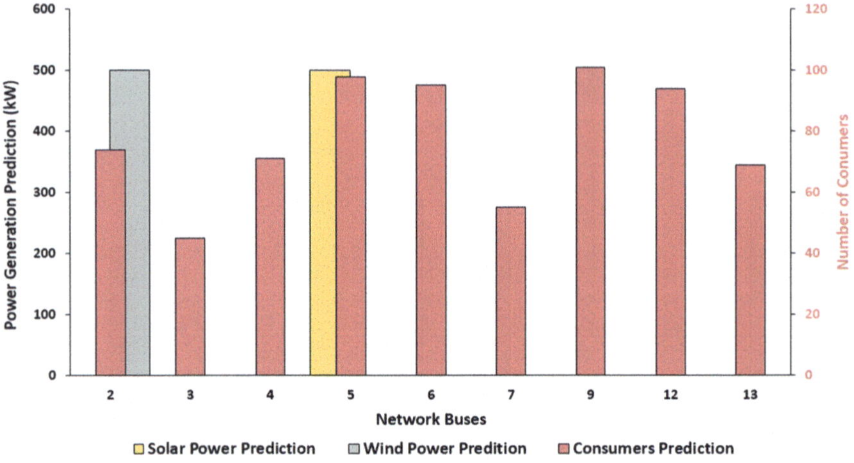

Fig. 13.6 Intermittent energy resources and consumers predictions

Table 13.1 Energy resource data

Energy resource		Capacity (MW) min–max	Prediction (MW) min–max	Units
Substation		0–10	–	1
Photovoltaic		0.75–0.75	0.2680–0.7470	1
Wind		0.75–0.75	0.1702–0.7707	1
Biomass		0–0.50	–	2
Storage	Available capacity for charge	0.20–0.20	–	2
	Available capacity for discharge	0.20–0.20	–	
Parking lots	Charge	1.20–1.20	–	1
Load demand		10–10	5.2859–8.0911	9

Table 13.2 Prediction for the number of consumers

	Expected minimum	Expected uncertain
Number of consumers	631	155

In this case study, the owners of energy storage systems (ESS) are external players. These owners have an agreement to keep a 20% reserve capacity for the network operator (this capacity should not be used by the ESS owner). This capacity can be used for instance to deal with excess or a lack of generation by the network operator. Two 1MW ESS units are available in the network. 0.4MW of capacity is reserved for the system operator (0.2MW for charge and 0.2MW for discharge). The other distributed energy resources belong to the network operator.

Average wind and solar power prediction, as well as the load demand prediction (considering 120 scenarios) are presented in Figs. 13.7 and 13.8, respectively.

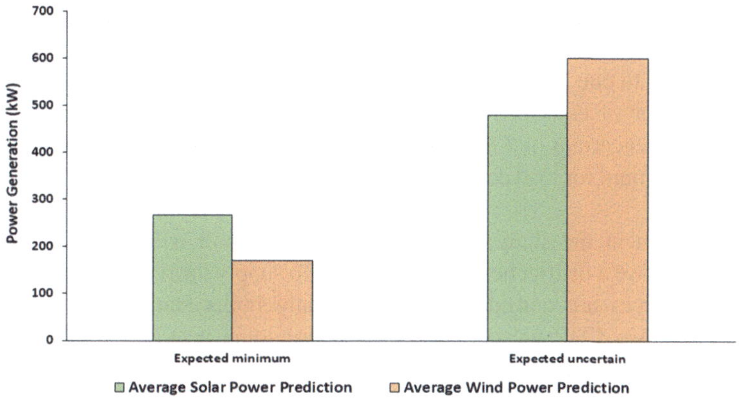

Fig. 13.7 Average solar and wind power prediction

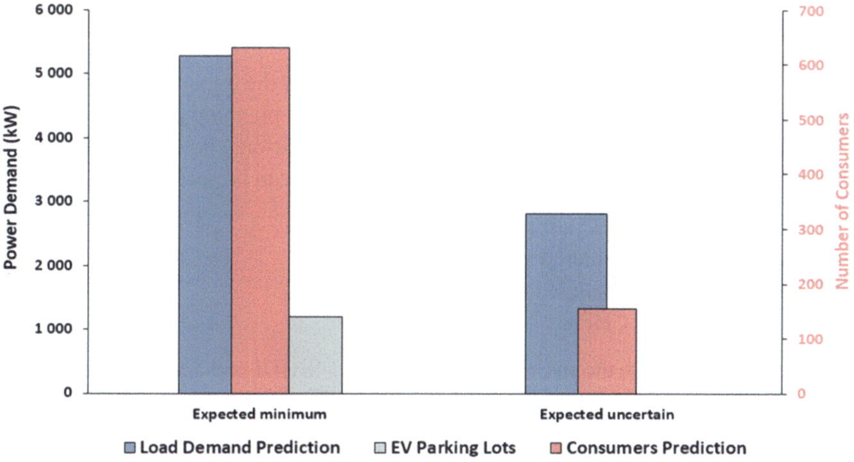

Fig. 13.8 Load demand and consumers prediction

The EVs parking lot is located at bus 3 and has 20 car places. The maximum charge capacity for each place is 60 kW. In this case study, a simultaneity factor equal to one is considered. Thus, the maximum charge capacity for the parking lot is 1200 kW.

Wind and solar power are average predicted values for the lifetime project over the year. Thus, these average values considering their standard deviations will be considered in the scenarios generation. Through Fig. 13.7 it is possible to see that the expected minimum for wind power and solar power are 0.1702MW and 0.2680MW, respectively. Considering the expected uncertain, the expected maximum for wind power and solar power are 0.7707MW and 0.7470MW, respectively.

Analyzing Fig. 13.8 the demand for EV parking lot considered in this case study is equal to the maximum charge capacity of the parking. As said above a simultaneity factor equal to one is also considered. The expected minimum for power demand and for number of consumers are 5.2859MW and 631 consumers, respectively, with expected uncertain of 2.8052MW and 155 consumers, respectively. Thus, the expected maximum for load demand is 8.0911MW and for the number of consumers is 786.

To complement the study presented in this chapter it will be considered the possibility to have a district heating (Fig. 13.9). To supply the required heat demand, four heat sources are considered: two heat-only boiler stations (or just boiler stations) and two CHP units. In addition to providing heat, the CHP units also provide electrical power, so this kind of unit can contribute to an improvement of the EENS. As a result, can also contribute to the reduction of the network investment costs, losses costs and EENS costs.

Thus, two CHP units and two heat-only boiler stations are carefully installed in the distribution network (Fig. 13.9). The heat demand points are in the same load demand buses.

The following four case studies are presented to show the impact of using storage units and the district heating in the distribution network planning problem.

District heating is only affected by CHP units and heat-only boiler stations. However, CHP heat and electricity supply are dependent as can be seen in Fig. 13.2.

- Case A—ESS and CHP are not considered;
- Case B—ESS is considered and CHP is not;
- Case C—CHP is considered and ESS is not;
- Case D—ESS and CHP are considered.

Table 13.3 presents the initial average values of SAIDI and SAIFI indexes, i.e., the values for the actual network considering and not considering the district heating (CHP units). For all analyses conducted in this case study it is intended to achieve a reduction at least 30% in SAIDI and 15% in SAIFI. In this case study, the way to achieve these reductions is investing in new lines construction. Two lines options are considered. Tables 13.4 and 13.5 present the lines thermal limits, the basic reliability indexes (failure rate—λ and repair time—r) for the investment opinion 1 (line AA90) and option 2 (line AA160), respectively. The bold values represent the possibilities of new connections between buses. It is possible to see in Table 13.4 costs and maintenance costs equal to zero. This means that the respectively AA90 line type exist in the actual network, thus its costs are considered zero in the proposed long-term planning method. This method considers also the possibility to change a line type for the other (AA90 by AA160). These tables also show the construction line cost (line cost plus installation cost). The maintenance cost for each new line is also presented.

Tables 13.6 and 13.7 present the heat resource and demand data. In Fig. 13.10 is depicted the expected minimum (2.8802MWth) and the expected uncertain (0.9027MWth).

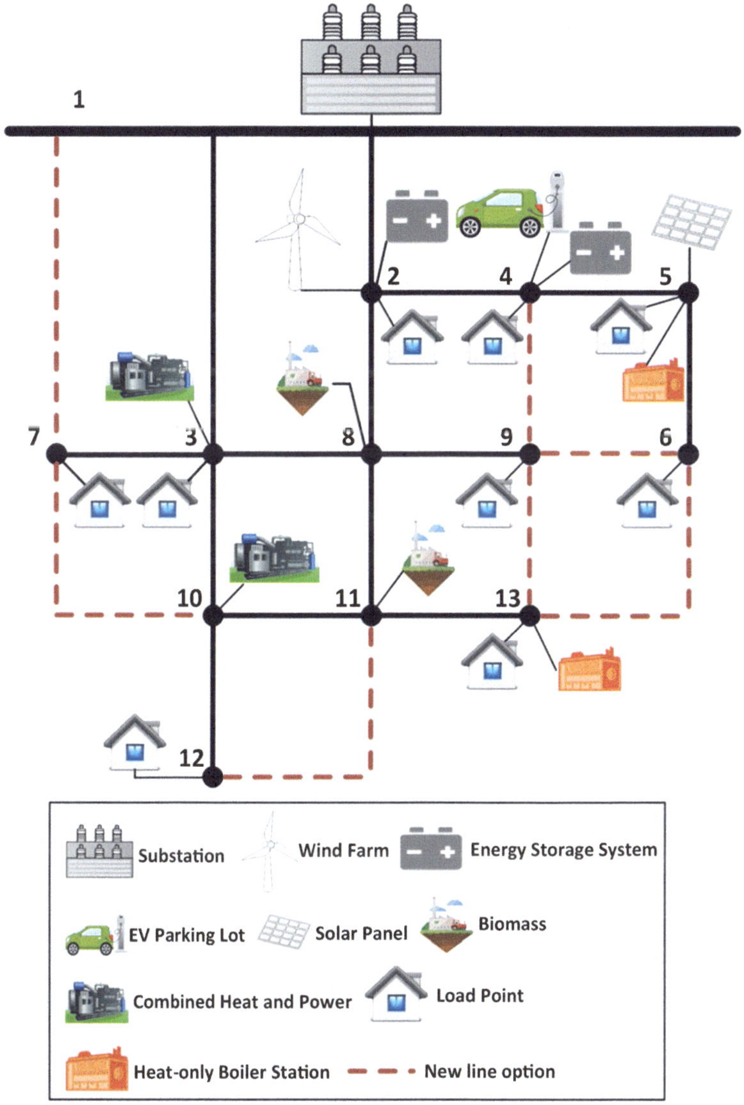

Fig. 13.9 13-bus distribution network with district heating

Table 13.3 Initial reliability indexes

District heating	SAIDI (h/customer × year)	SAIFI (interruption/customer × year)
Yes	7.1555	0.6887
No	8.3232	0.6561

Table 13.4 Lines data for option 1 (line AA90)

Bus out	Bus in	AA90				
		Limit (MVA)	Failure rate (failures/year)	Repair time (h)	Cost (m.u.)	Maintenance cost (m.u.)
1	2	4	0.1000	50.9600	0	0
1	3	4	0.3222	49.8230	0	0
1	7	4	0.1706	29.9800	220,250	44,050
2	4	4	0.5000	2.6490	0	0
2	8	4	0.1000	6.6886	0	0
3	7	4	0.3000	2.5900	0	0
3	8	4	0.1000	6.6886	0	0
3	10	4	0.2000	3.1537	0	0
4	5	4	0.3000	3.3631	0	0
4	9	4	0.0891	1.3546	4500	900
5	6	4	0.3000	2.6590	0	0
6	9	4	0.0891	1.3546	3000	600
6	13	4	0.0891	1.3546	18,000	3600
7	10	4	0.1200	1.2286	45,000	9000
8	9	4	0.1000	6.3954	0	0
8	11	4	0.1000	3.0476	0	0
9	13	4	0.0891	1.3546	7500	1500
10	11	4	0.3000	2.4480	0	0
10	12	4	0.4000	2.7211	0	0
11	12	4	0.1200	1.2286	9000	1800
11	13	4	0.2000	2.9631	0	0

Table 13.5 Lines data for option 2 (line AA160)

Bus out	Bus in	AA160				
		Limit (MVA)	Failure rate (failures/year)	Repair time (h)	Cost (m.u.)	Maintenance cost (m.u.)
1	2	6	0.1000	50.9600	210,125	42,025
1	3	6	0.3222	49.8230	213,500	42,700
1	7	6	0.1462	24.9834	224,750	44,950
2	4	6	0.5000	2.6490	13,125	2625
2	8	6	0.1000	6.6886	10,500	2100
3	7	6	0.3000	2.5900	7875	1575
3	8	6	0.1000	6.6886	15,750	3150
3	10	6	0.2000	3.1537	10,500	2100
4	5	6	0.3000	3.3631	18,375	3675
4	9	6	0.0764	1.1289	7875	1575
5	6	6	0.3000	2.6590	13,125	2625
6	9	6	0.0764	1.1289	5250	1050
6	13	6	0.0764	1.1289	31,500	6300
7	10	6	0.1029	1.0238	78,750	15,750
8	9	6	0.1000	6.3954	21,000	4200
8	11	6	0.1000	3.0476	13,125	2625
9	13	6	0.0764	1.1289	13,125	2625
10	11	6	0.3000	2.4480	5,250	1,050
10	12	6	0.4000	2.7211	36,750	7,350
11	12	6	0.1029	1.0238	15,750	3150
11	13	6	0.2000	2.9631	3,936	788

Table 13.6 Heat resource data

Energy resource	Power capacity (MW) min–max	Heat capacity (MWth) min–max	Units
CHP	0–1.5	0–1.0	2
Heat-only boiler	–	10–10	2

Table 13.7 Heat demand data

Energy resource	Heat capacity (MWth) min–max	Heat prediction (MWth) min–max
Heat demand	5–5	2.8802–3.7829

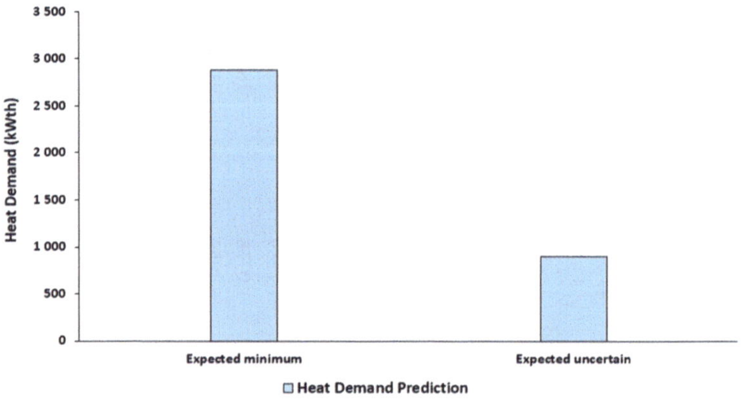

Fig. 13.10 Heat demand prediction

The most commonly used cogeneration units are the single-cycle gas or steam turbine units. In this chapter, the cogeneration unit used is the single-cycle steam turbine without condensation. Considering this kind of unit, the feasibility region is convex, which means that any point belonging to a straight line drawn between two distinct points belongs to the feasibility region presented. The considered CHP units present the following convex feasibility region (Fig. 13.11).

The EENS cost is 3 m.u/kWh, and of 0.12 m.u./kWh for loss cost. For the expected energy not supplied cost, investment cost, loss cost, a discount rate of 5% is considered for a 30 years lifetime project, which leads to a Capital Recovery Factor equal to 15.37. In this case study, the considered value for T_e is 4500 h and all the terms of the objective function (13.22) have the identical importance for the distribution system operator.

The proposed work was developed in MATLAB R2014b and TOMLAB 8.1 64 bits with CPLEX solver (version 12.5) using a computer with one Intel Xeon E5-2620 v2 processor and 16 GB of RAM running Windows 10 Pro.

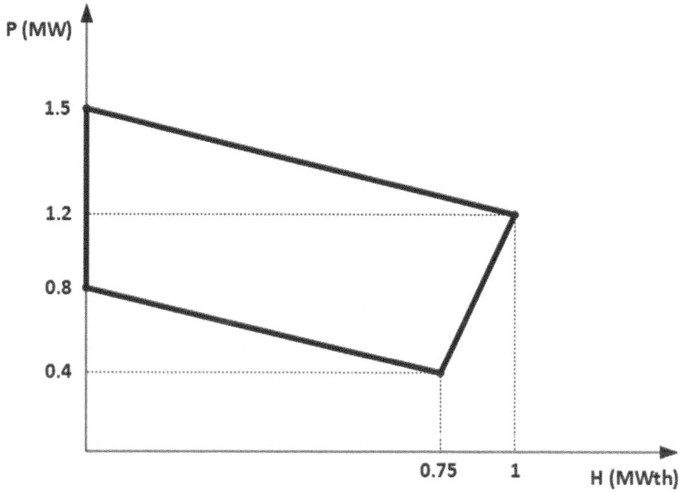

Fig. 13.11 Feasibility region for cogeneration units

Table 13.8 Peak memory and execution time for each case

Case	Peak memory (MB)	Execution time (s)
A	244	2152
B	192	1832
C	225	1246
D	140	625

13.5 Results and Discussion

Two-stage stochastic method is applied to solve a long-term planning problem in the considered case study. This optimization problem considered 120 scenarios and deals with 167,009 variables and 86,492 constraints. Table 13.8 presents the peak memory and the execution time for the two-stage stochastic long-term planning problem.

The execution times are less than an hour, so they are compatible for the available timeframe in the planning-making process. A memory test was made to evaluate the impact on computer system resources through MATLAB memory profiler. This command reports the peak memory for each function used in the methodology developed code. As can be seen in Table 13.8 the higher peak memory was verified in case A. Even the peak memory doesn't exceed 300 MB in this case. Thus, the proposed work in this chapter is compatible with a wide range of available computers in the market.

Figures 13.12 and 13.13 present the optimal radial topology for the two-stage stochastic method (Z^{S*}) without district heating and considering district heating, respectively. In other words, it is being considered the uncertainty in load and heat

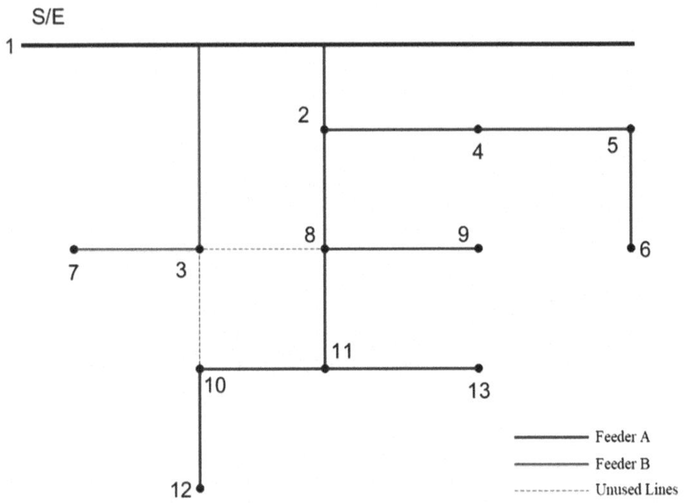

Fig. 13.12 Initial working radial topology without district heating

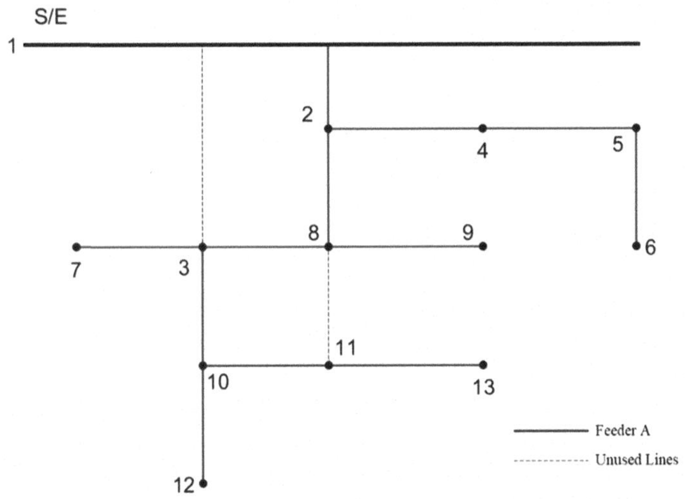

Fig. 13.13 Initial working radial topology considering district heating

demand, in the number of consumers, and in the wind and solar power in the actual distributed power network (without any option of line construction).

Through Table 13.9 it is possible to see the costs for power losses and EENS when the two-stage stochastic method is applied to the actual network with and without district heating. CHP units as distributed generator can contribute to the reduction of power losses and EENS costs (as can be seen in Table 13.9).

Table 13.9 Initial costs with and without district heating

District heating	Initial loss cost (m.u.)	Initial EENS cost (m.u.)
YES	626,970	1,131,900
NO	695,480	3,069,500

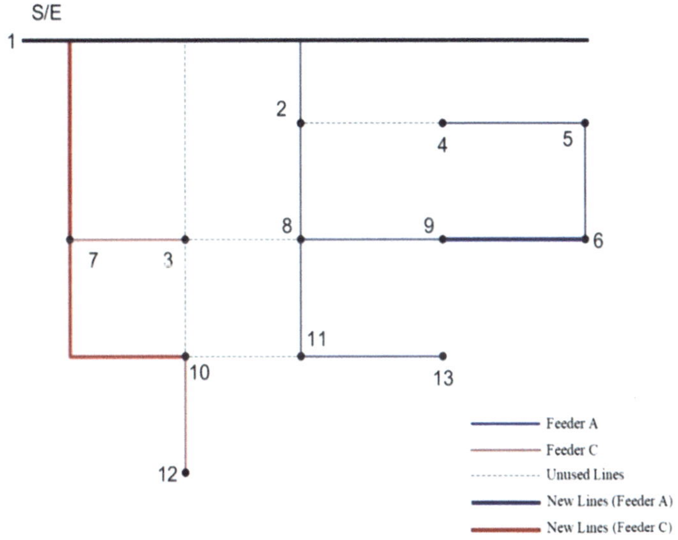

Fig. 13.14 Z^{S^*} radial topology for case A

The next two figures (Figs. 13.14 and 13.15) present the studies referred to case A and B using the two-stage stochastic optimization model. Optimal investment (construction of new lines) to be applied in network to improve the reliability indexes and at the same time minimizing the power losses cost, expected energy not supplied cost, and the investment cost is obtained. These studies also present the optimal radial topology to be chosen to operate in considered conditions (taking into account all scenarios). For case A, three new lines are chosen, one AA90 connected between bus 7 and bus 10, and two AA160 connected between bus 1 and bus 7 and bus 6 and bus 9. The total cost associated with this case is 3,516,065 m.u. and the total benefit of this investment is 1,604,200 m.u. for the lifetime project.

Regarding case B four new lines are selected, three AA90 between buses 7–10, 6–9, and 11–12 and one AA160 between busses 1–7. In this case the total cost is 3,565,618 m.u. and the total benefit is 1,740,050 m.u.

Figures 13.16 and 13.17 are related to the studies made for case C and D. Also, the optimal topology to be operated for each case is also obtained. For case C and D three new lines are chosen, one AA90 connected between bus 7 and bus 10, and two AA160 connected between bus 1 and bus 7 and bus 6 and bus 9. Case C presents a total cost of 2,863,415 m.u. and a total benefit of 69,060 m.u. Regarding case D the total cost is 2,701,645 m.u. and the total benefit is 177,710 m.u.

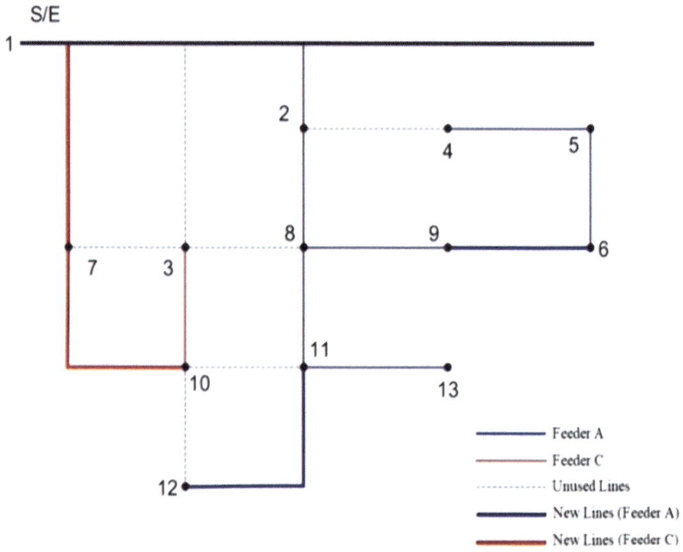

Fig. 13.15 Z^{S^*} radial topology for case B

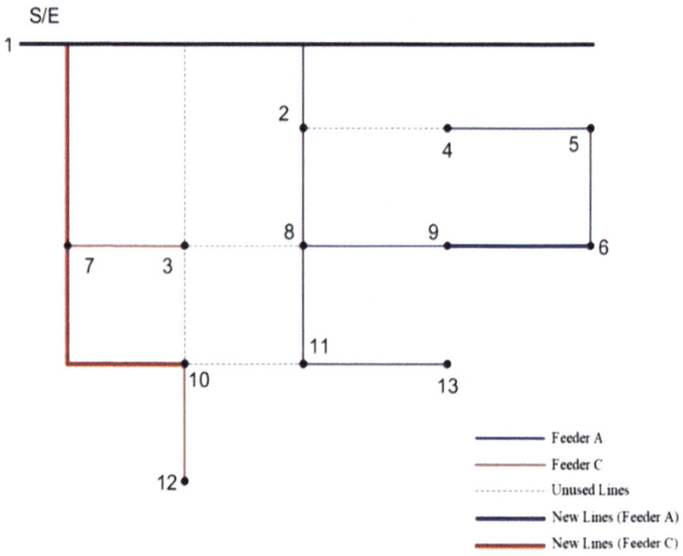

Fig. 13.16 Z^{S^*} radial topology for case C

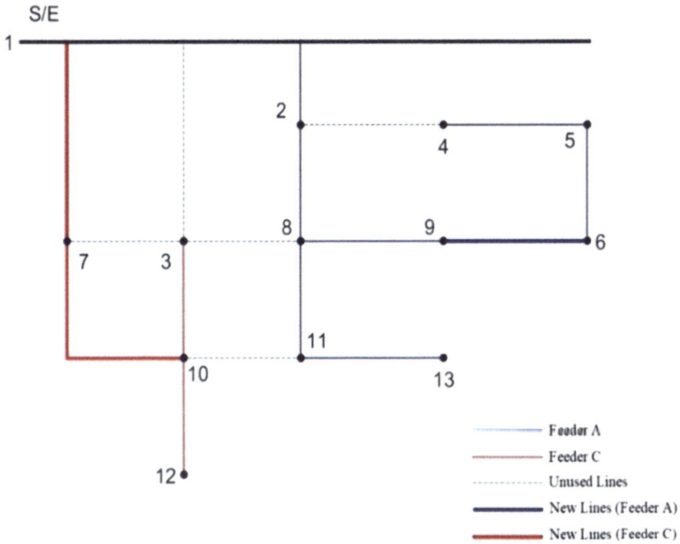

Fig. 13.17 Z^{S^*} radial topology for case D

Table 13.10 New lines construction for cases A, B, C, and D

	New line connections (Bus out–Bus in)		Substitution
	AA90	AA160	AA90 by AA160
Case A	7-10	1-7 9-6	–
Case B	7-10 9-6 11-12	1-7	–
Case C	7-10	1-7 9-6	–
Case D	7-10	1-7 9-6	–

A summary of the new lines construction for each case is presented in Table 13.10. Through this table it is possible to see that the substitution of the existence lines wasn't selected by proposal model.

Tables 13.11 and 13.12 present the result costs for each objective term as well as the total costs and the monetary benefits achieved in each case. Once CHP units are used in the district heating cases they also contribute as distributed generators to the distribution power network, thus the EENS costs and power losses costs are lower than the cases without district heating. Hence, the total costs for the cases that include CHP units are lower. It can be said that with the necessary investment to achieve the desired values of SAIDI and SAIFI the total monetary benefit is small when compared with the cases without CHP.

It can be seen in Table 13.13 that the paybacks for cases C and D are greater than the lifetime project and present an IRR negative. So, this means that the investment will not be recovered in the lifetime project. Thus, the investment in new lines construction to improve the SAIDI and SAIFI will not be economically feasible.

Table 13.11 Cost results for cases A, B, C, and D

	Case	EENS cost (m.u.)	Loss cost (m.u.)	Investment cost (m.u.)	Excess of power supply cost (m.u.)	Total cost (m.u.)
Z^{S*}	A	1,582,700	812,880	1,120,485	0	3,516,065
	B	1,525,800	891,830	1,147,988	0	3,565,618
	C	1,089,400	653,530	1,120,485	0	2,863,415
	D	974,610	606,550	1,120,485	0	2,701,645

Table 13.12 Benefit for cases A, B, C, and D

	Case	EENS cost benefit (m.u.)	Loss cost benefit (m.u.)	Total cost benefit (m.u.)
Z^{S*}	A	1,486,800	117,400	1,604,200
	B	1,543,700	196,350	1,740,050
	C	42,500	26,560	69,060
	D	157,290	20,420	177,710

Table 13.13 Economic evaluation

	Case	Payback (years)	IRR (%)	NPV (m.u.)
Z^{S*}	A	12.58	22.21	16,192
	B	13.10	17.37	12,969
	C	>30	<0	71,852
	D	>30	<0	61,329

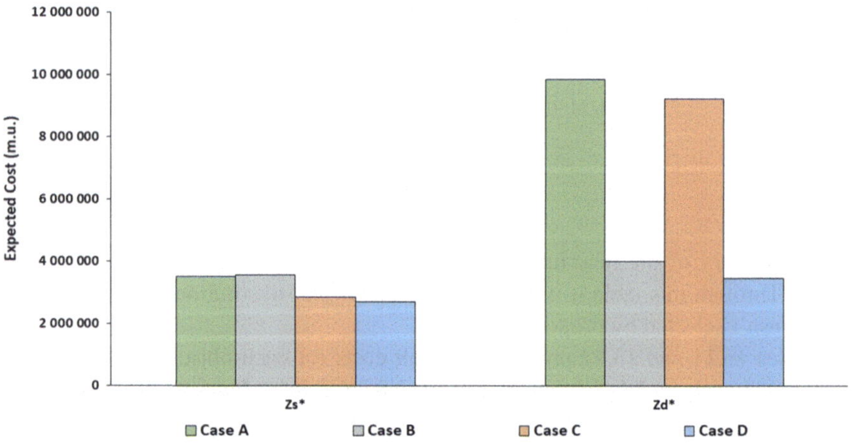

Fig. 13.18 Expected total costs for each case

In Fig. 13.18 is depicted a comparison between the total costs obtained by the two-stage stochastic method (Z^{s*}) and deterministic method (Z^{d*}). The lower costs presented by the two-stage stochastic method for each case is evident. The higher costs are present in cases A and C of deterministic method. This is due to the existence generation power excess and the nonexistence of ESS. Results suggest that ESS contributes to avoid a higher cost when the deterministic model is used and

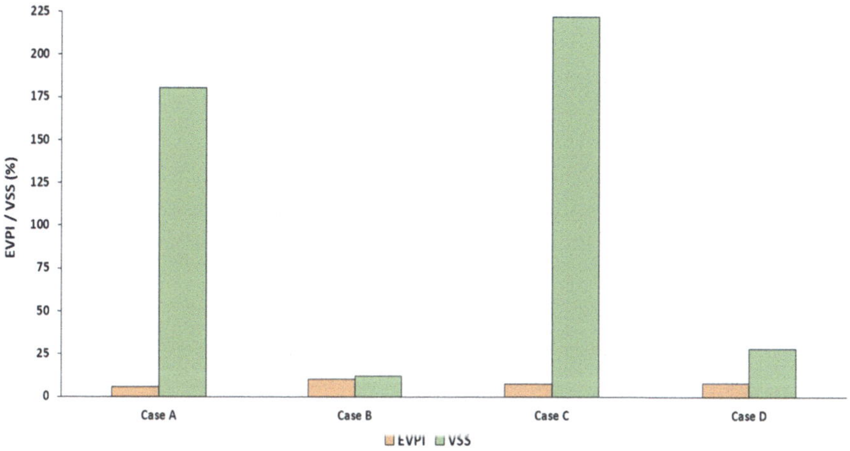

Fig. 13.19 EVPI and VSS for the considered cases

shows the two-stage stochastic method advantage (even without ESS the generation power excess is zero). The values of the quality indices are shown in Fig. 13.19. Case A and case C are good proofs of the previous statement, where the VSS is higher in cases A and C (180% and 222%) which means that without ESS the stochastic model is more important to achieve lower expected costs mitigating the uncertainty. In fact, these high VSS values for cases A and C are related to the existence of generation power excess.

For cases A, B, C, and D a reduction of 64%, 11%, 69%, and 22% is obtained when Z^{s*} is used.

The new reliability indexes when the two-stage stochastic model is used are shown in Figs. 13.20 and 13.21. As can be seen the obtained values are lower when compared with the initial values of SAIDI and SAIFI (Table 13.3). The reliability indexes values in the case B has more considerable changes when compared with the other three cases; this is related to the new lines constructions that the two-stochastic model has chosen (Table 13.10).

13.6 Conclusion

A two-stage stochastic model for a distribution power network long-term planning model was proposed to solve the challenging problem of considering several sources of uncertainty associated with the renewable generation and electric vehicles integration considering the network technical constraints. The problem complexity was reduced by the adequate aggregation of EVs instead of decentralized control. Therefore, it is possible to increase the scalability of the model and consider several uncertainty sources. The results also reveal that the increasing levels of uncertainty

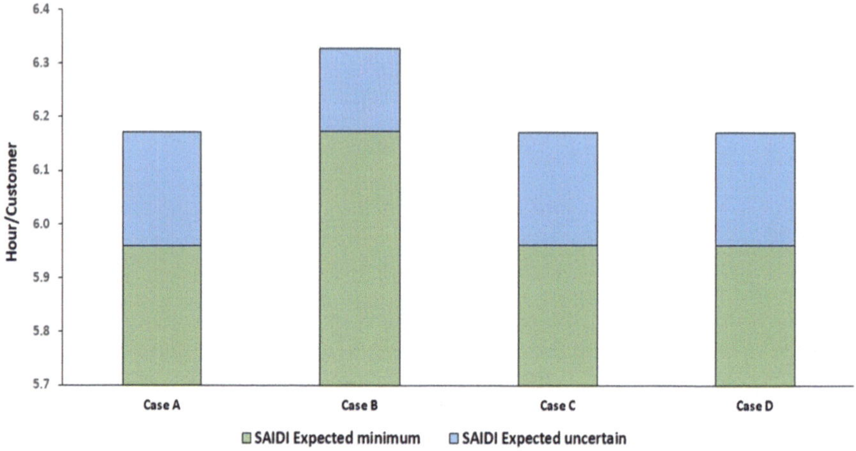

Fig. 13.20 Expected SAIDI index

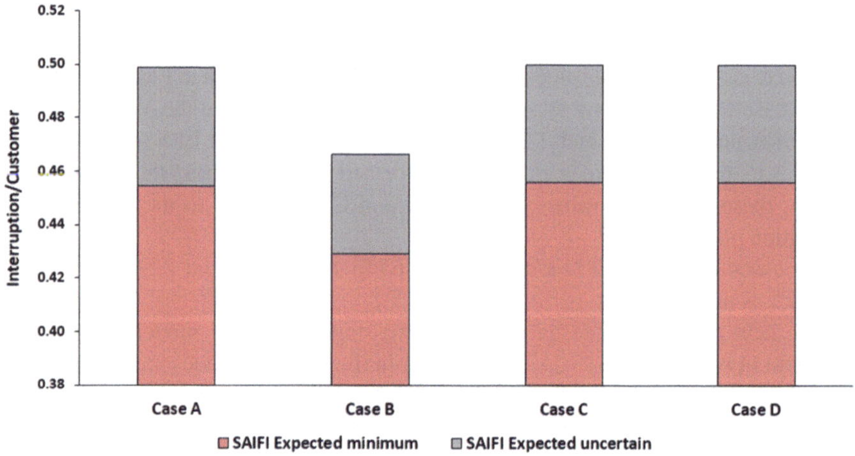

Fig. 13.21 Expected SAIFI index

can be mitigated by ESS. In fact, the deterministic model when the ESS are not used presents excess of generation power leading to high costs.

The district heating is also considered in the long-term planning problem. To deal with that some heat generators were considered (CHP units and heat-only boiler units). Results demonstrate that CHP, together with heat-only boiler units, can supply the district heating demand also contributing to network reliability reducing expected energy supplied and power losses costs avoiding the need to invest in new power lines for the considered lifetime project.

The method proved to be adequate to support the distribution network operator for future network expansion planning.

Acknowledgments This work has received funding from the EU's Horizon 2020 research and innovation programme under the Marie Sklodowska-Curie grant agreement No 641794 (project DREAM-GO) and from FEDER Funds through COMPETE program and from National Funds through FCT under the project UID/EEA/00760/2013. Bruno Canizes is supported by FCT Funds through SFRH/BD/110678/2015 PhD scholarship and M. Ali Fotouhi Ghazvini is supported by FCT Funds through SFRH/BD/94688/2013 PhD scholarship.

Nomenclature

Indices

c	Line options
e	Energy storage systems (ESSs)
g	Distributed generator (DG) unit
h	Heat-only boiler unit
hl	Heat load
hp	CHP heat power
i	Electrical buses
j	Electrical buses
l	Loads
s	Scenarios
sp	External suppliers
v	Electric vehicles parking lot (EV)
w	Transfer buses

Parameters

λ	Failure rate
ρ	Line resistivity at operating temperature ($\Omega \times \text{mm}^2/\text{km}$)
BNF	Benefit from the solution applied (€)
CostEENS	Expected energy not supplied cost (€)
CostGCP	Generation curtailment power cost (€)
CostINV	Initial investment in new lines (€)
CostM	Maintenance cost (€)
CostPL	Power losses cost (€)
dr	Discount rate
EENS	Expected energy not supplied
EVPI	Expected value of perfect information
FOR	Forced outage rate
$\text{FOR}_{(i,j,c)}$	Forced outage rate between bus i and bus j according to the chosen line option c
h	Number of service hours for the electric conduits per year
I	Current that flow in the line (A)

Investment	Total investment for the planning project (€)
J_{eco}	Economic current density (A/mm^2)
k' and k''	Constants that depend on the type of service (one or three phases)
L	Line length (km)
n	Number of active conductors
NB	Number of buses
nDG	Number of DG units
NL	Number of distribution network lines
NO	Number of line options
NPV	Net present value
NS	Number of scenarios
NW	Number of transfer buses
p	Energy price (€/kWh)
$P_{ChargeLimit(e)}$	Maximum charge rate of energy storage systems (kW)
$P_{DGMaxLimit(g)}$	Maximum active power of DG (kW)
$P_{DGMinLimit(g)}$	Minimum active power of DG (kW)
$P_{DGScenario(g,s)}$	Forecasted generation of DG (kW)
$P_{DischargeLimit(e)}$	Maximum discharge rate of energy storage systems (kW)
$P^{máx}{}_{(i,j,c)}$	Maximum admissible line flow between bus i and bus j according to the chosen line option c
$P_{SMaxLimit}$	Maximum active power of suppliers (kW)
$P_{SMinLimit}$	Minimum active power of suppliers (kW)
$P_{Supplier(sp)}$	Active power of external suppliers
q	Constant value dependent of the line/cable type
r	Repair time (h)
R	Line resistance (Ω/km)
S	Load (kVA)
SAIDI^{max}	Maximum Limit to System Average Interruption Duration Index Limit (h/consumer × year)
SAIFI^{max}	Maximum Limit to System Average Interruption Frequency Index (interruption/consumer × year)
S_{cc}	Line section (mm^2)
sf_v	Simultaneity factor
t	Project lifetime (years)
T	Number of total hours of a year
T_e	Time equivalent (h)
U	Unavailability

Variables

$a_{ESS(e,s)}$	Discharging status of the energy storage systems
$D_{(g)}$	Fictitious load of each distributed generator g
$d_{(i,j,c)}$	Fictitious flow associated with branch i,j for c line option

$\text{hb}_{(h,s)}$	Heat power for boiler unit h in scenario s
$\text{hchp}_{(\text{hp},s)}$	Heat power for CHP unit hp in scenario s
$\text{hload}_{(\text{hl},s)}$	Heat demand for hl heat load in scenario s
$P_{(i,j,c)}$	Power flow between bus i and bus j according to the chosen line option c
PC^1	Expected planning cost for the first stage
PC^2	Expected planning cost for the second stage
$P_{\text{Charge}(e,s)}$	Active power charging of energy storage systems (kW)
$P_{\text{Discharge}(e,s)}$	Active power discharge of energy storage systems (kW)
$P_{\text{Charge}(v,s)}$	Active power charging of EV parking lot (kW)
$P_{\text{GCP}(g,s)}$	Generation curtailment power of non-dispatchable DG units (kW)
$P_{\text{Load}(l,s)}$	Active power load for l load scenario s
SAIDI	System Average Interruption Duration Index (h/consumer × year)
SAIFI	System Average Interruption Frequency Index (interruption/consumer × year)
VSS	Value of stochastic solution
$x_{\text{ESS}(e,s)}$	Charging status of the energy storage systems
$z_{(w)}$	Binary variable related to the transfer buses

Sets

Ω_B	Set of buses
Ω_{BS}	Set of substation buses
Ω_{BT}	Set of transfer buses
Ω_{DG}	Set of DG
${\Omega^d}_{\text{DG}}$	Set of dispatchable DG
${\Omega^{\text{nd}}}_{\text{DG}}$	Set of non-dispatchable DG
Ω_E	Set of ESS
${\Omega^b}_E$	Set of ESS bus
$\Omega_{\text{heatboiler}}$	Set of heat boiler
Ω_{heatload}	Set of heat load
Ω_{hp}	Set of CHP heat power
Ω_L	Set of loads
${\Omega^b}_L$	Set of load buses
Ω_l	Set of lines
Ω_{SP}	Set of external suppliers
${\Omega^b}_{\text{SP}}$	Set of external supplier buses
Ω_V	Set of EV
${\Omega^b}_V$	Set of EV buses

References

1. Fitiwi DZ, Santos SF, Bizuayehu AW, Shafie-khah M, Catalao JPS (2016) A new dynamic and stochastic distributed generation investment planning model with recourse. In: 2016 IEEE power and energy society general meeting (PESGM). IEEE, pp 1–5
2. Soares J, Fotouhi Ghazvini MA, Borges N, Vale Z (2017) A stochastic model for energy resources management considering demand response in smart grids. Electr Power Syst Res 143:599–610. https://doi.org/10.1016/j.epsr.2016.10.056
3. Wu Z, Zeng P, Zhang X (2016) Two-stage stochastic dual dynamic programming for transmission expansion planning with significant renewable generation and N-k criterion. CSEE J Power Energy Syst 2:3–10. https://doi.org/10.17775/CSEEJPES.2016.00003
4. Gangammanavar H, Sen S (2016) Two-scale stochastic optimization framework for controlling distributed storage devices. IEEE Trans Smart Grid:1–1. https://doi.org/10.1109/TSG.2016.2616881
5. Cao Y, He M, Wang Z, Tao J, Zhang J (2012) Multiple resource expansion planning in smart grids with high penetration of renewable generation. In: 2012 IEEE third international conference on smart grid communications (SmartGridComm). IEEE, pp 564–569
6. Sauhats A, Petrichenko L, Beryozkina S, Jankovskis N (2016) Stochastic planning of distribution lines. In: 2016 13th international conference on the european energy market (EEM). IEEE, pp 1–5
7. Rong A, Lahdelma R (2007) Efficient algorithms for combined heat and power production planning under the deregulated electricity market. Eur J Oper Res 176:1219–1245. https://doi.org/10.1016/j.ejor.2005.09.009
8. Zhang X, Che L, Shahidehpour M, Alabdulwahab AS, Abusorrah A (2017) Reliability-based optimal planning of electricity and natural gas interconnections for multiple energy hubs. IEEE Trans Smart Grid 8:1658–1667. https://doi.org/10.1109/TSG.2015.2498166
9. Konstantelos I, Strbac G (2015) Valuation of flexible transmission investment options under uncertainty. IEEE Trans Power Syst 30:1047–1055. https://doi.org/10.1109/TPWRS.2014.2363364
10. Soares J, Canizes B, Gazvhini MAF, Vale Z, Venayagamoorthy GK (2017) Two-stage stochastic model using benders decomposition for large-scale energy resources management in smart grids. IEEE Trans Ind Appl:1. https://doi.org/10.1109/TIA.2017.2723339
11. Ghazvini MAF, Canizes B, Vale Z, Morais H (2013) Stochastic short-term maintenance scheduling of GENCOs in an oligopolistic electricity market. Appl Energy 101:667–677. https://doi.org/10.1016/j.apenergy.2012.07.009
12. Gong C, Wang X, Weiqiang X, Tajer A (2013) Distributed real-time energy scheduling in smart grid: stochastic model and fast optimization. IEEE Trans Smart Grid 4:1476–1489. https://doi.org/10.1109/TSG.2013.2248399
13. Su W, Wang J, Roh J (2014) Stochastic energy scheduling in microgrids with intermittent renewable energy resources. IEEE Trans Smart Grid 5:1876–1883. https://doi.org/10.1109/TSG.2013.2280645
14. Eajal AA, Shaaban MF, Ponnambalam K, El-Saadany EF (2016) Stochastic centralized dispatch scheme for AC/DC hybrid smart distribution systems. IEEE Trans Sustain Energy:1–14. https://doi.org/10.1109/TSTE.2016.2516530
15. Heitsch H, Römisch W (2009) Scenario tree reduction for multistage stochastic programs. Comput Manag Sci 6:117–133
16. Wu H, Shahidehpour M (2016) A game theoretic approach to risk-based optimal bidding strategies for electric vehicle aggregators in electricity markets with variable wind energy resources. IEEE Trans Sustain Energy 7:374–385. https://doi.org/10.1109/TSTE.2015.2498200
17. Fotouhi Ghazvini MA, Faria P, Ramos S, Morais H, Vale Z (2015) Incentive-based demand response programs designed by asset-light retail electricity providers for the day-ahead market. Energy 82:786–799

18. Wu H, Shahidehpour M, Alabdulwahab A, Abusorrah A (2015) Thermal generation flexibility with ramping costs and hourly demand response in stochastic security-constrained scheduling of variable energy sources. IEEE Trans Power Syst 30:2955–2964. https://doi.org/10.1109/TPWRS.2014.2369473
19. Bakirtzis EA, Ntomaris A V., Kardakos EG, Simoglou CK, Biskas PN, Bakirtzis AG (2014) A unified unit commitment – economic dispatch model for short-term power system scheduling under high wind energy penetration. International conference on the European energy market, EEM. https://doi.org/10.1109/EEM.2014.6861258
20. Nasri A, Kazempour SJ, Conejo AJ, Ghandhari M (2016) Network-constrained AC unit commitment under uncertainty: a benders' decomposition approach. IEEE Trans. Power Syst 31:412–422. https://doi.org/10.1109/TPWRS.2015.2409198
21. Gröwe-Kuska N, Heitsch H, Römisch W (2003) Scenario reduction and scenario tree construction for power management problems. In: 2003 IEEE Bologna PowerTech – conference proceedings, pp 152–158
22. Momber I, Siddiqui A, Roman TGS, Soder L (2014) Risk averse scheduling by a PEV aggregator under uncertainty. IEEE Trans Power Syst 30:882–891. https://doi.org/10.1109/TPWRS.2014.2330375
23. Majidi-Qadikolai M, Baldick R (2016) Stochastic transmission capacity expansion planning with special scenario selection for integrating N−1 contingency analysis. IEEE Trans Power Syst 31:4901–4912. https://doi.org/10.1109/TPWRS.2016.2523998
24. Nahman JM, Peric DM (2008) Optimal planning of radial distribution networks by simulated annealing technique. IEEE Trans Power Syst 23:790–795. https://doi.org/10.1109/Tpwrs.2008.920047
25. Pereira Junior BR, Mantovani JRS, Cossi AM, Contreras J (2014) Multiobjective multi-stage distribution system planning using tabu search. IET Gener Transm Distrib 8:35–45. https://doi.org/10.1049/iet-gtd.2013.0115
26. Gallo A (2015) A refresher on net present value. Harv Bus Rev:1–7
27. Bruni AL, Famá R, Siqueira JDO (1998) Análise Do Risco Na Avaliação De Projetos De Investimento: Uma Aplicação Do Método De Monte Carlo. Caderno de Pesquisas em Administração 1–na 6:62–75
28. Gan L, Low SH (2014) Optimal power flow in direct current networks. IEEE Trans Power Syst 29:2892–2904. https://doi.org/10.1109/TPWRS.2014.2313514
29. Warren CA (2003) IEEE Guide for Electric Power Distribution Reliability Indices (2012). In IEEE Std 1366-2012 (Revision of IEEE Std 1366-2003) , pp. 1–43. doi: 10.1109/IEEESTD.2012.6209381
30. Subcommittee D (2012) IEEE guide for electric power distribution reliability indices. Distribution. https://doi.org/10.1109/IEEESTD.2012.6209381
31. Canizes B, Soares J, Vale Z, Lobo C (2017) Optimal approach for reliability assessment in radial distribution networks. IEEE Syst J 11:1846–1856. https://doi.org/10.1109/JSYST.2015.2427454
32. Wenyuan L, Jiaqi Z, Kaigui X, Xiaofu X (2008) Power system risk assessment using a hybrid method of fuzzy set and Monte Carlo simulation. IEEE Trans Power Syst 23:336–343. https://doi.org/10.1109/TPWRS.2008.919201
33. Conejo AJ, Carrión M, Morales JM (2010) Decision making under uncertainty in electricity markets. Springer, New York
34. Venikov VA, Asambaev SN (1986) Quick estimation of the stability of a process on the basis of its initial stage. Energetika i Transport (Translation available from Allerton Press) 24:23–29
35. Gustafson MW, Baylor JS, Mulnix SS (1988) The equivalent hours loss factor revisited. IEEE Trans Power Syst 3:1502–1508. https://doi.org/10.1109/59.192959

Chapter 14
A Joint Energy Storage Systems and Wind Farms Long-Term Planning Model Considering Voltage Stability

Saman Nikkhah and Abbas Rabiee

14.1 Introduction

Radially nature and high ratio of R/X in distribution systems (DSs) cause several operational and security problems such as high power losses and low voltage profile in the grid. Recently, several solutions have been suggested to improve the reliability and stability of DSs. In spite of their costs, distributed generations (DGs) are considered to be one of the best viable solutions for these problems. Undeniable advantages of wind energy in today's power systems have resulted in rapid increase in penetration level of this kind of renewable energy sources into local and regional utility grids. However, despite various advantages of wind power technology, intermittent and stochastic nature of such DG resource can cause noticeable challenges for distribution system operators (DSOs), especially from the voltage stability point of view.

Recently, the integration of wind energy in DSs with energy storage systems (ESSs) has become a new solution to ensure the stability and reliability of a power system with facilitating increased penetration of wind energy. The dispatchable storage technologies can also provide additional benefits for distribution utilities, including better load management [1], mitigating power quality concerns [2], and overall reduction of energy costs [3]. In the following subsection, a review is made on previous researches in this regard and a background is made on planning of ESS to solve the uncertainty problem of renewable energy sources.

S. Nikkhah · A. Rabiee (✉)
Faculty of Electrical Engineering, University of Zanjan, Zanjan, Iran
e-mail: s.nikkhah@znu.ac.ir; rabiee@znu.ac.ir

B. Mohammadi-Ivatloo, F. Jabari (eds.), *Operation, Planning, and Analysis of Energy Storage Systems in Smart Energy Hubs*,
https://doi.org/10.1007/978-3-319-75097-2_14

14.1.1 Review of the Existing Literature

The ESS planning in DSs has been addressed in several research works. In [4], the authors propose a stochastic planning framework for storage systems to optimally site the battery ESS, aiming to maximize wind power penetration and minimize operation and investment costs. The uncertainty of wind energy is modeled by Monte Carlo simulations. In [5], a methodology is proposed for optimal allocation of ESS for a system with high penetration of wind energy. Both perspectives of the utility and the DG owner are taken into consideration by sizing the ESS to accommodate all amount of spilled wind energy. The authors of [6] consider a tradeoff between budget for investment and the daily operation cost in the ESS planning. The information gap decision theory is used in [7] for handling wind power uncertainty. The authors investigate the effect of storage devices on the uncertainty handling of wind energy. In [8], a multi-objective optimization model is proposed for scheduling of ESSs. In [9], the system robust operation ensured by using robust optimization technique, while the investment costs of storage units are minimized. The authors in [10] focus on the loss payment minimization using ESS and demand response in an uncertain environment, while electricity price is considered as an uncertain parameter. The work in [11] suggests a comprehensive framework for ESS allocation, aiming to increase wind power penetration and voltage stability enhancement. The authors consider economic requirements such as cost and profit obtained by sizing and siting of the ESSs.

Different objectives are addressed in the existing literature for the ESS scheduling problem. The economic objective in [12] is minimization of electricity usage cost and battery operation costs. In [13], a stochastic planning framework is proposed from the perspective of independent system operator aiming to maximize several objectives including: total expected net present value (NPV), cost and benefit of electricity utilization, power generation, etc. In [14], an economic dispatch model is proposed to increase wind utilization by utilization of ESSs with the objective of minimizing the composite operating costs of the system. The authors in [15] consider a coordinated wind power and ESS model for decreasing wind energy forecast errors. In [16], a probabilistic optimal power flow is introduced for optimal placement of ESSs in a system with the objective of minimizing the hourly social cost. In [17], the impact of ESS specific costs on the NPV of ESS installation investment is investigated considering the relationship between wind power penetration and daily load profile. In [18], the size of ESS installed in a wind-diesel power system is determined via a two-stage stochastic optimization model, with the objectives of fuel cost and operating cost minimization.

Although voltage stability is considered in some wind power planning research works [19–21], and improvement of voltage stability with application of ESS has been investigated in the literatures [22–26], the point which is not considered in previous works is consideration of voltage stability as a constraint of joint ESS and wind energy planning models. In [19], system loading margin (LM) is considered as constraints in the proposed corrective voltage control scheme for a system under the influence of wind energy. The work in [20] proposes voltage stability constrained optimal power flow which the relationship between the LM and uncertainty of wind energy is investigated. In [21], the improvement of voltage stability of power system under the influence of wind power generation is investigated while L-index considered as the voltage stability index. The work in [22], proposes a combinational photovoltaic and ESS model to improve voltage stability and decrease active and reactive power losses by optimal dispatching of load power factor. The customer-side ESSs are used in [23] to solve the voltage fluctuations of DSs with high penetration of photovoltaic systems, by giving the permission to DSO to control the output of ESSs. In [24], a coordinated control approach is proposed for decreasing the voltage and frequency deviations. Also, in [24] the voltage profile of a real DS is improved by coordinated operation of ESS and photovoltaic system. Due to the potential of battery ESS and STATCOM, the work in [25] is focused on the improvement of power quality and stability of DSs under the influence of high wind power penetration. In [26], grid voltage stability is improved while acceptable wind power penetration obtained by using ESS to control the intermittent nature of wind energy.

While the aim of the proposed models in the existing literature is improving voltage stability, this chapter considers the voltage stability as a constraint in the proposed model and optimal capacity of ESSs and wind turbines (as a kind of DGs) obtained subject to secure operation of power system from voltage stability point of view.

14.1.2 Chapter Contributions

It is concluded from the above literature survey that various planning frameworks have been proposed in the area of sizing and scheduling of ESS to mitigate the problems associated with the uncertainty of renewable DG units, and optimizing several objectives. However, the voltage stability considerations have not been included in the formulation of proposed models. Furthermore, the concept of integrating ESS in the system under the influence of wind energy from the perspectives of both DG owners and DSO has not been considered simultaneously. Therefore, the main focus of this chapter is to propose a voltage stability constrained wind-storage planning model (VSC-WSPM) which considers the perspectives of both DG owner and DSO. Due to the fact that one of the system operator's goals is minimizing power generation costs while preserving the system stability [27], the objective of the proposed model is to minimize power generation costs and charge/discharge costs of ESSs and to maximize the profit obtained by DG owner from wind energy procurement.

The main contributions of this chapter are outlined as follows:

- A comprehensive model is proposed for simultaneous planning of wind energy and ESS.
- The welfare of both DSO and DG owners is considered simultaneously in the long-term planning horizon.
- Voltage stability considered in the planning model.
- The impact of voltage stability on the capacity of DGs and ESSs is investigated.
- Optimal capacity of ESSs and DGs is determined in the entire planning horizon.

14.2 The Concept of Loading Margin Index for Voltage Stability Evaluation

LM is defined as the amount of power generation that can be increased to meet the additional demand increase in PQ buses prior to violation of operational limits or happening voltage collapse [28]. In order to guarantee the secure operation of a power system, it is necessary to preserve a desired level of LM. This level of LM specifies the distance between normal operation point and voltage collapse point of the system [29]. In this regard, power flow equations at the current operation point (COP) considered along with power flow equations in loadability limit point (LLP) (e.g., the nose point of PV curve), simultaneously [20].

For better description of the LM concept, consider the PV curve of an arbitrary load bus of a system that is depicted in Fig. 14.1. The points *A* and *B* are the system COP and the corresponding LLP, respectively. The distance (in MW or MVA) between points *A* and *B* is called LM, which could be characterized by loading parameter, λ. In order to have a proper safety margin, the system operator considers a desired level for the LM in which the system LM should be greater than it. The amount of system LM is an important factor for secure operation of the system, since the voltage instability can be prevented in the post-contingency conditions, if a sufficient LM is considered.

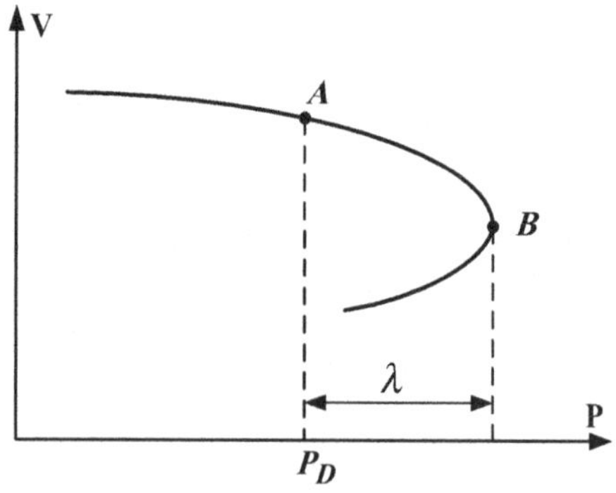

Fig. 14.1 System loading margin of P-V curve [20]

The generator's reactive power support is one of the most important factors which directly influences the system LM. The system LM will be large, when enough reactive power support exists in the system. Consequently, insufficient reactive power support or reaching the reactive power limit of generators could cause voltage collapse.

In this chapter, the LM is included in VSC-WSPM and power flow equality and inequality constraints considered in both LLP and COP points, simultaneously, in order to characterize the LM.

14.3 VSC-WSPM Problem Formulation

14.3.1 Objective Functions

The objective function of the problem is optimized by considering the minimization of power generation and ESS operation costs and maximization of wind energy profit obtained by DG owners.

14.3.1.1 Minimization of Power Generation Costs

Minimizing the total power generation cost in DSs is critical objective and should be considered in long-term planning models for improving total energy efficiency and economic reasons.

$$F_1 = \sum_{n=1}^{N_T} \sum_{t=1}^{24} 365 \times \left(\frac{1+\vartheta}{1+\varepsilon} \right)^n \cdot \mathrm{EC}_{n,t} \times \left(\sum_{i=1}^{N_G} P^G_{i,n,t} + \sum_{b=1}^{N_{\mathrm{DG}}} p^{\mathrm{DG}}_{b,n,t} \right) \tag{14.1}$$

where F_1 is the total power generation cost in planning horizon, $P^G_{i,n,t}$ and $p^{\mathrm{DG}}_{b,n,t}$ are active power generation and injected wind power to the grid at bus b in year n and time t, respectively. Also, $\mathrm{EC}_{n,t}$ is the energy cost in year n and time t, ϑ and ε are inflation and interest rates, respectively.

14.3.1.2 Minimization of ESS Charge/Discharge Costs

The ESS charge/discharge cost is a critical operation objective that should be minimized during the planning horizon.

$$F_2 = \sum_{n=1}^{N_T} \sum_{t=1}^{24} \sum_{b=1}^{N_{\mathrm{ESS}}} 365 \times \left(\frac{1+\vartheta}{1+\varepsilon} \right)^n \times \mathrm{CHC}_{n,t} \times \Delta t \left(p^{\mathrm{CH}}_{b,n,t} \cdot \eta^{\mathrm{ch}}_{b,t} + p^{\mathrm{DISCH}}_{b,n,t} / \eta^{\mathrm{disch}}_{b,t} \right) \tag{14.2}$$

where F_2 is the total ESS charge/discharge cost in planning horizon, $p_{b,n,t}^{\mathrm{CH}}/p_{b,n,t}^{\mathrm{DISCH}}$ are charge/discharge power of ESS at node b in year n and time t, $\eta_{b,t}^{\mathrm{ch}}/\eta_{b,t}^{\mathrm{disch}}$ are charging/discharging efficiencies, $\mathrm{CHC}_{n,\,t}$ is the operation cost of ESS.

14.3.1.3 Maximization of the Profit Obtained from the Wind Energy Procurement

The objective of DG owners is to maximize the net present value (NPV) of profit based on the annual cash flow over the time horizon of the investment.

$$F_3 = \left(\frac{1+\vartheta}{1+\varepsilon}\right)^n \times \left(\sum_{t=1}^{24}\sum_{n=1}^{N_T}\sum_{b=1}^{N_W}\left(\mathrm{IN}_{b,n,t}^{\mathrm{inc}} - C_{b,n,t}^{\mathrm{O\&M}}\right) - \sum_{n=1}^{N_T}\sum_{b=1}^{N_W} C_{b,n}^{\mathrm{inv}}\right) \quad (14.3)$$

where $\mathrm{IN}_{t,d,b}^{\mathrm{inc}}$ is the total annualized incoming of wind energy selling to the costumers and $C_{t,d,b}^{\mathrm{O\&M}}$ corresponds to the operation and maintenance cost of DGs, whereas $C_{t,b}^{\mathrm{inv}}$ denotes the DGs investment cost. These costs are formulated as follows:

$$\mathrm{IN}_{b,n,t}^{\mathrm{inc}} = p_{b,n,t}^{\mathrm{DG}} \times \mathrm{EC}_{n,t} \quad (14.4)$$

$$C_{b,n,t}^{\mathrm{O\&M}} = p_{b,n,t}^{\mathrm{DG}} \times \mathrm{DGC}_{\mathrm{O\&M}} \quad (14.5)$$

$$C_{b,n}^{\mathrm{inv}} = P_{b,n}^{\mathrm{DG}} \times \mathrm{DGC}_{\mathrm{inv}} \quad (14.6)$$

where $P_{b,n}^{\mathrm{DG}}$ is the added wind energy capacity to the grid at bus b in year n, $\mathrm{DGC}_{\mathrm{O\,\&\,M}}$ and $\mathrm{DGC}_{\mathrm{inv}}$ are operation and maintenance cost ($/MWh) and investment cost ($/MW) of DGs, respectively.

14.3.2 The Overall Objective Function

In order to consider all mentioned objective functions in one objective, two coefficients defined as the weighting coefficients which basically amounted in the interval [0, 1]. These coefficients are called w_1 and w_2. Due to this explanation, the total objective function which is the minimization of social welfare is defined as follows:

$$\mathrm{OF} = \min\left(w_1 \times (F_1 + F_2) - w_2 \times F_3\right) \quad (14.7)$$

14.3.3 Constraints

The VSC-WSPM is subject to following operation constraints.

14.3.3.1 Power Balance Constraints at COP

In order to optimize the objective function of proposed VSC-WSPM, it is necessary to consider the power flow constrains, operational and physical limits. Due to consideration of LM as the voltage stability index, the equality and inequality constraints at LLP should be considered additionally. In the following, a detailed description of equality and inequality constrains at COP are given.

$$\begin{aligned}&\left(\sum_{i=1}^{N_G} P_{i,n,t}^G\right) + P_{b,n,t}^{DG} + P_{b,n,t}^{DISCH} - P_{b,n,t}^{D} - P_{b,n,t}^{CH}\\&= V_{b,n,t}\sum_{j=1}^{N_B} V_{j,n,t}Y_{bj}\cos\left(\theta_{b,n,t} - \theta_{j,n,t} - \phi_{bj}\right)\end{aligned} \tag{14.8}$$

$$\left(\sum_{i=1}^{N_G} Q_{i,n,t}^G\right) + q_{b,n,t}^{DG} - Q_{b,n,t}^{D} = V_{b,n,t}\sum_{j=1}^{N_B} V_{j,n,t}Y_{bj}\sin\left(\theta_{b,n,t} - \theta_{j,n,t} - \phi_{bj}\right) \tag{14.9}$$

$$P_{G_i}^{\min} \le P_{i,n,t}^G \le P_{G_i}^{\max} \tag{14.10}$$

$$Q_{G_i}^{\min} \le Q_{i,n,t}^G \le Q_{G_i}^{\max} \tag{14.11}$$

$$V_b^{\min} \le V_{b,n,t} \le V_b^{\max} \tag{14.12}$$

$$|S_{l,n,t}(V,\theta)| \le S_l^{\max} \tag{14.13}$$

where constraints (14.8) and (14.9) are the power flow equations at COP, $P_{i,n,t}^G$, $Q_{i,n,t}^G$ are active and reactive power production of generator at bus i, in year n and time t, $P_{b,n,t}^D$, $Q_{b,n,t}^D$ are active and reactive power demand of b-th bus in year n and time t, Y_{bj}, ϕ_{bj} magnitude/angle of bj-th element of system admittance matrix, $V_{b,\,t,\,d}/\theta_{b,\,t,\,d}$ voltage magnitude/angle of bus b in year n and time t. Also, constraints (14.10)–(14.12) show the active and reactive power of the generators and voltage of system buses, respectively. Also, (14.13) shows the limit on power flowing through the branches.

14.3.3.2 Power Balance Constraints at LLP

As it is aforementioned, in the proposed VSC-WSPM, it is necessary to consider operational constraints at LLP in addition to considering those of COP. The following constraints represent the proposed constraints that considered in LLP.

$$\left(\sum_{i=1}^{N_G}\widehat{P}^G_{i,n,t}\right)+p^{\mathrm{DG}}_{b,n,t}+p^{\mathrm{DISCH}}_{b,n,t}-\widehat{P}^D_{b,n,t}-p^{\mathrm{CH}}_{b,n,t} = \widehat{V}_{b,n,t}\sum_{j=1}^{N_B}\widehat{V}_{j,n,t}Y_{bj}\cos\left(\widehat{\theta}_{b,n,t}-\widehat{\theta}_{j,n,t}-\phi_{bj}\right) \quad (14.14)$$

$$\left(\sum_{i=1}^{N_G}\widehat{Q}^G_{i,n,t}\right)+q^{\mathrm{DG}}_{b,n,t}-\widehat{Q}^D_{b,n,t}=\widehat{V}_{b,n,t}\sum_{j=1}^{N_B}\widehat{V}_{j,n,t}Y_{bj}\sin\left(\widehat{\theta}_{b,n,t}-\widehat{\theta}_{j,n,t}-\phi_{bj}\right) \quad (14.15)$$

$$\widehat{P}^D_{b,n,t}=(1+K_{D,b}\lambda)\,P^D_{b,n,t} \quad (14.16)$$

$$\widehat{Q}^D_{b,n,t}=(1+K_{D,b}\lambda)\left(Q^D_{b,n,t}\right) \quad (14.17)$$

$$\widehat{P}^G_{i,n,t}=\min\left(P^{\max}_{G_i},(1+K_{G,i}\lambda)\,P^G_{i,n,t}\right) \quad (14.18)$$

$$P^{\min}_{G_i}\le\widehat{P}^G_{i,n,t}\le P^{\max}_{G_i} \quad (14.19)$$

$$Q^{\min}_{G_i}\le\widehat{Q}^G_{i,n,t}\le Q^{\min}_{G_i} \quad (14.20)$$

$$V^{\min}_b\le\widehat{V}_{b,n,t}\le V^{\max}_b \quad (14.21)$$

$$\left|\widehat{S}_{l,n,t}(V,\theta)\right|\le S^{\max}_l \quad (14.22)$$

$$\lambda\ge\lambda_{\mathrm{des}}>0 \quad (14.23)$$

where (14.14) and (14.15) show the power flow equations at LLP, (14.16) and (14.17) correspond to the active and reactive power increment pattern of loads to meet the load increased from COP to LLP. $\widehat{P}^G_{i,n,t}$, $\widehat{Q}^G_{i,n,t}$ are active and reactive power production of generator at bus i, in year n and time t, at LLP, $\widehat{P}^D_{b,n,t}$, $\widehat{Q}^D_{b,n,t}$ are active and reactive power consumption of load connected to b-th bus in year n and time t, at LLP, $\widehat{V}_{j,n,t}/\widehat{\theta}_{b,n,t}$ is voltage magnitude/angle of bus b in year n and time t, at LLP. Also, (14.18) shows the increment of generators active power to cover the demand increment from the COP to the LLP. Also, constraints (14.19)–(14.22) correspond to active/reactive power limits, voltage magnitude limits, and the limits of apparent power flowing through branches at LLP, respectively. Finally, as it is aforementioned, desired value of LM that is defined by DSO should be lower than loading parameter, which is considered in (14.23).

14.3.3.3 System Load Growth

This chapter deals with the planning of ESSs and DGs in long-term planning horizon. Therefore, it is necessary to consider daily load model which considers the annual demand growth. This concept is mathematically expressed as follows:

$$P_{b,n,t}^{D} = (1 + \beta_{b,n})\, P_{b,n-1,t}^{D} \tag{14.24}$$

$$Q_{b,n,t}^{D} = (1 + \beta_{b,n})\, Q_{b,n-1,t}^{D} \tag{14.25}$$

where $\beta_{b,\,n}$ is the annually load growth for system load buses.

14.3.3.4 DG Capacity Constraints

Intermittency of wind energy is one of the main barriers against the high penetration of wind power in a grid. Therefore, it is necessary to limit the active and reactive capacity of each DG as follows:

$$0 \leq \sum_{b} \pi_{b,n}^{\mathrm{DG}} \leq \rho \times \sum_{b} P_{b,n,t}^{D} \tag{14.26}$$

$$\pi_{b,n}^{\mathrm{DG}} = \pi_{b,n-1}^{\mathrm{DG}} + P_{b,n}^{\mathrm{DG}} \tag{14.27}$$

$$P_{b,n}^{\mathrm{DG}} \leq I_{b,n}^{\mathrm{DG}} \cdot P_{\min}^{\mathrm{DG}},\ \left(I_{b,n}^{\mathrm{DG}} \in \{0,1\}\right) \tag{14.28}$$

$$P_{b,n}^{\mathrm{DG}} \geq I_{b,n}^{\mathrm{DG}} \times P_{\max}^{\mathrm{DG}} \tag{14.29}$$

$$0 \leq p_{b,n,t}^{\mathrm{DG}} \leq \mathrm{CF}_{t}^{\mathrm{DG}} \times \pi_{b,n}^{\mathrm{DG}} \tag{14.30}$$

$$q_{b,\min}^{\mathrm{DG}} \leq q_{b,n,t}^{\mathrm{DG}} \leq q_{b,\max}^{\mathrm{DG}} \tag{14.31}$$

where (14.26) gives the cumulative wind energy limit in n-th year of the planning horizon, which is increased in each year over the previous year due to the added wind capacity to the grid as shown in (14.27). Due to the economic and operational limits, added wind energy to the grid should be limited by (14.28) and (14.29) which binary variable $I_{b,n}^{\mathrm{DG}}$ denotes the years that wind energy needed to be added to the grid for a DG connected to bus b. Whereas, the DGs actual power output that is used in (14.8) and (14.9) is limited by (14.30) due to the capacity factor (CF) of DGs. Besides, (14.31) denotes the reactive power generation limits of the DGs.

14.3.3.5 ESS Constraints

The ESSs have technical operation constraints and should be considered in planning model. The ESS constraints proposed in this chapter are expressed as follows:

(a) Charging/discharging power constraints

$$0 \leq p_{b,n,t}^{\mathrm{CH}} \leq \delta_{b,t}^{\mathrm{ch}} \times p_{b,t}^{\mathrm{ch,max}} \tag{14.32}$$

$$0 \leq p_{b,n,t}^{\mathrm{DISCH}} \leq \delta_{b,t}^{\mathrm{disch}} \times p_{b,t}^{\mathrm{disch,max}} \tag{14.33}$$

$$\sum_{b}^{N_{\mathrm{ESS}}} \sum_{t=1}^{24} p_{b,n,t}^{\mathrm{CH}} \geq \sum_{b}^{N_{\mathrm{ESS}}} \sum_{t=1}^{24} p_{b,n,t}^{\mathrm{DISCH}} \tag{14.34}$$

$$\delta_{b,t}^{\mathrm{ch}} + \delta_{b,t}^{\mathrm{disch}} \leq 1, \left(\delta_{b,t}^{\mathrm{disch}}, \delta_{b,t}^{\mathrm{ch}} \in \{0, 1\}\right) \tag{14.35}$$

where (14.32) and (14.33) show the charging/discharging power limit. Also, due to the operation schedule of ESS, charging capacity of storage should be greater than discharging capacity of storage, which is modeled in (14.34). Also, (14.35) denotes that in each time interval t, only charge or discharge of ESSs is allowed.

(b) State of charge constraints

$$\mathrm{SOC}_{b}^{\min} \leq \mathrm{SOC}_{b,n,t}^{\mathrm{ESS}} \leq \mathrm{SOC}_{b}^{\max} \tag{14.36}$$

$$\mathrm{SOC}_{b,n,t}^{\mathrm{ESS}} \leq E_{b,n-1}^{\mathrm{ESS}} \tag{14.37}$$

$$E_{b,n+1}^{\mathrm{ESS}} = E_{b,n}^{\mathrm{ESS}} + e_{b,n}^{\mathrm{ESS}} \tag{14.38}$$

$$0 \leq e_{b,n}^{\mathrm{ESS}} \leq e_{\max}^{\mathrm{ESS}} \tag{14.39}$$

$$\mathrm{SOC}_{b,n+1,t}^{\mathrm{ESS}} = \mathrm{SOC}_{b,n,t}^{\mathrm{ESS}} + \Delta t \times \left(p_{b,n,t}^{\mathrm{CH}} \eta_{b,t}^{\mathrm{ch}} - p_{b,n,t}^{\mathrm{DISCH}} / \eta_{b,t}^{\mathrm{disch}}\right) \tag{14.40}$$

where constraint (14.36) shows the allowable state of charge of storage. Also according to (14.37) $\mathrm{SOC}_{b,n,t}^{\mathrm{ESS}}$ is the state of charge (SOC) for ESS installed at bus b, in year n and time t which is lower than total capacity of ESS installed till the n-th year ($E_{b,n-1}^{\mathrm{ESS}}$). The total capacity of ESSs increased each year over the previous year due to the added capacity of ESSs ($e_{b,n}^{\mathrm{ESS}}$) in each year which is shown by (14.38). Also, the annual capacity expansion of ESSs is limited by (14.39). Furthermore, the relationship between charging/discharging power and SOC of ESSs is modeled in (14.40), where Δt is the time slot of daily schedule which is assumed to be 1 h.

14.4 Simulations on a Standard Test System

In this section, the proposed VSC-WSPM is examined on a standard test system. In order to show the different aspects of the proposed model, different cases are considered. The following subsection gives the system data and the model parameters.

14.4.1 System Description

In order to evaluate the performance of the proposed VSC-WSPM, simulations are performed on the IEEE 33-bus standard distribution feeder. This system consists of 33 buses and 32 branches. The single line diagram of this system is depicted in Fig. 14.2. The proposed VSC-WSPM model, which is a mixed integer nonlinear programming problem (MINLP), is implemented in General Algebraic Modeling System (GAMS) [30] using DICOPT [31] and IPOPT [32] solvers. It is assumed that the demand of load buses varies in 24 h of a day with a pattern given in Fig. 14.3, while it is increased by 2% per year in the next years. It is assumed that three DGs are installed at buses 6, 14, and 32 whereas the penetration level (ρ) of DGs is supposed to be 50%. A planning horizon of 5 years is considered. It is assumed that DG units' yearly added capacity limited within 1–2 MW. Also, it is assumed that there is no injected wind energy to the grid during the first year to cope with budgeting delays and possible changes in policies. The proposed VSC-WSPM has been solved using parameters of Table 14.1. Table 14.2 summarizes the characteristics of dispatchable DG units. Also, Table 14.3 provides the daily variation of DGs' CF.

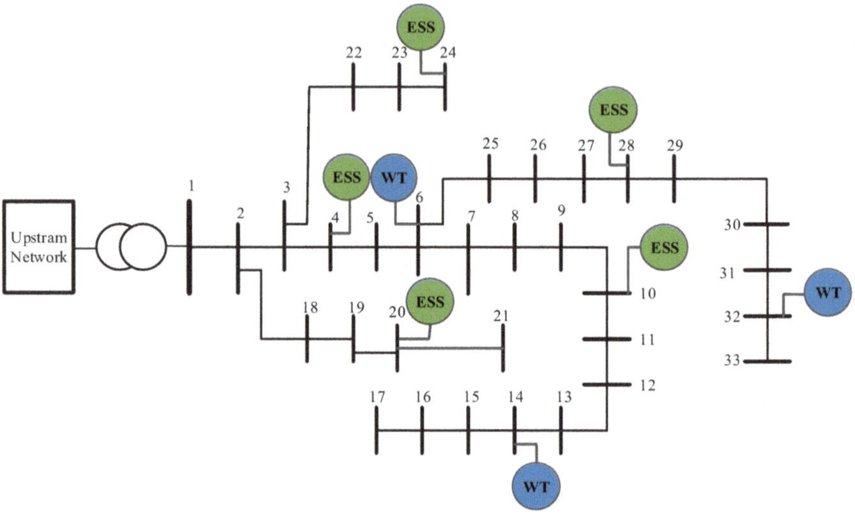

Fig. 14.2 Single line diagram of IEEE 33-bus distribution feeder

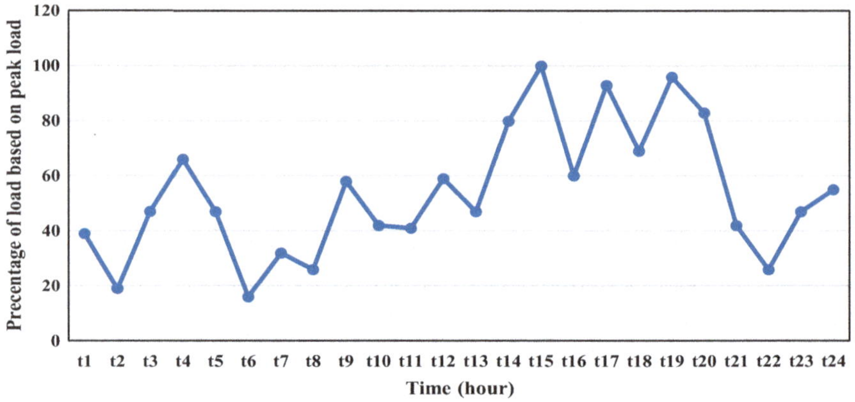

Fig. 14.3 Variation of load at load buses for 24 h of a day based on the peak value

Table 14.1 The simulation parameters

Parameter	Values (%)
$\beta_{b,t}$	2
ρ	50
λ_{des}	5
ITR	6
IFR	1

Table 14.2 Characteristics of dispatchable DG units [33]

Parameter	Unit	Value
DGs investment cost	\$/MW	318,000
DGs operation and maintenance cost	\$/MWh	10

Table 14.3 Daily variation of $\mathbf{CF}_t^W$

t_1	t_2	t_3	t_4	t_5	t_6	t_7	t_8	t_9	t_{10}	t_{11}	t_{12}
0.84	0.73	0.51	0.14	0.48	0.47	0.87	0.86	0.32	0.82	0.55	0.75
t_{13}	t_{14}	t_{15}	t_{16}	t_{17}	t_{18}	t_{19}	t_{20}	t_{21}	t_{22}	t_{23}	t_{24}
0.26	0.96	0.31	0.34	0.44	0.43	0.16	0.42	0.8	0.29	0.82	0.36

Table 14.4 Characteristics of ESS

$p_{b,t}^{\text{ch,max}}$ (MW)	$p_{b,t}^{\text{disch,max}}$ (MW)	$\eta_{b,t}^{\text{ch}}$	$\eta_{b,t}^{\text{disch}}$	$e_{\max}^{\text{ESS}}$ (MWh)
1	1	0.88	0.88	7

Table 14.5 Energy price ($\text{EC}_{n,t}$(\$/MWh))

t_1	t_2	t_3	t_4	t_5	t_6	t_7	t_8	t_9	t_{10}	t_{11}	t_{12}	t_{13}	t_{14}	t_{15}	t_{16}	t_{17}	t_{18}	t_{19}	t_{20}	t_{21}	t_{22}	t_{23}	t_{24}
40	36	48	54	48	36	38	36	52	44	42	52	47	60	60	52	60	57	60	60	44	36	48	50

Table 14.6 Charge/discharge cost of ESS ($\text{CHC}_{n,t}$(\$/MWh))

t_1	t_2	t_3	t_4	t_5	t_6	t_7	t_8	t_9	t_{10}	t_{11}	t_{12}	t_{13}	t_{14}	t_{15}	t_{16}	t_{17}	t_{18}	t_{19}	t_{20}	t_{21}	t_{22}	t_{23}	t_{24}
7	6	8	9	8	6	7	6	9	7	7	9	7.8	10	10	9	10	10	10	10	7	6	8	8

The buses 4, 10, 20, 24, and 28 selected for installation of ESSs. It is assumed that the ESSs will not charge or discharge in the first year of planning. Also, it is assumed that the SOC of each year equals to the amount of SOC at the end of previous year and the final energy stored in the ESSs at the end of a day, considered to be the initial state of ESSs in the next day. It is worth mentioning that each day simply models the peak load condition of a year, and since it is assumed the 5 years planning horizon, the 5 consecutive days considered accordingly. Characteristics of ESS, energy price, and charging cost of ESS are tabulated in Tables 14.4, 14.5, and 14.6, respectively.

In the following, the results obtained by implementing the proposed VSC-WSPM on IEEE-33 bus distribution test system are presented. The problem is examined in three case studies namely: Case-I: from the perspective of DSO (i.e., $w_1 = 1$, $w_2 = 0$, in Eq. (14.7)), Case-II: from the perspective of DG owner (i.e., $w_1 = 0$, $w_2 = 1$, in Eq. (14.7)), Case-III: from the perspective of both DG owner and DSO, simultaneously (i.e., $w_1 = w_2 = 0.5$, in Eq. (14.7)). In these cases, in order to show the effect of voltage stability constraints on the scheduled capacity of DGs and ESSs, Cases I and II solved with and without voltage stability constraints (i.e., without Eqs. (14.14)–(14.23)). For the sake of comparison, the results obtained for different cases are compared in Case-III.

14.4.2 Case-I: From the Perspective of DSO

In this case the proposed VSC-WSPM is implemented with and without the voltage stability constraints, from the perspective of DSO. Power generation cost with and without voltage stability constraints in this case is $4103451.9 and $4112595.9, respectively. Therefore, including voltage stability constraints imposes more cost to the DSO, which is reasonable. The annual added capacity of ESSs and DGs in this case for the entire planning horizon is depicted in Figs. 14.4 and 14.5, respectively. As it is observed from these figures, when voltage stability constraints

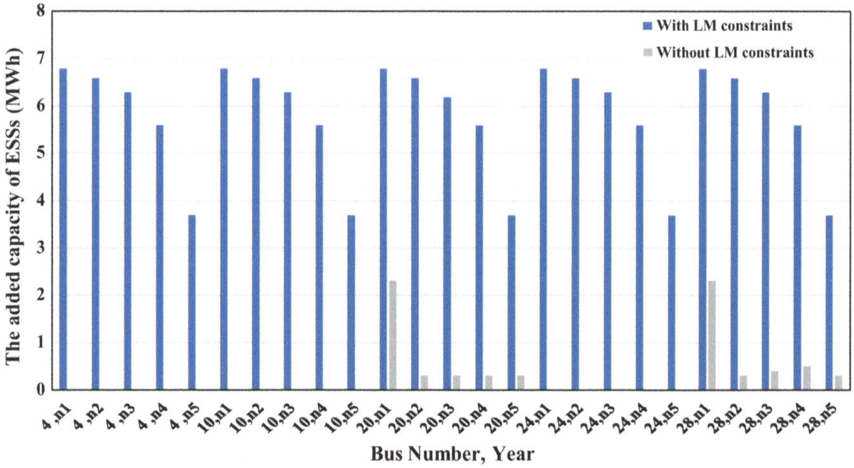

Fig. 14.4 The added capacity of ESSs in each bus number for planning horizon, in Case-I

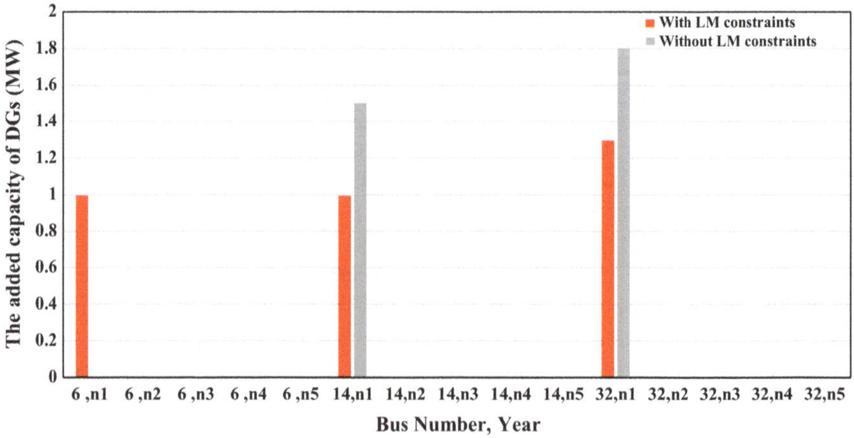

Fig. 14.5 The added capacity of DGs in each bus number for planning horizon in Case-I

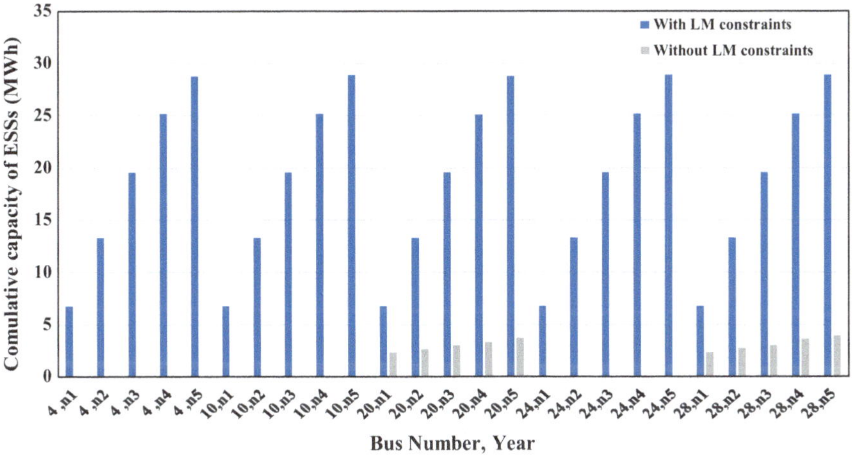

Fig. 14.6 Total capacity of ESSs in each bus for the planning horizon in Case-I

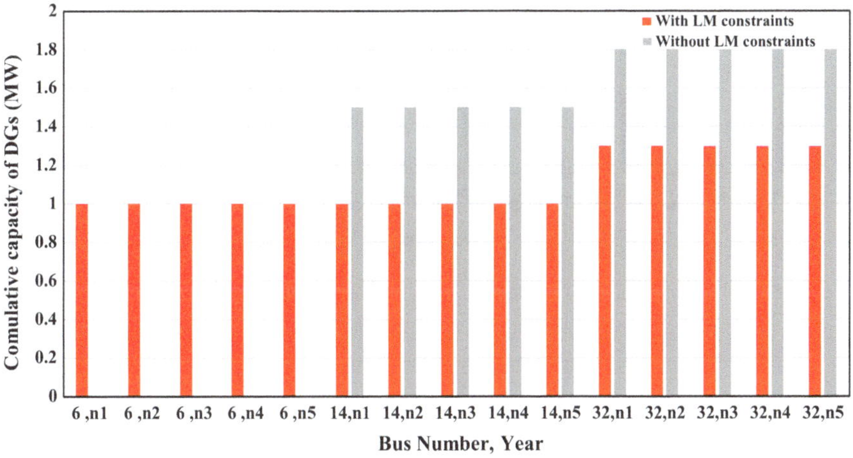

Fig. 14.7 Total capacity of DGs in each bus number for planning horizon, in Case-I

are considered in the VSC-WSPM, more wind power and ESS power are added to the grid in order to prevent voltage instability.

Besides, the total installed capacity of ESSs and DGs are depicted in Figs. 14.6 and 14.7, respectively. As it is observed from these figures, total capacities of wind energy and ESSs are affected by voltage stability constraints. It is evident from Fig. 14.6 that total capacity of ESSs increases yearly because additional capacity is added to the grid in each year. Also, for the DGs, no new capacity is scheduled since first year and total capacity is fixed in planning horizon for each installed DG bus number.

14.4.3 Case-II: From the Perspective of DG Owner

In this section the problem is solved from the DG owner's perspective and the effect of voltage stability constraints on the capacity of DGs and ESSs investigated. The net profit obtained from sharing wind energy with and without voltage stability constraints in this case is $875527.7 and $897296.7, respectively. It is worth to note that considering voltage stability decreases the DG owner's profit. The annual added capacity of ESSs and DGs in this case for the entire planning horizon is given in Figs. 14.8 and 14.9, respectively. It is inferred from these figures that adding wind

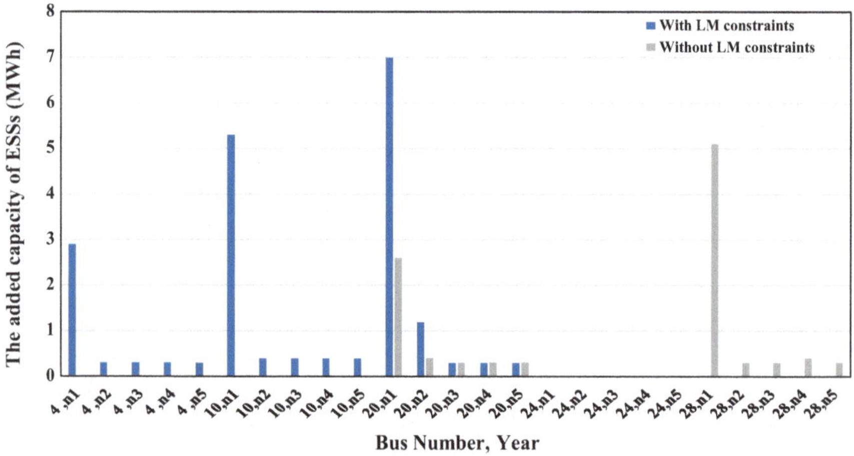

Fig. 14.8 The added capacity of ESSs in each bus number for planning horizon in, Case-II

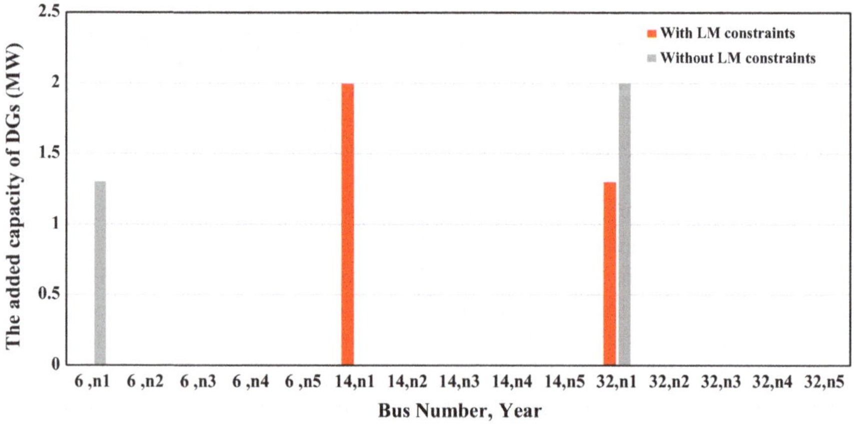

Fig. 14.9 The added capacity of DGs in each bus number for planning horizon, in Case-II

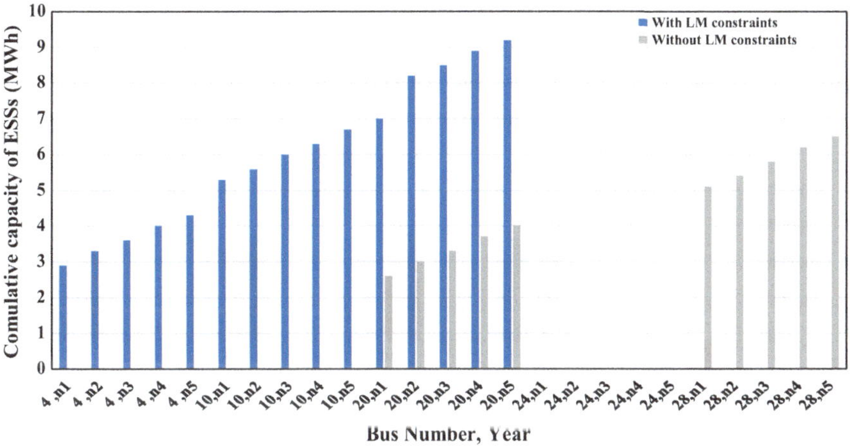

Fig. 14.10 Total capacity of ESSs in each bus number for planning horizon, in Case-II

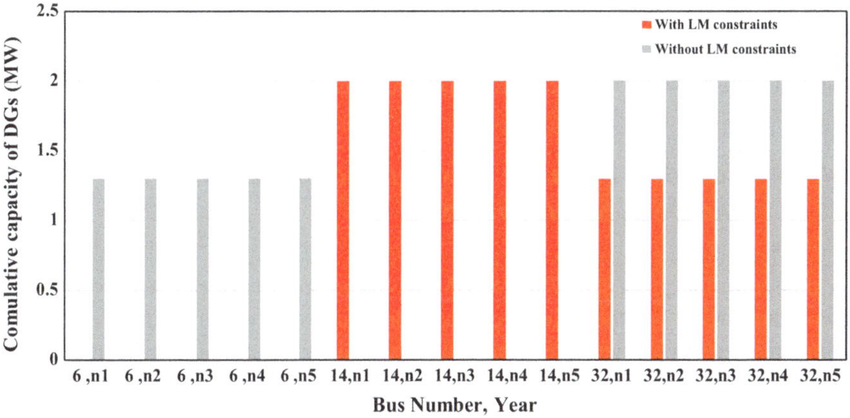

Fig. 14.11 Total capacity of DGs in each bus number for planning horizon, in Case-II

and ESSs energy to the system is affected by the voltage stability constraint. Also, more capacity is needed in order to guarantee the system security, when the voltage stability is considered in planning horizon.

Also, Figs. 14.10 and 14.11 show the total installed capacity of ESSs and DGs for this case, respectively. As it is evident from these figures total capacity of DGs and ESSs are affected by LM constraints in such a way that capacity of DGs and ESSs with the voltage stability constraints is bigger than those of without LM constraints.

Table 14.7 DG owner profit and power generation cost for all cases

Case #	DG owner profit ($)	Power generation cost ($)
Case-I	0	4103451.9
Case-II	875527.7	0
Case-III	906035.2	4260600.0

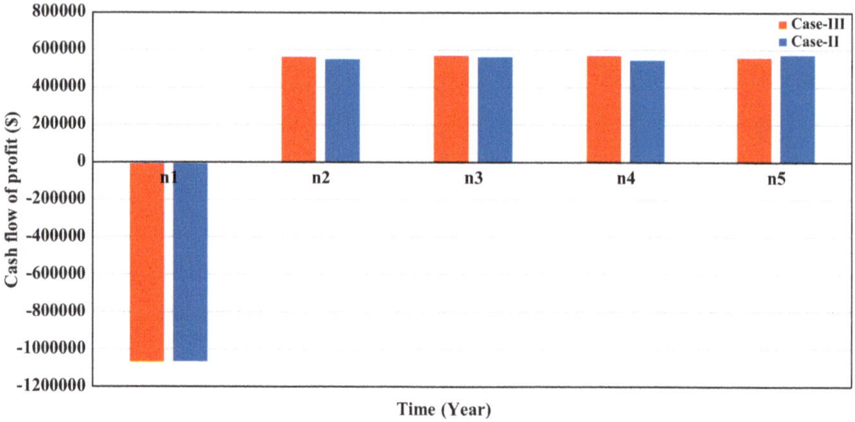

Fig. 14.12 Cash flow of DG owner's profit in different cases for the entire planning horizon

14.4.4 Case-III: From the Perspective of Both DG Owner and DSO, Simultaneously

In this case, the proposed VSC-WSPM is solved from both perspectives of DG owner and DSO. For the sake of comparison, the results obtained in this case are compared with two other cases when voltage stability constraints considered for all cases. It is assumed that desired LM is 5%. The net profit obtained from selling wind energy and power generation cost in all cases are tabulated in Table 14.7. It is evident from this table that considering both perspectives simultaneously provides more profit for the DG owner contrary to DSO. Also, Fig. 14.12 depicts the cash flow of profit obtained by DG owner in planning horizon for Case-II and Case-III. As it is observed from this figure that in the first year of planning the annual profit is negative, which means that the investment is not yet profitable in this year. Also, in the last 4 years, the profit becomes positive and is different for two cases.

The obtained annual capacity of ESSs and DGs which will be installed in the entire horizon is depicted in Figs. 14.13 and 14.14 for all cases, respectively. As it is observed in these figures, the added capacity of ESSs and DGs depends on the objective of decision maker and changes in different years due to the demand growth.

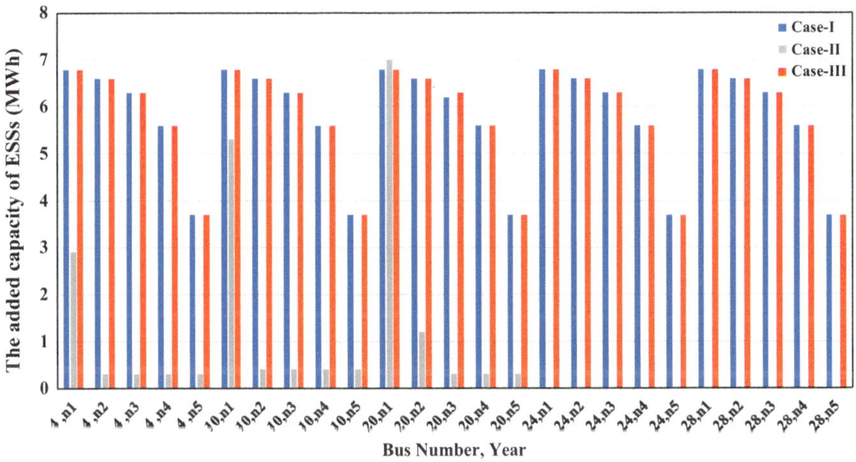

Fig. 14.13 The added capacity of ESSs in each bus number for planning horizon for different cases

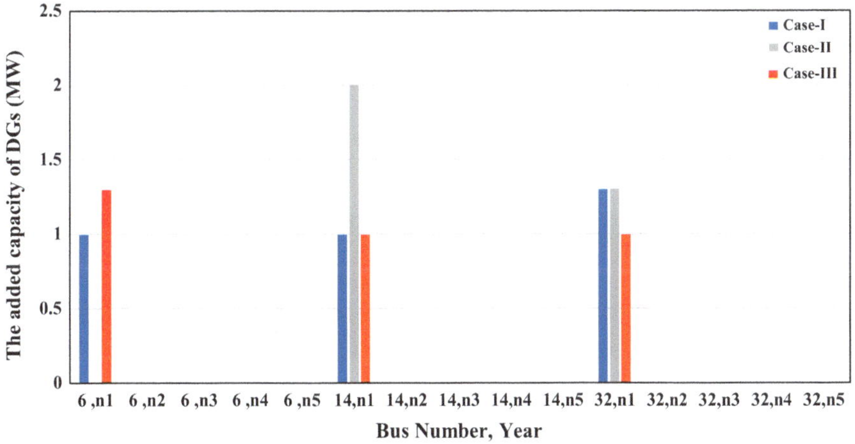

Fig. 14.14 The added capacity of DGs in each bus number for planning horizon for different cases

The cumulative capacity of ESSs and DGs in the planning horizon is given in Figs. 14.15 and 14.16 for all cases, respectively. As it is observed from these figures, the total capacities of ESSs and DGs are not the same in different cases. In other words, the capacity of DGs and ESSs depends on the goals of DG owner and DSO. Correspondingly, DSO and DG owner should specify their strategies when they make a planning decision.

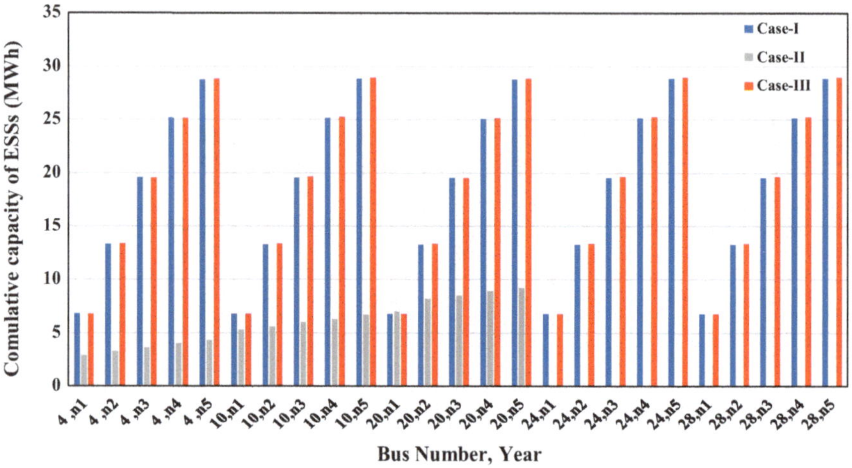

Fig. 14.15 Total capacity of ESSs in each bus number for planning horizon for different cases

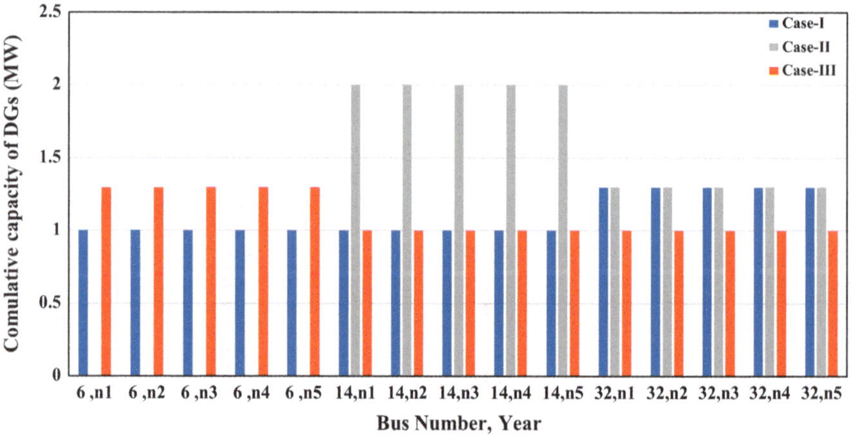

Fig. 14.16 Total capacity of DGs in each bus number for planning horizon for different cases

As it is aforementioned, one of the advantages of ESS is increasing wind power penetration. In this part, the impact of ESSs on capacity of DGs is investigated. In this regard, the proposed model is solved with and without ESS and total capacity of DGs is compared. It is evident from Fig. 14.17 that including ESSs to the grid increases the wind energy penetration in planning horizon.

Due to the relationship of SOC and ESS charge/discharge capacity, SOC can be used to investigate the charge/discharge states of ESS. The SOC of ESSs at buses 10 and 20 for third and fifth year of planning is depicted in Figs. 14.18 and 14.19,

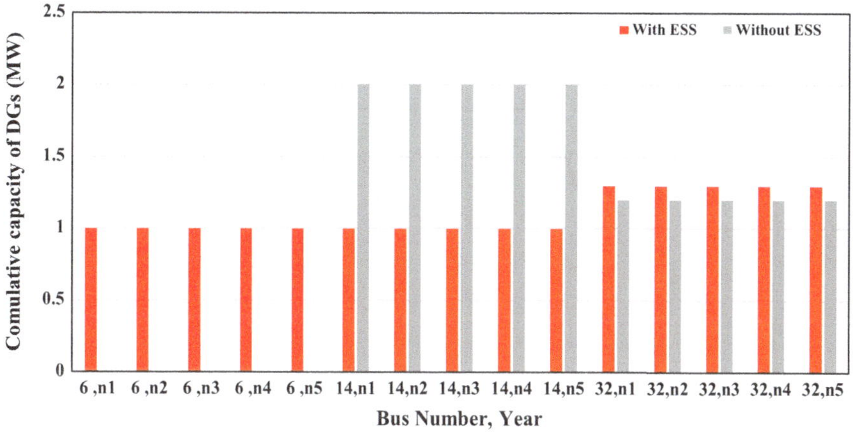

Fig. 14.17 The added capacity of DGs with and without ESS

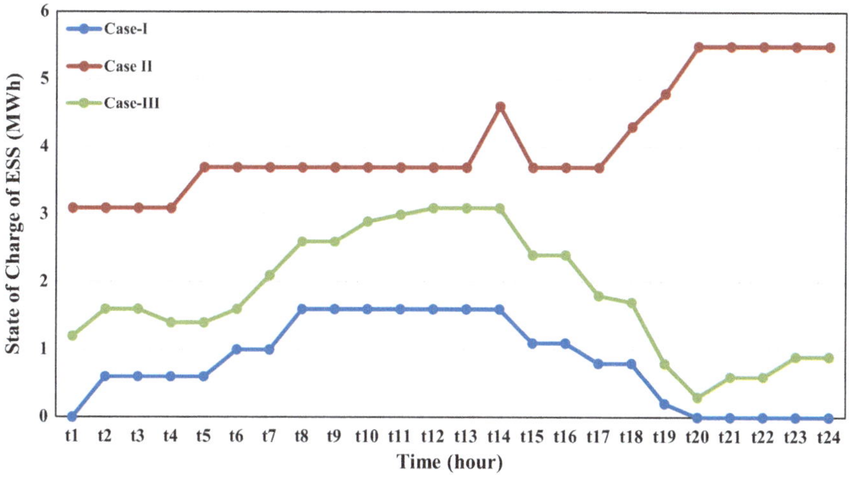

Fig. 14.18 SOC of ESS for bus 10 in third year of planning for a day in different cases

respectively. As it is observed from these figures, SOC increases and decreases in different time intervals of a day. Also, it is observed that in Case-II SOC is greater than other two cases, which is because of objective function of this case that ESS charge and discharge cost have not been considered in objective function. Furthermore, due to the topology of grid, SOC is different in ESS installed buses.

In this part, sensitivity of OF (i.e., in Eq. (14.7)) with respect to variation of interest rate in Case-III is investigated. Figure 14.20 depicts the variation of OF when the interest rate increases from 4% to 8% and inflation rate is kept 1%.

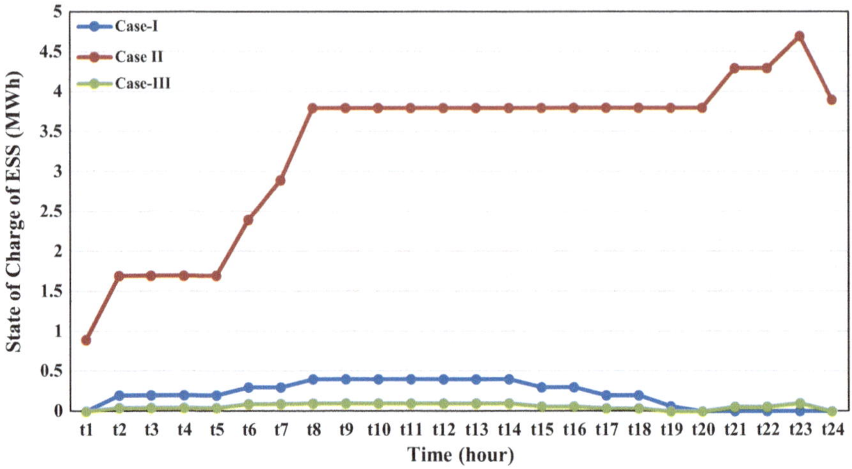

Fig. 14.19 SOC of ESS for bus 20 in fifth year of planning for a day in different cases

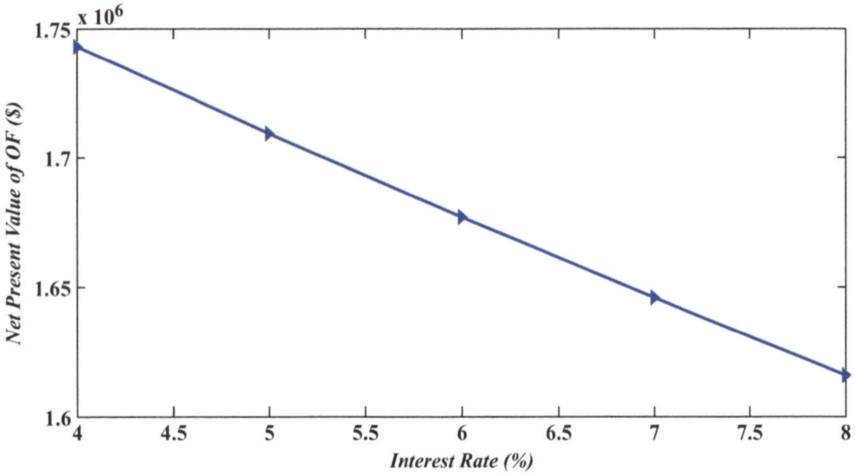

Fig. 14.20 Variation of net present value of OF versus the interest rate changes

It is inferred from this figure that if the interest rate increases, the OF reduces accordingly. Therefore the DG investor and DSO should consider proper value for the interest rate of their investments.

Finally, the P-V curve of an arbitrary load bus, i.e., bus 17 in the last year of the planning (i.e., fifth year) is depicted in Fig. 14.21. The P-V curves are plotted at the peak loading condition (i.e., t_{15}) for all three cases.

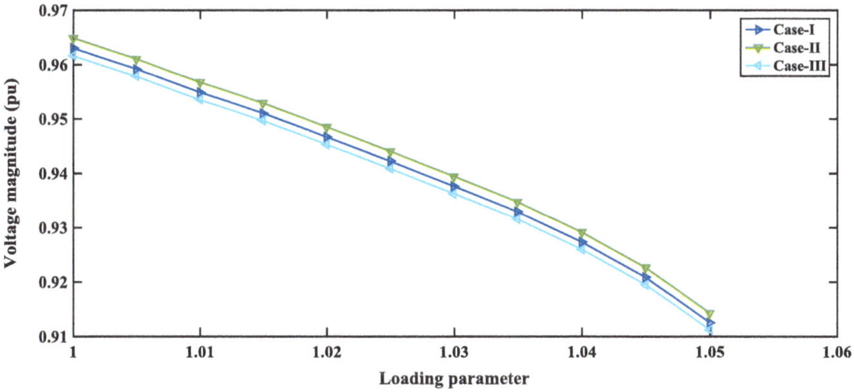

Fig. 14.21 Voltage profile of bus 17 in fifth year of planning horizon, at peak loading condition (t_{15})

14.5 Conclusion

In this chapter an approach is proposed for joint ESS and DG long-term planning, considering voltage stability constraints. Among the voltage stability indices, loading margin (LM) is considered in the formulation of the VSC-WSPM model. The proposed VSC- WSPM approach offers a decision-making tool for both DSO and DG owner to optimally determine their own strategies for utilization of wind energy and ESS.

The proposed VSC-WSPM is implemented on the IEEE 33-bus distribution test system in different cases. In the first case, the problem is solved from the perspective of DSO with the aim of power generation and ESS charge/discharge costs minimization, and the effect of voltage stability constraints on capacity of DGs and ESSs is investigated. In the second case, the problem is solved from the perspective of DG owner with the objective of maximizing his/her profit from wind energy procurement. In this case the capacity of ESSs and DGs is determined, with and without voltage stability constraints. In the third case, the problem solved from the perspective of both DSO and DG owner and the results compared with the former two cases when voltage stability constraints are taken into consideration. The following conclusions can be drawn:

- At the presence of voltage stability constraints the scheduled capacity of DGs and ESSs increases in order to preserve the voltage stability by ensuring the desired value of LM.
- It is necessary to charge the ESSs when DG owner wants to inject more wind energy to the grid. In such case, the SOC of ESSs is higher than the other cases.
- The added capacity of wind energy and ESS depends on the aims and priorities of the decision makers.

- The scheduled capacity of ESSs and DGs is affected by voltage stability constraints.
- DG owner and DSO should consider proper interest rate for their long-term investments.

Nomenclature

Sets

N_B	System buses
N_G	Generating units
N_L	Transmission lines
N_T	Planning horizon
N_{DG}	DG installed buses
N_{ESS}	ESS installed buses

Indices

b	System buses index
DG	DGs index
ESS	ESSs index
i	Thermal generating units' index
l	Transmission lines index
n	Index of planning years
t	Time index

Variables and Parameters

ϑ	Inflation rate
ε	Interest rate
ρ	Wind energy penetration factor
λ	Loading parameter
$\beta_{b,n}$	Demand growth rate at bus b in year n.
λ_{des}	Desired LM
$\eta_{b,t}^{ch/disch}$	Efficiency of charging and discharging of ESS (%)

$\delta_{b,t}^{\mathrm{ch/disch}}$	Charge/discharge binary indicators of ESSs
Δt	Timeslot duration
$\pi_{b,n}^{\mathrm{DG}}$	Cumulative wind power capacity of DG connected to bus b up to year n
$\mathrm{CHC}_{n,\,t}$.	Charge/discharge cost in year n and time t (\$/MWh)
$\mathrm{CF}_{t}^{\mathrm{DG}}$	Capacity factor of DG in time t
$\mathrm{DGC}_{\mathrm{inv}}$	Investment cost of DGs (\$/MW)
$\mathrm{DGC}_{\mathrm{O\,\&\,M}}$	Operation and maintenance cost of DGs (\$/MWh)
$E_{b,n}^{\mathrm{ESS}}$	Actual capacity of the ESS connected to bus b, in year n
$e_{b,n}^{\mathrm{ESS}}$	Added ESS capacity to bus b, in year n
$e_{\max}^{\mathrm{ESS}}$	Maximum annual added ESS capacity to the grid
$\mathrm{EC}_{n,\,t}$.	Energy price in year n and time t (\$/MWh)
$I_{b,n}^{\mathrm{DG}}$	Binary indicators of DGs
$K_{G,\,i}$	Rate of change in active power generation of unit i
$K_{D,\,b}$	Rate of load change at bus b
$P_{b,n}^{\mathrm{DG}}$	The added wind power capacity for DG connected to bus b in year n
$P_{\max/\min}^{\mathrm{DG}}$	Maximum/minimum wind energy added to the grid
$p_{b,n,t}^{\mathrm{CH/DISCH}}$	Charge/discharge power of ESS at node b in year n, at time t
$p_{b,t}^{\mathrm{ch/disch,max}}$	Maximum power charge/discharge of ESS at node b and time t
$(P/Q)_{G_i}^{\max/\min}$	Maximum/minimum active/reactive power of ith thermal generation unit
$P_{i,n,t}^{G}/Q_{i,n,t}^{G}$	Active/reactive power generation by ith thermal generation unit in year n, at time t
$\overset{\frown}{P}{}_{i,n,t}^{G}/\overset{\frown}{Q}{}_{i,n,t}^{G}$	Active/reactive power production of generator i in year n and time t at LLP
$P_{b,n,t}^{D}/Q_{b,n,t}^{D}$	Active/reactive power load of bus b in year n, at time t
$\overset{\frown}{P}{}_{b,t,d}^{D}/\overset{\frown}{Q}{}_{b,t,d}^{D}$	Active/reactive power consumption of load connected to bus b in year n and time t at LLP
$p_{b,n,t}^{\mathrm{DG}}/q_{b,n,t}^{\mathrm{DG}}$	Active/reactive power of DG injected to bus b in year n, at time t
$q_{b,\max/\min}^{\mathrm{DG}}$	Maximum/minimum reactive power of DG injected to bus b
$S_{l,\,n,\,t}(V,\theta)$	Power flow through l-th transmission line in year n, at time t
$S_{l}^{\max}$	Maximum transferable power through line l
$\mathrm{SOC}_{b,n,t}^{\mathrm{ESS}}$	State of charge (SOC) for the ESS connected at bus b in year n, at time t
$\mathrm{SOC}_{b}^{\max/\min}$	Maximum/minimum value of SOC
$V_{b,\,n,\,t}/\theta_{b,\,n,\,t}$	Voltage magnitude/angle of bus b in year n, at time t
$\overset{\frown}{V}_{b,t,d}/\overset{\frown}{\theta}_{b,t,d}$	Voltage magnitude/angle of bus b in year n and time t at LLP
$V_{b}^{\max/\min}$	Maximum/minimum voltage in bus b
Y_{bj}/ϕ_{bj}	Magnitude/angle of bj-th element of system admittance matrix

References

1. Gyuk IP, Eckroad S (2004) Energy storage for grid connected wind generation applications, US Department of Energy, Washington DC, EPRI-DOE Handbook Supplement, vol 1008703
2. Chen C, Duan S, Cai T, Liu B, Hu G (2011) Optimal allocation and economic analysis of energy storage system in microgrids. IEEE Trans Power Electron 26(10):2762–2773. https://doi.org/10.1109/TPEL.2011.2116808
3. Barton JP, Infield DG (2004) Energy storage and its use with intermittent renewable energy. IEEE Trans Energy Convers 19(2):441–448. https://doi.org/10.1109/TEC.2003.822305
4. Zhang Y, Dong ZY, Luo F, Zheng Y, Meng K, Wong KP (2016) Optimal allocation of battery energy storage systems in distribution networks with high wind power penetration. IET Renew Power Gener 10(8):1105–1113. https://doi.org/10.1049/iet-rpg.2015.0542
5. Atwa YM, El-Saadany E (2010) Optimal allocation of ESS in distribution systems with a high penetration of wind energy. IEEE Trans Power Syst 25(4):1815–1822. https://doi.org/10.1109/TPWRS.2010.2045663
6. Xiong P, Singh C (2016) Optimal planning of storage in power systems integrated with wind power generation. IEEE Trans Sustainable Energy 7(1):232–240. https://doi.org/10.1109/TSTE.2015.2482939
7. Maghouli P, Soroudi A, Keane A (2016) Robust computational framework for mid-term techno-economical assessment of energy storage. IET Gener Transm Distrib 10(3):822–831. https://doi.org/10.1049/iet-gtd.2015.0453
8. Farzin H, Fotuhi-Firuzabad M, Moeini-Aghtaie M (2017) A stochastic multi-objective framework for optimal scheduling of energy storage systems in microgrids. IEEE Trans Smart Grid 8(1):117–127. https://doi.org/10.1109/TSG.2016.2598678
9. Jabr RA, Džafić I, Pal BC (2015) Robust optimization of storage investment on transmission networks. IEEE Trans Power Syst 30(1):531–539. https://doi.org/10.1109/TPWRS.2014.2326557
10. Soroudi A, Siano P, Keane A (2016) Optimal DR and ESS scheduling for distribution losses payments minimization under electricity price uncertainty. IEEE Trans Smart Grid 7(1):261–272. https://doi.org/10.1109/TSG.2015.2453017
11. Le HT, Santoso S, Nguyen TQ (2012) Augmenting wind power penetration and grid voltage stability limits using ESS: application design, sizing, and a case study. IEEE Trans Power Syst 27(1):161–171. https://doi.org/10.1109/TPWRS.2011.2165302
12. Malysz P, Sirouspour S, Emadi A (2014) An optimal energy storage control strategy for grid-connected microgrids. IEEE Trans Smart Grid 5(4):1785–1796. https://doi.org/10.1109/TSG.2014.2302396
13. Murillo-Sanchez CE, Zimmerman RD, Anderson CL, Thomas RJ (2013) Secure planning and operations of systems with stochastic sources, energy storage, and active demand. IEEE Trans Smart Grid 4(4):2220–2229. https://doi.org/10.1109/TSG.2013.2281001
14. Chen H, Zhang R, Li G, Bai L, Li F (2016) Economic dispatch of wind integrated power systems with energy storage considering composite operating costs. IET Gener Transm Distrib 10(5):1294–1303. https://doi.org/10.1049/iet-gtd.2015.0410
15. Luo F, Meng K, Dong ZY, Zheng Y, Chen Y, Wong KP (2015) Coordinated operational planning for wind farm with battery energy storage system. IEEE Trans Sustainable Energy 6(1):253–262. https://doi.org/10.1109/TSTE.2014.2367550
16. Ghofrani M, Arabali A, Etezadi-Amoli M, Fadali MS (2013) A framework for optimal placement of energy storage units within a power system with high wind penetration. IEEE Trans Sustainable Energy 4(2):434–442. https://doi.org/10.1109/TSTE.2012.2227343
17. Chacra FA, Bastard P, Fleury G, Clavreul R (2005) Impact of energy storage costs on economical performance in a distribution substation. IEEE Trans Power Syst 20(2):684–691. https://doi.org/10.1109/TPWRS.2005.846091

18. Abbey C, Joós G (2009) A stochastic optimization approach to rating of energy storage systems in wind-diesel isolated grids. IEEE Trans Power Syst 24(1):418–426. https://doi.org/10.1109/TPWRS.2008.2004840
19. Rabiee A, Soroudi A, Mohammadi-Ivatloo B, Parniani M (2014) Corrective voltage control scheme considering demand response and stochastic wind power. IEEE Trans Power Syst 29(6):2965–2973. https://doi.org/10.1109/TPWRS.2014.2316018
20. Rabiee A, Nikkhah S, Soroudi A, Hooshmand E (2016) Information gap decision theory for voltage stability constrained OPF considering the uncertainty of multiple wind farms. IET Renew Power Gener 11(5):585–592. https://doi.org/10.1049/iet-rpg.2016.0509
21. Mohseni-Bonab SM, Rabiee A, Mohammadi-Ivatloo B (2016) Voltage stability constrained multi-objective optimal reactive power dispatch under load and wind power uncertainties: a stochastic approach. Renew Energy 85:598–609. https://doi.org/10.1016/j.renene.2015.07.021
22. Hung DQ, Mithulananthan N, Bansal R (2014) Integration of PV and BES units in commercial distribution systems considering energy loss and voltage stability. Appl Energy 113:1162–1170. https://doi.org/10.1016/j.apenergy.2013.08.069
23. Sugihara H, Yokoyama K, Saeki O, Tsuji K, Funaki T (2013) Economic and efficient voltage management using customer-owned energy storage systems in a distribution network with high penetration of photovoltaic systems. IEEE Trans Power Syst 28(1):102–111. https://doi.org/10.1109/TPWRS.2012.2196529
24. Lee SJ, Kim JH, Kim CH, Kim SK, Kim ES, Kim DU, Mehmood KK, Khan SU (2016) Coordinated control algorithm for distributed battery energy storage systems for mitigating voltage and frequency deviations. IEEE Trans Smart Grid 7(3):1713–1722. https://doi.org/10.1109/TSG.2015.2429919
25. Arulampalam A, Barnes M, Jenkins N, Ekanayake JB (2006) Power quality and stability improvement of a wind farm using STATCOM supported with hybrid battery energy storage. IEE Proc-Gener Transm Distrib 153(6):701–710. https://doi.org/10.1049/ip-gtd:20045269
26. Le HT, Santoso S (2007) Analysis of voltage stability and optimal wind power penetration limits for a non-radial network with an energy storage system. In: IEEE General Meeting in Power Engineering Society, IEEE, pp 1–8. https://doi.org/10.1109/PES.2007.385735
27. Rabiee A, Soroudi A, Keane A (2015) Information gap decision theory based OPF with HVDC connected wind farms. IEEE Trans Power Syst 30(6):3396–3406. https://doi.org/10.1109/TPWRS.2014.2377201
28. Rabiee A, Parniani M (2013) Voltage security constrained multi-period optimal reactive power flow using benders and optimality condition decompositions. IEEE Trans Power Syst 28(2):696–708. https://doi.org/10.1109/TPWRS.2012.2211085
29. Rabiee A, Soroudi A, Keane A (2015) Risk-averse preventive voltage control of ac/dc power systems including wind power generation. IEEE Trans Sustainable Energy 6(4):1494–1505. https://doi.org/10.1109/TSTE.2015.2451511
30. Brooke A, Kendrick D, Meeraus A, Raman R, Rosenthal R (1998) GAMS: a user's guide. GAMS Development Corporation, Washington, DC
31. Brooke A, Kendrick D, Meeraus A, Raman R, Rosenthal R (1998) GAMS: the solver manuals. GAMS Development Corporation, Washington, DC
32. Bussieck MR, Vigerske S (2010) MINLP solver software. In: Wiley encyclopedia of operations research and management science. Wiley, Hoboken. https://doi.org/10.1002/9780470400531.eorms0527
33. Ameli A, Bahrami S, Khazaeli F, Haghifam MR (2014) A multiobjective particle swarm optimization for sizing and placement of DGs from DG owner's and distribution company's viewpoints. IEEE Trans Power Delivery 29(4):1831–1840. https://doi.org/10.1109/TPWRD.2014.2300845

Chapter 15
Optimal Design, Operation, and Planning of Distributed Energy Systems Through the Multi-Energy Hub Network Approach

Syed Taha Taqvi, Azadeh Maroufmashat, Michael Fowler, Ali Elkamel, and Sourena Sattari Khavas

15.1 Introduction

Fossil fuels have been the primary source of energy globally for several decades. Infrastructures have been developed, almost everywhere, with a strong dependency on these nonrenewable resources. From internal combustion engines (ICE) for transport to centralized conventional power plants for electricity, these depleting means have played a vital role in meeting the increasing universal energy demand, yet, at a weighty economic, social and environmental cost [1]. Thus, concerns pertaining to the depletion of these valuable resources and adverse effects onto the environment need to be addressed. Renewable energy, on the other hand, has received a lot of attention in the recent years as promising "clean" alternatives. Utilization of these "green" resources has been studied using Distributed Energy Systems (DES) with focus on their economic and environmental impact [2–9].

Distributed Energy Systems embrace several advantages over conventional centralized power plants. In contrast to the large capacity central units, these smaller units are closer to their consumers. This prevents the 20% energy loss due to transmission of electricity from remote locations to end users as well as the need for continuous reliability improvement measures [10]. In general, DES are relatively easier to scale, flexible, and adaptable to various renewable and nonrenewable energy sources [11]. Its application continues to grow with rising interest in renewable energy generation as global focus orients towards the development of "smart energy systems" [12]. DES can be utilized to address future energy networks

S. T. Taqvi · A. Maroufmashat (✉) · M. Fowler · A. Elkamel
Department of Chemical Engineering, University of Waterloo, Waterloo, ON, Canada

S. S. Khavas
Department of Energy Engineering, Sharif University of Technology, Tehran, Iran

B. Mohammadi-Ivatloo, F. Jabari (eds.), *Operation, Planning, and Analysis of Energy Storage Systems in Smart Energy Hubs*,
https://doi.org/10.1007/978-3-319-75097-2_15

in a more holistic manner, accounting energy demand in almost all forms (i.e., electrical, thermal, material) [6, 7]. Due to their flexibility and their ease to network, a wide range of energy vectors can be integrated within the system with multiple nodes in order to make energy cheaper and cleaner. Its spectrum of compatible storage technologies (i.e., thermal, batteries, and compressed air) makes it a suitable candidate [12]. In addition, with progressive control technologies, operational costs have been observed to be lower when compared to centralized power plants [13, 14]. Despite these benefits, integrating DES to conventional energy infrastructure is a difficult task. The energy hub approach, conceptualized by Geidl et al. [15], has been identified as a significant methodology in designing such systems. Multi-objective problems can be formulated, using this approach, for the operation and planning of DES to reduce costs and GHG emissions.

15.2 Literature Review on Distributed Energy Systems Based on Energy Hub

Several studies have been carried out on Distributed Energy Systems, covering its various aspects in applications. Certain studies examine fossil-based DES whilst others study the integration of renewables to preexisting systems. Rieder et al. [16] developed a model for a small-scale DES that utilized hydrogen and natural gas as "clean" energy vectors. Another group of researchers built a model for combined heat and power (CHP) systems to meet the requirements for an urban settlement [17]. This group further investigated the operation of a DES with storage technology leading to reduced operational costs and emissions [18]. Yang [19] illustrated the significance of forming a hydrogen economy by considering compressed hydrogen as an energy vector in addition to electricity. A couple of comprehensive reviews, conducted by Chicco and Mancarella [8, 20], outlined the need for distributed energy systems whilst emphasizing on the importance of a multi-generation framework. The energy hub approach was classified as one of the notable strategies to model such a framework [8, 20].

Energy hubs, as the name suggests, are the central cores of an energy network. It is the place where the energy from different carriers comes together with the ability of being converted and stored [21]. They can also be perceived as interfaces between different energy generation and loads [10]. For many years, electricity and natural gas networks have been operating independently for separate applications. However, recent studies have outlined benefits of integrating renewable energy resources to these systems to achieve economic and environmental gains [22–24]. Thus, researchers have been actively seeking solutions to the energy crisis and global warming problems through these means in the residential, industrial, and transport sectors. Some have modeled the entire energy system as a single energy hub, whilst others have observed each entity as individual energy hubs, linked together in a network.

Bozchalui et al. [25] studied the operation of energy hubs in a residential building, whereas Syed et al. [26] designed a simulation model for a fleet of fuel cell vehicles and a commercial building. Based on the energy hub approach, Fabrizio et al. [27] developed a model to minimize renewable energy dependency and costs at the design concept stage of a building. Sheikhi et al. [28] presented a cost-benefit analysis on a hotel building by modeling combined cooling, heating, and power (CCHP) system. Moghaddam et al. [29] developed a Mixed Integer Nonlinear Programming (MINLP) model to self-schedule an energy hub over a short term (i.e., 24 h) to provide heating, cooling, and electricity demand of a hypothetical building, using electricity and natural gas vectors. Prices of these resources were used in order to make decisions in the scheduling process. On the other hand, Parisio et al. [30] employed the robust optimization technique to solve a Mixed Integer Linear Programming (MILP) problem, pertaining to energy hub scheduling. Hydrogen and electricity were the two energy vectors considered in this study. Sharif et al. [31] developed an energy hub model to replace a coal-fired power plant with a natural gas one, considering natural gas and renewable energy vectors (i.e., solar and wind). In another study, Maniyali et al. [32] presented an energy hub comprising of nuclear, wind, solar, biomass renewable energy vectors, as well as electrolyzers and fuel cells. The study aimed at fulfilling the electricity and hydrogen demands by industrial and transportation sectors, mainly through renewable energy.

Among the various papers on the network of energy hubs, the paper by Schulze et al. [33] is the one that considers different renewables to optimize power flow through a network. Yet, the case study presented in the paper lacks application on the network level. Maroufmashat et al. [10] developed a generic framework to help exhibit design and implementation of DES in urban areas. In another study, Maroufmashat et al. presented a network of energy hubs, based on a MILP formulation, to support hydrogen economy in such a setting [7]. This urban setting comprised of a network of 4 energy hubs: a school, food distribution center, residential complex, and a hydrogen refueling station. Hajimiragha et al. [34] used a 3 energy-hub system to demonstrate development towards a hydrogen economy. The comprehensive energy hub, presented in this study, comprised of natural gas, electricity, heat, and hydrogen energy vectors. In addition to conversion technologies, the modeled energy hub facilitated heat and hydrogen storage.

15.3 Description of Framework Development

This section aims at describing a generic framework, helpful in designing, operation, and planning of DES, using the multi-energy hub network approach. Three different case studies will be presented, in the following sections, with further details relevant to each. Figure 15.1 illustrates the route map to framework development with key elements.

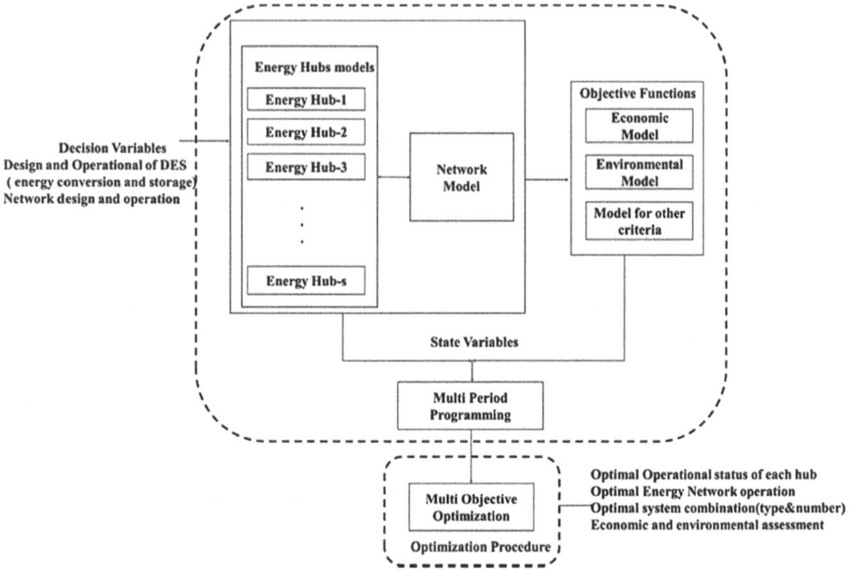

Fig. 15.1 Route map to framework development (adopted from Maroufmashat et al. [10])

As seen in Fig. 15.1, the framework comprises of two sub-models: (1) an energy hub model, and (2) a network model. A number of energy hubs can be considered within a network that encompass conversion and storage technologies. These can run for multiple periods with a single or multiple objective functions, based on technical, economic, and/or environmental criteria. A set of decision variables will be the basis used to define all energy hub and network relations. Finding an optimal solution to the overall problem will yield in the best design and operation of each energy hub and the entire network.

Figure 15.2 shows the superstructure of the generic energy hub model and the development at the optimal design and operational stages. The superstructure comprises of all possible units of DES which may exist within an energy hub. A certain set of technologies may be selected that may be seen as viable options. At the operation stage, some of those chosen technologies may or may not be utilized, at a particular period, depending on the objective function. The model will aim to yield the overall optimal value for the entire process.

The model, as stated earlier, is presented in the generic state. A number of renewable and nonrenewable energy vectors may be considered for each hub along with a number of storage technologies. A set of data may be required by the model concerning the topology of each energy hub (i.e., energy demand, available supply, and location), technical limitations (i.e., performance), economic (i.e., tariffs, capital, operation, and maintenance), and environmental factors (i.e., GHG emission factors). Based on this information, the model would be able to define the optimal design and operation of the energy hubs. It would state the type and number

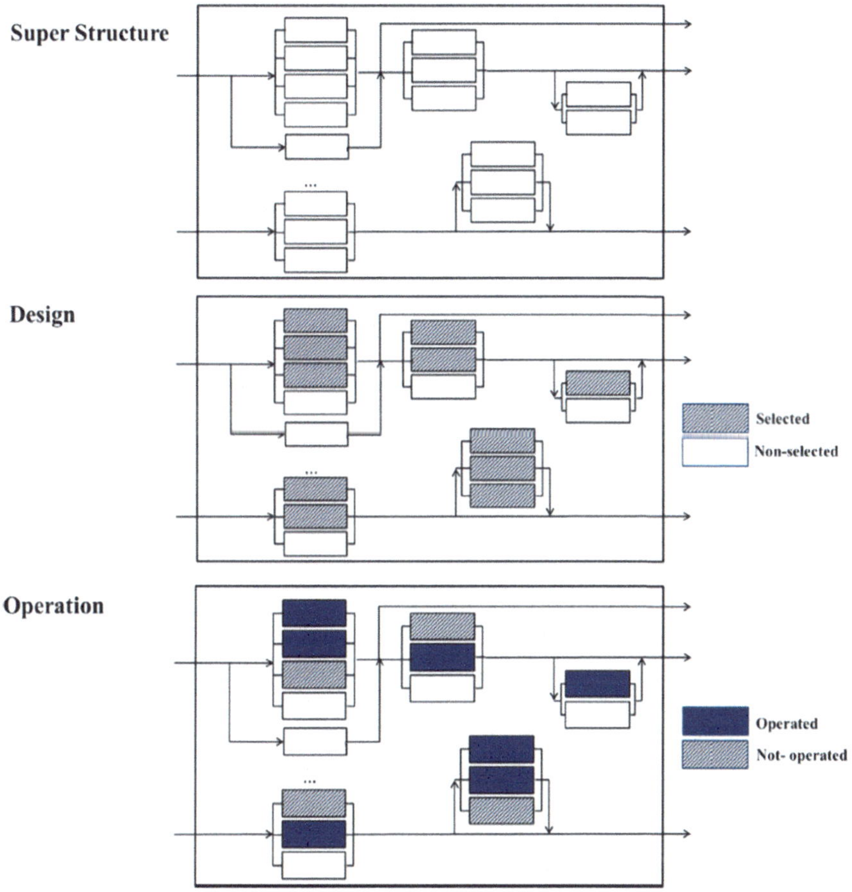

Fig. 15.2 Schematic depicting the superstructure of the generic energy hub model (adopted from Maroufmashat et al. [10])

of conversion and storage technologies needed for optimal design and operation. The model would be able to inform about the energies utilized during the process. Moreover, it would be able to calculate the economic and environmental costs in monetary and emission terms, respectively.

15.4 Modeling

Energy hubs, in addition to optimal multi-energy carrier systems, have also been identified as interfaces between different energy generation and loads, as depicted in Fig. 15.3 [15, 35]. The unit commonly comprises of three types of elements:

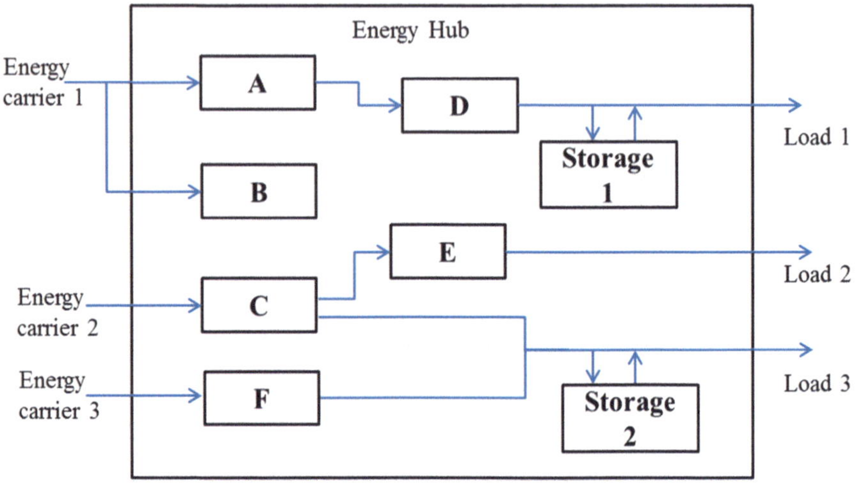

Fig. 15.3 Illustration of a simple energy hub (adopted from [6])

direct connections, converters, and storage. The connections include the different energy carriers (i.e., electricity grid, natural gas, etc.) that enter the system as well as the outputs to the consumer. Within the energy hub, there exist a set of conversion technologies to condition into the desired form. Additionally, energy storage systems can be considered in the hub for scheduled dispatch.

Among the various pros of this methodology, added reliability, load flexibility, and enhanced performance of the system are some of the notable ones [36]. Using the energy hub approach, a wide spectrum of energy-related problems can be addressed within the residential, commercial, and industrial areas [6]. As illustrated in Fig. 15.3, energy from carrier 1 is split between conversion technologies A and B. In contrast, energy from carrier 2 is split into two further energy vectors after passing through conversion technology C. D and E may represent other required for further conversion. For example, in the case where C may be a co-generation system, E may represent a chiller cascaded with to meet the demand of Load 2.

15.4.1 Generic Framework

One of the main aims of the VoFEN research project was to develop a generic modeling and analysis framework where the economical, ecological, and technical effects concerning energy systems could be studied [37]. This generic structure would allow high flexibility in modeling without posing any constraint on the size of the system. Hence, to model the energy conversion by each technology, as described in the previous section, Geidl et al. [35] proposed to use a coupling matrix C that would transform the input energy to the required energy vectors. Maroufmashat et al. [6] modified this formulation as shown in the following equation. Equation (15.1) shows a mathematical expression used to define the overall energy mapping process.

$$\overbrace{\begin{bmatrix} L_1 \\ L_2 \\ \cdot \\ L_i \\ \cdot \\ L_I \end{bmatrix}}^{L(t)} = \overbrace{\begin{bmatrix} C_{11} & C_{12} & .. & C_{1j} \\ C_{21} & C_{22} & .. & C_{2j} \\ \cdot & \cdot & \cdot & \cdot \\ \cdot & C_{ij} & \cdot & \cdot \\ \cdot & \cdot & \cdot & \cdot \\ C_{I1} & & & C_{IJ} \end{bmatrix}_{I\times J}}^{C(t)} .\mathrm{I}_{J\times J}.\mathrm{b}. \overbrace{\begin{bmatrix} P_1 \\ P_2 \\ \cdot \\ P_j \\ \cdot \\ P_J \end{bmatrix}_{J\times 1}}^{P(t)} \quad (15.1)$$

L and P, in the above equation, denote the load demand i and the input energy by the carrier j, respectively. b is a vector that converts the units of energy from the input to power, being consistent with that of the load. $I_{J\times J}$ is added to the equation which allows uniformity for matrix multiplication. The entities of the coupling matrix C represent the efficiency with which energy is converted. If a particular entity within the coupling matrix is zero, it depicts that no conversion of energy is taking place. If a single conversion technology is utilized, the efficiency of that conversion process is considered as the coupling factor. Additionally, if the load demand is met using one or more energy conversion technologies, the product of the efficiencies of each process is considered as the coupling factor. On the other hand, the input energy carriers may possess certain operational limits based on their capacity. Thus, their power needs to be constrained by lower and upper boundaries (i.e., min/max), as expressed by Eq. (15.2).

$$P^{\min} \leq P(t) \leq P^{\max} \qquad \forall t \quad (15.2)$$

In all, this simple model can be either utilized under steady state conditions or further developed to tackle dynamic systems with control strategies while including energy storage and losses. Moreover, unidirectional as well as bidirectional flow of power can be considered based on energy hub configuration [38]. For example, an electrical transformer would be able to realize reverse power flow whilst a turbine may not [36]. Based on this generic structure, the model opens a wide range of possibilities for optimization [33, 38, 39]. Stochastic models can be collated alongside for planning and operation of energy sources [8, 40, 41]. In addition, interactions between the energy carriers can be studied to assess reliability and performance [42, 43].

15.4.2 Energy Storage Modeling

Energy storage is one of the key elements of the energy hub considered by Geidl et al. [15, 36, 37]. More than half of the publications, adhering to multi-energy systems, have incorporated energy storage within their models, as evident from the classification. It is essential to account time dependency when energy storage is considered as energy accumulates over a certain period of time. Hence, the conversion technologies are perceived as discrete temporal systems [6].

$$\dot{M}_q = \alpha_q^{ch} Q_q^{ch} - \frac{1}{\alpha_q^{dis}} Q_q^{dis} \tag{15.3}$$

Equation (15.3) shows energy balance on the storage technology, accounting for energy entering the storage (i.e., charging) and leaving the battery (i.e., discharging). Q_q^{ch} represents the power in-flow through the storage technology q at an efficiency α_q^{ch} whilst Q_q^{dis} represents the power flowing out of it at an efficiency of α_q^{dis}.

$$\dot{M}_q = M_q(t) - M_q\,(t-1) + M_q^{stdby} \tag{15.4}$$

$M_q(t)$ and $M_q(t-1)$ represent the energy stored time periods (t) and $(t-1)$, respectively. In order to account for losses, the M_q^{stdby} term is added to the expression to express energy loss when the storage system is in its standby state. By compiling Eqs. (15.3 and 15.4), the overall equation for the qth storage device at time period (t) can be written as illustrated in Eq. (15.5).

As mentioned earlier, dynamic modeling is required when considering storage systems. Thus, the storage function needs to be discretized into separate time periods. This has been done using the forward difference formula, as seen in Eq. (15.4).

$$M_q(t) = M_q\,(t-1) + \alpha_q^{ch} Q_q^{ch}(t) - \frac{1}{\alpha_q^{dis}} Q_q^{dis}(t) - M_q^{stdby} \quad \forall q, \forall t \tag{15.5}$$

In matrix representation, Eq. (15.5) may be expressed as Eq. (15.6).

$$M(t) = M\,(t-1) + A^{ch} Q^{ch}(t) - A^{dis} Q^{dis}(t) - M^{stdby} \qquad \forall t \tag{15.6}$$

As written, A^{ch} and A^{dis}, in Eq. (15.6), are diagonal matrices representing charging and discharging efficiencies to allow matrix multiplication. In addition to the above model equations, technical constraints need to be structured to define the inability of the storage technology. For instance, simultaneous charging and discharging of a storage system is not possible. Hence, Eq. (15.7) comprises of two binary variables $\delta_q^{dis}(t)$ and $\delta_q^{ch}(t)$ that are introduced for each storing technology at each time period t to define the situation.

$$\delta_q^{dis}(t) + \delta_q^{ch}(t) \leq 1 \qquad \forall q, \forall t \tag{15.7}$$

Equation (15.8) shows the additional limitations on the capacity and exchange energy of each storage system.

$$
\begin{aligned}
&M_q^{\min} \leq M_q(t) \leq M_q^{\max} && \forall q, \forall t \\
&\delta_q^{\mathrm{ch}}(t) \times Q_q^{\mathrm{ch,min}} \leq Q_q^{\mathrm{ch}}(t) \leq \delta_q^{\mathrm{ch}}(t) \times Q_q^{\mathrm{ch,max}} \\
&\delta_q^{\mathrm{dis}}(t) \times Q_q^{\mathrm{dis,min}} \leq Q_q^{\mathrm{dis}}(t) \leq \delta_q^{\mathrm{dis}}(t) \times Q_q^{\mathrm{dis,max}}
\end{aligned}
\tag{15.8}
$$

$M_q^{\min}$ and $M_q^{\max}$ represent the minimum and the maximum level of energy stored in the qth storage system. Moreover, $Q_q^{\mathrm{ch,min}}$, $Q_q^{\mathrm{dis,min}}$, $Q_q^{\mathrm{ch,max}}$, and $Q_q^{\mathrm{dis,max}}$ represent the minimum and maximum energy that can flow through the qth storage technology during the energy charging and discharging process.

15.4.3 Network Modeling

In many cases, a single energy hub model suffices to represent the entire energy system. Yet, for large-scale planning and operational problems, a network of energy hubs is considered [6, 35, 44, 45]. These energy hubs are interconnected, facilitating energy transfer between each other.

Figure 15.4 shows a network of energy hubs with the focus on energy hub s. Each energy hub within the network either receives energy from outside the network (i.e.,

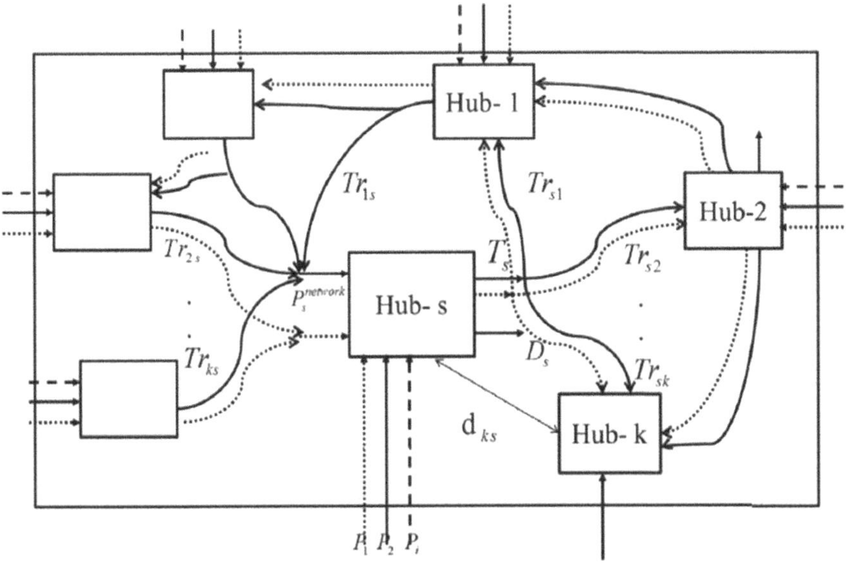

Fig. 15.4 Diagram depicting the interconnected energy hubs with energy hub s (adopted from Maroufmashat et al. [6])

grid, renewable energy sources, etc.) or from other energy hubs in the network. Likewise, each energy hub produces energy to meet energy demand within the energy hub or supply to other interconnected energy hubs. As evident in Fig. 15.4, three energy carriers have a flow of power into energy hub s. The total energy from hub s supplied to other connected energy hubs is represented by T_s. This total is the summation of individual energy output, Tr_{sk}, to each connect energy hub, k, from energy hub s. This relationship can be expressed mathematically in the following way:

$$T_s = \sum_{k \in S - \{s\}} Tr_{sk} \tag{15.9}$$

Similar to the coupling factors in the coupling matrix as well as energy storage efficiencies, a coefficient may be multiplied with Tr_{sk} to account for the losses due to transmission of energy from energy hubs s to k. All the energy vectors that exist between the interconnected energy hubs can be written in the matrix form, as shown below.

$$\begin{bmatrix} T_1 \\ T_2 \\ \bullet \\ \bullet \\ \bullet \\ T_s \\ \bullet \\ \bullet \\ Ts \end{bmatrix}_{S\times 1} = \begin{bmatrix} 0 & Tr_{12} & Tr_{13} & \cdots & Tr_{1k} \\ Tr_{21} & 0 & Tr_{23} & \cdots & Tr_{2k} \\ \bullet & \bullet & \cdots & & \\ \bullet & \bullet & \cdots & & \\ \bullet & \bullet & \cdots & & \\ Tr_{s1} & Tr_{s2} & \cdots & & Tr_{sk} \\ \bullet & & & & \\ \bullet & & & & \\ Tr & & \cdots & & 0 \end{bmatrix}_{S\times S} \bullet \begin{bmatrix} 1 \\ 1 \\ \bullet \\ \bullet \\ \bullet \\ 1 \\ \bullet \\ \bullet \\ 1 \end{bmatrix}_{S\times 1} \tag{15.10}$$

The first column vector contains the sum of all energy vectors leaving a particular energy hub (i.e., T_s). The Tr matrix contains each vector that leaves a particular energy hub s and enter energy hub k. The column vector on the right-hand side of the expression is a vector with each element equal to 1 to allow matrix multiplication.

15.5 Case Studies

This section discusses four case studies, starting with a simple predefined 2-energy hub system, considering electricity and natural gas only, and concluding with a 4-energy hub system with the addition of solar and hydrogen energy vectors. The aim of presenting these case studies is to outline a roadmap of evolution in modeling and optimization of smart energy systems, through the energy hub approach.

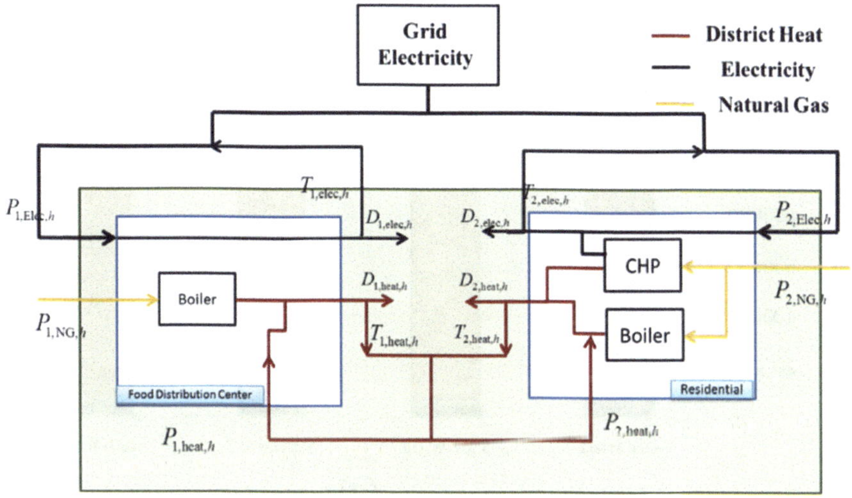

Fig. 15.5 Illustration of the two energy hubs with technologies considered and energy vectors involved (adopted from Maroufmashat et al. [6])

15.5.1 *Case Study 1*

In this study, an urban energy system was designed using two energy hubs: (1) a residential complex and (2) a light commercial building. The residential complex considered was a 10-floor building with a total area of 7765 m^2 whereas the commercial building is a 75,000 m^2 food distribution center. Figure 15.5 shows an illustration of the two buildings as energy hubs with the technologies and energy vectors considered in this study.

An hourly energy demand was considered for both energy hubs and three different time-of-use prices for electricity (i.e., on-peak, off-peak, and mid-peak) were applied. Additionally, the carbon dioxide emissions, arising from the consumption of energy from grid electricity and natural gas, were included in the model. A 1% energy loss was assumed for transmission of energy over every 200 m. In order to observe the effect of energy exchange between the two energy hubs, the following four scenarios were considered:

- No interaction between the energy hubs is assumed and no CHP technology is employed
- No interaction between the energy hubs is assumed but CHP technology is employed
- Energy hubs are allowed to interact but without the CHP technology
- Energy hubs are allowed to interact whilst employing distributed CHP technology in one of the hubs (as seen in Fig. 15.5).

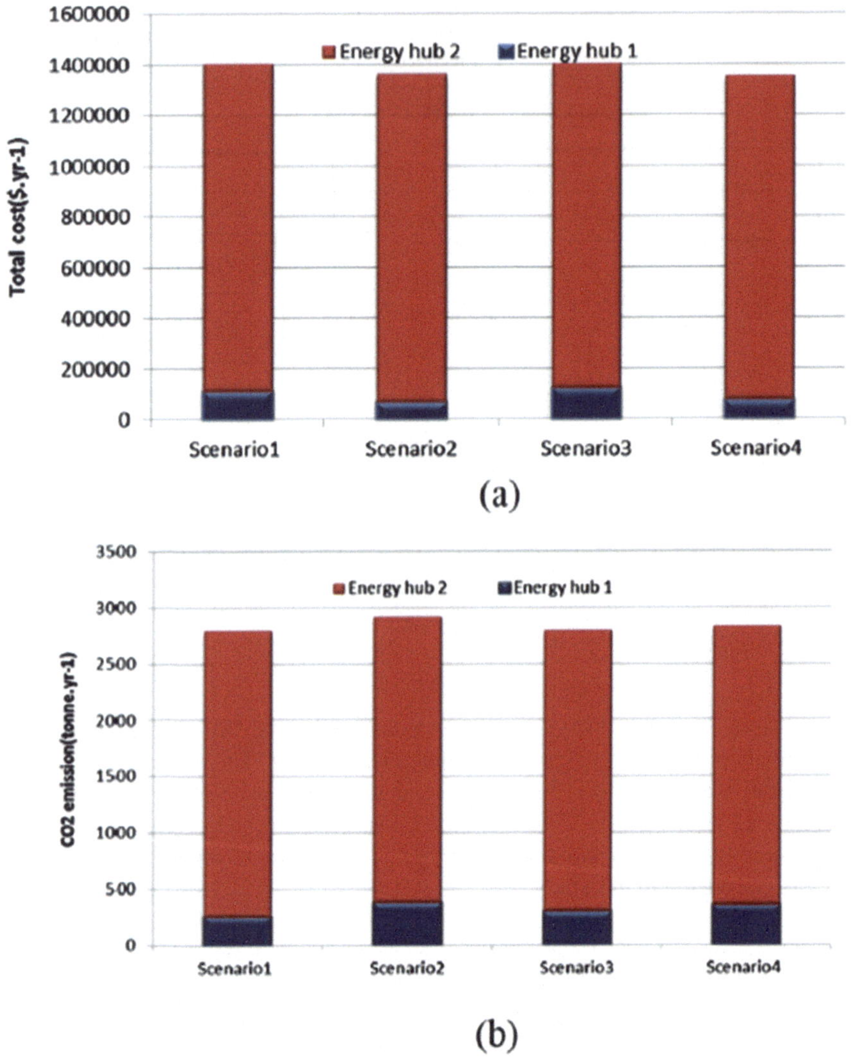

Fig. 15.6 Economic cost and carbon dioxide emissions resulting from the four scenarios for case study 1 (adopted from Maroufmashat et al. [6]). Results from the four scenarios for case study 1 depicting (**a**) economic cost and (**b**) carbon dioxide emissions, associated with each scenario

The economic cost due to purchasing of primary energy from sources incurred and the amount CO_2 emissions produced are shown in Fig. 15.6 for each scenario.

As evident from Fig. 15.6a, the least total economic cost was observed for scenario 4. Allowing interaction between the energy hubs and employing the CHP technology yielded 0.5% lower costs than scenario 2. Another interesting observation, perceived from the results of scenario 1 and 3, is that networking of energy hubs did not yield any economic benefit without considering the CHP technology. On the other hand, carbon dioxide emissions were the least for these two scenarios (i.e., 1 and 3), as shown in Fig. 15.6b. Scenario 2 resulted in the production of the most CO_2 emissions while scenario 4 showed 3% lower emissions, implying the effect of interaction of energy hubs on greenhouse gas emissions.

15.5.2 *Case Study 2*

A network of 3 energy hubs, comprising of a school, corner shop, and residential complex, were considered in this case study. The area of each of these structures was 20,000 m^2, 25,000 m^2, and 7765 m^2, respectively. The following figure represents these facilities as energy hubs, showing the distance between each of them.

Figure 15.7 also shows the energy vectors involved as well as the technologies considered within each energy hub. The assumptions made and operational scenarios observed were similar to those of case study 1.

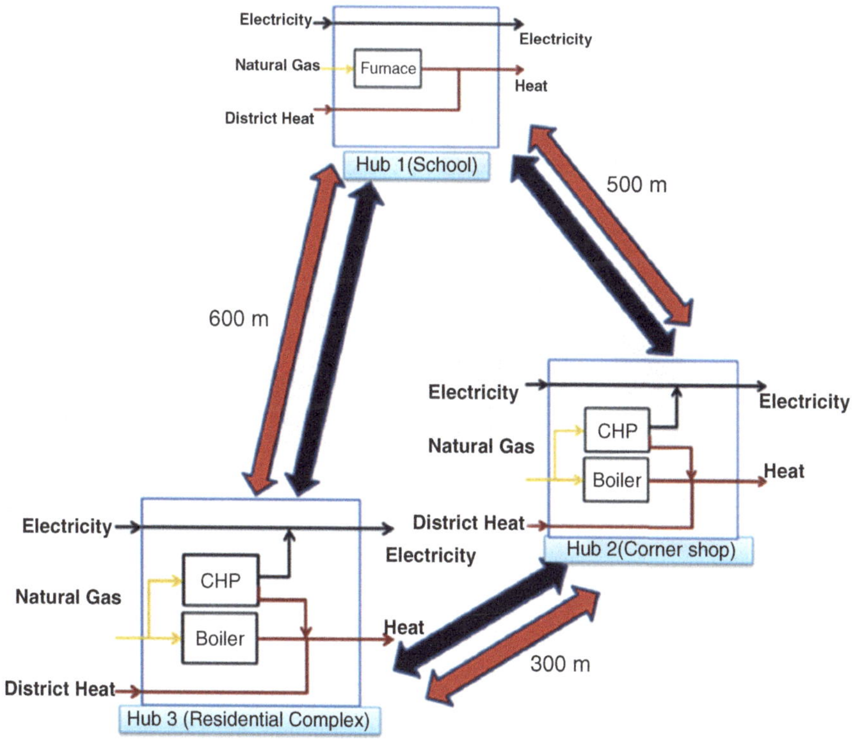

Fig. 15.7 Diagram of the three energy hubs with the distance between them (adopted from Maroufmashat et al. [6])

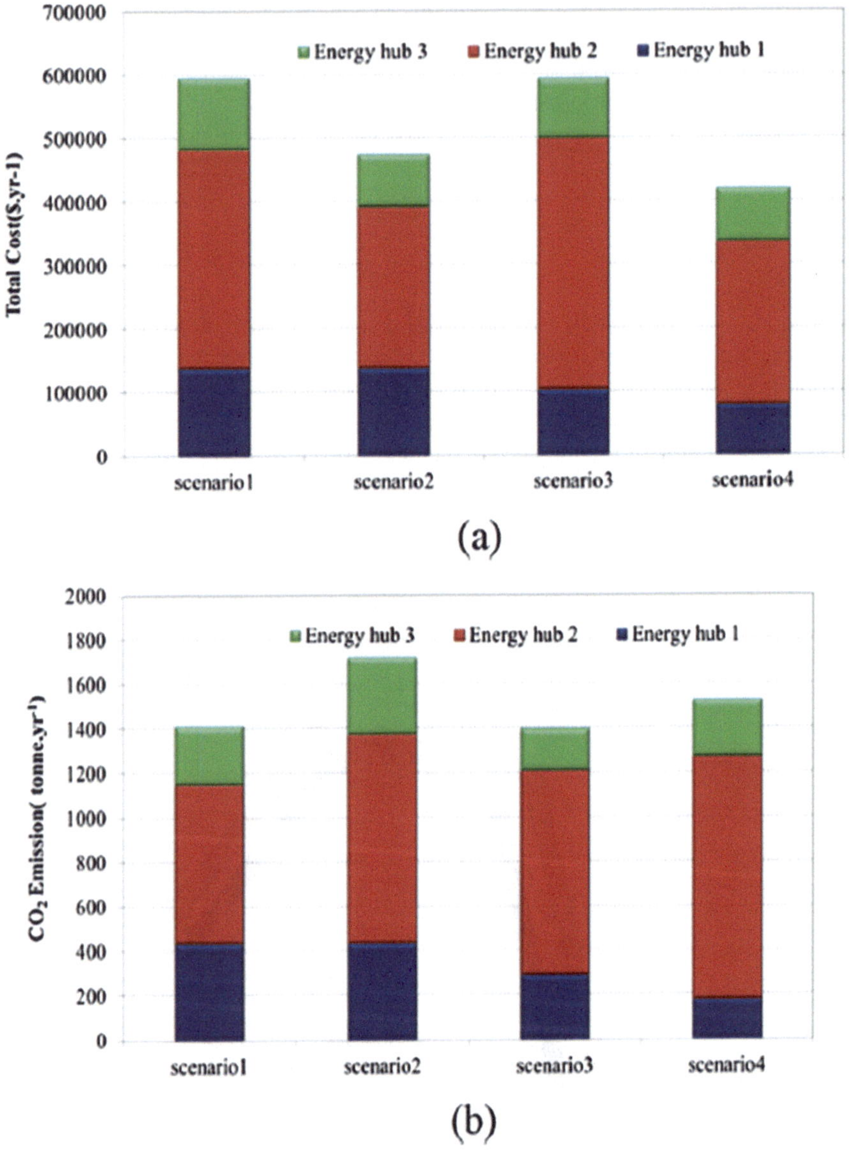

Fig. 15.8 Economic cost and carbon dioxide emissions resulting from the four scenarios for case study 1 (adopted from Maroufmashat et al. [6]). Results from the four scenarios for case study 1 depicting (**a**) economic cost and (**b**) carbon dioxide emissions, associated with each scenario

The results obtained from this analysis, as illustrated in Fig. 15.8, appear to have a similar trend in comparison to the results obtained from the previous case study. A much lower economic cost was realized in scenario 4 as compared to the other

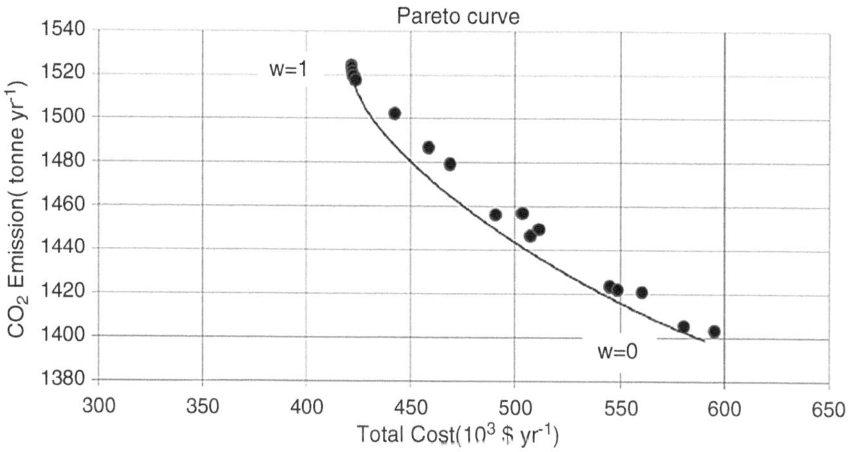

Fig. 15.9 Pareto optimality curve depicting the relationship between the economic cost and carbon dioxide emissions considering scenario 4 (adopted from Maroufmashat et al. [6])

3 scenarios, as seen in Fig. 15.8a. The economic cost in scenario 4 was 29% and 11% lower than scenarios 1 and 2, respectively. Moreover, similar to the results of case study 1, the economic costs incurred in scenarios 1 and 3 are observed to be the same. This emphasizes the need for employing coupled technologies, like CHP, in addition to networking of energy hubs, to increase economic gains. In Fig. 15.8b, the CO_2 emissions in scenario 4 appear to be 11% lower than scenario 2. Whereas, scenarios 1 and 3 appear to have the least carbon dioxide emissions among the 4 scenarios. In order to understand the relation between the economic cost and CO_2 emissions, bi-objective optimization was carried out. A Pareto front was constructed by assigning weights to these two objectives while considering scenario 4, as shown in Fig. 15.9.

It is evident from the figure above that the total economic cost and carbon dioxide emissions are inversely related. As more weight is assigned to economic cost (i.e., model solved for minimizing total cost), the amount of CO_2 emissions increases. Such an analysis can help design energy systems, using the energy hub approach, considering the availability of economic and environmental resources.

15.5.3 Case Study 3

In this case study, a network of 4 energy hubs is studied with the addition of solar energy as an energy vector to the system. The four different structures, each perceived as an energy hub, considered in this study are: (1) residential complex, (2) office building, (3) commercial building, and (4) a restaurant. The areas of each of these structures are 7765 m^2, 1000 m^2, 75,000 m^2, and 1000 m^2,

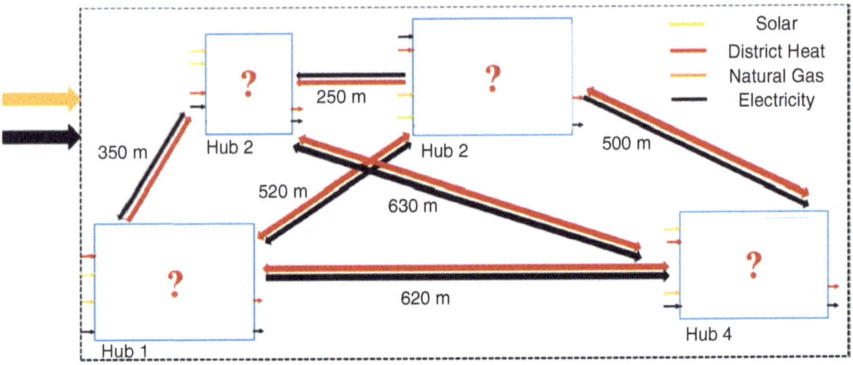

Fig. 15.10 Representation of the 4-energy hub network (adopted from Maroufmashat et al. [10])

respectively. In contrast to the previous studies, the configuration of each energy hub is not predefined, as shown in Fig. 15.10. The operational schedule of the energy conversion and storage technologies as well as networking disputes are addressed by the optimal model.

With regard to the solar energy vector, the solar irradiation data collected was from a region, located in the south of Ontario. Moreover, a nominal discount rate of 8% and a project lifetime of 20 years were assumed to conduct a thorough economic analysis of this potential distributed energy system (DES). Moreover, four scenarios, listed below, were considered to study the effect of networking of energy hubs and DES allocation within the energy hubs.

- No interaction between the energy hubs is allowed. Electricity and heat are only supplied from the utility grid (no DES). Moreover, no storage technology is considered.
- No interaction between the energy hubs is allowed. However, DESs comprising of difference energy conversion and storage technologies are allowed within the energy hubs.
- Energy exchange between energy hubs is allowed along with energy conversion technology within energy hubs. However, no storage is allowed in the energy hubs.
- Energy exchange between energy hubs as well as energy conversion and storage technologies in energy hubs are allowed.

Figure 15.11 shows how DES comprising of energy conversion and storage technologies are allocated within each energy hub, as determined by the optimal model. The values along the y-axis represent power in kW whereas the segment between each bar shows the number of technologies employed.

It is observed, from Fig. 15.11, that the highest number of technologies (i.e., conversion and storage) utilized were in scenario 4 while the least were considered in scenario 1. In scenario 1, four 300 kW and two 100 kW boilers were installed

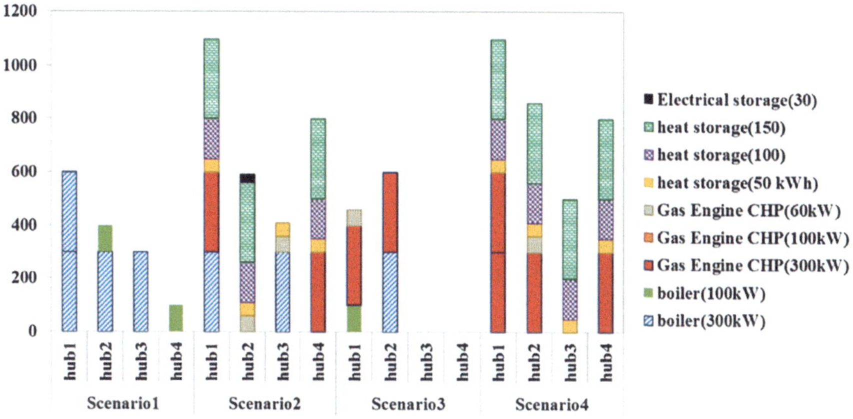

Fig. 15.11 DES allocation within each energy hub for the four scenarios (adopted from Maroufmashat et al. [10])

in order to meet the energy demand of the energy system. In scenario 4, four 300 kW gas engine CHPs were employed along with a total of 1200 kW heat storage technologies. A number of heat and electrical storage technologies are chosen in scenario 2, while in scenario 3, no technologies are chosen for hubs 3 and 4. Again, this outlines the gains achieved allowing energy exchange between energy hubs and storage. However, in all, solar energy is not selected in the optimal model due to the high capital cost and lack of incentive for power generation through renewable energy sources.

The total annual cost as well as CO_2 emission were studied in all four scenarios and the results are shown in Fig. 15.12.

Despite the least number of employed technologies, the highest annual cost and lowest carbon dioxide emissions were observed in Scenario 1, as seen from Fig. 15.12a, b, respectively. On the contrary, the least economic cost and highest carbon dioxide emissions were seen in scenario 4. The total annual cost incurred in scenario 4 was 12%, 6%, and 7% lower, as compared to scenarios 1, 2, and 3, respectively. These results imply that considering DES (i.e., energy conversion and storage) and networking of energy hubs, lower economic costs can be attained. Another significant finding is that the economic cost of scenario 3 is 3% higher than that of scenario 2. This indicates that utilizing energy storage yields in higher economic gains than networking of energy hubs.

A multi-objective analysis, similar to that carried in the previous case study, was conducted, considering scenarios 2, 3, and 4. The Pareto curve, illustrating the relationship between the annual economic cost and carbon dioxide emissions, for these scenarios are shown in Figs. 15.13, 15.14, and 15.15.

As evident from these figures above, the highest total economic cost yields in the least amount of carbon dioxide emissions. Looking at results obtained for scenario

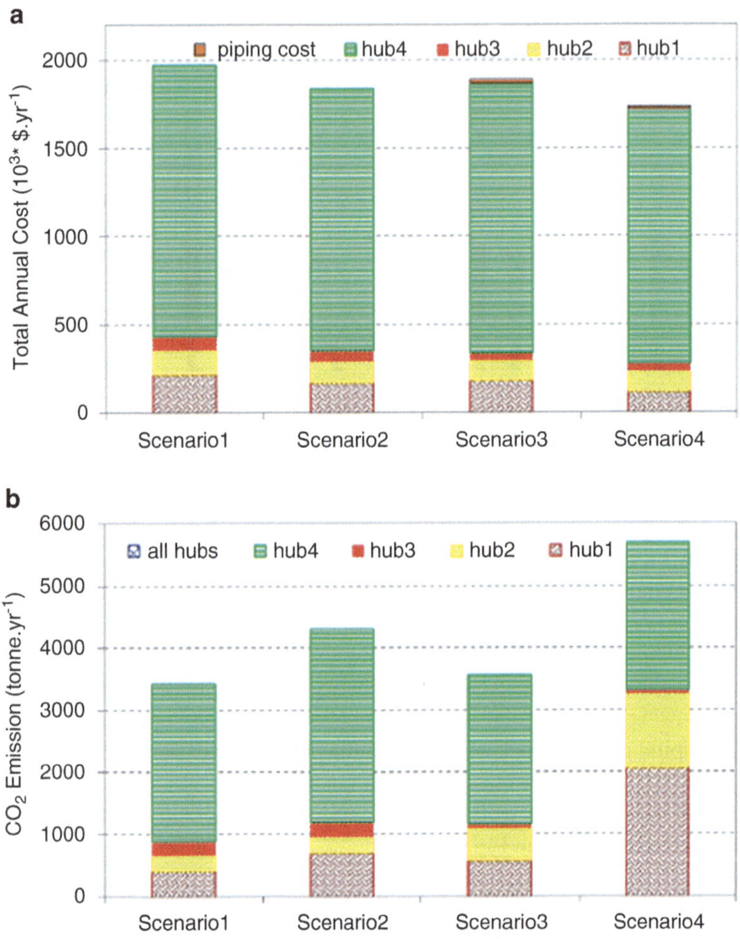

Fig. 15.12 Economic cost and carbon dioxide emissions resulting from the four scenarios for case study 3 (adopted from Maroufmashat et al. [10]). Results from the four scenarios for case study 1 depicting (**a**) economic cost and (**b**) carbon dioxide emissions, associated with each scenario

2 (Fig. 15.13), an increase of 2% in the total costs would result in 11% decrease in CO_2 emissions, in the lower part of the optimality curve (i.e., between points A and B). On the other hand, an 8% increase in cost yields a 14% decrease in carbon dioxide emissions in the upper part of the curve (i.e., between points A and B). Thus, making it more advantageous to operate within the lower region of the curve. It is also observed that the DES share in meeting energy demand is higher in the lower part of the curve than in the upper segment.

In scenario 3, it can be observed, from the Pareto optimality curve, that a 2% increase in total annual cost may lead to 2% and 5% decrease in CO_2 emissions, represented by points B and C, respectively. Additionally, the pie charts,

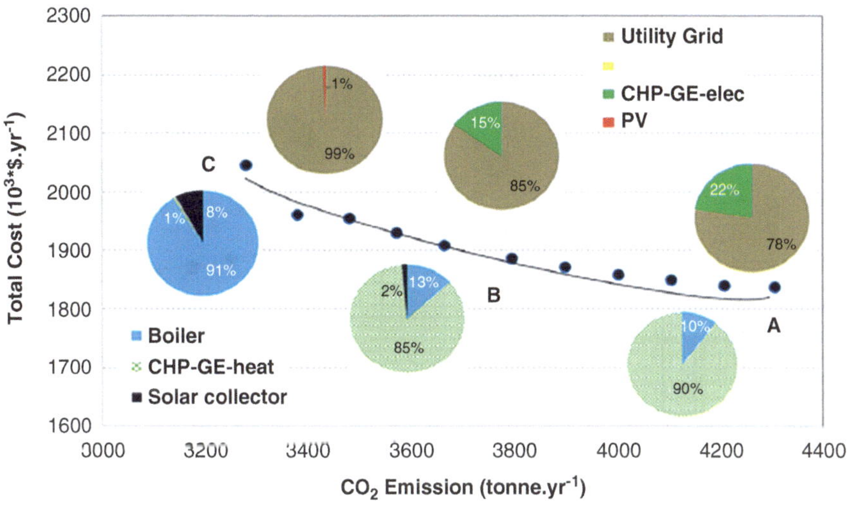

Fig. 15.13 Pareto optimality curve depicting the relationship between the economic cost and carbon dioxide emissions considering scenario 2 (adopted from Maroufmashat et al. [10])

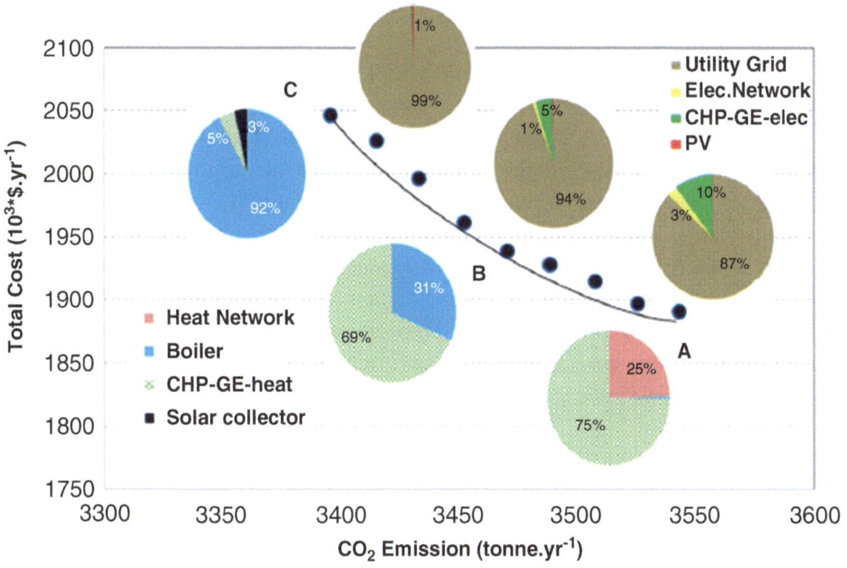

Fig. 15.14 Pareto optimality curve depicting the relationship between the economic cost and carbon dioxide emissions considering scenario 3 (adopted from Maroufmashat et al. [10])

in Fig. 15.14, depict 1% of electricity demand and 3% of heat demand is met using the solar energy technology (i.e., solar PV and collector). Whereas, the least cost is incurred when almost 99% of energy demand is met through purchasing electricity from the grid. From the comparison between scenarios 2 and 3, it can be

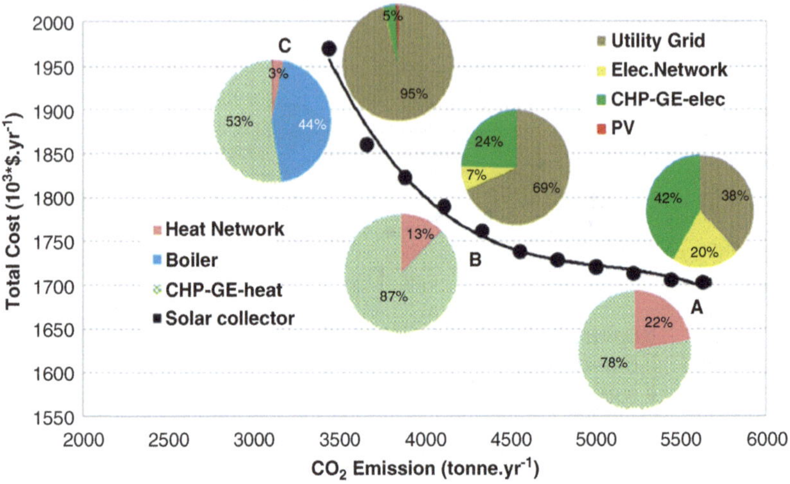

Fig. 15.15 Pareto optimality curve depicting the relationship between the economic cost and carbon dioxide emissions considering scenario 4 (adopted from Maroufmashat et al. [10])

explicitly seen that employing a storage technology leads to more economic gains than networking of energy hubs.

The bi-objective optimization results of allowing DES (energy conversion and storage) as well as networking of energy hubs are illustrated in Fig. 15.15. A 4% increase in economic costs between points A and B leads to a 23% reduction in CO_2 emissions, making it the most favorable out of the three multi-objective optimization analyses. Furthermore, in the upper segment (i.e., between B and C), a 11% increase in total economic cost leads to a 20% decrease in CO_2 emissions. A higher renewable energy share in meeting energy demand through solar PV and collectors is observed in this scenario as compared to scenario 2 and 3. In all, all these studies showed utilization of DES and energy hub network resulted in a decrease in economic costs but a higher environmental impact as compared to grid-based electricity.

15.5.4 *Case Study 4*

In this last case study, a network of 4 energy hubs was studied, with the additional consideration of solar and hydrogen energy vectors. The aim of this study is to investigate how to optimally design a hydrogen refueling station in an urban area where energy hubs can exchange their surplus energy with one another. The energy hubs considered are: (1) a school, (2) food distribution center, (3) a residential complex, and (4) hydrogen refueling station for fuel cell vehicles. The configurations of the first three energy hubs were defined as follows:

- School (*Energy Hub 1*): 530 kW boiler and a 50 m^2 of solar PV area

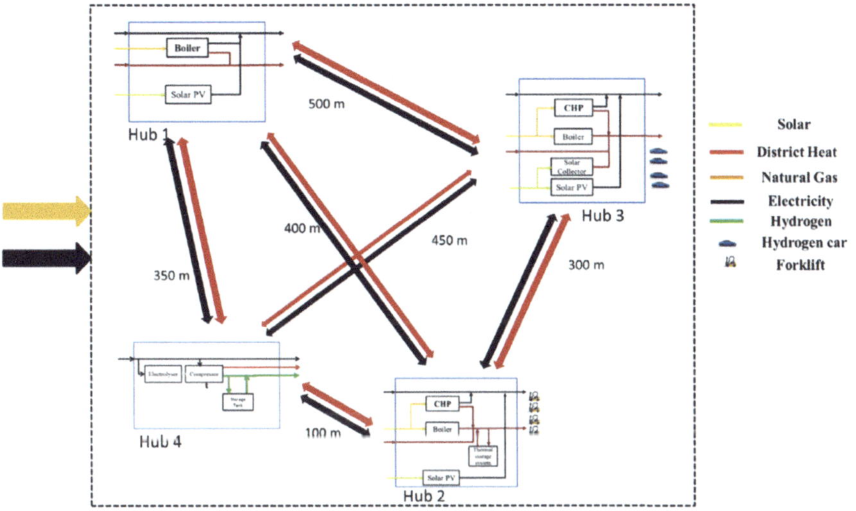

Fig. 15.16 Representation of the 4-energy hub network comprising of school, food distribution center, residential complex, and hydrogen refueling station (adopted from Maroufmashat et al. [7])

- Food distribution center (*Energy Hub 2*): 300 kW CHP, 147 kW boiler, heat storage tank, and a 100 m^2 solar PV area
- Residential complex (*Energy Hub 3*): 100 kW CHP, 300 kW boiler, an 80 m^2 solar thermal collector area, and an 80 m^2 solar PV area.

Energy hub 4 in Fig. 15.16 represents a hydrogen refueling station. The configuration of this hub was left to be determined optimally by the model. This includes the size of the electrolyzer and the hydrogen storage tank.

For this particular case study, the operating characteristics of a HySTAT-60 alkaline electrolyzer were considered. Moreover, in a course of a day, about 100 forklifts and 50 light duty fuel cell vehicles were considered to be fueled. Other data and assumptions pertaining to time-of-use of electricity and solar irradiation were similar to previous studies.

Similar to the previous case, four different scenarios were considered in this case study.

1. Distributed hydrogen production was allowed as well as energy exchange between energy hubs.
2. Distributed hydrogen production was allowed but no energy exchange between energy hubs was allowed.
3. Hydrogen was purchased (i.e., no distributed H_2 production) but energy exchange between energy hubs was allowed.
4. Hydrogen was purchased (i.e., no distributed H_2 production) but no energy exchange between energy hubs was allowed.

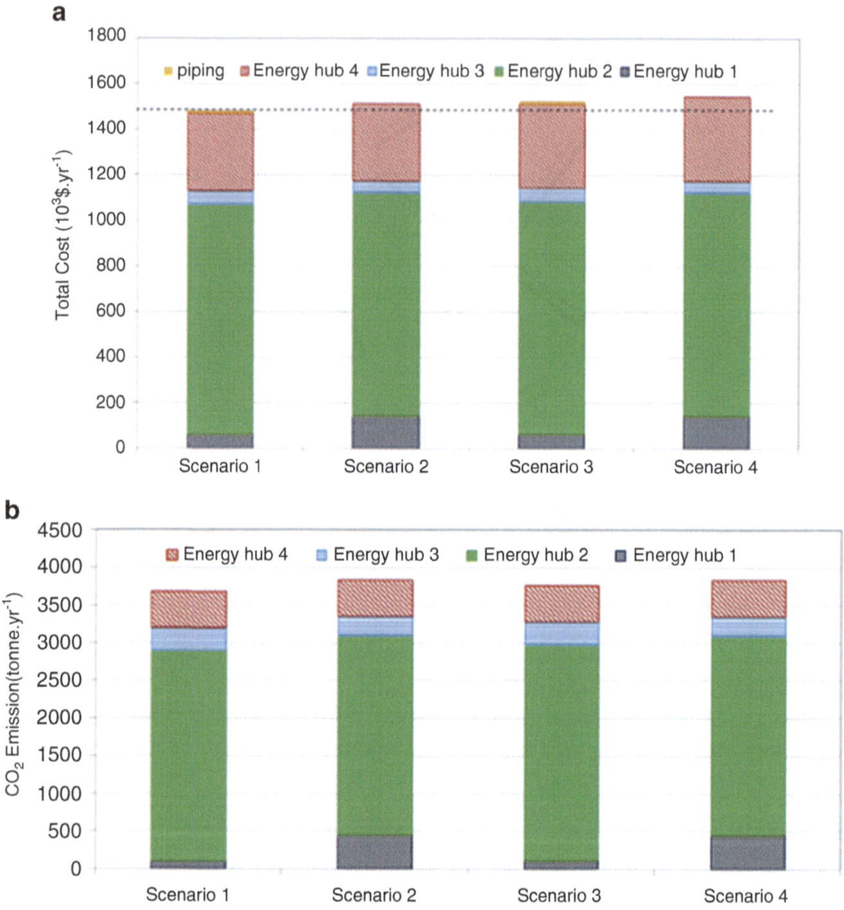

Fig. 15.17 Economic cost and carbon dioxide emissions resulting from the four scenarios for case study 2 (adopted from Maroufmashat et al [7]). Results from the four scenarios for case study 1 depicting (**a**) economic cost and (**b**) carbon dioxide emissions, associated with each scenario

Figure 15.17 shows the total economic costs and carbon dioxide emissions produced in each of the four scenarios.

It can be seen, from the figure above, that the least economic cost is incurred and the least carbon dioxide emissions are produced in scenario 1. Scenario 1 has 1.8% lower economic costs than scenario 2, implying that networking of energy hubs leads to economic gains. Additionally, it is evident that it is cheaper to produce hydrogen through this distributed setting (i.e., Scenario 1 and 2), within this smart energy network, in contrast to purchasing it (Scenario 3 and 4). When comparing scenario 1 to scenario 3, a 2% CO_2 emission reduction is also observed in the former scenario, indicating that distributed hydrogen can be relatively environmentally advantageous.

15.6 Conclusion

In this chapter, a brief introduction to the concept of optimal design of Distributed Energy Systems (DES) was presented. Relevant studies, focused on the modeling of DES through the energy hub approach, were outlined. A generic framework, based on the energy hub approach, was defined that can aid in the design, operation, and planning stages of Distributed Energy Systems. Four case studies were illustrated, starting from a simple 2-energy hub system to a more complex 4-energy hub DES with solar and hydrogen energy vectors. Findings from each study outlined that higher economic gains can be achieved when allowing energy hubs to interact with each as well as using distributed energy systems. Compared to scenarios with no networking between energy hubs, lower carbon dioxide emissions were observed in interacting energy hubs. However, in all case studies, purchasing of electricity from the grid was seen to result in the lowest CO_2 emissions. Since the emission factor of electricity grid is quite low in Ontario, employment of DES results in higher CO_2 emissions. However, in fossil-based economies, a lower carbon footprint can be expected when using DES and energy hub networks. This opens up an entire new area for application of DES, through multi-energy hub approach, in oil and gas based economies.

Nomenclature

α_q^{ch}	Charging efficiency of storage technology q
α_q^{dis}	Discharging efficiency of storage technology q
$L(t)$	Load demand/output energy i at time t
$\dot{M}_q$	Energy level in storage technology q at time t
M_q^{stdby}	Energy loss when the storage system q is in its standby state.
$P(t)$	Input energy by carrier j at time t
$P^{\min}$	Minimum input energy by carrier j
$P^{\max}$	Maximum input energy by carrier j
Q_q^{ch}	Power in-flow (charging) through the storage technology q
Q_q^{dis}	Power out-flow (discharging) through the storage technology q
T_s	Total energy from hub s supplied to other connected energy hubs
Tr_{sk}	Individual energy output to each connect energy hub, k, from energy hub s

References

1. Akella AK, Saini RP, Sharma MP (2009) Social, economical and environmental impacts of renewable energy systems. Renew Energy 34(2):390–396
2. Phdungsilp A, Martinac I (2016) Distributed energy resources systems towards carbon- neutral urban development: a review and application. Int J Energy Environ 7(1):77–88

3. Morvaj B, Evins R, Carmeliet J (2016) Optimising urban energy systems: simultaneous system sizing, operation and district heating network layout. Energy 116:619–636
4. Morvaj B, Evins R, Carmeliet J (2017) Decarbonizing the electricity grid: the impact on urban energy systems, distribution grids and district heating potential. Appl Energy 191:125–140
5. Maroufmashat A, Khavas SS, Bakhteeyar H (2014) Design and operation of a multicarrier energy system based on multi objective optimization approach. Int J Energy Power Eng 8(9):590–595
6. Maroufmashat A et al (2015) Modeling and optimization of a network of energy hubs to improve economic and emission considerations. Energy 93:2546–2558
7. Maroufmashat A, Fowler M, Khavas SS, Elkamel A, Roshandel R, Hajimiragha A (2015) Mixed integer linear programing based approach for optimal planning and operation of a smart urban energy networkto support the hydrogen economy. Int J Hydrog Energy 41(19):7700–7716
8. Chicco G, Mancarella P (2009) Distributed multi-generation: a comprehensive view. Renew Sust Energ Rev 13(3):535–551
9. Maroufmashat A, Elkamel A, Khavas SS, Fowler M, Roshandel R, Elsholkami M (2015) Development of the energy hub networks based on distributed energy technologies. In: Summer computer simulation conference, pp 1–8
10. Maroufmashat A, Sattari S, Roshandel R, Fowler M, Elkamel A (2016) Multi-objective optimization for design and operation of distributed energy systems through the multi-energy hub network approach. Ind Eng Chem Res 55(33):8950–8966
11. Alanne K, Saari A (2006) Distributed energy generation and sustainable development. Renew Sustain Energy Rev 10(6):539–558
12. Akorede MF, Hizam H, Pouresmaeil E (2010) Distributed energy resources and benefits to the environment. Renew Sustain Energy Rev 14(2):724–734
13. Mallikarjun S, Lewis HF (2014) Energy technology allocation for distributed energy resources: a strategic technology-policy framework. Energy 72:783–799
14. Pudjianto D, Pudjianto D, Ramsay C, Ramsay C, Strbac G, Strbac G (2007) Virtual power plant and system integration of distributed energy resources. IET Renew Power Gener 1(1):10–16
15. Geidl M, Koeppel G, Favre-Perrod P, Klöckl B, Andersson G, Fröhlich K (2007) The energy hub – a powerful concept for future energy systems. In: Third annual carnegie mellon conference on the electricity industry, pp 1–10
16. Rieder A, Christidis A, Tsatsaronis G (2014) Multi criteria dynamic design optimization of a small scale distributed energy system. Energy 74(C):230–239
17. Bracco S, Dentici G, Siri S (2013) Economic and environmental optimization model for the design andthe operation of a combined heat and power distributed generation system in an urban area. Energy 55:1014–1024
18. Bracco S, Delfino F, Pampararo F, Robba M, Rossi M (2014) A mathematical model for the optimal operation of the University of Genoa Smart Polygeneration Microgrid: evaluation of technical, economic and environmental performance indicators. Energy 64:912–922
19. Yang C (2008) Hydrogen and electricity: parallels, interactions, and convergence. Int J Hydrog Energy 33(8):1977–1994
20. Mancarella P (2014) MES (multi-energy systems): an overview of concepts and evaluation models. Energy 65:1–17
21. Mohammadi M, Noorollahi Y, Mohammadi-ivatloo B, Yousefi H (2017) Energy hub: from a model to a concept – a review. Renew Sustain Energy Rev 80:1512–1527
22. Zahedi A (2011) A review of drivers, benefits, and challenges in integrating renewable energy sources into electricity grid. Renew Sustain Energy Rev 15(9):4775–4779
23. Orehounig K, Mavromatidis G, Evins R, Dorer V, Carmeliet J (2014) Towards an energy sustainable community: an energy system analysis for a village in Switzerland. Energy Build 84:277–286
24. Liserre M, Sauter T, Hung JY (2010) Future energy systems: Inegrating renewable energy into the smart power grid through industrial electronics. IEEE Ind Electron Mag 4:18–37

25. Bozchalui MC, Hashmi SA, Hassen H, Cañizares CA, Bhattacharya K (2012) Optimal operation of residential energy hubs in smart grids. IEEE Trans Smart Grid 3(4):1755–1766
26. Syed F, Fowler M, Wan D, Maniyali Y (2010) An energy demand model for a fleet of plug-in fuel cell vehicles and commercial building interfaced with a clean energy hub. Int J Hydrog Energy 35(10):5154–5163
27. Fabrizio E, Corrado V, Filippi M (2010) A model to design and optimize multi-energy systems in buildings at the design concept stage. Renew Energy 35(3):644–655
28. Sheikhi A, Ranjbar AM, Oraee H (2012) Financial analysis and optimal size and operation for a multicarrier energy system. Energy Build 48:71–78
29. Moghaddam IG, Saniei M, Mashhour E (2016) A comprehensive model for self-scheduling an energy hub to supply cooling, heating and electrical demands of a building. Energy 94:157–170
30. Parisio A, Del Vecchio C, Vaccaro A (2012) A robust optimization approach to energy hub management. Int J Electr Power Energy Syst 42(1):98–104
31. Sharif A, Almansoori A, Fowler M, Elkamel A, Alrafea K (2014) Design of an energy hub based on natural gas and renewable energy sources. Int J Energy Res 38(3):363–373
32. Maniyali Y, Almansoori A, Fowler M, Elkamel A (2013) Energy hub based on nuclear energy and hydrogen energy storage. Ind Eng Chem Res 52(22):7470
33. Schulze M, Friedrich L, Gautschi M (2008) Modeling and optimization of renewables: applying the energy hub approach. In: 2008 IEEE international conference on sustainable energy technologies, ICSET 2008, pp 83–88
34. Hajimiragha H, Canizares C, Fowler M, Geidl M, Andersson G (2007) Optimal energy flow of integrated energy systems with hydrogen economy considerations. In: Bulk power system dynamics and control – VII, pp 1–11
35. Geidl M, Andersson G (2007) Optimal power flow of multiple energy carriers. IEEE Trans Power Syst 22(1):145–155
36. Geidl M, Andersson G (2005) A modeling and optimization approach for multiple energy carrier power flow. In: 2005 IEEE Russia PowerTech, pp 1–7
37. Geidl M, Koeppel G, Favre-Perrod P, Klockl B, Andersson G, Frohlich K (2007) Energy hubs for the future. IEEE Power Energy Mag 5(1):24–30
38. Hemmes K, Zachariah-wolff JL, Geidl M, Andersson G (2007) Towards multi-source multi-product energy systems. Int J Hydrog Energy 32:1332–1338
39. Adamek F (2008) Optimal multi energy supply for regions with increasing use of renewable resources. In: 2008 IEEE energy 2030 conference, energy 2008, pp 1–6
40. Del Real AJ, Arce A, Bordons C (2010) Solar-hydrogen residential power system considering seasonal weather variations. IFAC Proc Vol 43(1):169–174
41. Anastasiadis AG, Argyropoulou VM, Pagonis KD, Hatziargyriou ND (2013) Losses in a LV-microgrid with the presence of reactive power and CHP units. In: 8th mediterranean conference on power generation, transmission, distribution and energy conversion (MEDPOWER 2012), p 34
42. Schulze M, Del Granado P (2010) Optimization modeling in energy storage applied to a multi-carrier system. In: 2010 IEEE power and energy society general meeting, pp 1–7
43. Robertson E, Galloway S, Ault G (2012) The impact of wide spread adoption of high levels of distributed generation in domestic properties. In: IEEE power and energy society general meeting, pp 1–8
44. Galus MD, Andersson G (2009) Power system considerations of plug-in hybrid electric vehicles based on a multi energy carrier model. In: 2009 IEEE power and energy society general meeting, PES '09
45. Andrade F, Cardenas JJ, Romeral L, Cusido J (2012) Modeling and studying of power flow in a parking lot with plug-in vehicles and the impact in the public utility. In: 2012 IEEE PES innovative smart grid technologies, ISGT 2012, pp 1–7

Chapter 16
Joint Electricity and Heat Optimal Power Flow of Energy Hubs

Manijeh Alipour, Kazem Zare, and Heresh Seyedi

16.1 Introduction

Electricity is the utmost popular type of energy in recent days. All developed sectors of countries almost utilize electrical energy. In addition to the electricity, there are heating and cooling demands that should be supplied. Therefore, another transmission network, district heating network (DHN), is the other important system, which is very promising for carbon emission reduction and energy saving. District heating network is a well-extended system in several Northern European countries [1]. Conventionally, most energy service networks, i.e., electricity and local district heating systems have been scheduled separately without considering interdependency between various energy service infrastructures. Nevertheless, many welfares can be attained by envisaging the energy service system as cohesive. Energy flows provided from alternative resources could be administrated and consequently, safety of energy preparation can be improved. The energy could be supplied more efficiently and energy emissions, losses and costs would be reduced. However, in the case of separate planning and operation of energy systems an unlikely optimal solution will be reached, since optimization of each transmission network separately can obscure the optimal operation of the entire energy system. Henceforth, a unified study of energy networks is highly desirable [2, 3] and recent studies suggest integration of these networks so-called multi-carrier energy networks (MCENs) [2, 4, 5]. The important motivation behind aforesaid viewpoint is the growing exploitation of co-generation systems which creates a potent connection among various networks [2].

M. Alipour · K. Zare · H. Seyedi (✉)
Faculty of Electrical and Computer Engineering, University of Tabriz, Tabriz, Iran
e-mail: alipour@tabrizu.ac.ir; kazem.zare@tabrizu.ac.ir; hseyedi@tabrizu.ac.ir

B. Mohammadi-Ivatloo, F. Jabari (eds.), *Operation, Planning, and Analysis of Energy Storage Systems in Smart Energy Hubs*,
https://doi.org/10.1007/978-3-319-75097-2_16

The interdependency of these industries necessitates the integrated optimization of joint energy networks. For instance, a combined heat and power (CHP) unit generates electricity and heat by employing natural gas [6]. It connects the electrical network to the district heating and natural gas networks. Hence, these networks should be taken into account together as an integrated system for an optimization procedure that forms the so-called MCEN. The concept energy hub opens a new window on demonstration of a unified energy network which comprises several energy carriers such as heat, electricity, gas, etc. The concept of energy hub was first introduced in [7]. It can be specified as a mixture of energy conversion units which meets various kinds of energy demands [8]. Currently, multi-generation systems called as energy hub for combined and distributed generation of various energy sources can be established owing to advances in energy substructures. Energy hub can be envisioned as an integrated system where numerous energy carriers are stored, converted, and distributed [9, 10]. Compliance of multi-generation systems conveys noteworthy benefits in terms of improved energy efficiency, enhanced economy, and reduced CO2 emissions [11, 12]. The integration of several energy networks is investigated in some works [2, 4, 13]. These methodologies utilize energy hub model for the energy system. A heuristic optimization scheme is developed in [14] for multi-carrier systems.

Nowadays, the hubs are subject to more instable electricity prices as a result of liberalized electricity markets and are enthusiastic to alter their consumption pattern in order to diminish the costs. Demand response program (DRP) is one of the prevailing techniques of demand side management in which electricity end-users adapt their demand profile in response to operators request and/or electricity prices [15, 16]. The DRP is modeled and employed in various papers to evaluate the impact of DRP on electric demand profile characteristics [17, 18]. In Wu et al. [17], DRP is integrated in the scheduling model in order to manage the volatility of renewable energy. A new DRP has been proposed in [18] for distribution feeders with smart loads. An energy hub model in which distributed generations and electric load DRP are modeled and incorporated to gas and electricity substructures has been studied in [19]. Moreover, the DRP is formulated for the natural gas and electricity networks in [20]. The total daily heating and electricity demands of hub are supplied in [20].

On the other hand, the heating load profile of the MCEN can be modified in order to handle the interdependency of heat and power in CHP units and take more advantages of the units in producing power and heat with high efficiency. The thermal loads of a typical MCEN are responsive and flexible due to two motives. First, the human easement region is not a point but a span [21]. Second, warming can benefit present as well as adjoining future hours, since thermic insulation causes the thermal energy to be stocked. Therefore, regarding the proposed demand response program, in contrast to existing papers, the MCEN takes advantages of the curtailable and responsive heating demand of DHN. In addition, the hubs' thermal load will be modified regarding electric load profile in order to derive advantage of CHP units and alleviate total cost of provision of energy.

16.2 Basic Concepts of Multi-Carrier Energy Systems

This section introduces the energy hub concept and MCEN substructures model since the MCENs' prosperous role in future perspective of energy systems will precisely be distinguished through their basic components.

16.2.1 Energy Hub

Energy hub encompasses various technologies and devices. Firstly a typical energy hub model is described to clarify the optimization problem process. A generic energy hub model is presented in Fig. 16.1. As it is clear, the input energy carriers of hub are natural gas and electricity and the output side consists of thermal and electrical energies which will supply the heat and electric demands. The system converters are composed of a transformer, a micro-turbine, and a gas furnace. The input gas is dispersed between the micro-turbine and gas furnace. The micro-turbine uses natural gas and generates heat and electricity. In addition, the gas furnace generates heat from input natural gas.

The energy hub gets the information from day-ahead market and the hub's input and output states are liable for establishing optimal operation based on collected data. It is worth bearing in mind that the considered market in this chapter is a perfect

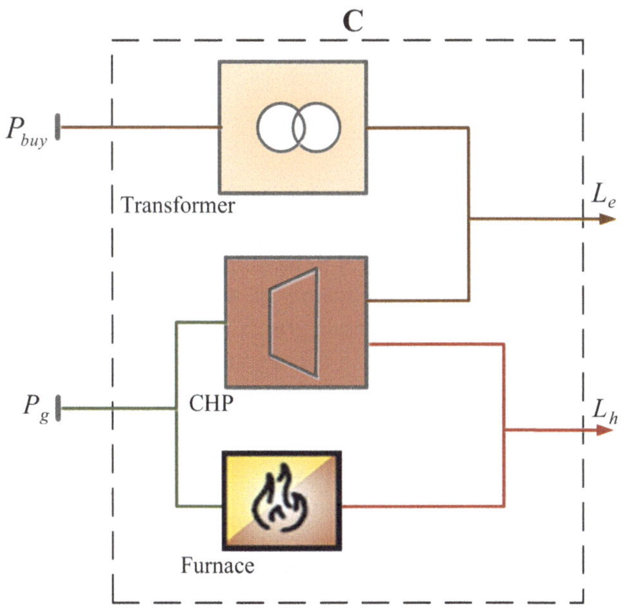

Fig. 16.1 Representation of a typical energy hub

market where all players are price takers. Therefore, the multi-carrier energy system only employs the market prices for optimal scheduling of hub and its strategies cannot affect the market price.

16.2.2 Multi-Carrier Energy System Structure

The carriers of the multi-carrier energy system are transferred to hubs via various transmission networks. The heat and electricity networks are typically connected through the coupling facilities of hubs (e.g., circulation pumps and CHP units). These coupling facilities permit the streams of energy among the networks. CHP units produce heat and power simultaneously; circulation pumps use electrical energy to circulate the water in the DHN. These hubs' converters enhance the flexibility of the heat and electricity systems for assisting the incorporation of uncertain renewable energy [22].

16.3 Problem Formulation

Integrated optimal thermal and electrical power flow constrained scheduling model of MCEN considering heat demand response is presented in this chapter. The aim of the optimal operational scheduling is minimizing the overall cost of hubs' power and heat procurement over a day-ahead period of time, satisfying several constraints.

16.3.1 Objective Function

The purpose of the MCEN scheduling is minimizing the cost of meeting hubs demands. The objective function in the thermal and electrical power flow constrained scheduling problem of MCEN to be minimized encompasses the expense of purchased power and gas from the main grid. Moreover, the MCEN is supposed to be capable of selling power to the main grid. Then, maximizing the income from selling the additional power to the grid is integrated in the objective function. The objective function to be optimized is as follows:

$$\underset{Pg_{h,t},P_t^{\text{grid}}}{\text{minimize}} \sum_t \left\{ P_t^{\text{grid}} \gamma_{e,t} + \sum_h Pg_{h,t} \gamma_{g,t} \right\} + H_{\text{Cur}} \gamma_{\text{Cur}} \tag{16.1}$$

It should be mentioned that, the interchanged power with the main grid, P_t^{grid}, would be positive in the case of buying power from the grid, else it would be negative. $H_{\text{Cur}} \gamma_{\text{Cur}}$ is the cost of curtailed load. In order to model the technical constraints related to various energy networks, the mathematical representations of district heat and electricity networks power flow will be addressed in the following.

16.3.2 Analysis of District Heating Networks

Heat can be produced at hubs by heat sources like CHPs or furnaces, and conveyed by water in supply pipe network. The water temperature drops at consumers' sites, owing to the heat consumption and supplies back to the hub through a return pipeline. Key DHS elements comprising heat sources, heat-exchangers, and the network of pipelines are modeled in Sects. 16.3.2.1–16.3.2.6.

16.3.2.1 Heat Sources

In general, heat sources include CHP units and gas furnaces that supply heat. CHP units and gas furnaces are modeled in the following:

$$Pg_{h,t} = Pg_{h,t}^{\mathrm{CHP}} + Pg_{h,t}^{F} \tag{16.2}$$

$$H_{h,t}^{F} = \eta^{F} Pg_{h,t}^{F} \tag{16.3}$$

$$H_{h,t}^{\mathrm{CHP}} = \eta_{q}^{\mathrm{CHP}} Pg_{h,t}^{\mathrm{CHP}} \tag{16.4}$$

$$P_{h,t}^{\mathrm{CHP}} = \eta_{e}^{\mathrm{CHP}} Pg_{h,t}^{\mathrm{CHP}} \tag{16.5}$$

$$H_{h,t}^{\mathrm{HS}} = H_{h,t}^{\mathrm{CHP}} + H_{h,t}^{F} \tag{16.6}$$

$$\underline{H}^{F} \le H_{h,t}^{F} \le \overline{H}^{F} \tag{16.7}$$

$$\underline{H}^{\mathrm{CHP}} \le H_{h,t}^{\mathrm{CHP}} \le \overline{H}^{\mathrm{CHP}} \tag{16.8}$$

Referring to (16.2), the purchased gas, $Pg_{h,\,t}$, is distributed between two streams. The $Pg_{h,t}^{\mathrm{CHP}}$ is fed into the CHP unit and the $Pg_{h,t}^{F}$ is fed into the gas furnace. Total heat output of hub h ($H_{h,t}^{\mathrm{HS}}$) is produced using furnace and CHP units as expressed in (16.6), where η_{h}^{CHP} and η^{F} are heat efficiency of CHP unit and efficiency of furnace unit, respectively. The capacity restrictions of gas furnace and CHP unit can be described as given in (16.7) and (16.8), respectively.

16.3.2.2 Water Pumps

The pumping power, which affords the required energy to sustain the water flow in the pipelines at the hub, is proportionate to the pressure difference and mass flow rate:

$$P_{h,t}^{\text{pump}} = \frac{\dot{m}_{h,t}^{\text{HS}}\left(pr_{n,t}^{S} - pr_{n,t}^{R}\right)}{\eta_{h}^{\text{pump}} \cdot \rho} \tag{16.9}$$

where ρ and η_{h}^{pump} are water density and the efficiency of pump. The pumping power is restricted by its technical limits

$$\underline{P}_{h}^{\text{pump}} \leq P_{h,t}^{\text{pump}} \leq \overline{P}_{h}^{\text{pump}} \tag{16.10}$$

16.3.2.3 Heat Production

The constraint defining the heat output of a hub which is employed to heat the flow is:

$$H_{h,t} = c\dot{m}_{h,t} \cdot \left(T_{n,t}^{S} - T_{n,t}^{R}\right) \tag{16.11}$$

where c is specific heat. As for the supply temperature of the heat sources, there are lower and upper limits, stated as:

$$\underline{T}_{n}^{S} \leq T_{n,t}^{S} \leq \overline{T}_{n}^{S} \tag{16.12}$$

16.3.2.4 Heat-Exchange Station

Thermal energy of heat-exchangers can be modeled as follows:

$$H_{t}^{\text{HES}} = c \cdot \dot{m}_{t}^{\text{HES}}\left(T_{n,t}^{S} - T_{n,t}^{R}\right) \tag{16.13}$$

The heat exchanger pressure should be above a firm level to make sure the sustainability of mass flow:

$$pr_{n,t}^{S} - pr_{n,t}^{R} \geq \underline{pr}^{\text{HES}} \tag{16.14}$$

The return temperature of the heat exchanger should be within its lower and upper bounds as well:

$$\underline{T}_{n}^{R} \leq T_{n,t}^{R} \leq \overline{T}_{n}^{R} \tag{16.15}$$

16.3.2.5 DHN Constraints

The DHN constraints including continuity of mass flow, node temperature, heat losses from a pipe, etc., are offered in this section:

Continuity of Mass Flow

The overall mass flow rate into any DHN node would be zero:

$$\sum \dot{m} = 0 \tag{16.16}$$

Node Temperature

The combination temperature at a node is equivalent to the temperature at the start of each pipeline leaving that node [23]:

$$\sum_{l_n^-=1}^{N_1} \left(T_{l_n^-,t}^{S,\text{out}} \cdot \dot{m}_{l_n^-,t}^{S,\text{pipe}} \right) = T_{l_n^+,t}^{S,\text{in}} \cdot \sum_{l_n^-=1}^{N_l} \dot{m}_{l_n^-,t}^{S,\text{pipe}} \tag{16.17}$$

$$\sum_{l_n^+=1}^{N_1} \left(T_{l_n^+,t}^{R,\text{out}} \cdot \dot{m}_{l_n^+,t}^{R,\text{pipe}} \right) = T_{l_n^-,t}^{R,\text{in}} \cdot \sum_{l_n^+=1}^{N_l} \dot{m}_{l_n^+,t}^{R,\text{pipe}} \tag{16.18}$$

Moreover, the temperatures of mixed mass at a DHN node are equivalent to mass flowing from that node:

$$T_{l_n^+,t}^{S,\text{in}} = T_{n,t}^{S} \tag{16.19}$$

$$T_{l_n^-,t}^{R,\text{in}} = T_{n,t}^{R} \tag{16.20}$$

Heat Losses from a Pipe

The temperature reduces exponentially during water flow in the pipe [24].

$$T_{l,t}^{S,\text{out}} = T_t^a + \left(T_{l,t}^{S,\text{in}} - T_t^a \right) \cdot e^{-\frac{\lambda_l \cdot L}{c \cdot \dot{m}_{l,t}^{S,\text{pipe}}}} \tag{16.21}$$

$$T_{l,t}^{R,\text{out}} = T_t^a + \left(T_{l,t}^{R,\text{in}} - T_t^a \right) \cdot e^{-\frac{\lambda_l \cdot sL}{c \cdot \dot{m}_{l,t}^{R,\text{pipe}}}} \tag{16.22}$$

where $T_{l,t}^{S,\text{out}}$, $T_{l,t}^{R,\text{out}}$, and T_t^a are the outlet supply, outlet return, and inlet temperatures of a pipe, respectively, λ_l is heat transfer coefficient unit length, and L and $\dot{m}_{l,t}^{\text{pipe}}$ are the length and the water flow rate of a pipe, respectively.

This relation can be approximately written as:

$$T_{t,l}^{S,\text{out}} = \begin{cases} T_t^a + \left(T_{l,t}^{S,\text{in}} - T_t^a\right) \cdot \left(1 - \frac{\lambda_l \cdot L_l}{c\dot{m}_{l,t}^{S,\text{pipe}}}\right) & \text{if} \quad \frac{\lambda_l \cdot L_l}{c\dot{m}_{l,t}^{S,\text{pipe}}} \leq 1 \\ T_t^a & \text{if} \quad \frac{\lambda_l \cdot L_l}{c\dot{m}_{l,t}^{S,\text{pipe}}} \geq 1 \end{cases} \tag{16.23}$$

$$T_{t,l}^{R,\text{out}} = \begin{cases} T_t^a + \left(T_{l,t}^{S,\text{in}} - T_t^a\right) \cdot \left(1 - \frac{\lambda_l \cdot L_l}{c\dot{m}_{l,t}^{R,\text{pipe}}}\right) & \text{if} \quad \frac{\lambda_l \cdot L_l}{c\dot{m}_{l,t}^{R,\text{pipe}}} \leq 1 \\ T_t^a & \text{if} \quad \frac{\lambda_l \cdot L_l}{c\dot{m}_{l,t}^{R,\text{pipe}}} \geq 1 \end{cases} \tag{16.24}$$

Mass Flow Rate Limit

Typically, increasing the flow rate of fluid causes reduction in the ultimate natural frequency of a pipeline. With large velocity of fluid flow, the pipeline can be unstable as the pipeline comes to be exposed to fatigue failure or resonance if its natural frequency is lower than certain limits [25]. Therefore, to avoid pipeline vibrations, the mass flow rates should not surpass their upper boundaries.

$$\underline{\dot{m}_l^S} \leq \dot{m}_{l,t}^{S,\text{pipe}} \leq \overline{\dot{m}_l^S} \tag{16.25}$$

$$\underline{\dot{m}_l^R} \leq \dot{m}_{l,t}^{R,\text{pipe}} \leq \overline{\dot{m}_l^R} \tag{16.26}$$

Pressure Loss

The static pressure drop between two nodes of a pipe is proportionate to the square of mass flow rate. The pressure drop can be stated by (16.27), [24]:

$$pr_{n^+,t}^S - pr_{n^-,t}^S = r_l \cdot \left(m_{l,t}^{S,\text{pipe}}\right)^2 \tag{16.27}$$

$$pr_{n^-,t}^R - pr_{n^+,t}^R = r_l \cdot \left(m_{l,t}^{R,\text{pipe}}\right)^2 \tag{16.28}$$

where r_l is hydraulic resistance of the pipe.

16.3.2.6 Thermal Storage

In this chapter, it is assumed that the hubs with distributed generation are equipped with thermal storages and the heat that the storage unit will be supplied could be presented as following:

$$Es_{h,t} = (1 - \eta_s)\, Es_{h,t-1} + H_{h,t}^{\mathrm{HS}} - H_{h,t}^{D} \tag{16.29}$$

The capacity of heat storage is limited as:

$$\underline{Es} \le Es_{h,t} \le \overline{Es} \tag{16.30}$$

16.3.3 Load Flow Equations

The electrical power flow constraints in the MCEN scheduling problem are modeled in order to simulate more realistic and precise framework. The flow of power through the power system can be expressed by the following equations which present the active and reactive power flow calculations and characterized by Kirchhoff's laws:

$$P_t^{\mathrm{grid}} + P_{i,t}^{g} - P_{i,t}^{l} = \sum_{j=1}^{N_{\mathrm{bus}}} \left(|V_{i,t}| \left|V_{j,t}\right| \left|Y_{ij}\right| \cos\left(\theta_{ij,t} - \delta_{i,t} + \delta_{j,t}\right)\right) \tag{16.31}$$

$$Q_t^{\mathrm{grid}} + Q_{i,t}^{g} - Q_{i,t}^{l} = -\sum_{j=1}^{N_{\mathrm{bus}}} \left(|V_{i,t}| \left|V_{j,t}\right| \left|Y_{ij}\right| \sin\left(\theta_{ij,t} - \delta_{i,t} + \delta_{j,t}\right)\right) \tag{16.32}$$

16.3.3.1 Voltage Limits

The voltage magnitude of substation buses, V_s, should be kept at nominal value V_S^n. Moreover, the bus voltages magnitude, $V_{i,\,t}$, should be maintained at permissible range.:

$$V_{\min} \le |V_{i,t}| \le |V_{\max}| \tag{16.33}$$

$$|V_S| = V_S^n. \tag{16.34}$$

16.3.3.2 Exchangeable Power Limit

In order to have the stable operation, interchangeable apparent power with the main grid should be in its admissible range [26].

$$\sqrt{P_t^{\mathrm{grid}^2} + Q_t^{\mathrm{grid}^2}} \leq \overline{S^{\mathrm{grid}}} \tag{16.35}$$

16.3.3.3 Apparent Power Flow Limits for Branches

It is indispensable to preserve the apparent power flowing from each branch, $S_{br,\,t}$, of the network in a limited bound:

$$\sqrt{P_{br,t}^2 + Q_{br,t}^2} \leq \overline{S_{br}} \tag{16.36}$$

16.3.4 Demand Response Program

The power and heat generations of hubs are almost correlated as CHP units' heat and power generations are interdependent. The sources of heat provision in the presented model for MCEN are gas furnaces and CHP units and these sources will be feed through bought gas. Hence, despite the constant price of gas, an efficient DRP is essential in MCENs to reduce the total cost. In addition, since the thermal load of a hub can be presumed more responsive and flexible than electrical load, the proposed DRP would be applied to the heat load. It is worth mentioning that, during the scheduling period, the thermal load is assumed to be shiftable and curtailable owing to the thermal load nature. The presented DRP for thermal load can be expressed as following:

$$H_t^{\mathrm{HES}} = \left(1 - \mathrm{cur}_t^{\mathrm{HES}}\right) \times H_t^{\mathrm{HES0}} + sl_t^{\mathrm{HES}} \tag{16.37}$$

$$H_{\mathrm{inc},t}^{\mathrm{HES}} = sl_t^{\mathrm{HES}} - \left(\mathrm{cur}_t^{\mathrm{HES}} \times H_t^{\mathrm{HES0}}\right) \tag{16.38}$$

$$0 \leq H_{\mathrm{inc},t}^{\mathrm{HES}} \leq \mathrm{inc}_t^{\mathrm{HES}} \times H_t^{\mathrm{HES0}} \tag{16.39}$$

$$\mathrm{cur}_t^{\mathrm{HES}} \leq \overline{\mathrm{cur}} \tag{16.40}$$

$$\text{inc}_t^{\text{HES}} \le \overline{\text{inc}} \tag{16.41}$$

$$H_{\text{Cur}} = \sum_t \sum_{\text{HES}} \left\{ \left(\text{cur}_t^{\text{HES}} \times H_t^{\text{HES0}} \right) - sl_t^{\text{HES}} \right\} \tag{16.42}$$

Equations (16.37) and (16.38) characterize the heat load and incremented heat load after employing the DRP. Equations (16.39), (16.40), and (16.41) restrict the increased load, conveyed load, and percentage of increased load. Equation (16.42) computes the total quantity of curtailed thermal load of MCEN. It should be mentioned that $H_{e,t}^{\text{HES}}$ in the above equations indicates the distribution network exchangers that are assumed as thermal loads.

16.4 Numerical Analysis

To scrutinize the validity and outperformance of the proposed model, a multi-carrier energy system consisting of a district heating and an electrical sub-networks has been employed in this section.

16.4.1 Multi-Carrier Energy System Structure

Figure 16.2 illustrates the structure of this test system. Configuration and characteristics of the multi-carrier energy system units [27] are presented in Tables 16.1 and 16.2, respectively. In the studied MCEN, bus 1 is connected to the main grid and the system is able to procure the electricity from the grid according to the day-ahead market prices. There are three energy hubs in the MCEN and their configurations are accordant with Fig. 16.3. Detailed characteristics of hubs' facilities are provided in Table 16.2. The predicted hourly active and reactive loads for of all buses are depicted in Fig. 16.4. Other network data comprising the impedance of branches are taken from [28]. The minimum and maximum values for voltage magnitude are assumed to be 0.95 p.u. and 1.05 p.u., respectively. Moreover, the gas price is considered 30 \$/MWh [27]. Minimum and maximum limits of hot water supply temperature are 70 °C and 100 °C, respectively. The specific heat of water, C, and the ambient temperature, T_a are 4.182×10^{-3} MJ kg^{-1} °C^{-1} and 10 °C, respectively. Two case studies have been studied to evaluate the proposed model. The first case schedules the MCEN without applying DRP, whereas the second case scrutinizes the impact of heat DRP in the scheduling procedure.

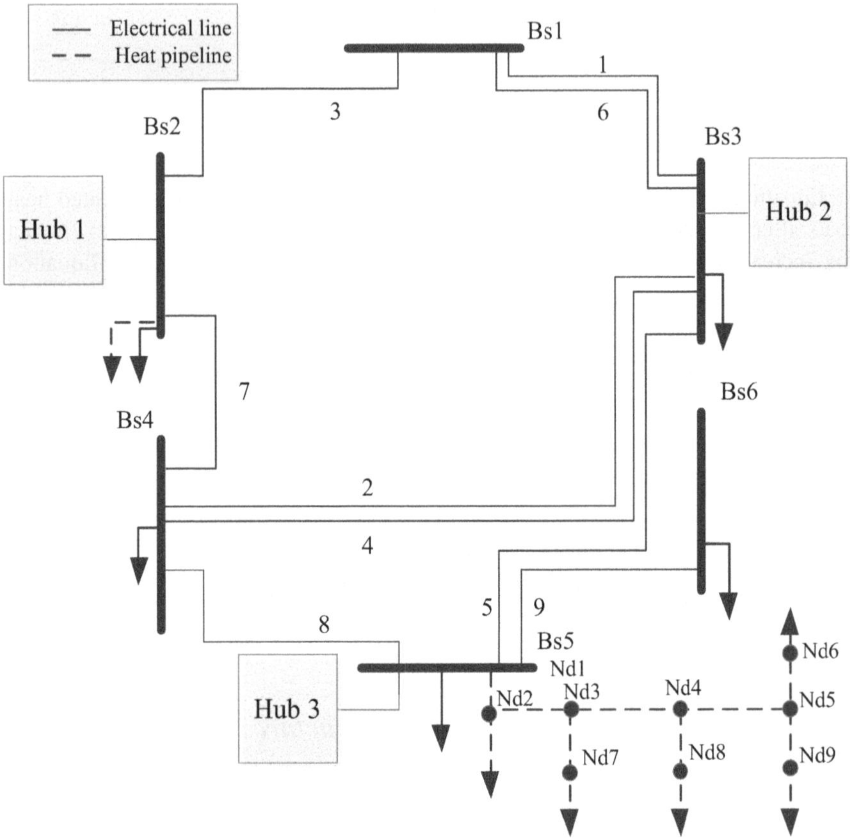

Fig. 16.2 The 3-hub test system under study

Table 16.1 MCEN configuration

Unit\Location	Electrical bus no.	Heat node no.	Hub no.
Generator 1	3	–	2
Gas furnace 1	2	–	1
Gas furnace 2	5	1	3
CHP unit 1	2	–	1
CHP unit 2	5	1	3
Heat storage	2	–	1

Table 16.2 Operational constraints of energy hubs generation units

Device	Efficiency	$\overline{P}$ (kW)	$\underline{P}$ (kW)
Generator of hub 2	$\eta^G = 0.6$	150	0
Gas furnace of hubs 1 & 3	$\eta^F = 0.75$	10	0
CHP units of hubs 1 & 3	$\eta_e^{\text{CHP}} = 0.35$, $\eta_q^{\text{CHP}} = 0.45$	100	0
Heat storage of hub 1	$\eta_s = 0.01$	50	0

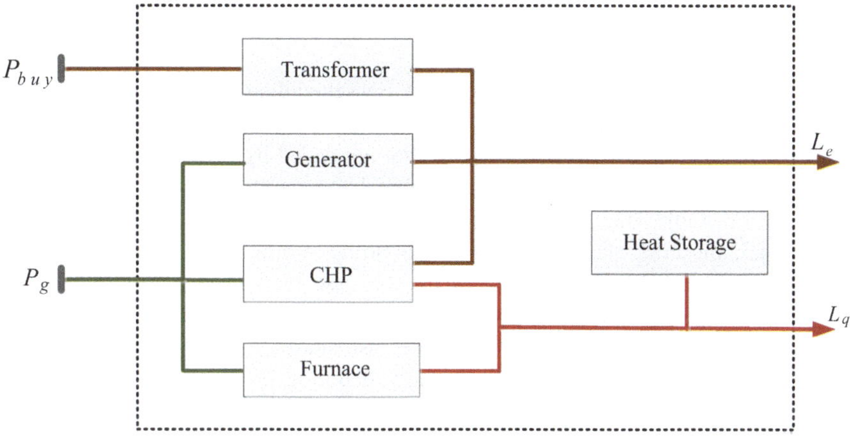

Fig. 16.3 A typical structure of an energy hub

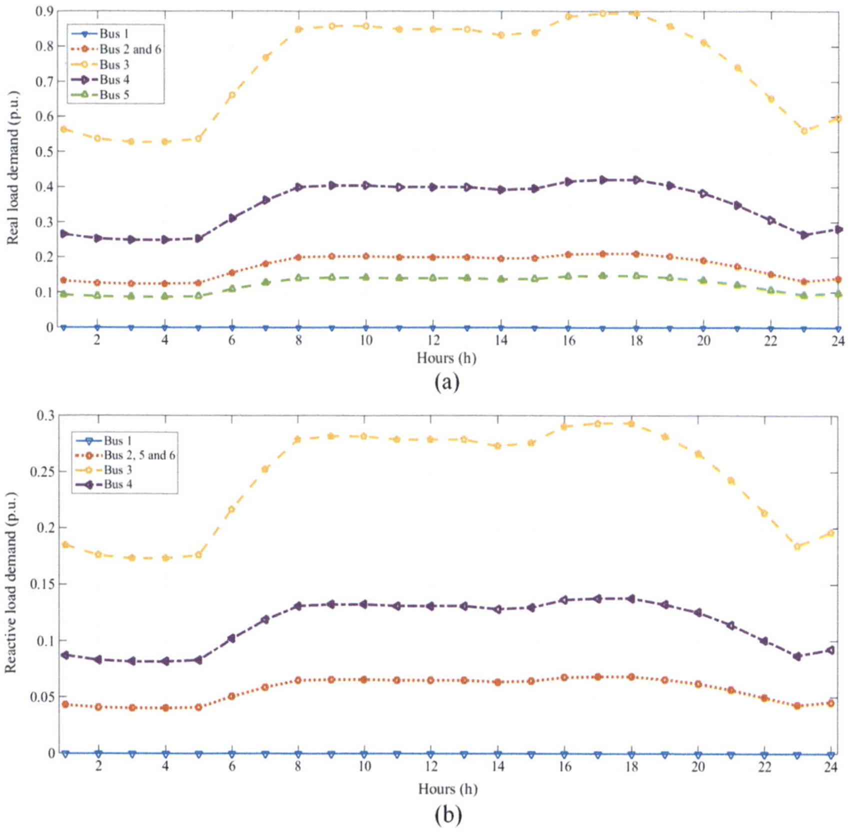

Fig. 16.4 Bus data (**a**) active loads (**b**) reactive loads

16.4.2 Simulation Results

This subsection is developed to study the MCEN scheduling problem employing the proposed framework. The mixed integer non-linear programming (MINLP) model has been applied in GAMS [29] environment unraveled by SBB/CONOPT solver. In the first case, all the economic and technical constraints are taken into account except DRP. Table 16.3 provides the summary of simulation results. Regarding Table 16.3, the cost of MCEN energy providing would be $239.81 for case 1 which has been reduced to $230.225 for the case with DRP. Furthermore, the system revenue from the electricity market participation over 24-hour time interval is about $30.08 for case 1 and $38.264 for case study 2. It can be inferred from the results that implementing DRP in the scheduling process has increased the revenue approximately 27.2% and reduced the total cost about 4%. The voltage magnitude of all buses is presented in Fig. 16.5 for case study 2. Regarding Fig. 16.5, the voltage magnitude of all buses is restricted between 0.95 and 1.05 p.u. Thermal load of DHN and the thermal load with distributed generation are depicted in Fig. 16.6. Fig. 16.7 shows the supply temperature of node 1 in DHN. According to Fig. 16.7, the temperature is decreased when the thermal demand is low in order to diminish losses. Moreover, the temperature is enhanced once the thermal demand is incremented to decrease the power expended by the pump. However, the temperature is reduced in some intervals that the thermal demand is high. This fact is due to the interdependency of heat and power generations of CHP facilities and the

Table 16.3 Summarized simulation results of MCEN

	Generation cost	Revenue from the sale of power	Cost of buying power	Value of objective function
Case study 1	$221.471	$30.081	$48.420	$239.810
Case study 2	$217.198	$38.264	$51.291	$230.225

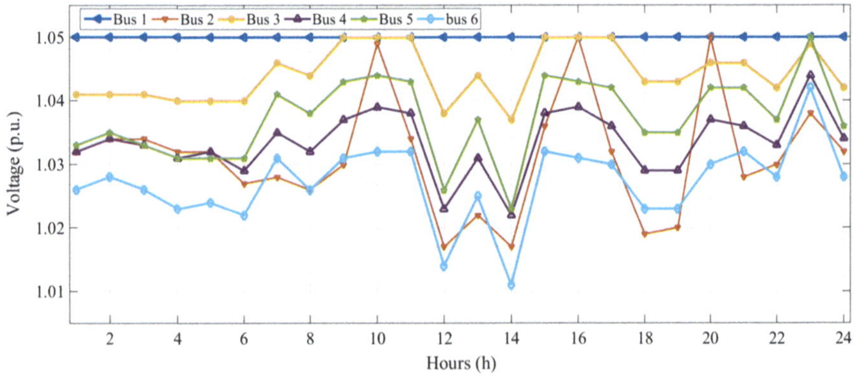

Fig. 16.5 Voltage profile of some electrical buses during 24 h

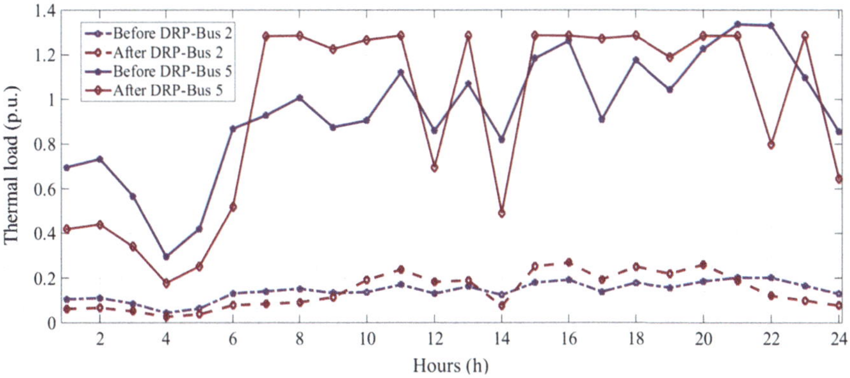

Fig. 16.6 Hourly heat demand of the hubs

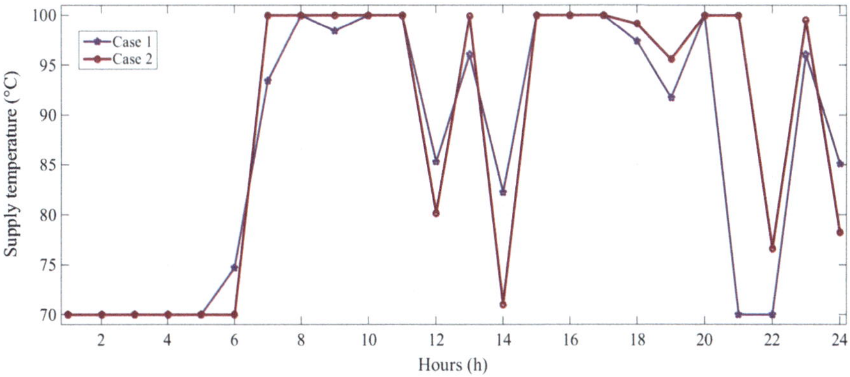

Fig. 16.7 Supply temperature of node 1

active power demand of power network. It is worth mentioning that the temperature in hours 21 and 22 has been improved by applying DRP. According to the simulation results, the temperature has been increased in these hours to decrease the consumed power by the pump.

Figure 16.8 depicts the gas distribution among DHN and hub 2's converters for case 1. Referring to this figure, it could be perceived that the gas furnace of DHN will contribute in providing thermal energy only when the CHP unit's capacity is reached, i.e. in hours 21:00 and 22:00. In addition, the CHP unit 1 is the only supplier of hub 1. According to the simulation results, the heat storage would be discharged till hour 24:00 to reduce the total cost. In addition, gas furnaces will not participate in providing thermal energy after applying DRP. The simulations indicate the similar results for case 2, except that by applying the DRP, the heat demand profile will be modified in a way that there will be no need to gas furnaces.

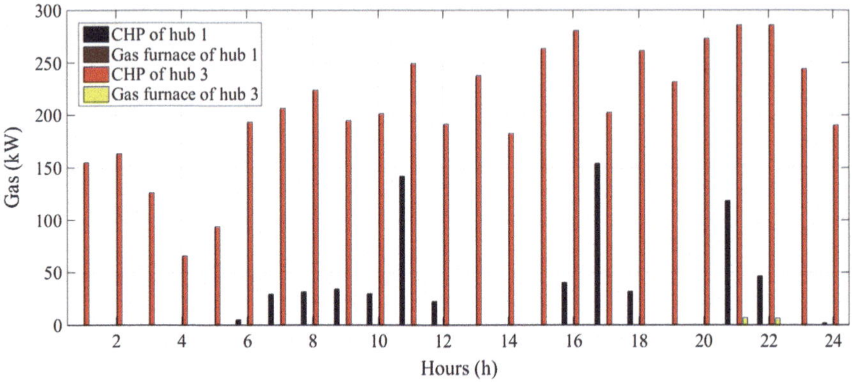

Fig. 16.8 Gas distribution among the hubs converters for case 1

16.5 Conclusion

In recent days, energy systems' optimal operation is a fundamental issue in system management scrutiny. This chapter has proposed a model for optimal scheduling of MCENs consisting of district heating and electrical networks. In the presented energy hub framework, the energy and continuity laws as well as the characteristic of district heating system's major elements for DHN, voltage magnitude of buses and line flow limits of electric network are modeled successfully. In addition, since the heating load profile of the MCEN can be modified, heat DRP has been implemented in order to handle the interdependency of heat and power in CHP units. The simulation outcomes have verified the usefulness and efficiency of the entire MCEN model and the capability of DRP, which can be employed to optimize the model. According to the simulation results, applying the heat DRP to the DHN reduces the total cost about 4% in the studied case. The results also indicate that the optimal operating strategy can improve the optimal temperature of nodes and decrease the consumed power by the pump.

Nomenclature

Indices

h	Index of hubs.
n	Index of nodes in the heating network.
l	Indices of pipelines in the heating network.
s	Superscript of supply in the heating network.
R	Superscript of return in the heating network.

n^+ Index of starting node of pipeline l.
n^- Index of ending node of pipeline l.
i Index of buses in the electricity network.

Variables

H_t^{HS} Total heat output of hub h.
T Temperature in the heating network.
pr Pressure in the heating network.
$P_{i,t}^g$ Active power flow of hubs positioned on bus i.
$Q_{i,t}^g$ Reactive power flow of hubs positioned on bus i.
$V_{i,t}$ Voltage of bus i.
$Pg_{h,t}$ Pumping power.
P_t^{grid} Active power bought from the Utility.
Q_t^{grid} Reactive power bought from the Utility.
$P_{h,t}^{\mathrm{pump}}$ Pumping power.
$\dot{m}$ Mass flow rate.
$\mathrm{cur}_t^{\mathrm{HES}}$ The participation factor of load in DRP.
sl_t^{HES} The amount of transferred load from other hours to hour t.
$\mathrm{inc}_t^{\mathrm{HES}}$ Incremental load factor.
H_{Cur} Total quantity of curtailed heat load.
$H_{\mathrm{inc},t}^{\mathrm{HES}}$ The increased load.
H_t^{HS} Total heat output of hub h.
T Temperature in the heating network.

Parameters

N_{bus} Number of buses of the power system.
H_t^{HES0} The primitive hub's load.
$P_{i,t}^l$ Active load of bus i.
$Q_{i,t}^l$ Reactive load of bus i.
Y_{ij} Magnitude of feeder's admittance.
$\theta_{ij,t}$ Phase angle of feeder's admittance.
$\gamma_{g,t}$ Gas price.
$\gamma_{e,t}$ Predicted day-ahead electricity market price.
η Efficiency.
N_{bus} Number of buses of the power system.
H_t^{HES0} The primitive hub's load.
$P_{i,t}^l$ Active load of bus i.

References

1. Gebremedhin A (2014) Optimal utilisation of heat demand in district heating system—a case study. Renew Sust Energ Rev 30:230–236
2. Geidl M, Andersson G (2007) Optimal power flow of multiple energy carriers. IEEE Trans Power Syst 22(1):145–155
3. Flexible energy delivery systems. http://www.cardiff.ac.uk/ugc/flexibleenergy-delivery-system-seminar-series-for-postgraduates-and-researchers. Accessed 25 Mar 13
4. Moeini-Aghtaie M, Abbaspour A, Fotuhi-Firuzabad M, Hajipour E (2014) A decomposed solution to multiple-energy carriers optimal power flow. IEEE Trans Power Syst 29(2):707–716
5. Geidl M, Andersson G (2006) Operational and structural optimization of multi-carrier energy systems. Eur T Electr Power 16(5):463–477
6. Barbieri ES et al (2014) Optimal sizing of a multi-source energy plant for power heat and cooling generation. Appl Therm Eng 71(2):736–750
7. Geidl M, Koeppel G, Favre-Perrod P, Klockl B, Andersson G, Frohlich K (2007) Energy hubs for the future. IEEE Power Energ Mag 5(1):24–30
8. Salimi M, Ghasemi H, Adelpour M, Vaez-ZAdeh S (2015) Optimal planning of energy hubs in interconnected energy systems: a case study for natural gas and electricity. IET Gener Transm Distrib 9(8):695–707
9. Anders GJ, Vaccaro A (2011) Innovations in power systems reliability. Springer, London
10. Krause T, Andersson G, Frohlich K, Vaccaro A (2011) Multiple-energy carriers: modeling of production, delivery, and consumption. Proc IEEE 99(1):15–27
11. Alipour M, Zare K, Mohammadi-Ivatloo B (2016) Optimal risk-constrained participation of industrial cogeneration systems in the day-ahead energy markets. Renew Sust Energ Rev 60:421–432
12. Parisio A, Del Vecchio C, Vaccaro A (2012) A robust optimization approach to energy hub management. Int J Electr Power Energy Syst 42(1):98–104
13. Martinez-Mares A, Fuerte-Esquivel CR (2012) A unified gas and power flow analysis in natural gas and electricity coupled networks. IEEE Trans Power Syst 27(4):2156–2166
14. Shabanpour-Haghighi A, Seifi AR (2015) Energy flow optimization in multicarrier systems. IEEE Trans Ind Inf 11(5):1067–1077
15. Medina J, Muller N, Roytelman I (2010) Demand response and distribution grid operations: opportunities and challenges. IEEE Trans Smart Grid 1(2):193–198
16. Mathieu JL, Price PN, Kiliccote S, Piette MA (2011) Quantifying changes in building electricity use, with application to demand response. IEEE Trans Smart Grid 2(3):507–518
17. Wu H, Shahidehpour M, Al-Abdulwahab A (2013) Hourly demand response in day-ahead scheduling for managing the variability of renewable energy. IET Gener Transm Distrib 7(3):226–234
18. Mosaddegh A, Canizares CA, Bhattacharya K (2017) Optimal demand response for distribution feeders with existing smart loads. IEEE Trans Smart Grid:1
19. Pazouki S, Haghifam MR, Olamaei J (2013) Economical scheduling of multi carrier energy systems integrating renewable, energy storage and demand response under energy hub approach. In: Smart Grid Conference (SGC), IEEE, pp 80–84
20. Bahrami S, Sheikhi A (2016) From demand response in smart grid toward integrated demand response in smart energy hub. IEEE Trans Smart Grid 7(2):650–658
21. Garage G Human comfort zone. [Online]. http://www.greengaragedetroit.com/index.php?title=HumanComfortZone
22. Liu X, Wu J, Jenkins N, Bagdanavicius A (2016) Combined analysis of electricity and heat networks. Appl Energy 162:1238–1250
23. Zhao H (1995) Analysis, modelling and operational optimization of district heating systems. PhD Thesis, Technical University of Denmark

24. Awad B, Chaudry M, Wu J, Jenkins N (2009) Integrated optimal power flow for electric power and heat in a microgrid. In: 20th international conference and exhibition on electricity distribution-part 1, 2009. CIRED 2009, IET, pp 1–4
25. Grant I (2010) Flow induced vibrations in pipes, a finite element approach. M.S. thesis, Department of Mechanical Engineering, Cleveland State University, Cleveland, OH, USA
26. Aghaei J, Alizadeh MI (2013) Multi-objective self-scheduling of CHP (combined heat and power)-based microgrids considering demand response programs and ESSs (energy storage systems). Energy 55:1044–1054
27. Alipour M, Zare K, Abapour M (2017) MINLP probabilistic scheduling model for demand response programs integrated energy hubs. IEEE Trans Ind Inf 14:79. https://doi.org/10.1109/TII.2017.2730440
28. Alipour M, Zare K, Seyedi H (2017) Power flow constrained short-term scheduling of CHP units. In: Sustainable development in energy systems. Springer, Cham, pp 147–165
29. Brooke AD, Kendrick AM, Roman R (1998) GAMS: a user's guide. GAMS Development, Washington, DC

Chapter 17
Power-to-Gas: A New Energy Storage Concept for Integration of Future Energy Systems

Azadeh Maroufmashat, Ushnik Mukherjee, Michael Fowler, and Ali Elkamel

17.1 Introduction

Energy storage is essential and well-accepted principle to SMART GRID. Currently Ontario has excess power; in 2016, $1 billion dollars of power was "curtailed" in addition to many billions sold outside of Ontario to the USA as a significant loss. In this work, power-to-gas has been shown to be one of the best alternatives for energy storage based on Ontario's grid profile. Hydrogen is generated with excess CO_2 free nuclear and wind power and used in a number of pathways. There are no real other alternatives for Ontario at this time: Best sites for pumped hydro are used now; Compressed Air Energy Storage (CAES) has little power density, on seasonal storage, efficiencies are low; and batteries have little power density and still higher cost, and repurposed batteries are not yet available in the market.

"Power-to-Gas" as a technology using commercialized electrolyzer has a lot of advantages and will be introduced below. First off, among all the currently available energy storage technologies it has the highest energy storage density, it has many different forms of storage such as compressed gas and liquefied hydrogen in storage tanks as well as storing in natural gas infrastructure, which is a great option for its storage and distribution since it efficiently uses the existing infrastructure and that brings better economic efficiency. Once they get stored, they can be stored for a long period and that allows for delay and offsetting for additional power generation. What's more power-to-gas is also a well-known clean technology because it reduces emissions, while being mixed with natural gas and it increases end-use petroleum fuels' renewable content without changing vehicle type or refueling infrastructure; more significantly when it gets combined with biogas generation more renewable

---

A. Maroufmashat (✉) · U. Mukherjee · M. Fowler · A. Elkamel
Department of Chemical Engineering, University of Waterloo, Waterloo, ON, Canada

B. Mohammadi-Ivatloo, F. Jabari (eds.), *Operation, Planning, and Analysis of Energy Storage Systems in Smart Energy Hubs*,
https://doi.org/10.1007/978-3-319-75097-2_17

natural gas will be developed through methanation to give lower CO_2 emission. Power-to-gas also has the ability to supply auxiliary electrical services what's more its incremental implementation property gives it the ability to adjust changing infrastructure needs. Last but not least, it has great ability in transporting energy over long distances and acting as a transportation fuel for lift trucks in the form of hydrogen, both prove its great commercialized potential [1–7].

Hydrogen as an ideal long-term energy vector can be created from many different resources such as fossil fuel, renewables, and carbon-free nuclear. The concept of "hydrogen economy" has developed in a fast face and it focuses based on the how hydrogen will be produced, distributed, and utilized in energy system. Hydrogen can be used to create electricity and has very various production pathways, when it gets used in transportation related area lower pollution and lower greenhouse gases emissions are in favor. It is well-known that the great price gap between peak and lower price hours has always been a concern for electricity markets, while the appearance of using hydrogen as an energy carrier in electrical grid solves this problem by storing energy generated by some seasonal power such as wind, solar and GHG free nuclear power and distributing them based on needs. It is obvious to see that the concept of "hydrogen economy" actually stands for an ideal fossil fuel free economy [8].

17.2 Different Alternative of Power-to-Gas Applications

Power-to-gas application can offer the most efficient usage of surplus power at all-time due to its gradual and incremental implement, and the pathways of this application are listed below:

Power to Hydrogen to Natural Gas End-users via hydrogen-enriched natural gas (HENG);

1. Power to Renewable Content in Petroleum Fuels;
2. Power to Power;
3. Power-to-gas—Seasonal Energy Storage to Electricity;
4. Power to Zero Emission Transportation;
5. Power to Seasonal Storage for Transportation;
6. Power to Microgrid;
7. Power to Renewable Natural Gas (RNG) to Pipeline ("Methanation"); and
8. Power to Renewable Natural Gas (RNG) to Seasonal Storage.

As shown in Fig. 17.1, electrolyzers are used in different pathways to convert the surplus power to hydrogen that will then be directly converted to methane with low carbon content. This low-carbon methane will go through another process to achieve higher cost and lower. Every process possesses losses, which can be accepted if the surplus electricity is required to reduce or cannot be used anywhere else. For those that have lower CO_2 emissions compared to current conventional natural gas such

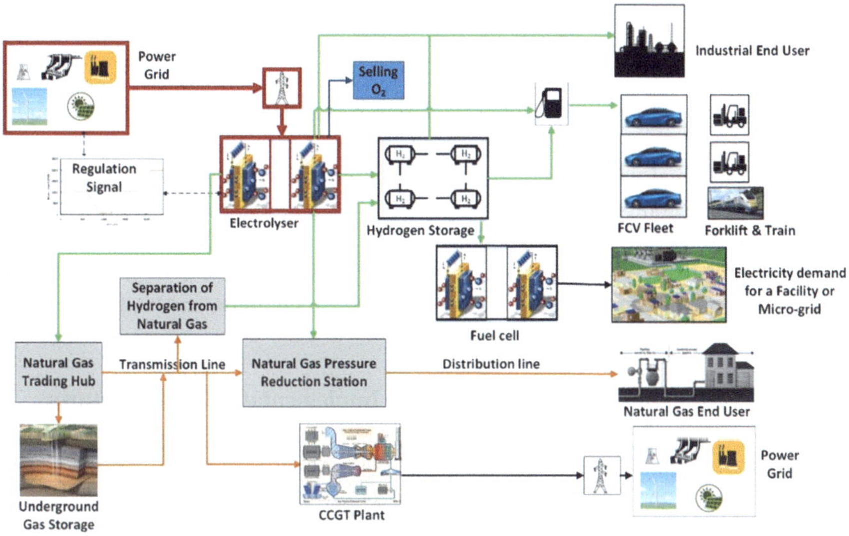

Fig. 17.1 Schematic view of Power-to-gas (adopted from Maroufmashat and Fowler [9])

as residential heat purposes, micro-CHP and large-scale gas turbine, low-carbon methane will be injected to natural gas pipeline. Either what's more it can also be used for the utilization of low-carbon transportation applications on seasonally or daily basis.

17.2.1 Power to Hydrogen to Natural Gas End-Users Pipeline Blending (HENG)

Hydrogen-enriched natural gas (HENG) can be made from decarbonizing natural gas by injecting generated hydrogen from surplus power including renewable energies to natural gas pipeline. This concept has its own limit, only when the composition of hydrogen blending to natural gas no larger than 10% the existing natural gas infrastructure or end-use equipment can function normally. HENG can be used to generate heat, electricity or as a fuel for transportation with no modification to the equipment of HENG systems due to the fact that it has lower CO_2 emissions compared to natural gas.

When this electrolytic hydrogen is injected into natural gas transportation or distribution pipelines, certain limits are required, which are between 5 and 20% based on different types of applications. Many electrolyzers and hydrogen storage tanks are required to build and install even with these concentrations. This kind of storage system has a challenging problem on its optimum capacity and many literatures suggest that this hydrogen storage system improves the performance of the whole systems even though it adds complexity and costs more to power-to-gas system.

In modern society converting diesel Class 8 vehicles to natural gas can certainly accomplish more benefits because no change needed for natural gas infrastructure and vehicles fueled with hydrogen-enriched natural gas becomes more renewable with low-carbon contents.

This pathway can show immediate effect for energy storage with minimal investment cost. What's more it does not need to shed or sell energy for loss as much to avoid high amount of Global Adjustment expenditures in Ontario.

17.2.1.1 Technology Issues Regarding the Implementation of this Pathway

One of the most significant limitations is the allowable fraction of hydrogen into natural gas pipeline. The composition has great effect on end-user systems, safety and risk issue, and durability of pipeline material and leakage of hydrogen. The end-user systems that get affected are furnaces, boilers, and power generators. They are affected by natural gas composition, type of appliances, engine and their ages with an acceptable range of hydrogen between 5 and 20%, higher composition may cost more to the system.

Compression stations and compressed natural gas tanks only have a limitation of 2% for hydrogen concentration which is small comparing to dried compressed blended hydrogen that has 20% and preferred due to higher performance. Current installed gas turbine only has a limitation up to 1%, it can be somehow increased to 5–15% if turbine gets adjusted and upgraded. Gas engine has a limitation of 2% preferably if higher concentration wants to be achieved the simple upgraded control systems shall be used [10].

Safety and risk analysis focuses on hydrogen concentration, pipeline types, material, and failure mode conditions. Comparing to large-scale coal and nuclear plants natural gas systems have a lower risk of severe accident while comparing to some renewable systems such as solar PV and wind, natural gas systems have a higher risk. Some of the risks are the possibility of ignition and the severity of explosion while close to urban areas. One of the exceptions is that during initial implementation due to low concentration of hydrogen in the natural gas system no significant risk appears. Under higher pressure and higher concentrations of hydrogen, the addition of hydrogen may cause material of pipelines degrading faster. This is mainly an issue for transmission pipelines not for steel pipelines. Hydrogen leaks 4 to 5 times faster from fittings than methane due to its lighter density. Once the concentration exceeds 20%, it will have the same order of the leakage [11, 12].

17.2.2 Power to Increased Renewable Content in Petroleum Fuels

Biofuels, typically ethanol gets blended into the distributed gasoline with the range of 5–10% in order to reduce the dependency of imported oil, promote renewable fuel industry, and reduce the carbon emissions released from cars. Under the condition of not changing the quality of the fuel itself and encounter the limitation of ethanol caused by its low energy output renewable electrolytic hydrogen can also be seen as a potential method to increase the renewable energy. Traditional petroleum fuels normally have the following stages for life-cycle emissions: crude extraction, crude transport, crude refining, petroleum fuels transportation, distribution and vehicle consumption, which all contribute to the carbon intensity of gasoline and diesel. In Ontario province unlike Steam Methane Reforming (SMR), the production of hydrogen via electrolysis has a significant low-carbon footprint but meanwhile it costs more than SMR. Well refiners will also implement electrolysis hydrogen to meet the carbon intensity reduction target regulated by government. This pathway uses power-to-gas for oil refining to reduce the carbon intensity as well as decarbonize transportation sector on the life-cycle basis without converting current infrastructure. It is also complimentary with the addition of ethanol to gasoline so that both methods of renewable content can be implemented at the same time.

17.2.3 Power to Power

Hydrogen can be converted from surplus power via electrolyzer, then pressurized, stored, and utilized through fuel cell or hydrogen gas turbine. The drawback is that it might cause potential loss of energy because of the need of fuel cells in the facility. What's more the round trip efficiency is lower than battery than battery energy storage but it is favorable in some remote applications or emergency situations, which provides extended power applications.

17.2.4 Power-to-Gas to Seasonal Energy Storage to Electricity

Produced hydrogen can be stored in the underground facility; moreover, along with natural gas storage electricity can be generated in large-scale natural gas-based power plant. This pathway can properly use wind energy in its off-season as well as for daily and weekly variation in energy demand. It is also very useful for load leveling of baseload nuclear power in Ontario.

17.2.5 Power to Hydrogen for Zero-Emission Transportation

Hydrogen can also be used as a transportation fuel, which gets compressed and stored at high pressure ranging from 300 to 700 bar for hydrogen vehicles and lift trucks. While the concentration is highly required for this application, that is the appearance of 99.995% pure hydrogen.

Using hydrogen separated via electrolysis from nuclear, hydro, wind, or solar sources as fuel to vehicles can truly achieve zero emissions. Comparing to gasoline internal combustion engine (ICE) vehicles, fuel cells vehicles using hydrogen made from natural gas still have a significantly less GHG reductions. How well hydrogen as a future transportation fuel will develop really depends on fuel cell vehicle availability and the development of hydrogen refueling stations. This pathway can integrate electrical and transport energy sectors without the need to upgrade electricity distribution systems to possibly achieve zero emission transportation in an urban area with consumers' preference. In order to improve urban air quality and associated benefits in society's health outcomes using hydrogen FCV in urban areas is a good approach. For different specific transportation applications battery electric vehicles and fuel cell vehicles can be a desirable complementary technology in both the short and long terms.

17.2.6 Power to Seasonal Storage for Transportation

Salt caverns or depleted oil and gas reservoirs can be used to store pressurized hydrogen produced from surplus power via electrolyzer. It can then be separated from other gases via Pressure Swing Adsorption (PSA) and sent to the end-users once it is needed by transportation. It has similar benefits as "Power to Hydrogen for Zero Emission Transportation" with additional benefit, that is, the hydrogen can be produced while renewable energy is plentiful and used all year round. It requires very high penetration of wind energy and baseload nuclear and large capacities of sessional energy storage.

17.2.7 Power to Microgrid

Due to the nature of the intermittency of renewable energies the mismatch between electricity grid congestion at peak demand and under-utilized excess power distribution infrastructure during off-peak hours are all great technical concerns for urban communities. Power-to-gas which stores energy in the form of hydrogen within micro grid is an alternative to utilize for variety of microgrid energy requirements such as transportation demand or to be used for community. It can also be helpful for remote off-grid communities and mining sites with larger needed storage capacities.

17.2.8 *Power to Renewable Natural Gas (RNG) to Pipeline ("Methanation")*

A stream of renewable natural gas made from combination of hydrogen and carbon dioxide can be mixed into natural gas distribution system, this methane production has a higher energy loss and cost but with no limitation of blending into the natural gas distribution systems comparing to simple hydrogen production. It can even be complimentary with the first pathway to inject hydrogen into natural gas pipeline up to allowable limit and convert remaining hydrogen to methane via methanation. It is not a fully developed technology occurred inside a chemical reactor, biological reactor, or natural methanation in underground storage so that the purity of carbon dioxide and the quality of synthesis methane both shall be taken into account. Sometimes if the synthetic methane has low quality an additional gas cleaning process will be needed.

Qualitative benefit of this pathway is carbon sequestration from biogas production or industrial processes such as cement production.

17.2.9 *Power to Renewable Natural Gas (RNG) to Seasonal Storage*

Underground storage can be used to store renewable methane (RNG) once it is produced from surplus electricity. This pathway considering methanation can be matched to an ongoing industrial or agricultural operation for carbon sequestration and independent on natural gas demand profiles.

17.3 Key Technologies in Power-to-Gas

The core technology of power-to-gas system is electrolyzer that converts electricity into fuel. Alkaline, polymer electrolyte membrane (PEM), and solid oxide membrane (SO) are all different types of electrolyzers. Comparing to the most commercial one alkaline electrolyzer PEM electrolyzer has higher potential for cost reduction, durability, and efficiency improvement in future. One of the other electrolyzers that have potential for greater efficiency gain is solid oxide electrolyzer but they require high operating temperature and still in research phase of development. Speaking of higher current density and operational flexibility in terms of dynamic response and frequency regulation that are preconditions for future capital cost reduction PEM electrolyzers have great benefits. When load should be immediately ramp up or down from the point of the normal operation the operation flexibility for the utility electricity grid becomes a key advantage of PEM electrolyzers. Meanwhile it also has the potential to provide auxiliary services that increase the technology's availability factor. If the electrolyzers can be maintained at elevated pressure it will be a benefit for future power-to-gas systems because they can reduce the compression requirement for storage systems, PEM electrolyzer is one of those while alkaline ones are not. Nowadays PEM electrolyzers are limited by the rate of hydrogen production per stack and cell lifetime, even with those limitations they are still expected to surpass alkaline technology in the near future.

The following table lists out the technical, operational, and economic information of the two most applicable electrolyzers that are alkaline and PEM electrolyzers. The information has been collected from different manufacturers' data and existing literature. Commercial or pre-commercial applications with a time frame of up to 10 years are represented by "current perspective," while the long-term planning depends on future technologies with the improvement of their cost and performance as well as a period of more than 10 years ahead [13–15].

The investment costs of alkaline electrolyzers and PEM electrolyzers are \$1000 and \$2000 per kW, respectively. This cost might be changed based on the specific size and thermodynamic condition and based on E&E consultant the cost of PEM electrolyzers is expected to be \$1300 per kW [9].

The other important technologies considering in power-to-gas applications are hydrogen storage and compression systems. Different types of storing hydrogen exist, including underground compressed gas, metal hydride, and liquid hydrogen. Based on the applications, the type of storage may vary. The compressed gas storage is the simplest one, while the issue about the storing liquid hydrogen is boil off losses that results in limited time of storage. For long-term, large-scale energy storage, the underground hydrogen storage is desirable. The information regarding the energy storage systems are summarized in the following table (Table 17.1).

17.4 Technical and Economic Assessment of Power-to-Gas Pathways

Different pathways of power-to-gas applications are presented in the following figure (Fig. 17.2). The overall efficiencies of each pathway along with the economic benefits of them are presented in Tables 17.2 and 17.3.

Since there is a wide range of efficiencies for some technologies, the allowable range of pathway efficiency is mentioned in Table 17.2. In order to calculate the levelized cost of product, as shown in Table 17.3, the hourly Ontario electricity prices (HOEP) are used. The calculation is classified for two groups of HOEP when it is less than 2.5 cent per kW and when it is more than 2.5 per kW.

Results in Tables 17.2 and 17.3 show that in terms of economic and technical points of view, the power-to-gas pathways with hydrogen to the end-users are better options compared to other alternatives that have additional energy conversion technologies leading to more energy losses and lower efficiency and more cost.

Table 17.1 Technical and economic data for hydrogen storage, compression and purification technologies (adopted from [9])

Technology	Explanation	Efficiency		Cost (CAN per kg or kg h^{-1})		Lifetime
		Current	Long term	Current	Long term	
Low pressure hydrogen storage	3–300 kg	Almost 100% (without compression)		260–430	15	20
Compressor—for low pressure storage	Until 180 bar	88–95%	88–95%	3000	3000	20
Compressor—for refueling station	Until 700 bar	80–91%	80–91%	8700–17,000	13,000	20
Injection to pipeline compression	Including compression	95%	95%	–	–	
Underground storage	GWh to TWh (including compression)	90–95%	95%	300–350	40	30
Hydrogen purification system	PSA	80–95%	85%	4000	4000	20

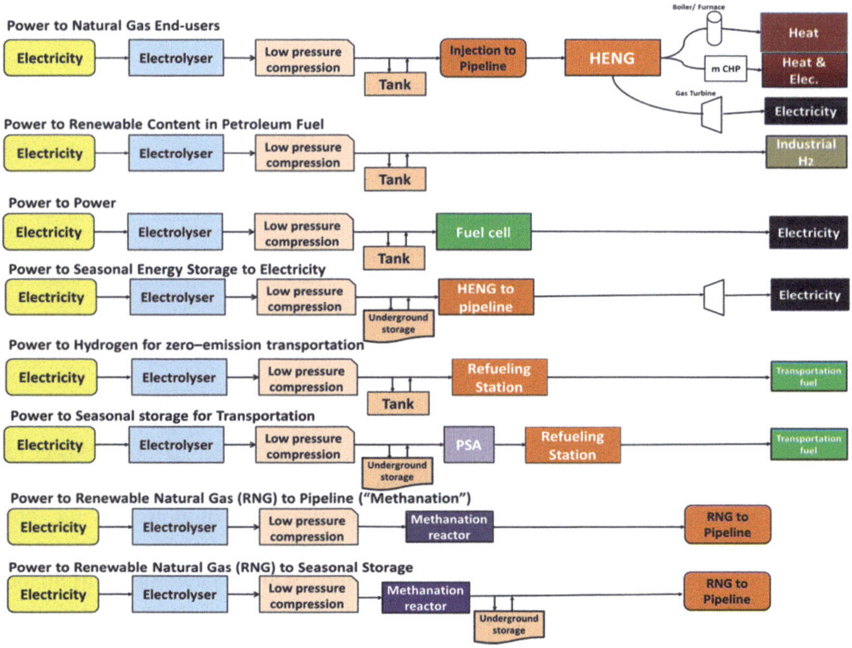

Fig. 17.2 Different pathways of power-to-gas (adopted from [9])

Table 17.2 Technical comparison of the power-to-gas pathways

P2G pathways	Technologies	Current (%)	Long term (%)
Power to natural gas end-users	Electrolyzer, low pressure hydrogen storage/compression, injection to pipeline	59–83	64–86
	To heat for residential	52–76	56–79
	To micro-CHP	40–72	55–74
	To large-scale gas turbines	18–26	23–31
Power to renewable content in petroleum fuel	Electrolyzer, low pressure hydrogen storage/compression	55–83	59–86
Power to power	Electrolyzer, low pressure hydrogen storage/compression, fuel cell	17–40	27–43
Power to seasonal energy storage to electricity	Electrolyzer, low-pressure compression, underground storage, transmission pipelines, natural gas-based power plants	16–24	22–29
Power to hydrogen for zero–emission transportation	Electrolyzer, low-pressure compression and storage, high-pressure compression for refueling station	50–79	54–82
Power to seasonal storage for transportation	Electrolyzer, low-pressure compression, underground storage, hydrogen separation technologies, high-pressure compression	36–68	43–66
Power to renewable natural gas (RNG) to pipeline ("Methanation")	Electrolyzer, low-pressure energy storage and compression, methanation reactor, gas clean-up, injection of renewable natural gas to the natural gas pipeline	40–63	45–65
Power to renewable natural gas (RNG) to seasonal storage	Electrolyzer, low-pressure compression, methanation reactor, gas clean-up, underground storage, injection of RNG to the natural gas pipeline	34–60	43–58

From technical point of view, the power to hydrogen for heat purposes and for transportation has the highest energy efficiency, which is around 50–80%. Power to power has a lowest efficiency specially when utilizing from underground energy storage. Seasonal storage can lower the overall efficiency of power-to-gas, but not more than 10% lower. Power to renewable content in petroleum fuels has an average efficiency of 68% (current) and 72% (future), while a large Steam Methane Reforming (SMR) has an efficiency in a range of over 70%. Results indicate that in future years power-to-gas can be competitive to the conventional SMR from

Table 17.3 Economic comparison of the power-to-gas pathways

$ per kWh	0 < HOEP<2.5		2.5 < HOEP<7.5	
P2G pathways	Current	Long term	Current	Long term
Power to natural gas end-users	0.15–0.27	0.02–0.04	0.19–0.37	0.07–0.1
Power to power	0.53–0.57	0.07–0.09	0.76–78	0.19–0.23
Power to hydrogen for zero–emission transportation	0.18–0.27	0.03–0.04	0.23–0.34	0.08–0.09
Power to seasonal storage for transportation	0.27–0.33	0.09–0.1	0.53–1.06	0.22
Power to renewable natural gas (RNG) to pipeline ("Methanation")	0.22–0.35	0.04–0.05	0.3–0.47	0.11–0.13
Power to renewable natural gas (RNG) to seasonal storage	0.42–0.45	0.08–0.09	0.53	0.15–0.16

technical point of view. From economic point of view, power to power has the highest cost, regarding different types of electrolyzer and fuel cell technologies; the cost is between 0.38 and 0.53 $ per kWh when the HOEP is less than 2.5 cent per kWh, while that of is 0.6 and 0.8 $ per kWh when the HOEP is more than 2.5.

The levelized cost of power to hydrogen pathways is less than the others. Seasonal storage technologies can increase cost by double. The levelized cost of product for hydrogen is around 17–29 cent per kWh, while in the long term the cost must be as low as 4–5. The reason is that the price of storage technologies will be decreased significantly in the long-term. The levelized cost of methane for natural gas vehicles is in the range of 28–41 cent per kWh. Improvements in technologies can increase the overall efficiency of power-to-gas pathways in future years that make them more economically and technically feasible for implementation.

17.5 Case Studies

Different applications of power-to-gas in different projects are summarized as below: a deterministic energy system model for the production of electrolytic hydrogen from off-peak grid and intermittent wind power is developed. The interaction of the energy system with the existing: Power grid infrastructure (in Ontario);—Natural Gas grid infrastructure (for hydrogen distribution in Ontario) is studied [1]. A Pricing Mechanism for Valuing Ancillary, Transportation and Environmental Services Offered by a Power-to-Gas Energy System is developed [16]. The benefit of accounting for uncertainty in electricity pricing and future zero-emission transportation sector by considering hydrogen demand that influence the operation of the power-to-gas energy hub is demonstrated in Ref. [17]. Benchmarking and selection of power-to-gas utilizing electrolytic hydrogen as an energy storage alternative is carried out in Ref. [17]. Decarbonizing transportation through the use of power-to-gas for oil refining operations is investigated in Ref. [18].

Assessing the feasibility of Methanation (Renewable Natural Gas) as a viable energy recovery pathway for power-to-gas energy systems and the integration of Renewable Natural Gas into the Natural Gas Distribution System as Renewable Natural Gas potential are the subjects of future projects.

17.6 Summary and Concluding Points

Power-to-Gas is a part of transition plan to sustainable low-carbon energy systems in order to respond the climate change. In this work, different potential alternatives of power-to-gas are presented and some limitations regarding the technology readiness are discussed. Hydrogen generated from clean sources of energy can be mixed with natural gas to make hydrogen-enriched natural gas to be injected in natural gas pipelines and utilized for different application. The hydrogen-enriched natural gas can be separated in to hydrogen and natural gas at the end-user to supply the hydrogen demand of transportation sector in the urban areas. Hydrogen can also be stored in underground in the existing infrastructure for seasonal energy storage. Different pathways of power-to-gas are discussed from economic and technical points of view. With this information, policy maker is able to develop energy policy transition plans and strategies towards a fossil-free economy. The use of electrolytic hydrogen from intermittent renewable energy sources and baseload nuclear power will provide needed energy storage and clean emissions free transportation fuels for the energy requirement of the future. More importantly, through the gradual implementation of electrolysis capacity, current energy needs and issues can be immediately addressed, while developing infrastructure capacity for the future.

References

1. Mukherjee U, Elsholkami M, Walker S, Fowler M, Elkamel A, Hajimiragha A (2015) Optimal sizing of an electrolytic hydrogen production system using an existing natural gas infrastructure. Int J Hydrog Energy 40(31):9760–9772
2. Peng D (2013) Enabling utility-scale electrical energy storage through underground hydrogen-natural gas co-storage. University of Waterloo
3. Qadrdan M, Abeysekera M, Chaudry M, Wu J, Jenkins N (2015) Role of power-to-gas in an integrated gas and electricity system in Great Britain. Int J Hydrog Energy 40(17):5763–5775
4. Mazloomi K, Gomes C (2012) Hydrogen as an energy carrier: prospects and challenges. Renew Sust Energ Rev 16(5):3024–3033
5. Wallbrecht J (2006) International gas union triennium 2003e2006. Amsterdam
6. Fogelson J (2014) California plans for a hydrogen future. Forbes. http://www.forbes.com/sites/jasonfogelson/2014/04/15/california-plans-for-a-hydrogen-future/
7. de Santoli L, Basso GL, Nastasi B (2017) The potential of hydrogen enriched natural gas deriving from power-to-gas option in building energy retrofitting. Energ Buildings 149:424
8. Ball M, Weeda M (2015) The hydrogen economy–vision or reality? Int J Hydrog Energy 40(25):7903–7919

9. Maroufmashat A, Fowler M (2017) Transition of future energy system infrastructure; through power-to-gas pathways. Energies 10(8):1089
10. Penev M, Melaina M, Bush B, Muratori M, Warner E, Chen Y (2016) Low-carbon natural gas for transportation: well-to-wheels emissions and potential market assessment in California, NREL (National Renewable Energy Laboratory (NREL), Golden, CO (United States))
11. Melaina MW, Antonia O, Penev M (2013) Blending hydrogen into natural gas pipeline networks: a review of key issues. Contract 303:275–3000
12. Judd R, Pinchbeck D (2013) Power to gas research roadmap. Offering a solution to the energy storage problem. Gas for energy, p 2
13. Körner A, Tam C, Bennett S, Gagné J (2015) Technology roadmap-hydrogen and fuel cells. International Energy Agency (IEA), Paris
14. Vanhoudt W, et al (2016) Power-to-gas: short term and long term opportunities to leverage synergies between the electricity and transport sectors through power-to-hydrogen. Hinicio and LBST • Ludwig-Bölkow-Systemtechnik GmbH. http://www.lbst.de/download/2016/Hinicio-LBST_2016_PtH2-study_Fondation-Tuck.pdf. Retrieved Oct 2016
15. Benjaminsson G, Benjaminsson J, Rudberg RB (2013) Power-to-Gas: a technical review. Svenskt Gastekniskt Center
16. Mukherjee U, Maroufmashat A, Narayan A, Elkamel A, Fowler M (2017) A stochastic programming approach for the planning and operation of a power to gas energy hub with multiple energy recovery pathways. Energies 10(7):868
17. Walker SB, Mukherjee U, Fowler M, Elkamel A (2016) Benchmarking and selection of Power-to-Gas utilizing electrolytic hydrogen as an energy storage alternative. Int J Hydrog Energy 41(19):7717–7731
18. Al-Subaie A, Maroufmashat A, Elkamel A, Fowler M (2017) Presenting the implementation of power-to-gas to an oil refinery as a way to reduce carbon intensity of petroleum fuels. Int J Hydrog Energy 42(30):19376–19388

Chapter 18
Multi-Objective Optimization Framework for Electricity and Natural Gas Energy Hubs Under Hydrogen Storage System and Demand Response Program

Majid Majidi, Sayyad Nojavan, and Kazem Zare

18.1 Introduction

Multi-carrier energy systems or so-called energy hub systems have been focused to be further extended in future power systems [1]. Imported natural gas and electric power from natural gas and electric power are usually the main input resources injected to these systems [2]. Imported gas is consumed by boiler and cogeneration systems for heat and electric power generation [2]. In addition to the mentioned non-renewable generation units, renewable ones like wind turbine can also be integrated with other resources to supply local electric power for various applications [3].

18.1.1 Literature Review

Due to new concept that energy hub systems have provided in the planning and scheduling of power system, these systems have been studied in some papers and their brief summaries are presented in this section.

With the aim of reducing energy consumption in a hub energy system, an energy management system algorithm has been presented in [4]. A new design based on hub model has been presented in [5] to integrated renewable energy resources into CHP plants for reduction of emission. Optimal scheduling of electricity and gas networks of hub energy system in a smart environment has been presented in [6]. Using power to gas concept, optimal operation of hub energy system has been studied to satisfy economic goals in [7]. Using the concept of energy hub,

M. Majidi · S. Nojavan (✉) · K. Zare
Faculty of Electrical and Computer Engineering, University of Tabriz, Tabriz, Iran
e-mail: majidmajidi95@ms.tabrizu.ac.ir; sayyad.nojavan@tabrizu.ac.ir; kazem.zare@tabrizu.ac.ir

B. Mohammadi-Ivatloo, F. Jabari (eds.), *Operation, Planning, and Analysis of Energy Storage Systems in Smart Energy Hubs*,
https://doi.org/10.1007/978-3-319-75097-2_18

optimal sizing and operation of a combined cold, heat, and power system has been evaluated in [8]. Optimal operation of a residential hub energy system has been investigated considering flexible loads in [9]. As a seasonal storage system, power to gas technology has been employed to provide supporting services for natural gas and electrical infrastructures in [10]. Operation of hub energy system has been analyzed from reliability viewpoint in [11, 12]. An optimization model has been developed for a smart hub energy system in [13] to minimize total cost of system through optimal operation of appliances and generation units. Optimal operation of a hub energy system including combined heat and power system and electric vehicle has been investigated considering flexible loads in [14]. With the aim of minimizing total operation cost of hub energy system, evolutionary algorithms have been used to optimize heating network in [15]. Single and multi-objective power flow problem of an electrical and gas hub energy system has been investigated using time varying acceleration coefficient gravitational search algorithm in [16]. Optimal economic operation of hub energy system has been evaluated using robust optimization approach in [17]. Using energy hub concept, steady-states in microgrids have been studied in [18]. Uncertainty-based planning and operation of hub energy system has been studied in [19]. With the aim of improving environmental performance, multi-carrier energy system concept has been used for optimum dispatch of energy resources in [20]. In order to minimize operation cost of microgrid-based hub energy system, real time pricing mechanism has been employed in [21]. A decentralized energy system has been developed in a neighborhood using hub energy concept in [22]. By employing stochastic programming, risk-based economic operation problem of hub energy system has been solved in [23]. Economic dispatch problem of energy hub system is studied in [24]. Hub energy systems have been comprehensively reviewed from various viewpoints in [25]. Finally, in order to maximize total profit of hub energy system, an MINLP based model has been investigated in [26].

18.1.2 Novelty and Contributions of this Research

Utilization of energy storage systems in energy systems is vital for energy management purposes.

Various technologies of storage systems with different applications and efficiencies are available. One of the important and integrating types of these storage systems is hydrogen storage system (HSS). This storage is composed of two sections, namely electrolyzer and fuel cell unit. At the times that generation is more than consumption, excess energy is consumed by electrolyzer and therefore hydrogen molar is generated. Produced hydrogen molar is stored in special hydrogen tanks and it is later consumed by fuel cell unit at peak times that energy consumption is at its maximum level to generate electric power [27, 28].

As the main focus of this chapter, hydrogen storage system has been studied to be utilized in future hub energy systems for further efficiency improvement of these systems. Using a bi-objective optimization model in this chapter, optimum performance of hub energy system has been investigated from both economic and emission viewpoints in the presence of HSS and DRP. Demand response program has been available for participation of loads to reduce their payments as much as possible and improve their environmental performance. Therefore, presented contributions of this chapter can be expressed as follows:

- Bi-objective optimization model for eco-environmental operation of hub energy system.
- Employing ε-constraint and fuzzy satisfying approaches for solving proposed bi-objective model.
- Implementation of hydrogen storage system for optimum operation of hub energy system.
- Utilization of demand response program for reduction of emission and operation cost of hub energy system.

18.2 Problem Formulation

In order to solve optimal performance problem of hub energy system from economic and emission viewpoints, a bi-objective optimization model has been proposed in the presence of HSS and DRP which mathematical formulation is presented in the following.

18.2.1 Economic Objective Function

Total operation cost of hub energy system to be minimized (18.1) is composed of several individual costs which are presented through Eqs. (18.2)–(18.8).

$$\text{Min } \Phi_1 = \text{Total cost} = C_{\text{net}} + C_{\text{Wind}} + C_{\text{TS}} + C_{\text{DR}} + C_{\text{CHP}} + C_{\text{Bo}} + C_{\text{Wa}} \tag{18.1}$$

$$C_{\text{net}} = \sum_{t=1}^{H} \left(\lambda_t^e \times p_t^e\right) \tag{18.2}$$

$$C_{\text{Wind}} = \sum_{t=1}^{H} \left(\lambda^{\text{wi}} \times p_t^{\text{wi}}\right) \tag{18.3}$$

$$C_{\text{TS}} = \sum_{t=1}^{H} \left(\lambda_s^h \times \left(p_t^{\text{ch},h} + p_t^{\text{dis},h}\right)\right) \tag{18.4}$$

$$C_{\mathrm{DR}} = \sum_{t=1}^{H} \left(\lambda^{\mathrm{DR}} \times \left(p_t^{e,\mathrm{shdo}} + p_t^{e,\mathrm{shup}}\right)\right) \tag{18.5}$$

$$C_{\mathrm{CHP}} = \sum_{t=1}^{H} \left(\lambda^{g} \times g_t^{\mathrm{CHP}}\right) \tag{18.6}$$

$$C_{\mathrm{Bo}} = \sum_{t=1}^{H} \left(\lambda^{g} \times g_t^{B}\right) \tag{18.7}$$

$$C_{\mathrm{Wa}} = \sum_{t=1}^{H} \left(\lambda^{\mathrm{wa}} \times \mathrm{wa}_t\right) \tag{18.8}$$

18.2.2 Environmental Objective Function

Generation of electrical energy is sometimes results of burning fossil fuels which leads to environment pollution and hub system containing fossil fuel-burning equipment is not an exceptional case. Therefore, with the aim of having environmental performance, total generated emission of hub energy system should be minimized (18.9). Like cost total cost function, total emission of hub energy system is composed of several individual emission functions which are presented through Eqs. (18.10)–(18.13).

$$\mathrm{Min}\ \Phi_2 = \mathrm{Em} = \left(\mathrm{Em}^{\mathrm{CHP}} + \mathrm{Em}^{B} + \mathrm{Em}^{L} + \mathrm{Em}^{\mathrm{NET}}\right) \tag{18.9}$$

$$\begin{aligned}\mathrm{Em}^{\mathrm{CHP}} = {} & \left(\mathrm{EF}_{\mathrm{CO}}^{\mathrm{CHP}} \times g_t^{\mathrm{CHP}}\right) \\ & + \left(\mathrm{EF}_{\mathrm{SO}}^{\mathrm{CHP}} \times g_t^{\mathrm{CHP}}\right) + \left(\mathrm{EF}_{\mathrm{NO}}^{\mathrm{CHP}} \times g_t^{\mathrm{CHP}}\right)\end{aligned} \tag{18.10}$$

$$\mathrm{Em}^{B} = \left(\mathrm{EF}_{\mathrm{CO}}^{B} \times g_t^{B}\right) + \left(\mathrm{EF}_{\mathrm{SO}}^{B} \times g_t^{B}\right) + \left(\mathrm{EF}_{\mathrm{NO}}^{B} \times g_t^{B}\right) \tag{18.11}$$

$$\mathrm{Em}^{L} = \left(\mathrm{EF}_{\mathrm{CO}}^{L} \times g_t^{L}\right) + \left(\mathrm{EF}_{\mathrm{SO}}^{L} \times g_t^{L}\right) + \left(\mathrm{EF}_{\mathrm{NO}}^{L} \times g_t^{L}\right) \tag{18.12}$$

$$\mathrm{Em}^{\mathrm{NET}} = \left(\mathrm{EF}_{\mathrm{CO}}^{\mathrm{NET}} \times p_t^{e}\right) + \left(\mathrm{EF}_{\mathrm{SO}}^{\mathrm{NET}} \times p_t^{e}\right) + \left(\mathrm{EF}_{\mathrm{NO}}^{\mathrm{NET}} \times p_t^{e}\right) \tag{18.13}$$

18.2.3 Thermal Model

Studied multi-carrier energy system should supply heating demand through its available thermal units (18.14).

$$p_t^h = \left[\eta_{\mathrm{gh}}^B \times g_t^B\right] + \left[A^{\mathrm{CHP}} \times \eta_{\mathrm{gh}}^{\mathrm{CHP}} \times g_t^{\mathrm{CHP}}\right] + \left(p_t^{\mathrm{dis},h} - p_t^{\mathrm{ch},h}\right) \tag{18.14}$$

18.2.3.1 Model of Boiler

As one of thermal resources supplying heating demand, boiler should be operated within its nominal capacity for heat generation (18.15).

$$\eta_{\mathrm{gh}}^B \times g_t^B \le p_c^B \tag{18.15}$$

18.2.3.2 Model of CHP System

Some percentage of imported gas from gas network is injected to CHP system to generate electric power and heating. Since thermal generation of CHP system is usually proportional with its electrical generation, thermal operation limitation will be satisfied if electrical operation limitation is satisfied. Equation (18.16) has been used for safe electrical operation of CHP system.

$$\eta_{\mathrm{ge}}^{\mathrm{CHP}} \times g_t^{\mathrm{CHP}} \le p_c^{\mathrm{CHP}} \tag{18.16}$$

18.2.3.3 Model of TSS

In order to have no energy waste, various energy storage systems including HSS and TSS have been used in the hub energy system. Here, mathematical modeling of TSS has been presented through Eqs. (18.17)–(18.20).

Stored heat level of TSS is expressed by Eq. (18.17).

$$C_t^{\mathrm{st},h} = C_{t-1}^{\mathrm{st},h} + p_t^{\mathrm{ch},h} \times \eta_{\mathrm{ch}}^h - p_t^{\mathrm{dis},h} / \eta_{\mathrm{dis}}^h - p_t^{\mathrm{loss},h} \tag{18.17}$$

Limitations of stored heat level, input and output heat of TSS are expressed through Eqs. (18.18)–(18.20).

$$\alpha_{\min}^h \times C_c^{\mathrm{st},h} \le C_t^{\mathrm{st},h} \le \alpha_{\max}^h \times C_c^{\mathrm{st},h} \tag{18.18}$$

$$\frac{\alpha_{\min}^h \times C_c^{\mathrm{st},h} \times I_t^{\mathrm{ch},h}}{\eta_{\mathrm{ch}}^h} \le p_t^{\mathrm{ch},h} \le \frac{\alpha_{\max}^h \times C_c^{\mathrm{st},h} \times I_t^{\mathrm{ch},h}}{\eta_{\mathrm{ch}}^h} \tag{18.19}$$

$$\alpha_{\min}^{h} \times C_{c}^{\mathrm{st},h} \times I_{t}^{\mathrm{dis},h} \times \eta_{\mathrm{dis}}^{h} \leq p_{t}^{\mathrm{dis},h} \leq \alpha_{\max}^{h} \times C_{c}^{\mathrm{st},h} \times I_{t}^{\mathrm{dis},h} \times \eta_{\mathrm{dis}}^{h} \tag{18.20}$$

Heat loss of TSS is expressed in Eq. (18.21).

$$p_{t}^{\mathrm{loss},h} = \alpha_{\mathrm{loss}}^{h} \times C_{t}^{\mathrm{st},h} \tag{18.21}$$

In order to prevent TSS from simultaneous operation in charging and discharging modes, Eq. (18.22) is employed.

$$I_{t}^{\mathrm{ch},h} + I_{t}^{\mathrm{dis},h} \leq 1 \tag{18.22}$$

18.2.4 Electrical Model

Electrical demand capable of participating in DRP is due to be supplied through renewable and non-renewable units as well as upstream network and available hydrogen energy storage system (18.23).

$$\begin{aligned}\left(p_{t}^{l} + p_{t}^{\mathrm{shup}} - p_{t}^{\mathrm{shdo}} + P_{t}^{\mathrm{el}}\right) &= \left(A^{\mathrm{WIND}} \times \eta_{\mathrm{ee}}^{\mathrm{CON}} \times p_{t}^{\mathrm{wi}}\right) + \left(A^{\mathrm{NET}} \times \eta_{\mathrm{ee}}^{T} \times p_{t}^{e}\right) \\ &+ \left(A^{\mathrm{CHP}} \times \eta_{\mathrm{ge}}^{\mathrm{CHP}} \times g_{t}^{\mathrm{CHP}}\right) + \left(P_{t}^{\mathrm{fc}}\right)\end{aligned} \tag{18.23}$$

18.2.4.1 Model of Renewable Resources

Wind turbine speed should be within a predefined range to be able for power generation. So, wind turbine output can be formulated using Eq. (18.24) as follows:

$$p_{t}^{\mathrm{wi}} = \begin{cases} 0 & w < w_{\mathrm{ci}} \\ p_{r}\left(z - y \cdot w(t) + x \cdot w^{2}(t)\right) & w_{\mathrm{ci}} \leq w < w_{r} \\ p_{r} & w_{r} \leq w < w_{\mathrm{co}} \\ 0 & w \geq w_{\mathrm{co}} \end{cases} \tag{18.24}$$

18.2.4.2 Model of HSS

Hydrogen storage system can be extensively employed in future hub energy systems for energy management purposes. At the times of excess energy, HSS is operated and therefore excess energy is converted to hydrogen molar and stored in hydrogen tanks. Later, in other periods, stored hydrogen molar is used to generate electric power by fuel cell to be consumed in demand side.

Produced hydrogen molar by electrolyzer in HSS is expressed and limited through Eqs. (18.25) and (18.26), respectively.

$$N_{H2,t}^{\mathrm{EL}} = \frac{\eta^{\mathrm{EL}} P_t^{\mathrm{EL}}}{\mathrm{LHV}_{H2}} \tag{18.25}$$

$$N_{H2,t}^{\mathrm{EL}} \leq N_{H2,\max}^{\mathrm{EL}} \times U_t^{\mathrm{EL}} \tag{18.26}$$

Consumed hydrogen molar by fuel cell in HSS is expressed and limited through Eqs. (18.27) and (18.28), respectively.

$$N_{H2,t}^{\mathrm{FC}} = \frac{P_t^{\mathrm{FC}}}{\eta^{FC} \mathrm{LHV}_{H2}} \tag{18.27}$$

$$N_{H2,t}^{\mathrm{FC}} \leq N_{H2,\max}^{\mathrm{FC}} \times U_t^{\mathrm{FC}} \tag{18.28}$$

Dynamic model of pressure inside HSS is expressed and limited in Eqs. (18.29) and (18.30), respectively.

$$P_t^{H2} = P_{t-1}^{H2} + \frac{\Re T_{H2}}{V_{H2}} \left(N_{H2,t}^{\mathrm{EL}} - N_{H2,t}^{\mathrm{FC}} \right) \tag{18.29}$$

$$P_{\min}^{H2} \leq P_t^{H2} \leq P_{\max}^{H2} \tag{18.30}$$

Total power consumption of electrolyzer is limited by Eq. (18.31).

$$P_{\min}^{\mathrm{EL}} \times U_t^{\mathrm{EL}} \leq P_t^{\mathrm{EL}} \leq P_{\max}^{\mathrm{EL}} \times U_t^{\mathrm{EL}} \tag{18.31}$$

Generated power by fuel cell is limited through Eq. (18.32).

$$P_{\min}^{\mathrm{FC}} \times U_t^{\mathrm{FC}} \leq P_t^{\mathrm{FC}} \leq P_{\max}^{\mathrm{FC}} \times U_t^{\mathrm{FC}} \tag{18.32}$$

Finally, Eq. (18.33) is used to prevent HSS from simultaneous operation in charging and discharging modes.

$$U_t^{\mathrm{EL}} + U_t^{\mathrm{FC}} \leq 1 \tag{18.33}$$

18.2.4.3 Model of DRP

In this chapter, electrical loads have been considered to be capable of participating in demand response programs. One of the common types of programs available in DRP is time-of-use (TOU) rates of DRP in which loads are shifted from peak time periods to other time periods that leads to their cost reduction [29–32]. It should be mentioned that total amount of load is fixed. In other words, increased and decreased loads at the end of planning period are equal. Also, it is noteworthy that loads maximum participation limit in DRP has been considered to be 20% in this chapter. Based on the given explanation above, mathematical formulation of DRP is presented through Eqs. (18.34)–(18.37) in the following:

$$p_t^{\text{el,DRP}} = p_t^{\text{el}} + p_t^{\text{shup},e} - p_t^{\text{shdo},e} \tag{18.34}$$

$$0 \le p_t^{\text{shup},e} \le \text{LPF}^{\text{shup},e} \times p_t^l \times I_t^{\text{shup},e} \tag{18.35}$$

$$0 \le p_t^{\text{shdo},e} \le \text{LPF}^{\text{shdo},e} \times p_t^l \times I_t^{\text{shdo},e} \tag{18.36}$$

$$I_t^{\text{shup},e} + I_t^{\text{shdo},e} \le 1 \tag{18.37}$$

18.2.4.4 Model of Electrical, Gas and Water Networks

Total imported power from upper grid should satisfy transformers nominal capacity limitation (18.38).

$$\eta_{\text{ee}}^T \times p_t^e \le p_c^T \tag{18.38}$$

Total purchased gas from gas network is divided into several parts for various applications (18.39). It should be noted that total purchased gas should be in the defined nominal rage (18.40).

$$g_t^{\text{net}} = g_t^B + g_t^{\text{CHP}} + g_t^l \tag{18.39}$$

$$g_{\text{min}}^{\text{net}} \le g_t^{\text{net}} \le g_{\text{max}}^{\text{net}} \tag{18.40}$$

Total purchased water from gas network is used to satisfy water demand (18.41). It should be noted that total purchased water should be in the defined nominal rage (18.42).

$$\text{wa}_t^l = \text{wa}_t^{\text{net}} \tag{18.41}$$

$$\text{wa}_{\text{min}} \le \text{wa}_t^{\text{net}} \le \text{wa}_{\text{max}} \tag{18.42}$$

18.3 Multi-Objective Problem

In this section, methods used for solving multi-objective problem are briefly explained.

18.3.1 ε-Constraint Technique

This technique is used to solve multi-objective problems involving conflict objective functions. This method sets one of the objective functions as the main objective function of problem and sets the other objective functions as the limitation for the main objective function (18.43).

$$\begin{aligned} &\mathrm{OF} = \min(\Phi_1) \\ &s.t. \\ &\begin{cases} \Phi_2 \leq \varepsilon \\ \text{Equal } \& \text{ unequal constraints} \end{cases} \end{aligned} \tag{18.43}$$

Determining the minimum and maximum values of second objective functions, ε-constraint technique varies ε vector within this range. So, the first objective function changes in accordance with the changing of second objective function through varying ε vector and as a results of that Pareto front is obtained [1, 31].

18.3.2 Fuzzy Satisfying Approaches

This technique is another part of used methodology for solving multi-objective problem. This technique calculates the normalized forms of each objective function. Later, it makes a comparison between per unit values of each objective function in each iteration and then choosing the minimums in each iteration, it selects the maximum value between the chosen minimums which is the selected solution of multi-objective problem [33].

18.4 Case Study

In this section, optimal eco-emission operation of hub energy system has been studied in the presence of HSS and DRP. Studied hub energy system is illustrated in Fig. 18.1.

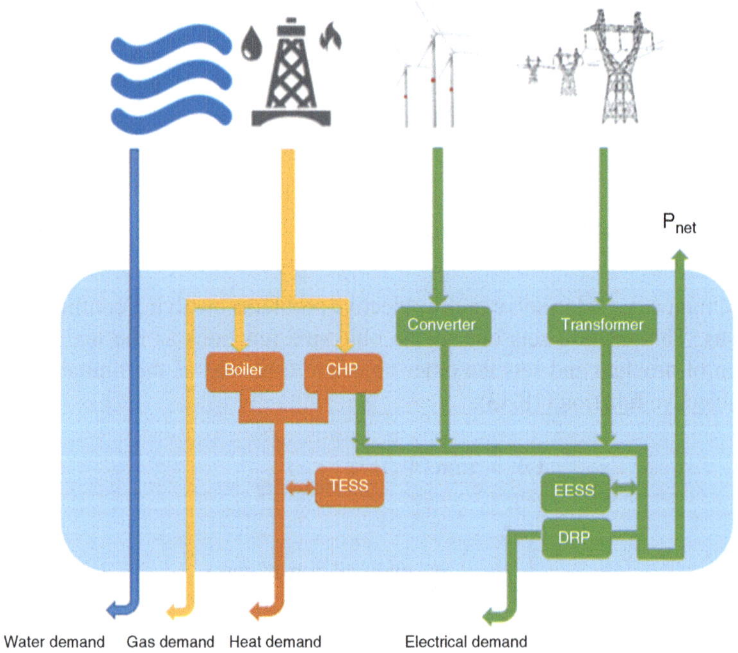

Fig. 18.1 Hub energy system with HSS and DRP

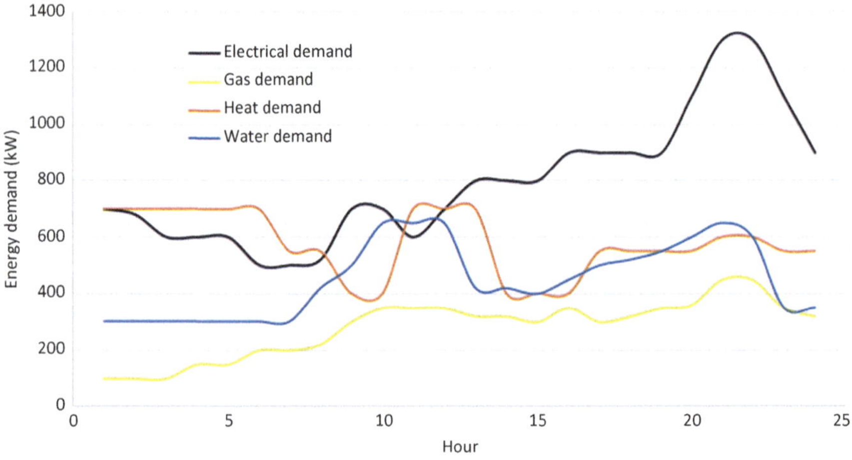

Fig. 18.2 Energy demands

18.4.1 *Input Data*

Energy demands to be supplied by hub energy system are presented in Fig. 18.2.

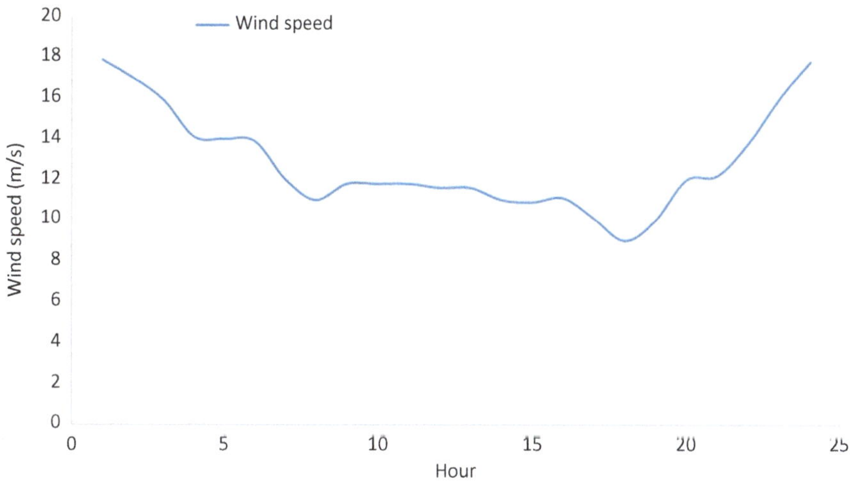

Fig. 18.3 Wind speed

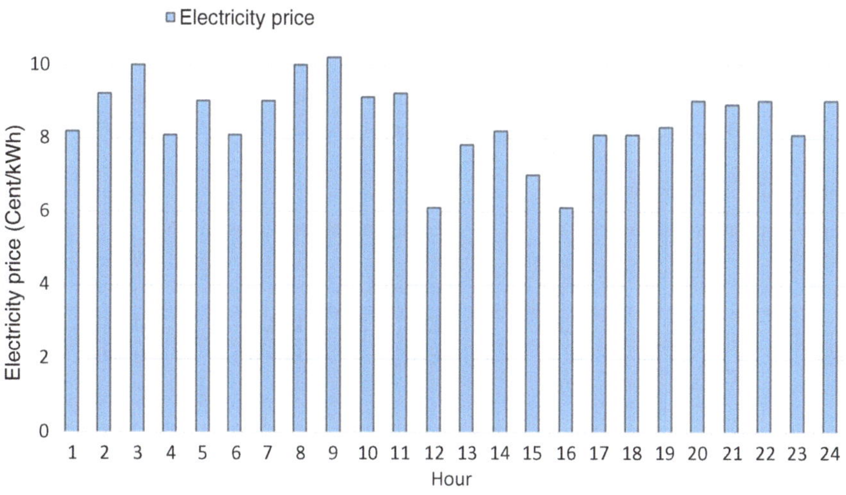

Fig. 18.4 Upper grid power price

Wind speed and upper grid price are illustrated in Figs. 18.3 and 18.4, respectively.

Parameters of thermal generation units are presented in Table 18.1.

Parameters of HSS and TSS are presented in Table 18.2.

Parameters of upper network are presented in Table 18.3.

Prices and operation cost of various networks and units are presented in Table 18.4.

Finally, parameters of wind turbine are presented in Table 18.5.

Table 18.1 Thermal generation unit's parameters

CHP parameter [34]			Boiler parameter [34]		
#	Unit	Value	#	Unit	Value
η_{ge}^{CHP}	%	40	η_{gh}^{B}	%	85
η_{gh}^{CHP}	%	35	p_c^B	kW	800
A^{CHP}	–	0.96	–	–	–
p_c^{CHP}	kW	800	–	–	–
Boiler emission [35]			CHP emission [35]		
#	Unit	Value	#	Unit	Value
EF_{CO}^{B}	kg/kWh	0.37	EF_{CO}^{CHP}	kg/kWh	0.37
EF_{SO}^{B}	kg/kWh	0.000003	EF_{SO}^{CHP}	kg/kWh	0.000003
EF_{NO}^{B}	kg/kWh	0.00009	EF_{NO}^{CHP}	kg/kWh	0.00009

Table 18.2 Parameters of HSS and TSS

HSS parameter [28]						TSS parameter [34]		
#	Unit	Value	#	Unit	Value	#	Unit	Value
LHV_{H2}	MJ/k mol	240	P_{min}^{EL}	KW	1.5	α_{min}^{h}	–	0.05
η^{EL}	%	90	P_{max}^{EL}	KW	6.2	α_{max}^{h}	–	0.9
η^{FC}	%	9	P_{min}^{FC}	KW	0.5	α_{loss}^{h}	–	0.2
$\Re$	J/^{0}K mol	8.314	P_{max}^{FC}	KW	6	η_{ch}^{h}	%	90
T_{H2}	^{0}K	313	$N_{H2,max}^{EL}$	Nm^3/h	1.05	η_{dis}^{h}	%	90
V_{H2}	m^3	4	$N_{H2,max}^{FC}$	Nm^3/h	3.90	$C_c^{st,h}$	kW	200
P_{min}^{H2}	Bar	2	P_{max}^{H2}	Bar	13.8	–	–	–

Table 18.3 Upper grid info

Upstream network parameter [34]			Upstream network emission [36]		
#	Unit	Value	#	Unit	Value
A^{NET}	–	0.99	EF_{CO}^{Net}	kg/kWh	0.368
p_{max}^{e}	kW	1000	EF_{SO}^{Net}	kg/kWh	0.0002
p_{min}^{e}	kW	0	EF_{NO}^{Net}	kg/kWh	0.0008
p_c^T	kW	800	–	–	–

Table 18.4 Prices and operation costs of different networks and units [34]

Parameter	Value	Unit
λ^{g}	6	Cent/kWh
λ^{wa}	4	Cent/kWh
λ^{wi}	0	Cent/kWh
λ_s^{e}	2	Cent/kWh
λ_s^{h}	2	Cent/kWh
λ^{DR}	2	Cent/kWh

Proposed bi-objective model for optimum operation of hub energy system from financial and environmental viewpoints has been molded using a mixed-integer linear programming and the whole simulations are carried out under CPLEX solver of GAMS [37].

Table 18.5 Parameters of wind turbine [34]

Parameters	Unit	Value
A^{WIND}	–	0.96
x, y, z	–	0.07, 0.01, 0.03
w_r	m/s	10
w_{ci}	m/s	4
w_{co}	m/s	22
p_r	kW	400

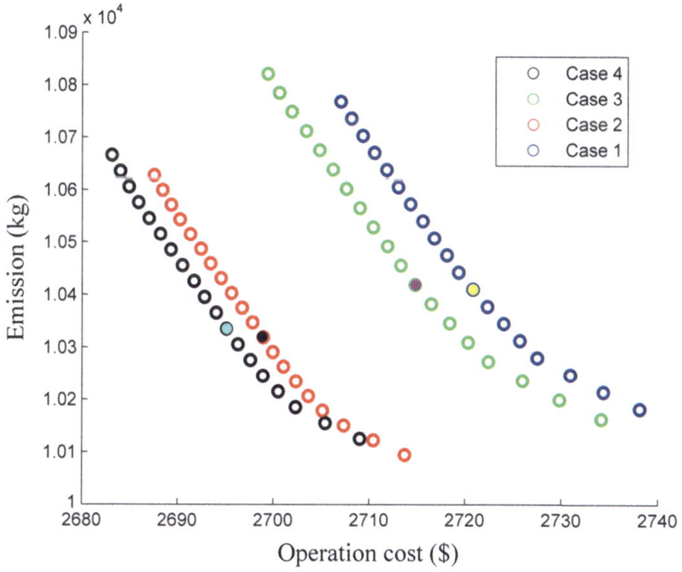

Fig. 18.5 Pareto front of 4 case studies

18.4.2 Simulation Results

Optimal eco-emission performance of hub energy system in the presence of HSS and DRP is investigated in 4 case studies as follows:

Case 1: Eco-emission performance of hub energy system without HSS & without DRP.
Case 2: Eco-emission performance of hub energy system without HSS & with DRP.
Case 3: Eco-emission performance of hub energy system with HSS & without DRP.
Case 4: Eco-emission performance of hub energy system with HSS & with DRP.

Solving the proposed multi-objective model in 4 cases, Pareto front id obtained for the whole cases, which is illustrated in Fig. 18.5.

It can be understood from Fig. 18.5 that by employing HSS and DRP in cases 2–4, total emission and operation cost of hub energy system have been reduced. Using min-max fuzzy satisfying technique, trade-off solutions are selected in each

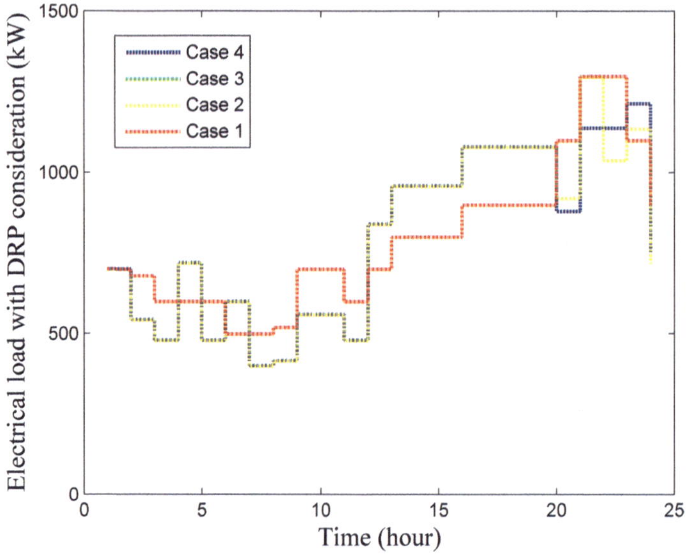

Fig. 18.6 Electrical load

case study. According to the selected solutions, total operation cost of hub energy system in cases 1–4 is 2720.80 $, 2698.86 $, 2714.82 $, and 2695.14 $, respectively. Evaluating the obtained results above, it can be seen that total operation cost of hub energy system in the presence of HSS and DRP in case 4 has been reduced 0.94%, 0.13%, and 0.72% in comparison with cases 1, 2, and 3, respectively. Also, total generated emission of hub energy system in cases 1–4 is 10,410.40 kg, 10,319.58 kg, 10,419.32 kg, and 10,335.33 kg, respectively. By analyzing achieved results above it can be observed that by employing HSS and DRP, generation of emission in the hub energy system is reduced in comparison with case 1 that no HSS and DRP has been used which convinces environmental concerns. As the outcome of analyzed results above, it can be concluded that utilization of HSS and DRP can provide financial and environmental benefits for hub energy system.

In order to understand additional benefits of HSS and DRP, further results have been presented in the following:

Electrical load which has been participated in DRP is shown in Fig. 18.6. According to this pattern which has been changed after implementation of DRP, total imported power from upper network has been changed in a way that financial benefits have been provided for hub energy system. Figure 18.7 illustrates pattern of imported power from upper grid.

Total imported gas from gas network to be used by CHP system and boiler is captured in Fig. 18.8. According to this figure, total imported gas in the presence of HSS and DRP in cases 2–4 has been reduced.

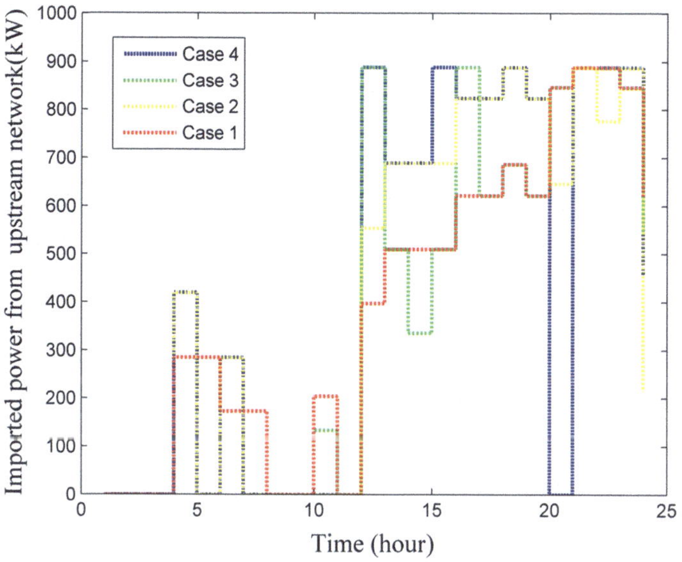

Fig. 18.7 Purchased power from upper grid

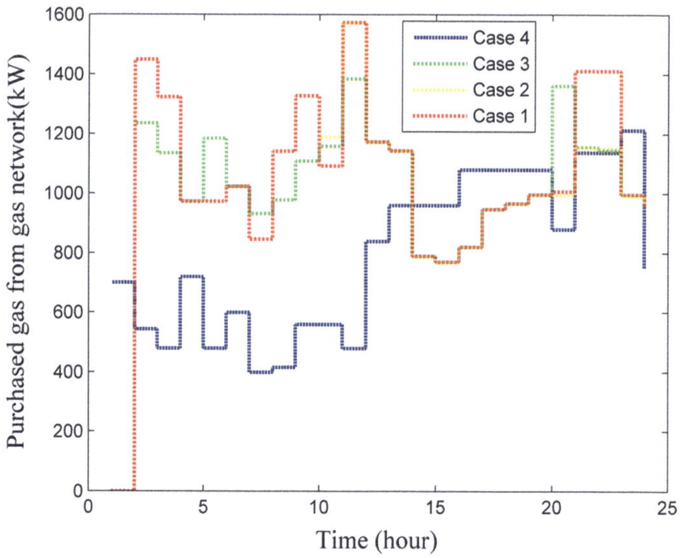

Fig. 18.8 Purchase gas from gas network

As results of reduction of total purchased gas, share of CHP system in gas consumption has been reduced in the presence of HSS and DRP which is shown in Fig. 18.9.

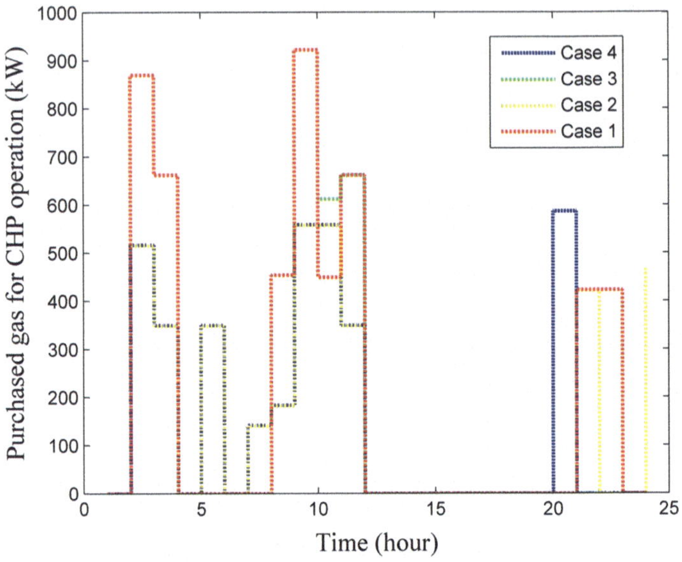

Fig. 18.9 CHP gas consumption

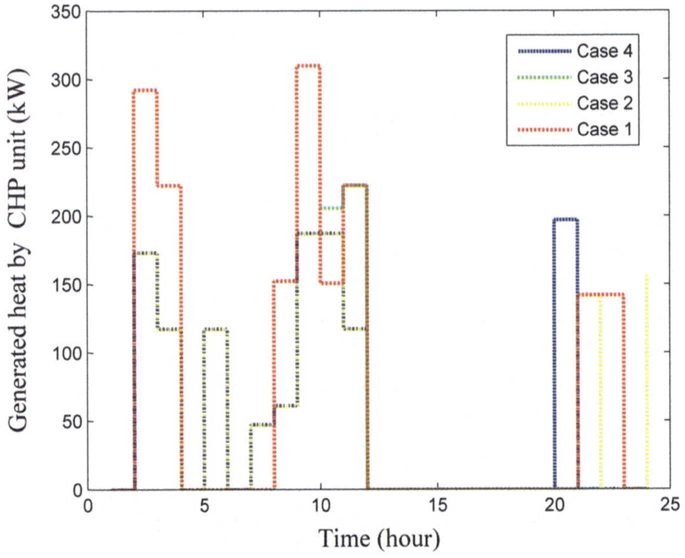

Fig. 18.10 Thermal generation of CHP system

As shown in below in Figs. 18.10 and 18.11, less gas consumption by CHP system consequently has led to less heat and electric power generation of CHP system in the presence of HSS and DRP.

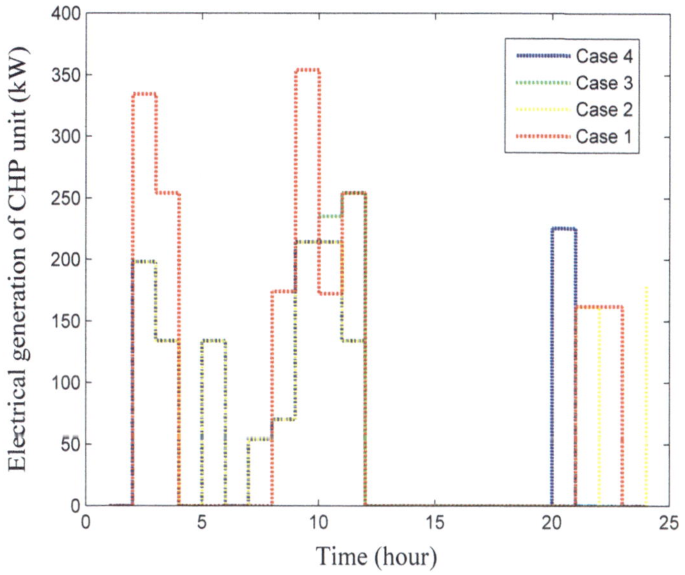

Fig. 18.11 Electrical generation of CHP system

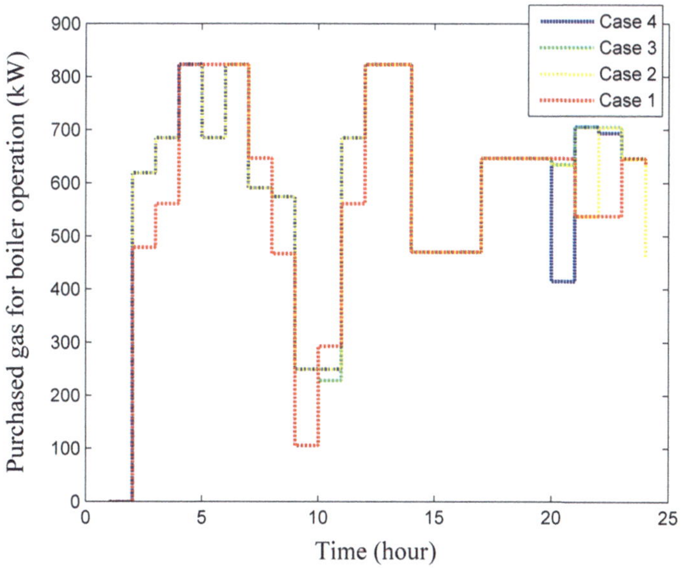

Fig. 18.12 Boiler gas consumption

Due to less gas consumption of CHP system, share of boiler in supplying thermal demand has been increased and the new pattern according to which boiler consumes gas is illustrated in Fig. 18.12.

18.5 Conclusion

In this chapter, a bi-objective optimization model has been proposed for optimum operation of hub energy system from financial and environmental viewpoints. Hydrogen storage system has been used for saving energy in off-peak periods and supplying energy demand in peak periods. Also, electrical load has been considered to be capable of participating in demand response program to reduce its financial payments and improve environmental performance of hub energy system. In order to evaluate efficiency of employed techniques, proposed bi-objective model is analyzed in 4 case studies. Extracted results from simulations revealed that total operation cost of energy hub system in case 4 in the presence of HSS and DRP has been reduced 0.94 %, 0.13 %, and 0.72 % in comparison with cases 1, 2, and 3,respectively. This means that implementation of both HSS and DRP satisfies financial goals. In addition to operation cost, total emission of hub energy system with HSS and DRP is reduced which provides a good situation for hub energy system from environmental-friendly operation viewpoint.

Nomenclature

Indices

t Time period index

Parameters

η_{ee}^{T}	Efficiency of transformer
η_{ge}^{CHP}	Gas to electricity efficiency of CHP system
η_{ee}^{CON}	Converter efficiency
η_{ch}^{h}, η_{dis}^{h}	Charge and discharge efficiency of TSS
α_{min}^{h}, α_{max}^{h}	Minimum and maximum limitation coefficient of TSS
η^{EL}	Efficiency of electrolyzer unit
η^{FC}	Efficiency of fuel cell unit
α_{loss}^{h}	Loss of heat coefficient in TSS
A^{NET}	Availability of upper grid power
A^{CHP}	Availability of CHP generation
A^{WIND}	Availability of wind generation
$C_{c}^{st,h}$	Limitation of available stored heat in TSS
EF_{CO}^{CHP}	CO_2 emission factor for CHP unit
EF_{SO}^{CHP}	SO_2 emission factor for CHP unit
EF_{NO}^{CHP}	NO_x emission factor for CHP unit

$\mathrm{EF}_{\mathrm{CO}}^{B}$	CO_2 emission factor for boiler
$\mathrm{EF}_{\mathrm{SO}}^{B}$	SO_2 emission factor for boiler
$\mathrm{EF}_{\mathrm{NO}}^{B}$	NO_2 emission factor for boiler
$\mathrm{EF}_{\mathrm{CO}}^{L}$	CO_2 emission factor for residential gas consumption
$\mathrm{EF}_{\mathrm{SO}}^{L}$	SO_2 emission factor for residential gas consumption
$\mathrm{EF}_{\mathrm{NO}}^{L}$	NO_x emission factor for residential gas consumption
$\mathrm{EF}_{\mathrm{CO}}^{\mathrm{Net}}$	CO_2 emission factor for provided power by upper network
$\mathrm{EF}_{\mathrm{SO}}^{\mathrm{Net}}$	SO_2 emission factor for provided power by upper network
$\mathrm{EF}_{\mathrm{NO}}^{\mathrm{Net}}$	NO_x emission factor for provided power by upper network
$g_{\min}^{\mathrm{net}}, g_{\max}^{\mathrm{net}}$	Gas network minimum and maximum limitations
g_t^l	Gas demand in residential section
LHV_{H2}	Hydrogen lower heating value
$N_{H2,\max}^{\mathrm{EL}}$	Maximum hydrogen molar generation limit of electrolyzer
$N_{H2,\max}^{\mathrm{FC}}$	Maximum hydrogen molar consumption limit of fuel cell unit
$p_{\min}^{e}, p_{\max}^{e}$	Upper network minimum and maximum limitations
p_c^T	Nominal limitation of transformer
p_c^{CHP}	Nominal limitation of CHP system
p_c^B	Nominal limitation of boiler
p_r	Wind-turbine rated power
p_t^{el}	Electrical energy demand
p_t^h	Heating demand
$P_{\min}^{\mathrm{EL}}$	Minimum power consumption limit in electrolyzer
$P_{\max}^{\mathrm{EL}}$	Maximum power consumption limit in electrolyzer
$P_{\min}^{\mathrm{FC}}$	Minimum power generation limit in fuel cell unit
$P_{\max}^{\mathrm{FC}}$	Maximum power generation limit in fuel cell unit
$P_{\min}^{H2}$	Hydrogen tank minimum pressure limitation
$P_{\max}^{H2}$	Hydrogen tank maximum pressure limitation
$\Re$	Constant of gas
T_{H2}	Mean temperature of the vessel
V_{H2}	Total volume of tank
wa_t^l	Water demand
$\mathrm{wa}_{\min}, \mathrm{wa}_{\max}$	Minimum and maximum limitations of water network
$w_{\mathrm{ci}}, w_{\mathrm{co}}, w_r$	Cut-in, cut-out and rated wind speeds
x, y, z	Coefficients modeling wind generation
λ_t^e	Price of power provided by upper grid
λ^{wi}	Price of power provided by wind turbine
λ^g	Price of imported gas from gas network
λ^{wa}	Price of imported water from gas network
λ_s^h	Operation cost of TSS

Variables

Cost	Total cost of hub energy system
$C_t^{\mathrm{st},h}$	Available stored heat in TSS
g_t^{CHP}	Injected gas to the CHP system
g_t^B	Injected gas to the boiler
g_t^{net}	Total injected gas to the hub energy system
$I_t^{\mathrm{ch},h}$, $I_t^{\mathrm{dis},h}$	Charge and discharge modeling binary variables of TSS
$N_{H2,t}^{\mathrm{EL}}$	Fuel cell hydrogen molar generation
$N_{H2,t}^{\mathrm{FC}}$	Fuel cell hydrogen molar consumption
p_t^e	Imported power from upper network
$p_t^{\mathrm{ch},h}$, $p_t^{\mathrm{dis},h}$	Charge and discharge heat of TSS
$p_t^{\mathrm{loss},h}$	Loss of heating energy in TSS
p_t^{wi}	Produced power by wind-turbine
P_t^{H2}	Pressure of hydrogen tank
P_t^{FC}	Power generation of electrolyzer
P_t^{EL}	Power consumption of electrolyzer
U_t^{FC}	Binary variable which is 1 if fuel cell unit is ON; Otherwise 0
U_t^{EL}	Binary variable which is 1 if electrolyzer is ON; Otherwise 0
$\mathrm{wa}_t^{\mathrm{net}}$	Total injected water to the hub energy system

References

1. Majidi M, Nojavan S, Zare K (2017) A cost-emission framework for hub energy system under demand response program. Energy 134:157
2. Nojavan S, Majidi M, Zare K (2017) Performance improvement of a battery/PV/fuel cell/grid hybrid energy system considering load uncertainty modeling using IGDT. Energy Convers Manag 147:29–39
3. Ghalelou AN, Fakhri AP, Nojavan S, Majidi M, Hatami H (2016) A stochastic self-scheduling program for compressed air energy storage (CAES) of renewable energy sources (RESs) based on a demand response mechanism. Energy Convers Manag 120:388–396. https://doi.org/10.1016/j.enconman.2016.04.082
4. Kamyab F, Bahrami S (2016) Efficient operation of energy hubs in time-of-use and dynamic pricing electricity markets. Energy 106:343–355. https://doi.org/10.1016/j.energy.2016.03.074
5. AlRafea K, Fowler M, Elkamel A, Hajimiragha A (2016) Integration of renewable energy sources into combined cycle power plants through electrolysis generated hydrogen in a new designed energy hub. Int J Hydrog Energy 41(38):16718–16728. https://doi.org/10.1016/j.ijhydene.2016.06.256
6. Sheikhi A, Bahrami S, Ranjbar AM (2015) An autonomous demand response program for electricity and natural gas networks in smart energy hubs. Energy 89:490–499. https://doi.org/10.1016/j.energy.2015.05.109
7. Mukherjee U, Walker S, Maroufmashat A, Fowler M, Elkamel A (2017) Development of a pricing mechanism for valuing ancillary, transportation and environmental services offered by a power to gas energy system. Energy 128:447–462. https://doi.org/10.1016/j.energy.2017.04.042

8. Sheikhi A, Ranjbar AM, Oraee H (2012) Financial analysis and optimal size and operation for a multicarrier energy system. Energ Buildings 48:71–78. https://doi.org/10.1016/j.enbuild.2012.01.011
9. Brahman F, Honarmand M, Jadid S (2015) Optimal electrical and thermal energy management of a residential energy hub, integrating demand response and energy storage system. Energ Buildings 90:65–75. https://doi.org/10.1016/j.enbuild.2014.12.039
10. Peng DD, Fowler M, Elkamel A, Almansoori A, Walker SB (2016) Enabling utility-scale electrical energy storage by a power-to-gas energy hub and underground storage of hydrogen and natural gas. J Nat Gas Sci Eng 35:1180–1199. https://doi.org/10.1016/j.jngse.2016.09.045
11. Shariatkhah M-H, Haghifam M-R, Parsa-Moghaddam M, Siano P (2015) Modeling the reliability of multi-carrier energy systems considering dynamic behavior of thermal loads. Energ Buildings 103:375–383. https://doi.org/10.1016/j.enbuild.2015.06.001
12. Koeppel G, Andersson G (2009) Reliability modeling of multi-carrier energy systems. Energy 34(3):235–244. https://doi.org/10.1016/j.energy.2008.04.012
13. Rastegar M, Fotuhi-Firuzabad M (2015) Load management in a residential energy hub with renewable distributed energy resources. Energ Buildings 107:234–242. https://doi.org/10.1016/j.enbuild.2015.07.028
14. Rastegar M, Fotuhi-Firuzabad M, Lehtonen M (2015) Home load management in a residential energy hub. Electr Power Syst Res 119:322–328. https://doi.org/10.1016/j.epsr.2014.10.011
15. Shabanpour-Haghighi A, Seifi AR (2016) Effects of district heating networks on optimal energy flow of multi-carrier systems. Renew Sust Energ Rev 59:379–387. https://doi.org/10.1016/j.rser.2015.12.349
16. Derafshi Beigvand S, Abdi H, La Scala M (2016) Optimal operation of multicarrier energy systems using Time Varying Acceleration Coefficient Gravitational Search Algorithm. Energy 114:253–265. https://doi.org/10.1016/j.energy.2016.07.155
17. Parisio A, Del Vecchio C, Vaccaro A (2012) A robust optimization approach to energy hub management. Int J Electr Power Energy Syst 42(1):98–104. https://doi.org/10.1016/j.ijepes.2012.03.015
18. Wasilewski J (2015) Integrated modeling of microgrid for steady-state analysis using modified concept of multi-carrier energy hub. Int J Electr Power Energy Syst 73:891–898. https://doi.org/10.1016/j.ijepes.2015.06.022
19. Pazouki S, Haghifam M-R (2016) Optimal planning and scheduling of energy hub in presence of wind, storage and demand response under uncertainty. Int J Electr Power Energy Syst 80:219–239. https://doi.org/10.1016/j.ijepes.2016.01.044
20. Orehounig K, Evins R, Dorer V, Carmeliet J (2014) Assessment of renewable energy integration for a village using the energy hub concept. Energy Procedia 57:940–949. https://doi.org/10.1016/j.egypro.2014.10.076
21. Ma T, Wu J, Hao L (2017) Energy flow modeling and optimal operation analysis of the micro energy grid based on energy hub. Energy Convers Manag 133:292–306. https://doi.org/10.1016/j.enconman.2016.12.011
22. Orehounig K, Evins R, Dorer V (2015) Integration of decentralized energy systems in neighbourhoods using the energy hub approach. Appl Energy 154:277–289. https://doi.org/10.1016/j.apenergy.2015.04.114
23. Najafi A, Falaghi H, Contreras J, Ramezani M (2016) Medium-term energy hub management subject to electricity price and wind uncertainty. Appl Energy 168:418–433. https://doi.org/10.1016/j.apenergy.2016.01.074
24. Beigvand SD, Abdi H, La Scala M (2017) A general model for energy hub economic dispatch. Appl Energy 190:1090–1111. https://doi.org/10.1016/j.apenergy.2016.12.126
25. Mancarella P (2014) MES (multi-energy systems): An overview of concepts and evaluation models. Energy 65:1–17. https://doi.org/10.1016/j.energy.2013.10.041
26. Moghaddam IG, Saniei M, Mashhour E (2016) A comprehensive model for self-scheduling an energy hub to supply cooling, heating and electrical demands of a building. Energy 94:157–170. https://doi.org/10.1016/j.energy.2015.10.137

27. Nojavan S, Zare K, Mohammadi-Ivatloo B (2017) Selling price determination by electricity retailer in the smart grid under demand side management in the presence of the electrolyser and fuel cell as hydrogen storage system. Int J Hydrog Energy 42(5):3294–3308
28. Nojavan S, Zare K, Mohammadi-Ivatloo B (2017) Application of fuel cell and electrolyzer as hydrogen energy storage system in energy management of electricity energy retailer in the presence of the renewable energy sources and plug-in electric vehicles. Energy Convers Manag 136:404–417
29. Nojavan S, Majidi M, Esfetanaj NN (2017) An efficient cost-reliability optimization model for optimal siting and sizing of energy storage system in a microgrid in the presence of responsible load management. Energy 139:89
30. Nojavan S, Majidi M, Zare K (2017) Risk-based optimal performance of a PV/fuel cell/battery/grid hybrid energy system using information gap decision theory in the presence of demand response program. Int J Hydrog Energy 42(16):11857–11867
31. Nojavan S, Majidi M, Najafi-Ghalelou A, Ghahramani M, Zare K (2017) A cost-emission model for fuel cell/PV/battery hybrid energy system in the presence of demand response program: ε-constraint method and fuzzy satisfying approach. Energy Convers Manag 138:383–392
32. Majidi M, Nojavan S, Zare K (2017) Optimal stochastic short-term thermal and electrical operation of fuel cell/photovoltaic/battery/grid hybrid energy system in the presence of demand response program. Energy Convers Manag 144:132–142
33. Majidi M, Nojavan S, Esfetanaj NN, Najafi-Ghalelou A, Zare K (2017) A multi-objective model for optimal operation of a battery/PV/fuel cell/grid hybrid energy system using weighted sum technique and fuzzy satisfying approach considering responsible load management. Sol Energy 144:79–89
34. Pazouki S, Haghifam M-R, Moser A (2014) Uncertainty modeling in optimal operation of energy hub in presence of wind, storage and demand response. Int J Electr Power Energy Syst 61:335–345
35. Elsied M, Oukaour A, Gualous H, Hassan R (2015) Energy management and optimization in microgrid system based on green energy. Energy 84:139–151
36. Elsied M, Oukaour A, Gualous H, Brutto OAL (2016) Optimal economic and environment operation of micro-grid power systems. Energy Convers Manag 122:182–194
37. The GAMS software website (2017). [Online]. http://www.gams.com/dd/docs/solvers/cplex.pdf

Index

B. Mohammadi-Ivatloo, F. Jabari (eds.), *Operation, Planning, and Analysis of Energy Storage Systems in Smart Energy Hubs*,
https://doi.org/10.1007/978-3-319-75097-2

S

T

The manufacturer's authorised representative in the EU is Springer Nature Customer Service Centre GmbH, Europaplatz 3, 69115 Heidelberg, Germany. If you have any concerns regarding our products, please contact ProductSafety@springernature.com

Printed and bound by CPI Group (UK) Ltd, Croydon, CR0 4YY
05/10/2026
02228141-0013